BRIEF CONTENTS

TABLE OF CONTENTS

Chapter 5

Fire Behavior and Building Construction

Chapter 6

Features of Fire Protection

Chapter 7

Chapter 8

Chapter 9

Chapter 12

Firefighting Concerns of Green Construction .349

Chapter 13

Specific Occupancy-Related Construction Hazards361

Chapter 14

INSTRUCTOR AND STUDENT RESOURCES

Instructor Resources
Instructor's ToolKit CD

Preparing for class is easy with the resources on this CD, including:

- **PowerPoint Presentations** Provides you with a powerful way to make presentations that are educational and engaging to your students. These slides can be modified and edited to meet your needs.
- **Lesson Plans** Provides you with complete, ready-to-use lesson plans that include all of the topics covered in the text. Offered in Word documents, the lesson plans can be modified and customized to fit your course.
- **Test Bank** Contains multiple-choice questions, and allows you to create tailor-made classroom tests and quizzes quickly and easily by selecting, editing, organizing, and printing a test along with an answer key, including page references to the text.
- **Image and Table Bank** Provides you with a selection of the most important images and tables found in the textbook. You can use them to incorporate more images into the PowerPoint presentations, make handouts, or enlarge a specific image for further discussion.

Student Resources
Student Workbook

This resource is designed to encourage critical thinking and aid comprehension of course material through:

- Case studies and corresponding questions
- Matching questions
- Fill-in-the-blank questions
- Multiple-choice questions

ACKNOWLEDGMENTS

Reviewers and Contributors

Robert "Jay" Alley
Blue Ridge Community College
Flat Rock, NC

Captain Raul A. Angulo
Ladder Company 6, Seattle Fire Department
Seattle, WA

Brian Bagwell, Psy.D.
Metropolitan State University of Denver
Denver, CO

Tim Baker
Fire Science Instructor
Lansing Community College
Lansing, MI

Brian R. Carlson
Anderson Township Fire & Rescue
Cincinnati, OH

Richard C. Carroll
Fire Protection Instructor
Shelby Fire & Rescue
Cleveland Community College
Shelby, NC

David L. Cates
Instructor – Fire Protection Technology
Wilson Community College
Wilson, NC

Sean DeCrane
Cleveland Fire Training Academy
Cleveland, OH

Kevin A. Gallagher
Acushnet Fire & EMS Department
Acushnet, MA

Ian K. Harrell
Surry Community College
Director of Fire and Rescue Programs /
Emergency Services Training Center
Dobson, NC

Bradley Iverson
Las Vegas Fire & Rescue,
College of Southern Nevada
Las Vegas, NV

Gary R. Johnson
Fire Science Instructor, Chief (ret.)
Central Ohio Technical College
Newark, OH

Shad Johnson
Cleveland Community College
Shelby, NC

James D. Lea
Training Officer
City of Spartanburg Fire & Rescue
Spartanburg, SC

Robert Lindstedt
Gray Fire Rescue
Gray, Maine

Mark Martin
Fire Chief, MPA
Stark State College
North Canton, OH

Wesley B. Rogers
Blue Ridge Community College
Flat Rock, NC

Steven F. Sawyer
Senior Fire Service Specialist
NFPA
Quincy, MA

John Shafer
Green Maltese
Greencastle Fire Department
Greencastle, IN

Richard M. Siebel
City of Raleigh Fire Department
Raleigh, NC

Landon Stallings
Fire Marshal
Fort Worth Fire Department
Fort Worth, Texas

Jimmy Talton
Fire Chief
Reece City V. F. D.
Attalla, Al.

Kenneth R. Willette
Division Manager, Public Fire Protection Division
NFPA
Quincy, MA

William Wren
NYS Fire Instructor
NYS Fire Academy – Oneida County
New Hartford Fire Department
Clinton, NY

This textbook is a record of our collective fire history. It contains a compendium of building fires, recent and long ago, large and small, which offer hard-earned lessons on firefighter safety as well as insights on firefighting strategies and tactics. Many of these lessons have cost the fire service dearly, in the form of the supreme sacrifice.

Since the 2007 publication of the last edition, the fire service has continued to suffer greatly in terms of reduced staffing in many departments. In terms of building construction, the incessant march of lightweight construction has made the fireground an even more dangerous place. The toxic mix of fewer hands on the fire scene, higher rates of heat release, and buildings that fail even more quickly than ever before presents perhaps one of the most challenging times in fire service history.

Although we've made improvements in our building and fire codes to stem this tsunami of risk, we still have a long way to go. As always, the fire service—local and national—needs to voice its objections to the public regarding dangerous building practices.

This new edition attempts to identify and analyze many of these changes. This includes the rapid growth of modular construction, lighter and newer engineered wood structural members, and, of course, the widespread permeation of green construction in our lives. In terms of fire incidents and experience with green buildings, it is an unknown and uncharted territory. The new green innovations coming about are greatly outpacing the codes' ability to regulate them. While the public is enthralled with green construction and consider it modern and environmentally friendly, for us, it presents a new set of problems. Remember, the words "green" and "safe" are not often interchangeable.

During the preparation of the last edition, we lost the fire service legend Frank Brannigan. Since the last edition, we lost his beloved wife Maureen. For those of you who knew her, you will remember her role as "student advocate." She always made sure Frank's projector slides were in order and ready to go. She also made sure Frank's students got their lecture breaks. Frank and Maureen were the dynamic duo, dedicated to fire fighter safety.

It's my hope that I have been able to properly carry the time-honored torch they lit more than 40 years ago with the publication of their first edition. I respectfully dedicate this edition to Maureen and her enduring legacy of fire fighter training.

Glenn Corbett
Waldwick, NJ
November 2013

Introduction

OBJECTIVES

At the conclusion of this chapter, you will be able to:

- Understand the importance of studying building construction as a fire fighter.
- Understand why using proper language is so important for comprehending building construction techniques and for fireground safety.
- Understand why evacuation protocols are critical.
- Identify the role of prefire plans and prefire analysis in firefighting operations.
- Understand the importance of looking for building deficiencies and dangerous conditions while conducting prefire planning walkthroughs.
- Explain why it is so important to study reports of significant fires, NIOSH Fire Fighter Fatality and Near Miss reports of the past.
- Understand the relationship between risk and benefit and its impact on fire fighter safety.

Case Study

Figure 1-1 The Worcester Cold Storage building fire.

© Paul Connors/AP Photos.

On December 3, 1999, a fire broke out in an abandoned cold storage warehouse in Worcester, Massachusetts (**Figure 1-1**). The large seven-story structure—actually two connected buildings—was built in the years preceding World War I. Since the structure had been used to process meat and store frozen foods, its interior surfaces were lined with asphalt-impregnated cork, supplemented in later years with polystyrene foam. In addition, the structure was essentially windowless, given the need to keep the interior cold. There was only a single exit from the upper floors.

Arriving fire fighters began searching for the fire and people in the maze-like structure. The building was a known domicile for squatters; rumors of homeless people being trapped in the building surfaced during the early stages of the fire.

As fire fighters operated in the building, conditions deteriorated rapidly. Two fire fighters became lost and began running low on air. They called for help, and succeeding sets of fire fighters began a frantic search for them. With conditions getting even worse, four additional fire fighters became lost in the building.

The district chief in charge of interior operations made the decision to not allow any more fire fighters into the building. He physically put himself in the stairway, preventing fire fighters from going up the stairs. The six missing fire fighters lost their lives.

1. How difficult was it for the district chief to decide to not allow any additional fire fighters into the building to search for their lost brother fire fighters?

Introduction

As you begin to read this text, you may be asking yourself this question: "Why do I need to study building construction?" The answer is simple: because your life depends upon it! Firefighting is conducted in, under, on top of, and around buildings. It is a fire fighter's place of business. As long as there have been fire fighters, burning buildings have presented the real risk of death and injury. It is important, then, that fire fighters know as much as possible about the hazards associated with the different types of buildings in their community.

It is essential that you have a working knowledge of how buildings are designed, constructed, renovated, maintained, altered, and fall into disrepair. You must know the language of construction and the general types of construction. You must understand how construction materials behave under the attack of fire, how different types of buildings fail in a fire, and how the type of occupancy impacts fire load and life safety. Through preplanning, you must collect as much information as possible about the buildings in which you will fight fires.

In short, you must understand how buildings are built. Even more importantly, you must understand how they can fall apart. In the United States, 100 fire fighters on average are killed annually in the line of duty. The largest building collapses in recorded history—the collapse of the twin towers of the World Trade Center in New York City on September 11, 2001—killed a total of 343 members of the New York Fire Department and one fire patrolman **Figure 1-2**. It was the worst loss of life in fire service history.

In 2012, a total of nine fire fighters were killed in collapses, by falling through floors, by running out of air, by being enveloped in an explosion inside of a structure, or by being struck by falling debris. More than 30,500 fire fighters were injured in 2012 on the fireground, sustaining injuries such as sprains, burns, and smoke inhalation.

Figure 1-2 The worst loss of life in fire service history.
© Shawn Baldwin/AP Photos.

Watch Your Language

Clear communication is a critical factor for all fire departments. It follows that a wide range of fire professionals, in trying to do their job at a facility or building site or when that site becomes the fireground, must rely on effective communications to succeed—and sometimes survive. Knowing how to use and understand proper language and industry jargon, and communicate with different agencies and professionals, is critical to all fire fighters.

In 1989, an Avianca airliner crashed on Long Island, New York. A major contributing factor was "language." The pilot kept telling the controller that he was low on fuel. The pilot should have used the proper term and declared an "emergency." The controller would have then given him an immediate priority for landing and the catastrophe could have been avoided.

In one building collapse, the fireground commander was heard ordering, "Pull your line out." This order might simply mean, to some, to reposition the lines. In this case, however, it meant, "Get out of there fast."

Many years ago, the aviators of the world settled on the French words *m'aider*, meaning "help me" (in English, *mayday*), as a universal distress signal. All fire departments should include this term in their radio language and develop protocols for dealing with such situations. A mayday message is used by fire fighters who are trapped, low on air, disoriented, or other similar situations to alert the incident commander and other fire fighters to their situation. It is also used in some departments to alert everyone that a collapse has occurred or a collapse is imminent. In other departments, the use of the term *emergency traffic* (based on national standards) is used to clear the "air" on the radio to issue an evacuation order. Such "collapse" messages are used to initiate an evacuation order. This should mean leaving any tools or equipment behind in order to get out fast. Fire fighters have been injured and killed pulling lines out when collapse was imminent. The radio order is sometimes supplemented with the use of air horn blasts and siren signals. It is important that you know the specific evacuation messages and protocols used in your department.

By nature and training, fire fighters are aggressive and often resist orders to back out. The overall gravity of the situation may not be apparent to those fighting the fire. "We've got a good shot at the fire" is often an excuse for delay. Very few fire departments drill on immediate evacuation. This is a critical oversight.

Some aggressive fire departments resent any intimation of retreat. However, they should remember that the U.S. Navy is a very aggressive organization, but it continually drills on "abandon ship." In similar fashion, fire departments should regularly drill on getting out of buildings. Fire fighters should respond immediately to such an order. They should not look back at the collapse while fleeing; rather, they should be trained to stay together and look where they are going.

Today, there remains a need for a specific change in dispatching language. Apparatus should be dispatched to a building fire. As soon as fire command determines that the fire involves the *structure*, as distinct from the *contents*, there is a radically different situation. When the structure is involved, the gravity resistance system of the building is under attack. All units should be advised of this fact—and they should acknowledge the emergency message, "This is now a *structural fire*." The word *structural* should be reserved for just such a situation. In the

case of lightweight trusses or wooden I-beams, a department's standard operating procedures (SOPs) should provide for evacuation of the affected area as soon as it is involved in fire.

In the past, the phrase "withdraw in good order" was used by fire departments. This is a military term, and its meaning is not appropriate in fire scenarios. When a military retreat is necessary, it is important to deceive the enemy. Rear guard troops may keep up a fierce fight while all others slip out. Finally the rear guard moves. This operation may succeed because the enemy can be deceived. In contrast, fire cannot be deceived or bluffed. When the time comes to get out, speed is of the essence. There is much truth in Francis Brannigan's humorous dictum, "Panic will save your life, provided you are the first to panic and are headed in the right direction."

As a fire fighter, it is also important for you to have a working knowledge of not only construction techniques and materials, but also the appropriate terminology related to these topics. For example, it is inappropriate to refer to a building as a "brick building" rather than a "building of ordinary construction."

Besides being a sign of professionalism, using the correct language is a sign of safety. By using proper terminology, you convey not only the characteristics of the building to other fire fighters, but also *the hazards*. Get into the habit of using the correct words to describe buildings, including the proper terms when preparing technical reports. The rest of this text is filled with construction terms, some of which may be new to you. Learn them and use them.

Prefire Planning and Analysis

Prefire planning is another key element for fire professionals. Its value cannot be overstated. Without it, fire fighters are reduced to just reacting to a fire rather than being prepared for its many potential hazards.

Francis Brannigan's great interest in prefire planning was forced by necessity. As a young naval officer, he was assigned as Chief of the Panama Canal District as part of his naval fire department duties. All of a sudden, he had responsibilities for lives and property for which he was totally unprepared. His first fire companies were composed of "kids" whom he had trained himself. Brannigan decided that to be effective and safe, fire fighters needed to minimize surprises on the fireground by learning everything possible about potential problems ahead of time.

Thus, the inspiration for the first article in the fire service press on prefire planning was born: "Surveys Aid in Preparation for Handling Large Fires" in the January 1948 issue of *Fire Engineering*. In it, Brannigan described the prefire survey of a huge wooden Navy warehouse full of vital spare airplane propellers.

Many fire departments do preplanning today. Preplans typically include diagrams of the buildings, notations on the location of stairs and utilities, hazard contents, fire hydrant locations, and the types of fire protection systems. Sharing prefire planning knowledge is very beneficial to all concerned, including fire departments responding on mutual aid. For years, these preplans were organized into notebooks and carried on fire apparatus to the fire scene. Today's practice is for fire departments to use computer software to collect the information and have mobile display terminals in apparatus to display the data **Figure 1-3**. The future also appears promising for "interactive preplans," where real-time data from a building for such issues as fire alarm device activations and individual elevator cab status will be transmitted directly to responding fire fighters, supplementing their preplans.

Geographic Information Systems (GIS) have become a useful tool for preplanning. From a firefighting perspective, GIS systems typically take the form of data-enhanced maps of neighborhood response areas that include information on building uses and types, street arrangements, and utility locations. The specific location of important features, such as a fire hydrant or a flammable liquid storage tank, can be incorporated into a GIS system through the use of portable Global Positioning System (GPS) devices (using the object's longitude and latitude coordinates). GIS technology can be integrated with specific electronic building preplans to provide responding firefighters with a more complete set of information for use on the fireground.

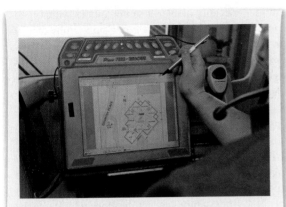

Figure 1-3 Information is vital to all fire fighters.
© Jones and Bartlett Learning. Courtesy of MIEMSS.

The word *plan*, though commonly used to describe this prefire effort, might be a misnomer. A more accurate term may be **prefire analysis**. Analysis goes a step beyond the basic plan; from our perspective, it involves studying the prefire plan information that has been gathered to identify specific construction issues/concerns and interactions. Analysis allows us to anticipate problems.

The use of the word *plan* sometimes leads to overemphasis on the planning of *action*. Strong efforts should be made to accumulate the necessary information on which sound judgments can be based at the time of the emergency, particularly if the emergency force is well trained. If the emergency force is less well trained, then more detailed planning against specific problems is necessary. If the plan is too specific, there is the potential problem that failure to follow the plan, for whatever reason, may be taken as evidence of negligence or incompetence.

⊕ Branniganism

"Undress the building when analyzing how and where the fire will spread in a building."

Perhaps unbelievably, at some locations the concept of preplanning is sadly lacking or even missing. At one government facility, there are 20–30 places where the fire department was instructed to take no action until approval was obtained from a department head or representative. Although this rule can be enforced against the fire department, it has *no effect whatsoever on the fire*. The fire does not recognize any authority. The fire chief is not in charge of the fire, but rather of the fire department's effort to contain the fire. The fire sets its own agenda.

Consider separating prefire planning and the inspection function. An inspection is essentially a policing function. In an inspection, the property should be examined to see whether it complies with standards. If it does not comply, penalties may result. The inspector, therefore, is principally a police officer. As Gilbert and Sullivan told us, "A policeman's lot is not a happy one." The reason is simple: you can't win with a police officer; the best you can do is come out even. It may not be wise to just walk into a facility and announce that you are there to develop a prefire plan. Your visit may happen to coincide with the worst possible day of the year for the building's management. Instead of showing up unannounced for prefire planning, we suggest that the department send a letter from the fire chief indicating the mayor's (or other city/town manager's) appreciation of the valuable contribution the facility is making to the economic life of the community. According to the Chubb Group of Insurance Companies, a large percentage of businesses that suffer a serious fire or other major disaster never reopen, so it is in everyone's interest that such a fire not happen. Any pertinent local experience should be cited in the letter. The letter can continue:

> The mayor has directed the fire chief to make every effort to keep local businesses as safe as possible in an attempt to minimize the effects of a fire. In order for the fire department to operate most effectively on a fire at your facility, there are a number of things that the fire department should know and record ahead of time. Because such surveys require mutual cooperation, they are best done by appointment. Please call the fire chief to arrange a mutually agreeable appointment time.

Wow! A business owner may never have received such a letter in his or her life from any government agency. Usually official letters begin with much more forceful language: "You are hereby directed to…" Imagine the talk at the country club, the business association, or the Rotary club. Although such enthusiasm may be a little overstated, this type of letter may produce cooperation that may be "contagious."

Few people have any idea that firefighting is a planned operation. *Firefighting* is a commonly used term for taking care of unplanned emergencies. For example, "We're so busy fighting fires around here, we can't get next year's budget organized." We have never seen a firefighting movie in which fire fighters were shown discussing a particular hazard or studying a book—but it happens, and it needs to happen more.

Prefire planning surveys are important—even critical. Sometimes the perceived importance of such planning is enhanced by the stature of the fire officer performing the planning survey **Figure 1-4** . If the client is uncooperative, then other measures may be legally available to help ensure cooperation.

Fire departments operate differently than other services or industries operating around the clock, such as police departments, prisons, hospitals, military bases, or manufacturing plants. In such organizations, the decision makers work the day shift. Serious events that happen on the night shift are often referred to these supervisors at home. By contrast,

fire departments deliver a senior officer with broad authority to make important decisions immediately to the scene of an emergency at any time of day. Many fire departments using three- or four-platoon systems operate almost as if each platoon were a separate fire department.

A company or entity that functions in a "normal" organizational manner may not be able to fully understand an organization that has three chiefs, all of whom regard themselves as equal. There are obviously many ways of handling a possible emergency situation (perhaps all equally good), so it is very likely that if each of these three senior officers develops an emergency plan, three different procedures might emerge. This may be the cause for a lack of confidence by an owner or facility manager in the ability of the fire department to handle such a fire situation. Consequently, all contact with an owner or manager should be through a single individual who has the lead responsibility for that particular risk. Differences on procedures should be ironed out within the fire department.

Although the basic objective of prefire planning is to consider potential emergency situations and develop the best possible plans for coping with them, preplanning serves another valuable purpose. If a credible disaster scenario is developed, and the best planning provides an inadequate solution, then there is a clear-cut problem. If the risk has been calculated and the answer is unsatisfactory, either a solution must be found or those responsible must face the distasteful fact of that risk. When planning indicates that such a disaster potential exists, the conclusion should be shared immediately with the management and appropriate officials.

Figure 1-4 Prefire planning surveys are important—even critical.
© Jones & Bartlett Learning. Photographed by Glen E. Ellman.

Expect the Unexpected

While we will study the role of building and fire codes in subsequent chapters of this text, it is important to understand that many—if not most—buildings are not totally compliant with all building (and fire) codes. On the one hand, people make honest mistakes, which can lead to noncompliance. On the other hand, some building designers and contractors take shortcuts, thereby creating dangers for

Figure 1-5 A very observant captain spotted this auto body shop "come-along" inside a library attic. A set of them—cabled across the roof span—were actually holding the roof of the building together!

Courtesy of Paul C. Kearns, Jr.

fire fighters. It is important for you to be alert to such conditions, and preplanning can reveal them before an emergency occurs. Having a basic knowledge of codes will help you identify such safety hazards. Remember—if it doesn't look right, it probably isn't.

Keep an eye out for problems as you conduct a prefire plan walkthrough. Problems may not be so obvious. Look above ceilings. Look for void spaces. Explore all areas of the building. Expect unusual and bizarre conditions from time to time Figure 1-5. In short, expect the unexpected.

Study Fire Reports

Over the last several years, there has been a significant increase in the number of "after-action" reports, particularly when fire fighters are killed. Most notable among these are the fire fighter fatality reports issued by the National Institute of Occupational Safety and Health (NIOSH). In addition, some states and cities have issued their own after-action reports. Most of these studies include important information about the building construction and its behavior in the fire. These reports provide critical information on fire fighter safety, albeit

obtained at a steep price—the supreme sacrifice. Take the time to read them and learn from them.

In addition to these after-action reports, websites have been developed to collect and disseminate information about fireground "near-misses." The two most well-known websites are www.firefighterclosecalls.com and www.firefighternearmiss.com. It is important to periodically visit these sites and review the reports there to learn about the experiences of fire fighters who were nearly injured or killed, including information about the role of building construction in the incident.

Why We Do What We Do

The fire service is full of traditions. It is full of history, including both tragedy and triumph. We cherish the traditions of service and sacrifice. We are proud of our history, and rightfully so. This text is filled with lessons from our real-world firefighting history. These lessons were learned—often through supreme sacrifice—over many years, in many buildings, at many fires. We fight fires in buildings large and small, old and new.

Traditions of heroism and bravery should be honored. And yet, some traditions are simply outdated and need to be abandoned. Changing a fireground philosophy into one of *fire fighter* safety and survival is critical for everyone involved. As you read the pages of this text, keep Phoenix Fire Chief Alan Brunacini's philosophy in mind: Take a significant risk to protect a savable life, a little very controlled risk to protect savable property, and no risk for what is already lost.

Changes in fire company staffing and the explosion of lightweight construction across the United States have tremendously increased the danger to today's fire fighters. Conducting a risk–benefit analysis under such conditions has brought Chief Brunacini's philosophy even more into focus. It is critical that incident commanders and rank-and-file fire fighters alike keep safety as their number one priority. Remember, you didn't start the fire or build the building—take only reasonable risks to protect the property.

TACTICAL CONSIDERATIONS

1. Fire officers should include smoke and fire conditions, location of the fire, and the type of building construction in their initial radio reports. Proper terms not only convey building characteristics, but also the associated hazards.
2. The trend toward lighter and lighter building construction has life-and-death implications for your safety. A risk–benefit analysis should be performed on any lightweight building with an advanced fire. Withdrawal of crews and a defensive strategy should be considered when the civilian life hazard has been accounted for.
3. Fire does not recognize any authority, and the fire chief is not in charge of the fire; the fire sets its own agenda. Fire cannot be bluffed or deceived. The gravity resistance system of the building is constantly under attack during a fire. When the time comes to get out, speed is of the essence.
4. Understanding building construction will help ensure the safety of every fire fighter on the scene. If you understand where the weaknesses in a building are and how they will react when under attack by fire, you and your crew can avoid those dangerous areas.

Chapter Summary

- Understanding building construction will ensure the safety of every fire fighter on the scene. If you understand where the weaknesses in a building are, you can make sure that you and your team steer clear of those areas.
- For fire fighters, using proper building construction terminology is critical to clarify communication, enhance professionalism, and ensure safety.
- Preplanning—and knowing your buildings—is an important fire department function that enables fire fighters to minimize surprises on the fireground by learning everything possible about potential problems ahead of time.
- Prefire analysis goes a step beyond the basic preplanning; it involves studying the prefire plan information that has been gathered to identify specific construction issues/concerns and interactions.
- Many—if not most—buildings are not totally compliant with all building (and fire) codes, so preplanning fire surveys can reveal dangerous conditions before an emergency occurs.
- Study reports of past fires and learn the lessons.
- Fire fighter safety is the number one priority when considering the risks of any particular situation.

Key Terms

Prefire analysis A study of information that has been gathered as part of a prefire plan to identify specific construction issues/concerns and interactions.

Prefire planning A document developed by gathering general and detailed data used by responding personnel to determine the resources and actions necessary to mitigate anticipated emergencies at a specific facility.

Case Study

Many fire service leaders have said that the two main subjects that fire fighters must thoroughly understand are fire behavior and building construction. Understanding how a building is put together and knowing how fire is going to react within that building will save fire fighters and civilians.

1. Without a _____, fire fighters are reduced to reacting to a fire rather than being prepared for it.

 A. size-up
 B. code enforcement plan
 C. prefire plan
 D. forward staging area

2. Which organization/agency provides information on fire fighter deaths and how they occurred?

 A. DEP
 B. NIOSH
 C. EPA
 D. ISO

3. Why is clear communication such a critical factor for all fire departments' operation?

 A. To convey the building's characteristics
 B. To notify fire fighters of hidden hazards
 C. To prepare for fire attack
 D. All of the above

4. In addition to lower fire company staffing levels, which other specific problem has surfaced over the last several years that poses a significant danger to fire fighters?

 A. An increase in drunken drivers
 B. The proliferation of lightweight construction
 C. Construction in flood zones
 D. The increased use of cast iron columns

Challenging Questions

1. Why is it important to study building construction?

2. What is a more appropriate term for a "brick building"?

3. What is a more accurate term for prefire planning?

4. Why should prefire planning and fire inspections be separate activities?

5. According to Phoenix Fire Chief Alan Brunacini, how much risk should be taken to save what is already lost?

Flames: © Dro/Dreamstime.com; Steel texture: © SharpshotDreamstime.com; Chapter opener photo: © Dennis Tokarzewski/ShutterStock, Inc.

Concepts of Construction

OBJECTIVES

At the conclusion of this chapter, you will be able to:

- Identify the variety of forces to which buildings are subjected.
- Identify the different types of loads and explain how they are applied to buildings.
- Define safety factor and explain how it affects the choice of materials used in a building assembly.
- Identify composite materials and explain how the combination of two different materials affects structural elements.
- Identify and define key types of structural members, such as beams, trusses, columns, walls, roofs, and arches.
- Identify the various types of structural connections and explain how they affect transmission of loads and behave in a fire.

Case Study

Figure 2-1 The Strand Theatre collapse in Brockton, Massachusetts, March 10, 1941.

Roof collapse has always been a danger for fire fighters. On March 10, 1941, in Brockton, Massachusetts, fire fighters responded to a report of a fire at the Strand Theatre (**Figure 2-1**). This structure was of ordinary construction, with brick load-bearing walls and wood joists. The theater included seats for more than 1,600 people, including a balcony that spanned the width of the building. To cover this large distance between the exterior walls (and to avoid columns in the balcony itself), the wood roof joists were supported by long-span unprotected steel trusses. A plaster and wire lath ceiling was installed under the trusses.

The fire began in the basement and quickly ran up the wall void spaces and vertical ductwork to the roof. Fire fighters ascended to the balcony to attack the roof fire; even though fire was breaking through the roof on the outside, little fire was visible on the inside. The roof suddenly collapsed onto the balcony, trapping fire fighters between the roof and the balcony seats. A portion of the balcony itself then collapsed. A total of 13 fire fighters died. Subsequent investigations revealed that the steel trusses, heated by the burning wood joists, had twisted and likely elongated, dropping their load. A moderate snow load on the roof was discounted as causing the collapse.

This incident raised serious questions about the relationship between construction and firefighting operations, which continue to plague modern-day fire fighters:

1. Is it possible that fire conditions can appear different on the outside of a building than they appear inside a building?
2. How does the span of a truss relate to load-carrying capacity?
3. When a truss fails, how much area is involved?

Introduction

Building construction has evolved over time. Although the overall rationale for a structure—to provide shelter—has remained the same, the way we build buildings has changed as new technologies and techniques emerge. From a firefighting perspective, it is important to realize that the built environment that you respond to in your community is actually an aging collection of different types of buildings, incorporating a myriad of construction methods, materials, and design concepts, along with the hazards associated with each one. In essence, you must become the local "historian of building construction" to understand how the buildings you protect—and the people in them—will behave under fire conditions.

One thing that hasn't changed, however, is the laws of physics. As a fire fighter, you must have a working knowledge of how the forces of nature and humans cause buildings to collapse. Anticipating the **reaction** of a particular building to these forces is essential to your safety. It is important, therefore, that you have an understanding of the basic concepts of construction. Despite the fact that learning from fire-ground experience is critical, you can't rely solely on this to keep you safe.

From experience, some fire officers have learned to make useful, but limited, judgments regarding the loss of structural stability in a fire. Many of the commonly taught indicators, such as sagging floors, strange noises, and the like, may be cases of "too little, too late." In addition, experience with one type of building element, such as solid sawn wood **joists**, is not valid for **trusses** or wooden I-beams.

Relying on experience alone is not sufficient for today's fire fighters; they must be aware of the theories and principles involved in building construction. We will begin our study by looking at the greatest of all fire fighter enemies—gravity.

Gravity

The force of gravity is the eternal enemy of every building. Twenty-four hours a day, seven days a week, gravity exerts a force on the building. Buildings may appear to be just quietly sitting on their **foundations**, but, in fact, they are under great stresses. Although this may not be apparent as you look at the structure, the stresses are there nonetheless. By one means or another, in assembling a structure, the builder has defied gravity. Not all builders recognize this fact.

The **gravity resistance system** in a building consists of all the **structural elements** and the connections that support and transfer the **loads**—that is, the forces attributable to gravity and other sources of stress on the structure. Fire weakens and destroys the structural elements and/or the connections in a building, and places loads on some structural elements that cannot handle them, which can cause a collapse. Once the collapse starts, the results are unpredictable.

Some buildings are designed to be fire resistive. This means that to some degree they will resist fire-caused collapse. The majority of buildings are non-fire-resistive, and thus may easily collapse as a result of assault by a significant fire on the gravity resistance system.

Definitions of Loads

A load in a building works to destroy the gravity resistance system of the structure. According to *NFPA 5000: Building Construction and Safety Code*, loads are "forces or other actions that result from the weight of all building materials, occupants and their possessions, environmental effects, differential movement, and restrained dimensional changes."

Specific terms are used to describe the various types of loads and the way they are placed on a building. It is important to understand and use these terms correctly. Dictionary definitions are not always identical to the definitions used in the construction field. In addition, the various terms are not always mutually exclusive. For example, a load may be a **live load** and an **impact load** at the same time.

Stress and Strain

Before we proceed any further, we must first understand the concepts of **stress** and **strain**, the result of forces applied to a structural member. In the common vernacular, stress and strain are considered synonyms; however, a clear understanding of the technical definitions of these terms is critical for fire fighters.

As we have discussed, an external force that acts on a structure is called a load. The internal forces that resist the load are called stress and strain. Stress is the force per unit area that produces a deformation. It is usually measured in pounds per square inch (psi), although occasional references to pounds per square foot (psf) are also found. The unit area measurement is at the discretion of the person making the calculations. Always be careful to note the unit area when examining any calculations. **KIP**, a term meaning 1,000 pounds, is used in engineering calculations where the number would be so large as to be unwieldy.

Strain, in contrast, is the actual percentage of elongation (deformation) that occurs when a material is stressed. It is measured in fractions of an inch of deformation per inch of original length of the material.

Compression, Tension, Torsion, and Shear Forces

Generally, four types of forces can be applied to a structural member: **compression**, **tension**, **torsion**, and **shear** Figure 2-2 . In essence, a compressive force is one in which the force squeezes a structural member, such as a concrete **column** supporting a floor. Tensile forces stretch a member such as a steel cable that is supporting a suspended walkway. A torsional force is a twisting force such as that created by turning a screwdriver on a screw; it is the result of **torque** (the measurable turning force applied to a structural member). Shear forces occur *within* a building member when opposing forces pull the member in opposite directions and are often the result of **wind loads**.

Figure 2-2 The design of a suspension bridge anticipates three separate forces: compression, torsion, and tension.

© Brandon Holmes/ShutterStock, Inc.

Figure 2-3 What will be the result of a fire burning underneath this roof with its dead loads?

© Donald King/AP Images.

Dead Loads

Dead load is the weight of the building itself and any equipment permanently attached to it or built in. A more accurate term is **self-weight**. Years ago, dead load was often piled on the building without regard for the consequences.

The modern designer more clearly understands that dead weight breeds dead weight. Consequently, efforts are now made to lighten all parts of a building. Removing a pound of dead weight at the top of a building enables the builder to reduce ounces at many points in the supporting structure. As proof of the criticality of this consideration, advertisements for building materials such as gypsum shaft liners now emphasize their weight saving over the traditional masonry enclosures. Today, buildings can be considered as being bought "by the pound."

Fire resistance is closely related to mass. All other things being equal, a heavier steel **beam** will take longer to fail due to heating than a lighter beam. A 4- by 10-inch (102- by 254-mm) wooden beam will support its load longer while burning than will a 2- by 10-inch (51- by 254-mm) beam. This point highlights a valid general principle: Any substitute structural element that is of less mass than the element previously used to carry an equivalent load is inherently less fire resistant.

Added Dead Loads

A structure's dead load is often increased during alteration of a building. For example, the addition of air conditioners to the roof of a building previously without air conditioning may mean that its dead load is increased without any strengthening of the structure **Figure 2-3**. At one fire in a large city, fire fighters were overhauling a fire in a restaurant, seeking out hidden pockets of fire in the overhead. They attacked the fire-weakened supports of the building. Above them, added-on air-conditioning units fell, causing several fatalities.

Even when a structure is strengthened to cope with the added load, the modifications may not prove adequate to avoid collapse in a fire. For instance, if **lightweight trusses** are strengthened by additional trusses, there is no improvement to resistance from fire collapse because all the trusses will burn and lose strength at the same time.

In one case, a supermarket was converted to a Japanese restaurant with many grills. Each grill required a heavy fume hood. These hoods were hung from the roof. When the building was a supermarket, the roof did not have to support such a dead load. Now the same roof must carry the hoods—and safety has been greatly reduced.

Dead loads can be deadly—especially dead loads for which the building has not been designed. It is critical that your local building department be notified and that a structural engineer assess the situation

when you suspect a dead load is a catastrophe waiting to happen or you see dead loads on a building.

Live Loads

Live loads are any loads other than dead loads. An elevated water tank, for example, is a dead load (the tank itself is a dead load). The water in the tank may be there for years—but it is still a live load. A concrete vault is a dead load. A movable safe, no matter how heavy, is a live load.

The differences between these types of loads are not just picky distinctions; they have practical implications for firefighting operations:

- Dead loads can be accurately calculated.
- Live loads are indeterminate. The live load must be estimated based on the projected use of the building and such variables as snow, wind, or rain.

All buildings are designed with a given use in mind. The building code, in turn, specifies the minimum live-floor load design for specific types of buildings. Typical building code minimum design load requirements for modern buildings are spelled out in *NFPA 5000: Building Construction and Safety Code*, **Table 2-1**. The uniform "live load" column refers to loads distributed over the entire area (in psf), while the **concentrated load** applies to specific large objects. An office building floor, for example, must be able to sustain a load of at least 50 psf over the entire floor but also have the ability to sustain a load of 2,000 lb (907 kg) at a specific location on the floor for individual or collected heavy objects like a heavy safe. Note that these are minimum design loads—if you were to find loading in areas of actual buildings that exceed these limits, it is best to contact your buildings department and advise it of the potentially overloaded conditions.

Table 2-1 *NFPA 5000: Building Construction and Safety Code*, Table A.35.6.1.2: Live Loads		
	Uniform	**Concentrated**
Occupancy or Use	**psf (kN/m²)**	**lb (kN)**
Apartments (see residential)		
Access floor systems		
Office use	50 (2.4)	2000 (8.9)
Computer use	100 (4.79)	2000 (8.9)
Armories and drill rooms	150 (7.18)[a]	
Assembly areas and theaters		
Fixed seats (fastened to floors)	60 (2.87)[a]	
Lobbies	100 (4.79)[a]	
Movable seats	100 (4.79)[a]	
Platforms (assembly)	100 (4.79)[a]	
Stage floors	150 (7.18)[a]	
Balconies and decks	1.5 times the live load for the occupancy served. Not required to exceed 100 psf (4.79 kN/m²)	
Catwalks for maintenance access	40 (1.92)	300 (1.33)
Corridors		
First floor	100 (4.79)	
Other floors, same as occupancy served, except as indicated		

Reprinted from: NFPA 5000, Table A.35.6.1.2: Live Loads.

(Continues)

Table 2-1 *NFPA 5000: Building Construction and Safety Code,* Table A.35.6.1.2: Live Loads, (*Continued*)

Occupancy or Use	Uniform psf (kN/m²)	Concentrated IB (kN)
Dining rooms and restaurants	100 (4.79)[a]	
Dwellings (see residential)		
Elevator machine room grating [on area of 2 in. by 2 in. (50 mm by 50 mm)]		300 (1.33)
Finish light floor plate construction [on area of 1 in. by 1 in. (25 mm by 25 mm)]		200 (0.89)
Fire escapes	100 (4.79)[a]	
On single-family dwellings only	40 (1.92)	
Fixed ladders	See Section 4.5 [of ASCE/SEI 7]	
Garages	40 (1.92)[a, b, c]	
Passenger vehicles only		
Trucks and buses		
Handrails, guardrails, and grab bars	See Section 4.5 [of ASCE/SEI 7]	
Helipads	60 (2.87)[d, e] Nonreducible	[e, f, g]
Hospitals		
Operating rooms, laboratories	60 (2.87)	1000 (4.45)
Patient rooms	40 (1.92)	1000 (4.45)
Corridors above first floor	80 (3.83)	1000 (4.45)
Hotels (see residential)		
Libraries		
Reading rooms	60 (2.87)	1000 (4.45)
Stack rooms	150 (7.18)[a, h]	1000 (4.45)
Corridors above first floor	80 (3.83)	1000 (4.45)
Manufacturing		
Light	125 (6.00)[a]	2000 (8.90)
Heavy	250 (11.97)[a]	3000 (13.40)
Office buildings		
File and computer rooms shall be designed for heavier loads based on anticipated occupancy		
Lobbies and first floor corridors	100 (4.79)	2000 (8.90)
Offices	50 (2.40)	2000 (8.90)
Corridors above first floor	80 (3.83)	2000 (8.90)
Penal institutions		
Cell blocks	40 (1.92)	
Corridors	100 (4.79)	
Recreational uses	75 (3.59)[a]	
Bowling alleys, poolrooms, and similar uses	100 (4.79)[a]	
Dance halls and ballrooms	100 (4.79)[a]	
Gymnasiums	100 (4.79)[a, k]	
Reviewing stands, grandstands, and bleachers	60 (2.87)[a, k]	
Stadiums and arenas with fixed seats (fastened to the floor)		

Reprinted from: NFPA 5000, Table A.35.6.1.2: Live Loads.

Table 2-1　*NFPA 5000: Building Construction and Safety Code,* Table A.35.6.1.2: Live Loads, (*Continued*)

Occupancy or Use	Uniform psf (kN/m²)	Concentrated IB (kN)
Residential		
One- and two-family dwellings		
Uninhabitable attics without storage	10 (0.48)[l]	
Uninhabitable attics with storage	20 (0.96)[m]	
Habitable attics and sleeping areas	30 (1.44)	
All other areas, except stairs	40 (1.92)	
All other residential occupancies		
Private rooms and corridors serving them	40 (1.92)	
Public rooms[a] and corridors serving them	100 (4.79)	
Roofs		
Ordinary flat, pitched, and curved roofs	20 (0.96)[n]	
Roofs used for roof gardens	100 (4.79)	
Roofs used for assembly purposes	Same as occupancy served	
Roofs used for other occupancies		
Awnings and canopies		
Fabric construction supported by a skeleton structure	5 (0.24) nonreducible	300 (1.33) applied to skeleton structure
Screen enclosure support frame	5 (0.24) nonreducible and applied to the roof frame members only, not the screen	200 (0.89) applied to supporting roof frame members only
All other construction	20 (0.96)	
Primary roof members, exposed to a work floor		
Single panel point of lower chord of roof trusses or any point along primary structural members supporting roofs over manufacturing, storage warehouses, and repair garages		2000 (8.9)
All other primary roof members		300 (1.33)
All roof surfaces subject to maintenance workers		300 (1.33)
Schools		
Classrooms	40 (1.92)	1000 (4.45)
Corridors above first floor	80 (3.83)	1000 (4.45)
First-floor corridors	100 (4.79)	1000 (4.45)
Scuttles, skylight ribs, accessible ceilings		200 (0.89)
Sidewalks, vehicular driveways, and yards subject to trucking	250 (11.97)[a, p]	8000 (35.60)[q]
Stairs and exit ways	100 (4.79)	300[r]
One- and two-family dwellings only	40 (1.92)	300[r]
Storage areas above ceilings	20 (0.96)	
Storage warehouses (shall be designed for heavier loads if required for anticipated storage)		
Light	125 (6.00)[a]	
Heavy	250 (11.97)[a]	

Reprinted from: NFPA 5000, Table A.35.6.1.2: Live Loads.

(*Continues*)

Table 2-1 *NFPA 5000: Building Construction and Safety Code*, Table A.35.6.1.2: Live Loads, (*Continued*)

Occupancy or Use	Uniform psf (kN/m²)	Concentrated IB (kN)
Stores		
Retail		
First floor	100 (4.79)	1000 (4.45)
Upper floors	75 (3.59)	1000 (4.45)
Wholesale, all floors	125 (6.00)[a]	1000 (4.45)
Vehicle barriers	See Section 4.5. [of ASCE/SEI 7]	
Walkways and elevated platforms (other than exit ways)	60 (2.87)	
Yards and terraces, pedestrian	100 (4.79)[a]	

Reprinted from: NFPA 5000, Table A.35.6.1.2: Live Loads.

[a] Live load reduction for this use is not permitted by Section 4.7 [of ASCE SEI 7] unless specific exceptions apply.

[b] Floors in garages or portions of buildings used for the storage of motor vehicles shall be designed for the uniformly distributed live loads of Table 4-1 [Table A.35.6.1.2] or the following concentrated load: (1) for garages restricted to passenger vehicles accommodating not more than nine passengers, 3000 lb (13.35 kN) acting on an area of 4.5 in. by 4.5 in. (114 mm by 114 mm); and (2) for mechanical parking structures without slab or deck that are used for storing passenger vehicles only, 2250 lb (10 kN) per wheel.

[c] Design for trucks and buses shall be per AASHTO LRFD Bridge Design Specifications; however, provisions for fatigue and dynamic load allowance are not required to be applied.

[d] Uniform load shall be 40 psf (1.92 kN/m²) where the design basis helicopter has a maximum take-off weight of 3000 lb (13.35 kN) or less. This load shall not be reduced.

[e] Labeling of helicopter capacity shall be as required by the authority having jurisdiction.

[f] Two single concentrated loads, 8 ft (2.44 m) apart shall be applied on the landing area (representing the helicopter's two main landing gear, whether skid type or wheeled type), each having a magnitude of 0.75 times the maximum take-off weight of the helicopter and located to produce the maximum load effect on the structural elements under consideration. The concentrated loads shall be applied over an area of 8 in. by 8 in. (200 mm by 200 mm) and shall not be concurrent with other uniform or concentrated live loads.

[g] A single concentrated load of 3000 lb (13.35 kN) shall be applied over an area 4.5 in. by 4.5 in. (114 mm by 114 mm), located so as to produce the maximum load effects on the structural elements under consideration. The concentrated load need not be assumed to act concurrently with other uniform or concentrated live loads.

[h] The loading applies to stack room floors that support nonmobile, double-faced library book stacks, subject to the following limitations: (1) the nominal book stack unit height shall not exceed 90 in. (2290 mm); (2) the nominal shelf depth shall not exceed 12 in. (305 mm) for each face; and (3) parallel rows of double-faced book stacks shall be separated by aisles not less than 36 in. (914 mm) wide.

[k] In addition to the vertical live loads, the design shall include horizontal swaying forces applied to each row of the seats as follows: 24 lb per linear ft of seat applied in a direction parallel to each row of seats and 10 lb per linear ft of seat applied in a direction perpendicular to each row of seats. The parallel and perpendicular horizontal swaying forces need not be applied simultaneously.

[l] Uninhabitable attic areas without storage are those where the maximum clear height between the joist and rafter is less than 42 in. (1067 mm), or where there are not two or more adjacent trusses with web configurations capable of accommodating an assumed rectangle 42 in. (1067 mm) in height by 24 in. (610 mm) in width, or greater, within the plane of the trusses. This live load need not be assumed to act concurrently with any other live load requirement.

[m] Uninhabitable attic areas without storage are those where the maximum clear height between the joist and rafter is 42 in. (1067 mm) or greater, or where there are two or more adjacent trusses with web configurations capable of accommodating an assumed rectangle 42 in. (1067 mm) in height by 24 in. (610 mm) in width, or greater, within the plane of the trusses. At the trusses, the live load need only be applied to those portions of the bottom chords where both of the following conditions are met:
 i. The attic area is accessible from an opening not less than 20 in. (508 mm) in width by 30 in. (762 mm) in length that is located where the clear height in the attic is a minimum of 30 in. (762 mm); and
 ii. The slope of the truss bottom chord is no greater than 2 units vertical to 12 units horizontal (9.5 percent slope).
The remaining portions of the bottom chords shall be designed for a uniformly distributed nonconcurrent live load of not less than 10 lb/ft² (0.48 kN/m³).

[n] Where uniform roof live loads are reduced to less than 20 lb/ft² (0.96 kN/m²) in accordance with Section 4.8.1 [of ASCE/SEI 7] and are applied to the design of structural members arranged so as to create continuity, the reduced roof live load shall be applied to adjacent spans or to alternate spans, whichever produces the greatest unfavorable load effect.

[o] Roofs used for other occupancies shall be designed for appropriate loads as approved by the authority having jurisdiction.

[p] Other uniform loads in accordance with an approved method, which contains provisions for truck loadings, shall also be considered where appropriate.

[q] The concentrated wheel load shall be applied on an area of 4.5 in. by 4.5 in. (114 mm by 114 mm).

[r] Minimum concentrated load on stair treads [on area of 2 in. by 2 in. (50 mm by 50 mm)] is to be applied noncurrent with the uniform load.

Reprinted from: NFPA 5000, Table A.35.6.1.2: Live Loads

Sources: Reprinted with permission from ASCE/SEI 7.

Added Live Loads

The water trapped in a building or on a roof can be a significant added live load. Flat-roof buildings, for example, often are built with little reserve strength, and the added weight of water—such as that delivered to the roof either from natural events, like rain or snow storms, or as part of a firefighting operation—could lead to their collapse. Similarly, an elevated water tank is designed for a certain live load—the water in the tank. If the tank overflows during freezing weather, the escaping water will form an ice coating on the structure—and the added load could overload it to the point of collapse.

In the past, both rainwater and snow have been the cause of many roof collapses. One of the most notable early collapses due to snow involved the Washington, D.C., Knickerbocker Theater in 1922; its roof caved in after a 28-inch (71 cm) snowfall **Figure 2-4**. A total of 98 people were killed during a movie at the theater, including a congressman, in what is still known today as the "Knickerbocker storm." Subsequent investigations pointed to an underdesigned steel truss and **wall** anchorage problems.

In 1978, the **spaceframe** roof (characterized by three-dimensional pyramid-like trusses) of the Hartford Civic Center in Hartford, Connecticut, collapsed

Figure 2-4 The Knickerbocker Theater roof collapsed in 1922 after a 28-inch snowfall.
Courtesy of the Library of Congress.

under a heavy snow load. This collapse was blamed on a variety of factors, including an underestimation of the dead load of the truss frame itself and inadequate bracing within the spaceframe.

As both the Knickbocker Theater and Hartford Civic Center collapses illustrate, **structural frame** design flaws may not be immediately obvious; indeed, they may not become apparent until a catastrophic failure occurs. Be wary when you see heavy rainwater and snow loads on roofs—the structures may be moments away from collapse.

When the use of a building changes (i.e., in "recycled buildings"), the design of the building should be reviewed to determine whether the structure should be structurally strengthened for its new use and whether its fire safety features should be upgraded. (This topic is discussed in detail in the chapter that reviews building and fire codes). Unfortunately, this need to update the building's infrasture is often ignored, sometimes owing to pressure to eliminate "bureaucratic obstructions" or to save a historic building from wreckers.

As mentioned earlier, live loads can also be added to a building by firefighting operations at the scene of a fire. In one example, water thrown on a theater marquee in freezing weather turned to ice and collapsed the marquee, trapping several fire fighters. Fortunately, one fire fighter had access to a construction crane and the marquee was lifted.

Keep in mind the tremendous amounts of weight added to a building by the use of heavy-caliber streams. Water weighs 8.34 pounds per gallon, so a 1,000-gallon per minute (gpm) (3785 Lpm) master stream will potentially add over 4 tons (3632 kg) of weight to the building in just 1 minute **Figure 2-5**. Be cognizant of the amount of water runoff: How much water is remaining and adding to an already fire-compromised structure?

Impact Loads

Impact loads are loads that are delivered in a short time. A load that the structure might resist, if delivered as a static load over time, may cause collapse if

Figure 2-5 A 1,000-gpm (3785 Lpm) master stream will add an additional live load of four tons every minute.

© Design Pics Inc./Alamy Images.

Figure 2-6 This tractor trailer ran off a highway, hit an office building (with an exposed structural frame in a parking area on the first floor), and burst into flames. Fire fighters must be kept clear of such a building in anticipation of a possible collapse as they extinguish the fire.

Courtesy of Glenn Corbett.

delivered as an impact load. The jetliner attacks on the twin towers of New York's World Trade Center on September 11, 2001, are examples of a tremendous impact load.

Buildings have a habit of standing up as long as they are left alone. However, fire can release different kinds of impact loads on a building. In the case of a fire, all sorts of interrelationships among building elements, which have mutually supported one another and provided strength not even calculated by the designer, are disturbed. A fire can cause innumerable *undesigned changes in loading*, many of which can cause collapse. An explosion; the overturning of heavy live loads, such as a big safe; the collapse of heavy, nonstructural, ornamental masonry; or even the weight of a fire fighter jumping onto a roof may be enough of an impact load to cause collapse.

The point, of course, is that there is no such thing as a "no impact" load. No matter how gingerly personnel and equipment are placed on a structure, there is a substantial increase in the stress on that structure, at least momentarily. In any event, assaults on a structure already under attack by fire are very dangerous.

Buildings may also experience impact loads unrelated to fires. Buildings have been hit by vehicles as well as falling trees, towers, and other buildings **Figure 2-6** . Most buildings are not designed

for these kinds of assaults and may suffer structural failure consequences as a result.

Lateral Impact Loads

<u>Lateral impact loads</u> can produce disastrously high stresses. A lateral impact load, such as from an explosion, can be delivered from a direction that has little or no stress resistance.

Some walls are designed as load bearing and to accept vertical compressive loads. However, even impressive masonry walls are not necessarily resistant to undesigned lateral loads. The lateral thrust of collapsing floors has brought down many brick walls.

The sudden ignition of trapped carbon monoxide or other fire-generated combustible gases has sometimes caused detonations that have destroyed substantial masonry walls. In 1968, for example, a cooking-gas explosion caused the collapse of a 22-story concrete-panel apartment house in England—the so-called "Ronan Point collapse." In this incident, the undesigned lateral impact load from the explosion blew out one bearing wall of the structure. The floor it was supporting then fell and became an impact load

on the floor below, and the successive collapse of the underlying floors followed. In the wake of this event, gas service in the Ronan Point building was replaced with electricity, and repairs were made to other units. In 1984, however, tenants were moved out of eight other buildings in the same group because of a fear of collapse. The repairs after the first collapse were apparently inadequate.

The **safety factors** built into ordinary buildings are rarely large enough to assure that there will not be progressive collapse in the wake of the first excessive impact load. Progressive collapse is a particular hazard in the construction of concrete frame buildings. Backdraft explosions can occur when the combination of products of combustion, including a mixture of carbon monoxide and oxygen, is exactly right, and they can blow the building apart.

Impact Loads Due to Terrorism

While terrorist attacks using explosives have occurred for many years, the last few decades have seen an increase in both their frequency and public awareness of these incidents. This has led to research efforts in an attempt to better fortify structures from such attacks. Locating buildings farther away from the street (i.e., increasing the "setback") and placing bollards (posts) and barriers in front of "terrorist targets" are two types of modifications that use distance as a means to minimize the effects of a bomb exploded near a building **Figure 2-7**. Structural engineers are now routinely hired to design blast resistance into high-profile buildings.

Broken and flying glazing (such as window glass and decorative glass) from an exploding bomb may kill and injure many people. Specially designed blast-resistant "security" windows were in the process of being installed in the Pentagon on September 11, 2001; the area of the building hit by the terrorist-hijacked plane had actually been retrofitted with these features. The windows in the immediate area of the impact remained intact—a factor that likely saved many lives. Unfortunately, some security-type windows present significant challenges for forcible entry and escape in a fire

Figure 2-7 The new "Freedom Tower" at the World Trade Center site in New York City uses setback distance and a reinforced concrete enclosed core for protection against a terrorist bomb attack.

© Jay Lazarin/iStockphoto.

incident. Glazing and window materials are covered in greater depth in the chapter that discusses the methods and materials of construction, renovation, and demolition.

Static and Repeated Loads

Static loads are loads that are applied slowly and remain constant. A heavy safe is an example of a live, static load; it is not a dead load. **Repeated loads** are loads that are applied intermittently. A rolling bridge crane in an industrial plant applies repeated loads to the columns as it passes over them.

Wind Loads

Wind load is the force applied to a building by the wind. The designer can counter this force in a variety of ways. **Shear walls** are installed in buildings to resist lateral loads from wind and earthquakes. In a small wood-frame structure, the plywood sheathing over the studs provides the needed shear strength. Larger structures also use shear walls constructed of steel, concrete, or **reinforced concrete**. The goal is to provide a stiff and strong enclosure.

Two closely related structural frames, the **braced frame** and the **moment frame**, are also used in buildings to resist lateral wind and earthquake loads (and, more recently, the lateral blast loads from a terrorist's bomb). Braced frame uses diagonal members for bracing purposes Figure 2-8 , while moment frames use special **moment** connections between columns and beams that resist rotation.

The term moment refers to a force causing rotation of a structural member around a fixed point of connection, where the force is applied some distance from the fixed point. An analogous example would be the force you apply to the edge of a book to actually open the book—the force you apply causes the book's cover to rotate at the book's spine and open to the first page. The force you applied multiplied by the distance between the location of the force to the point of rotation (cover edge to spine) gives the "moment."

Wood-frame homes built in coastal areas subject to hurricane-force winds demand more than simply wood-panel sheathing. Originally, small wood-frame buildings in these areas were built with only a **gravity connection** to the foundation. Today, **hurricane bracing** in wood-frame dwellings, using strategically placed metal straps to hold roofs in place and bolts to hold the frame to the foundation, have found great acceptance since Hurricane Andrew devastated Florida in 1992 Figure 2-9 .

Figure 2-8 Diagonal bracing in a steel frame high-rise.
Courtesy of Glenn Corbett.

Figure 2-9 These straps are designed to counter the uplift forces on roof rafters during a hurricane.
Courtesy of IIBHS – The Insurance Institute for Business & Home Safety.

Low-rise bearing wall buildings made of masonry usually have enough mass in the walls that special consideration of wind load is unnecessary. High-rise buildings, in contrast, must be reinforced against the substantial wind forces that can occur at greater heights, often with bracing as described previously. Some arrangements of braces between columns resemble the letter "K" and, therefore, are called **K-bracing**. In old buildings and railroad bridges, heavy riveting of **girders** to columns from top to bottom of the frame is called **portal bracing**.

The framing of lightweight, unprotected, noncombustible steel buildings, however, is tied together to resist wind forces. Fire-caused failure of one part may cause the collapse of other sections when unanticipated torsional loads, which the building design does not take into account, are transmitted through the **ties**. Most floors simply carry loads to the support system of columns and floors. **Diaphragm floors** are designed to stiffen the building against wind and other lateral loads such as earthquakes.

Masonry walls, which are needed to enclose vertical shafts, can do double duty as shear walls, resisting overturning during high winds. This method of construction is satisfactory only for low- or medium-rise buildings. If used in a **megastructure**, it would require shear walls, with no other purpose than wind resistance, to be erected above the floors where some elevators terminate.

Very tall high-rise buildings are built to take the wind load on the exterior walls rather than on the interior core. Sometimes diagonal columns are visible on the exterior. Externally braced buildings are known as **tube construction**, as contrasted with **core construction** Figure 2-10 . Unique giant **vierendeel trusses** (rectangular trusses with very rigid corner bracing), formed by exterior box columns and spandrels, provided the wind bracing of New York's World Trade Center and allowed the buildings to carry the vertical loads around the openings created by the planes after impact.

The wind load can be particularly hazardous to fire fighters when operating in or near buildings under construction. This is because the full benefit

Figure 2-10 Chicago's Willis Tower: A high-rise of tube construction.
© Medioimages/Photodisc/Thinkstock.

of interconnecting all of the parts of the completed building has not yet been realized. Temporary bracing may be inadequate.

In addition to its effect on structures, the wind is a force to be reckoned with in the movement of smoke in buildings. This force is often ignored by some whose solution to the high-rise fire and life safety problem rests first on "sophisticated, state-of-the-art" smoke-control systems.

Concentrated Loads

Concentrated loads are heavy loads located at one point in a building Figure 2-11 . A steel beam resting on a masonry wall is an example of a concentrated dead load. A safe is a concentrated live load.

Figure 2-11 Some buildings are designed to hold the additional weight of a concentrated load. However, some objects are placed without the addition of structural supports.

Courtesy of the estate of Francis L. Brannigan.

In a typical concrete block wall, the building designer may insert solid block, brick, reinforced concrete, or a "**wall column**" to stiffen the wall and carry the weight of the concentrated load. If a wall is being breached and the structure is found to be stronger than normal, choose another location. You are probably right under a concentrated load.

Years ago, floors that were considered "fireproof" (of course, there is no such thing as fireproof) were made of brick segmental **arches**. Many still exist. The wooden floor was leveled in some cases by supporting it on little **piers** (short columns). At some locations, there was a gap of several inches between the floor and the arch. When such a floor burned, a live load such as a safe would fall the few inches. The impact often proved too much for the arch, which had carried the safe for years, and the safe wound up in the basement. To this day, the testing and rating of safes involves dropping the heated safe a considerable distance.

Throughout the United States and Canada are existing examples of almost every feature or practice of building construction. Although you may not have every feature or problem in your area, you should be aware of them, even though you concentrate on those common to your locale.

Imposition of Loads

Loads are also classified according to the orientation in which they are placed on the structure. They can be classified as axial or eccentric. An **axial load** is a force that passes through the **centroid** of the section under consideration. An axial load is perpendicular to the plane of the section. In simpler words, an axial load is straight and true; the load is evenly applied to the bearing structure. All other conditions being equal, a structure will sustain its greatest load when the load is axial **Figure 2-12**.

Structural elements are not always loaded in the most efficient manner. Other considerations may be more important. For instance, a ladder is strongest when absolutely vertical, so that a person standing on the ladder represents an axial load. But human

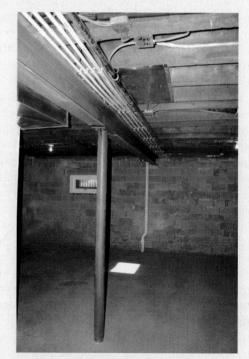

Figure 2-12 Loads are applied evenly in the case of this axially loaded column.

Courtesy of Glenn Corbett.

nature and mechanics work against axial loading and require slanting of the ladder.

An **eccentric load** is a force that is perpendicular to the plane of the section but does not pass through the center of the section, thus bending the supporting members. In simpler terms, the load is straight and true but is concentrated to one side of the center of the supporting wall or column **Figure 2-13**.

For an example of such a load, place a stack of books on the floor. Sit on the stack squarely. Note how evenly applied weight stabilizes the stack. Now shift your weight until it bears on only part of the stack. Note the tendency of the stack to bend outward. One side of the stack is being compressed,

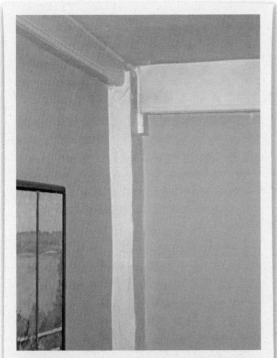

Figure 2-13 This wood post (column), in one of the oldest post and beam houses in the United States, is eccentrically loaded. The bracket for additional support was added since the beam had shrunken away from the post.

Courtesy of Glenn Corbett.

while the other is being pulled apart. These forces are called compression and tension.

Fire Loads

Some engineers do not recognize the term **fire load** or even consider the impact of a fire on a structure. Fire load represents the potential fuel available to a fire. The research work done on fire load was directed toward the contents of buildings. When the building is combustible, however, the building itself is part of the fire load.

Fire load represents the total amount of potential energy (heat) in the fuel; the term **heat release rate (HRR)** (also expressed as \dot{Q}) indicates the rate of available energy released. This distinction is important. All wood can generate approximately the same total amount of heat per pound, but the *rate* at which fuels burn varies dramatically—that is, heat is released more rapidly from certain materials. For example, a 5-pound (2.3 kg) solid chunk of wood will burn more slowly than 5 pounds of wood chips; a pound of plywood will burn many times faster than a pound of heavy timber.

For many years, fire fighters would assess the fire loading of a building solely in terms of pounds of fuel per square foot. To make this calculation, three pieces of information are needed: the weight of the combustible material, the **caloric value** for the material, and the floor area.

Each combustible material has its own caloric value. The basic measurement of caloric value is the **Btu** (British thermal unit), the amount of heat required to raise a pound of water one degree Fahrenheit. The metric equivalent is the **kilojoule (kJ)**. One Btu is approximately equal to 1 kJ. Two estimates for caloric values are commonly used. Wood, paper, and similar materials are estimated at 8,000 Btu/lb. For plastics and combustible liquids, 16,000 Btu/lb is a common estimate, though the caloric value for some of these fuels is much higher.

To measure the fire load in Btu/sq ft, the weight of the fuel is multiplied by the caloric value and divided by the floor area. The equivalent metric statement is kJ/m². In some cases, you may find fire

load expressed in psf, a practice that dates back to when there was only one basic fuel (i.e., wood or paper), estimated at 8,000 Btu/lb. Under this system, plastics were converted into "equivalent pounds" on the basis that 1 pound of plastics equals 2 pounds of wood.

A fire load of 80,000 Btu/sq ft or 10 lb of ordinary combustibles/sq ft is approximately the equivalent of a 1-hour exposure to the standard fire endurance test ASTM E-119, *Standard Test Methods for Fire Tests of Building Construction and Materials* (discussed in greater detail in subsequent chapters); 160,000 Btu/sq ft is equivalent to a 2-hour exposure!

Although this can give one a sense of the severity, it is more important to assess a fire in terms of its total HRR, Q̇. Q̇ is usually expressed in terms of **watts (W)**, **kilowatts (KW)**, or **megawatts (MW)**. For example, a typical polyurethane sofa fire has a peak HRR of approximately 3,120 KW **Table 2-2**. It is this HRR that is a primary determinant of whether a compartment will reach flashover. Fire protection engineers also use HRR to describe a fire's size (and resultant severity). To better understand this relationship, readers are encouraged to read Chapter 5, Basic Fire Science, in *NFPA 921: Guide for Fire and Explosion Investigations*. We'll review this in more detail in the chapter that discusses the effects of fire loads on a structure.

Suspended Loads

Slender tensile members can carry a load that would require a compressive member of much greater size. This feature currently is being used in the design of many buildings.

In many buildings, designers are eliminating interior columns at selected locations by hanging the ends of beams from the overhead structure—thereby creating a **suspended load**. A slender rod can replace a much larger column. When this is done, however, several fire problems develop. The tensile member, having less mass, has less fire resistance. In addition, the load cannot be delivered to the ground in tension; it must be converted into a compressive load. This requires one or more connections in the overhead area, where fire temperatures are often greater than on the floor. These connections may be hidden in the structure. In such a case, an attic fire might cause the collapse of the first floor.

Table 2-2 Representative Peak HRRs (Unconfined Burning)

Fuel	(kg)	(lb)	Peak HRR (kW)
Wastebasket, small	0.7–1.4	1.5–3	4–50
Trash bags, 42 L (11 gal) with mixed plastic and paper trash	1.1–3.4	2.5–7.5	140–350
Cotton mattress	11.8–13.2	26–29	40–970
TV sets	31.3–32.7	69–72	120 to over 1500
Plastic trash bags/ paper trash	1.2–14.1	2.6–31	120–350
PVC waiting room chair, metal frame	15.4	34	270
Cotton easy chair	17.7–31.8	39–70	290–370
Gasoline/kerosene in 0.185 m² (2 ft²) pool	19	—	400
Christmas trees, dry	6–20	13–44	3000–5000
Polyurethane mattress	3.2–14.1	7–31	810–2630
Polyurethane easy chair	12.2–27.7	27–61	1350–1990
Polyurethane sofa	51.3	113	3120
Wardrobe, wood construction	70–121	154–267	1900–6400

Reproduced with permission from NFPA 921-2011, Guide for Fire and Explosion Investigations, Copyright © 2011, National Fire Protection Association. This reprinted material is not the complete and official position of the NFPA on the referenced subject, which is represented only by the standard in its entirety.

Sources: Values are from the following publications:

Babrauskas, V., "Heat Release Rates," in SFPE Handbook of Fire Protection Engineering, 3rd ed., National Fire Protection Assn., Quincy MA (2002).

Babrauskas, V. and Krasny, J. (1985) Fire Behavior of Upholstered Furniture, NBS Monograph 173 Fire Behavior of Upholstered Furniture, National Bureau of Standards, Gaithersburg MD.

Lee, B.T. (1985), Heat Release Rate Characteristics of Some Combustible Fuel Sources in Nuclear Power Plants, NBSIR 85-3195, National Bureau of Standards, Gaithersburg MD.

NFPA 72, National Fire Alarm and Signaling Code, 2010 edition, Annex B.

In old heavy timber buildings, overloaded beams are sometimes restored by inserting a **tie rod**, which goes up through the building, generally to a truss or beam extending from wall to wall in the cockloft. This may be well hidden. Look for the telltale end of the tie rod about midpoint of the beam.

Suspended High-Rise Buildings

The concept of **suspended beams** has been applied in the construction of some high-rise buildings. Columns are replaced with cables suspended from beams **cantilevered** out from the top of the central reinforced concretecore. This provides some economy of construction, unobstructed floor space, and an open plaza at the entry level to the building. Possible fire problems of this type of structure are discussed further in the chapter that discusses fire-resistive construction.

Safety Factor

The safety factor represents the ratio of the strength of the material just before failure to the safe working stress. The term "safety factor" is sometimes misunderstood. It is not practical to use a material in a structure so that it will be loaded to its **ultimate strength**, as shown in tests. After all, the material used in a structure may not be as good as the sample tested, the workmanship may be inferior, or the material may deteriorate over the years. For these reasons, the design load is only a fraction of the tested strength of the material. If the design load is only one-tenth of the tested strength, the safety factor is 10; if the design load is half the tested strength, the factor of safety is 2.

Steel, which is made under controlled conditions, has a safety factor of 2; masonry constructed in place might have a safety factor of 10. The less that is known about the characteristics of a material and its role in a building assembly, the greater the factor of safety required.

The safety factor represents a measure of what is not known about the material. Lower safety factors made on the basis of knowledge of the material and its loading are not reductions in safety. In contrast, improper reductions in the safety factor that might survive normal loading may be significant under fire conditions, creating hazards for residents of the building and fire fighters alike.

The safety factor of steel used in excavation bracing is half what would be permitted in a structure because it is temporary; in other words, the permitted load is doubled while the building is under construction. If a fire occurs in such an unfinished structure, the steel will fail much sooner because the temperature at which heated steel fails decreases as the load increases.

Composite Materials

At times, two materials are combined to take advantage of the best characteristics of each. For instance, concrete is a relatively inexpensive material that is strong in compression but weak in tension; steel is strong in both ways but is more costly than concrete. By providing steel at the locations where tensile stresses develop, a **composite** material, called reinforced concrete, is developed **Figure 2-14**.

Figure 2-14 With the addition of steel rebar, the concrete will withstand loads under tension.

© Graham Prentice/ShutterStock, Inc.

All elements of a composite material must react together if there is to be no failure. If the materials separate, the composite no longer exists, and the two materials separately may be unable to carry the load successfully.

Composite Structural Elements

Two different materials may be combined in a structural element. Steel and concrete are combined in composite floors. In some cases, studs are welded to steel beams and then imbedded in the concrete of the floor. This is done to produce a diaphragm floor, which stiffens the structure. In one serious high-rise building fire, heat caused shear connectors, which joined the beams to columns, to fail. The beams did not fall because the imbedded studs kept the beams attached to the concrete.

Composite concrete-steel floors can be constructed with bar-joist trusses. The top **chord** of the truss is set below the tops of the **web**, allowing triangles of steel to project upwards. These are imbedded in the concrete.

A **Flitch plate girder** is made by sandwiching a piece of steel between two wooden beams. The girder is much stronger than a piece of wood of the same dimensions, yet it can be installed using ordinary carpentry techniques. A sheet of plywood may also be used as the "meat" of this "sandwich."

A **brick and block composite wall**, in which cheaper concrete block substitutes for brick where it will not be seen, is another common example of composite structural elements . Do not confuse this with a brick-veneered concrete block wall, in which the brick and block are not structurally united. In older construction, hollow tile is found in these types of **composite walls** instead of concrete block.

The term "composite construction" is sometimes used to describe buildings in which two different materials carry structural loads. Some concrete buildings, for instance, have a top floor or penthouse of lightweight steel.

Figure 2-15 This photo depicts a brick and block wall. These composite walls are used to cut the overall construction costs.

© Remi Cauzid/ShutterStock, Inc.

Structural Elements

Buildings are made up of structural elements, such as beams, trusses, columns, arches, and walls, that differ in how they carry the load and transfer it to the next element. The principles that govern each element are the same regardless of the material. Elements (or members) that are assembled together into a structure are often called a structural frame.

Beams

The beam is probably the oldest structural member **Figure 2-16**. It is not hard to imagine primitive humans dropping a tree across a stream to form the first bridge. A beam transmits forces in a direction perpendicular to such forces to the reaction points (points of support). Consider a load placed on a floor beam: The beam receives the load, turns it laterally, divides it, and delivers it to the reaction points.

The definition of a beam does not consider its **attitude**—that is, its vertical or horizontal orientation. Although beams are ordinarily thought of as horizontal members, this is not always the case. A vertical or diagonal member that performs the functions of a beam, although it may have another name such as a rafter, is structurally a beam.

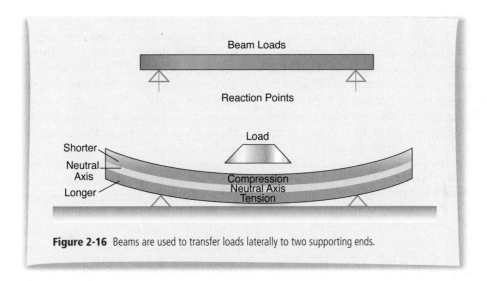

Figure 2-16 Beams are used to transfer loads laterally to two supporting ends.

When a beam is loaded, it deflects or bends downward. The initial load is its own dead weight. The load placed on it is a superimposed load. Some beams are built with a slight **camber**, or upward rise, so that when the design load is superimposed, the beam will be more nearly horizontal. A carpenter selects wood floor beams accordingly.

Deflection causes the top of a beam to shorten so that the top is in compression. The bottom of the beam elongates and thus is in tension. The line along which the length of the beam does not change is termed the **neutral axis** or plane. It is along this line that the material in the beam is doing the least work and material can be most safely removed.

The cable is an ideal beam. Fully in tension, it makes the most economical use of the available material. Cable-supported roofs are being used for some large open-area structures. Nevertheless, for a beam to be safe and economical is not always enough; it must also deflect so little that the deflection will not be noticed. This is accomplished by using additional material or by rearranging the shape of the material. In other words, **stiffness**, or reduced deflection, can be achieved by material mass or by geometry. The economics of using geometry (e.g., truss shapes) over mass has had a tremendous effect on structures.

Carrying Capacity and Depth of Beams

The load-carrying capacity of a beam increases by the *square* of its depth. Consider a 2- by 4-inch (51- by 102-mm) wood beam carrying a certain load. If another 2- by 4-inch (51- by 102-mm) beam is laid alongside the first beam, the carrying capacity is increased by a factor of two. If the same amount of wood is sawn into a 2- by 8-inch (51- by 203-mm) beam, thus multiplying the depth by two, the carrying capacity is increased by the square of two—that is, by four.

The principles governing beams are the same for all beams, regardless of the material. For examples, we will look at wooden beams. Standard carpentry manuals provide a table that details safe longtime loads, uniformly distributed, for 1,800-pound (816-kg) structural grade lumber. For convenience in determining loads for beams of different thicknesses, the information is arranged by size from 1- by 4-inch to 1- by 16-inch, (25- by 102-mm to 25- by 406-mm) even though no beams are as thin as 1 inch. To find the strength of a 3-inch-thick (76 mm) beam, the figure is multiplied by three **Figure 2-17** . A 1- by 4-inch (25- by 102-mm) beam spanning 6 feet (1.8 m) can carry 533 pounds (242 kg); a 1- by 8-inch (25- by 203-mm) beam spanning 6 feet (1.8 m) can carry 2,133 pounds (968 kg) (four times as much); and a 1- by 16-inch (25- by 406-mm) beam spanning 6 feet (1.8 m) can carry

8,533 pounds (3,871 kg). This latter beam is four times the depth of a 1- by 4-inch (25- by 102-mm) beam.

To repeat, the load-carrying capacity or strength of a beam increases as the square of the depth, but increases only in direct proportion to increases in width. Look at this in another way. A 4- by 8-inch (102- by 203-mm) beam spanning 10 feet (3 m) can carry 5,120 pounds (2322 kg). In standard terminology, the width is given first. Suppose the 4- by 8-inch (102- by 203-mm) beam is used as an 8- by 4-inch (203- by 102-mm) beam. The 8- by 4-inch beam can carry only half the load.

The greater efficiency of a deeper beam must be balanced against other considerations, such as the desired thickness of the floor and the deflection of flooring between widely spaced beams. Floor boards are themselves beams. For a variety of reasons, then, the almost universal spacing for sawn wooden beams in ordinary–construction is 16 inches (406 mm) and the depth of the beam is determined by the design load and the span.

The depth of the beam required is particularly significant in the case of trusses because of the greater void in which combustible gases can be generated or

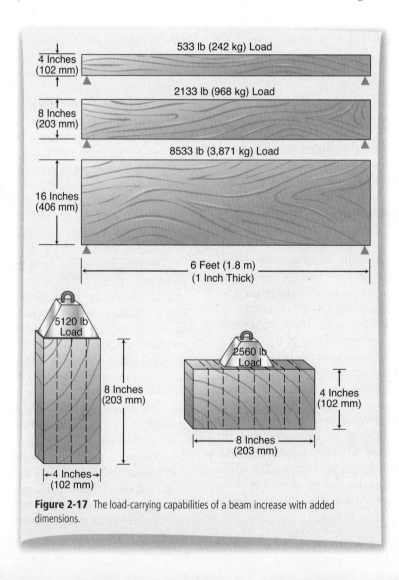

Figure 2-17 The load-carrying capabilities of a beam increase with added dimensions.

accumulated. The length of a span, or the distance between supports, is a determinant of the safe load of a beam. As the length of the span increases, the safe load capacity decreases in direct proportion. Table 2-3 demonstrates this principle.

The figures related to load also consider the beam's weight, which must be deducted to get the permissible superimposed or added load. The figures assume a **uniformly distributed load** Figure 2-18 .

If the load is concentrated at the center of a beam, then the permitted load is half the distributed load. The principles involved are the same for all beams, regardless of the material used.

Types of Beams

Different terms are used for various beams:

- A **simple beam** is supported at two points near its ends. In simple beam construction, the load is delivered to the two reaction points and the rest of the structure renders no assistance in an overload.
- A **continuous beam** is supported at three or more points. Continual construction is structurally advantageous because if the span between two supports is overloaded, the rest of the beam assists in carrying the load.
- A **fixed beam** is supported at two points and is rigidly held in position at both points. This rigidity may cause collapse of a wall if the beam collapses and the rigid connection does not yield properly.

Table 2-3 Safe Load of a Beam.

Span Distance (Feet)	Load Capacity (Pounds)
6 (1.8 m)	3,333 (1,512 kg)
12 (3.7 m)	1,666 (756 kg)
18 (5.5 m)	1,111 (504 kg)
24 (7.3 m)	833 (378 kg)
30 (9.1 m)	667 (303 kg)
36 (11 m)	555 (252 kg)

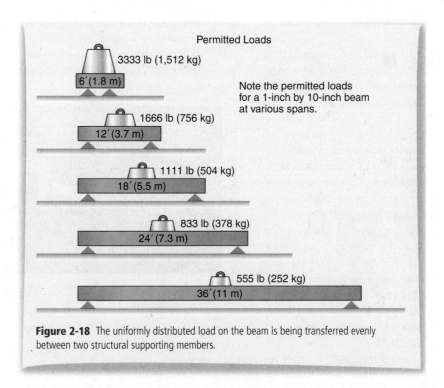

Permitted Loads

3333 lb (1,512 kg)
6′ (1.8 m)

Note the permitted loads for a 1-inch by 10-inch beam at various spans.

1666 lb (756 kg)
12′ (3.7 m)

1111 lb (504 kg)
18′ (5.5 m)

833 lb (378 kg)
24′ (7.3 m)

555 lb (252 kg)
36′ (11 m)

Figure 2-18 The uniformly distributed load on the beam is being transferred evenly between two structural supporting members.

- An **overhanging beam** projects beyond its support, but not far enough to be a cantilever.
- A **bracket** is a diagonal member that supports what would otherwise be a cantilever.
- A joist is a beam, often made of wood.
- A **steel joist** or **bar joist** is a lightweight steel truss joist.
- A girder is any beam, of any material (not just steel), that supports other beams.
- A **built-up girder** is made of steel plates and angles riveted together, as distinguished from a girder rolled from one piece of steel.
- A **spandrel girder** is a beam that carries the load on the exterior of a framed building between the top of one window and the bottom of the window above.
- A **lintel** is a beam that spans an opening in a masonry wall **Figure 2-19**. Stone lacks tensile strength so it can only be used for quite short lintels. Many **precast** concrete lintels have the word TOP cast into the top to be sure they are erected with the reinforcing rods that provide tensile strength in the bottom.
- A **grillage** is a series of closely spaced beams designed to carry a particularly heavy load.
- A **cantilever beam** is supported at only one end, but it is rigidly held in position at that end **Figure 2-20**. It projects out over a support point. Beyond the support point, the tension is in the top and the compression is in the bottom. Cantilever structures are very likely to be unstable in a serious fire because the fire may destroy the method by which the beam is held in place. The cantilever beam resembles a playground seesaw. Many of us can recall being lifted up in the air by a playmate who then ran off and dropped us to the ground. Failure of the interior section of the cantilever can cause collapse of the projecting section. As a cantilevered structure such as a balcony collapses, it is likely to pull down a wall. When this happened in a New Orleans incident, three fire fighters were killed. Cantilevers

Figure 2-19 Lintels are used to support the downward force of the wall above them. This appears to be a window replacement in an old wall; note the dangerous lack of mortar between bricks.

© Meredith Lamb/ShutterStock, Inc.

Figure 2-20 Cantilevered construction.

© Raymond Forbes/age fotostock

are used widely for both architectural and design economy considerations. Wooden cantilevered balconies are common. Temporary construction platforms, for example, on which a crane delivers material, are cantilevered out from a building. They are supported by wooden compression members delivering the load to the floor above. A fire could destroy the wood and cause the platform to fall.

- When any change is to be made in the foundation of an existing wall, the wall must be supported. Often holes are cut through the wall, and so-called **needle beams** are inserted and supported on both sides. They pick up the load of the walls.

- A suspended beam is a simple beam, with one or both ends suspended on a tension member such as a chain, cable, or rod. The typical theater marquee is a suspended beam. Fire may destroy the anchoring connection; the beam then becomes an undesigned cantilever. In a Washington, D.C., theater, the ends of the chains were connected to the roof trusses. When an attic fire destroyed the connections, alert junior officers, who were fire science students, noticed that the front brick wall had been pulled out almost to the point of collapse. Another old theater was converted into a furniture store. In a subsequent fire, its suspended marquee collapsed, killing six fire fighters. The collapsing marquee pulled down the masonry along the entire length of the building; in fact, one victim was about 100 feet (30.5 m) from the initial collapse.

- A **transfer beam** moves loads laterally when it is not convenient to arrange columns one above the other—the ideal arrangement. If it is necessary to change the vertical alignment, a transfer beam must be designed to receive the concentrated load of the column and deliver it laterally to supports. Improper alterations of buildings may produce unintended transfer beams that are points of weakness. In

1981, the Kansas City Hyatt Regency Hotel suffered a disastrous collapse of a walkway in which more than 100 people were killed. A construction change had created a weak connection and caused a short section of the upper walkway beam to act as an undesigned transfer beam; when placed under stress, it failed, with disastrous effects.

The critical nature of beams as part of building structure becomes evident when their failure occurs. Consider the nine-story Alfred P. Murrah federal office building in Oklahoma City, which was destroyed in a terrorist attack in 1995 **Figure 2-21**.

Figure 2-21 The Alfred P. Murrah federal office building in Oklahoma City was destroyed in a terrorist attack in 1995.
Courtesy of FEMA.

It used a reinforced concrete transfer girder at the third-floor level to allow larger spacings between exposed exterior columns on the first two floors. The blast from the bomb set off at the front of the building destroyed these exterior columns, resulting in a lack of support for the transfer girder and initiating a collapse of all of the upper floors.

Beam Loading

Beam loading refers to the distribution of loads along a beam. Assume a given simple beam, which can carry eight units of distributed load. If the load is concentrated at the center, it can carry only four units. If the beam is cantilevered and the load is distributed, two units can be carried. If the load is at the unsupported end of the cantilever, only one unit can be carried.

Particularly in buildings under construction or under demolition by fire or otherwise, excessive loads may be concentrated in one area. This may have been the case in New Orleans when scaffolding with stacks of bricks ready for use collapsed suddenly at a building being reconstructed. Do not increase the overload by the weight of fire fighters and equipment.

Reaction and Bending of Beams

The reaction of a beam is the result of force exerted by a beam on a support. The total of the reactions of all the supports of a beam must equal the weight of the beam and its load.

The bending moment of a beam can be simply described as that load that will bend or break the beam. The amount of bending moment depends not only on the weight of the load, but also on its position. The farther a load is removed from the point of support, the greater the moment. Heavy loads should be placed directly over or very close to the point of support.

Trusses

A truss is a type of beam, a framed structure consisting of a triangle or group of triangles arranged in a

⊕ Branniganism

It's been said people judge others by, amongst other things, the way they write. When it comes to building construction terms, fire fighters must traverse a spelling minefield. For example, a *lintel* is a beam that carries the load over an opening in a wall; a *lentil* is a bean in soup. The word is *masonry*, not *masonary*. Trusses have top and bottom *chords*, not *cords*.

Bolt: © iadams/Shutterstock, Inc.; Texture: © Eky Studio/Shutterstock, Inc; Steel: © Sharpshot/Dreamstime.com

single plane in such a manner that loads applied at the points of intersections of the members will cause only direct stresses (tension or compression) in the members. Loads applied between these points cause flexural (bending) stresses. Trusses can support axial loads but are not designed to handle rotational moments.

The rigidity of the truss rests in the geometric principle that only one triangle can be formed from any three lines. Thus, the triangle is inherently stable. An infinite number of quadrilaterals can be formed from four lines, so the rectangle is inherently unstable.

The economy of a truss is derived from the separation of compressive and tensile stresses so that a minimum of material can be used. There are many designs of trusses, each with its own name **Figure 2-22**.

The top and bottom members of the truss are called chords. The compressive connecting members are called **struts**. The tensile connecting members are called ties. Connections are called **panel points**. As a group, the struts, ties, and panel points are called the web.

Types of Trusses

The principal types of trusses can be described by their overall appearance. In a **parallel-chord truss**, the bottom and top chords are parallel. The steel bar joist. (a parallel-chord truss) used for both roofs and

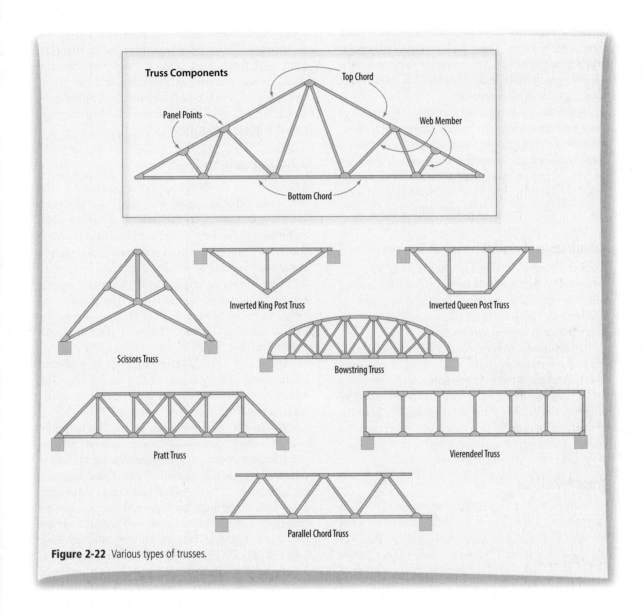

Figure 2-22 Various types of trusses.

floors has been around for some time. Very long-span parallel-chord roof trusses may have a slight upward pitch to the center to facilitate water runoff. Truss floors have been designed in which the trusses were buried in the ceilings or floors of apartments, so that the living space is within the truss. Wooden parallel-chord trusses are being used for floors and roofs in single-family homes, row houses, apartment houses, and smaller office buildings. Recently, builders have been using wooden parallel-chord trusses as roof rafters. This provides a peaked roof, with usable attic space. Dormer windows may be a clue to this type of roof.

Many roof trusses are triangular in shape to provide a peaked roof; these are known as **triangular trusses**. Lightweight triangular trusses must be closely spaced. As a result, they leave no clear space that might be used as an attic.

A serious and continuous problem in truss design has been connectors. In recent years, tremendous strides have been made in the development of connectors capable of transmitting heavy loads. Trusses can provide huge clear spans yet have a dead weight considerably less than that of a corresponding ordinary beam. In some cases, no ordinary beam can do the job.

Compression versus Tension in Trusses

The top chord of a truss is in compression. The bottom chord is in tension. When a truss is cantilevered out, the situation reverses. Beyond the point of support, the tension is in the top chord and the compression is in the bottom chord.

In sketches of trusses, compression members are often shown as thick lines, while tension members are shown as thin lines. This graphic device accounts for the fact that compressive loads are best resisted by columns whose material is as far as possible from the center (e.g., a cylinder). Tensile loads need only the required strength of material. Shape is unimportant.

Truss Principles

Consider a building span of 20 feet (6.1 m), in which two 10-foot (3 m) I-beams extend from the walls on opposite sides to a column support in the center. Assume that each beam can carry 1,000 pounds (454 kg). If the column is removed, the beam would have to be 20 feet (6.1 m) long. According to the principle that doubling the length of a beam decreases its carrying capacity by half, a beam that is 20 feet (6.1 m) long, of the same material and cross-section area, can carry a load of only 500 pounds (227 kg).

Suppose the column is cut off and a stub remains at the junction of the beams. If a tension member is tied from the stub to the beam ends at the wall, a triangle (truss) will be formed, in effect restoring the load-carrying capacity of the beam, yet

removing the obstruction of the column from the floor below. This type of truss is called an **inverted king post truss**. It is called a "king post" because of the single compression member and "inverted" because the compression member extends downward. A truss with two compression members is called a **queen post truss**.

Problems with Trusses

The truss can be designed and constructed as the minimum structure that will carry the designed load. Use of trusses, in fact, is emphasized in some engineering schools. Competitions are held in which the winner produces a truss with the highest ratio of superimposed load to weight of the truss.

"Our trusses are engineered" is sometimes presented as an unquestioned virtue. The question is, "Engineered to what objective?" Unfortunately, the answer is often, "To meet the lowest price in a competitive market."

Trusses can fail in a variety of ways. All parts and connections of a truss are vital to its stability. The failure of one element of a truss may cause the entire truss to fail.

Multiple truss failures can occur. The failure of one truss can have serious impacts on other parts of the structure, even parts far from the initial failure point. For example, steel trusses are tied together for stability to increase wind resistance. When one truss fails, unanticipated torsional stresses can be passed to other trusses by way of the ties, which may cause multiple failures. In one case, a 1,400-foot (427 m) TV tower collapsed as an antenna was being installed. Two diagonals had been removed to permit the antenna to be hoisted up the interior of the tower. Some trusses come with labels warning carpenters not to cut any part of the truss. Despite this, trusses are often cut to accommodate plumbing or other fixtures.

Rising roofs present a serious wood and metal plate truss problem. For example, some attic trusses bow upward, causing ceilings to separate from walls. The *Journal of Forestry* blames this defect on early harvested tree species. Juvenile wood is more prone

to shrinkage and warping. Another theory is that during the cold winter months, the bottom chord of the attic truss, buried in insulation, is warmer and loses more moisture. The upper chord, which is much cooler, retains moisture. The result is that the top chord expands slightly and the bottom chord shrinks slightly, bending the truss and making it "rise." From a firefighting point of view, the most serious effect of rising roofs is the weakened joints that can result from shrinking wood pulling away from **gusset plates**. Gas and electric lines located in such attics may be dangerously affected as well.

Designers rely on standard tables of allowable design stresses in structural components when developing their building plans. A timber truss can have many defects that may lead it to collapse under non-fire conditions such as high wind or snow loads. These same defects may precipitate early collapse under fire conditions. For example, shrinkage of wood may cause stresses that the design of the structure did not take into account. Chemical action, wet and dry rot, enzyme decay, and insect infestation can destroy the timber. Excessive loads may be carried on the trusses.

Defective design may lead to the truss not acting as a truss. Poor maintenance also affects trusses. For example, the building's owners may fail to tighten loose tension rods. Lightweight gusset plate trusses (gusset plates made of small, thin pieces of galvanized sheet steel with teeth punched into one side; the teeth are then pressed into two pieces of wood to hold them together) are also sometimes poorly made or mishandled in the field.

There is intense competition in the building materials supply business, as builders have a clear economic incentive to purchase their materials from the least expensive supplier. Clearly, these factors can lead to provision of less-than-desirable (or even outright shoddy) materials, especially given that the materials will be "hidden" in the building—that is, trusses and any of their potential defects are unseen in a completed structure. With luck, the defects may not cause failure for some time, if ever. When the defects do become apparent in a fire situation, however, the results can be disastrous.

Truss Failure in Fires

The economy of the truss is found in the fact that it separates compression and tension forces. However, this economy can also cause disaster. The bottom chord of a truss is under tension. Recall that a tension member is like a chain—one break precipitates failure.

Conversely, the top chord of a truss is under compression; therefore, it responds exactly like a column. Consider a truss with panel points 5 feet (1.5 m) apart. This is loaded like a column tied off every 5 feet, a series of 5-foot columns one atop the other. If a panel point or connection fails, two 5-foot segments join to become a 10-foot (3 m) segment. This segment can carry only 25% of the design load. Its failure can precipitate the failure of other trusses.

Compare a lightweight wooden parallel-chord truss with a sawn beam. The top of an ordinary beam is in compression and the bottom is in tension. The greatest compression is at the top, gradually lessening as the neutral plane is approached. The greatest tension is in the bottom, gradually lessening as the neutral plane is approached. There is a continuum in the loading of the beam. The loss of some material, therefore, is not necessarily fatal.

In the truss, all of the compressive load is carried on the top chord, typically a light member such as a 2 × 4 or 2 × 6. Loss of wood can precipitate crushing failure in compression.

All of the tensile load is carried on the bottom chord. Except for quite short trusses, the bottom chord is made up of two or more pieces of wood, joined end-to-end with metal gusset plates. The loss of a gusset plate would cause the tensile load to pull the truss apart.

Web members may be connected by gusset plates or by nailing and gluing. In one design, the ties or tension members are stamped from one piece of metal and flattened on the end to make the gusset plate. The steel tie could act as a heat collector, delivering heat to the gusset plate teeth, which can pyrolytically destroy the tensioned wood fibers that had been gripping the 0.375-inch (9.5 mm) deep metal teeth. Tubular steel and wood trusses can burn through at panel points, where there is little wood. A steel pin that carries heat into the wood can cause pyrolytic decomposition.

Large triangular trusses such as those used in apartment house roofs have bottom chords made up of two or more pieces of wood butted together and held by gusset plates. In the case of such trusses used in apartment house roofs, disaster to fire fighters operating below may be averted by the fact that room partitions support the failing trusses, so that only parts of the truss fall into the top floor. The same truss, used in a commercial building to provide a wide clear space, can and has failed catastrophically.

Columns

A column is a structural member that transmits a compressive force along a straight path in the direction of the member **Figure 2-23**. Columns are usually thought of as being vertical, but any structural member that is compressively loaded is governed by the laws of columns, despite its attitude.

Columns by themselves are often used for monuments. Nonvertical columns are often called by other names, such as struts or **rakers**, which are diagonal columns with brace foundation piling.

A **bent** is a line of columns in any direction. If a line of columns is specially braced to resist wind, it is called a wind bent. A bay is the floor area between any two bents. A pillar is a freestanding masonry load-carrying column, as in a cathedral.

Figure 2-23 Columns transmit compressive force along a straight path in the direction of the member.
© S.Borisov/ShutterStock, Inc.

Load-Carrying Capacity of Columns

Beams decrease in load-carrying capacity proportionately—a 12-foot (3.7 m) long beam can carry half the load of a similarly-sized 6-foot-long (1.8 m) beam. Columns, however, lose strength by the square of the change in length. Thus, a 12-foot column can carry only one-fourth the load of a 6-foot column of the same material and cross section.

Shapes of Columns

The most efficient shape for a column is one that distributes the material equally around the axis as far as possible from the center of the cylinder. The thin wall tubing used for the legs of a child's swing set is an example of an efficient use of steel in forming a column. In theory, column material could be paper thin, but it would be subject to local damage, as by a puncture or dent. Most codes provide for minimum wall thickness for columns to prevent local damage.

To obtain a free floor space 63 × 357 feet (19.2 × 108.8 m) in a 48-story high-rise, the designer provided super-columns. They consist of four 39-inch (99 cm) square steel box columns spaced so they form a super-column 21 feet (6.4 m) on a side. The box-column steel walls are as thick as 3.125 inches (79.4 mm) at the base and get narrower farther up the height of the column. Note that the principle of handling a compressive load—place the column material as far from the center as possible—was followed. The space inside the square is used for stairs and elevators.

It is more difficult to attach beams to round columns than to rectangular columns, so less efficient rectangular columns are often used. In cast-iron construction in buildings, interior columns are usually circular, while wall columns (within exterior walls) are rectangular. Rectangular columns often fit better into floor plans. Structural design is often a compromise between competing needs.

When beams were discussed, we explained how a 2- by 8-inch (51- by 203-mm) joist set on the narrow edge would be a much more efficient use of wood than the same amount of wood in a 4- by 4-inch (102- by 102-mm) shape. Conversely, the 4- by 4-inch shape

would be the most efficient use of materials under a compressive load in a column because it is most nearly circular in cross section.

Masonry walls under construction are often braced against high winds. Available scaffold planks, usually 2 by 8 inches (51 by 203 mm) or 2 by 10 inches (51 by 254 mm), are most often used for this purpose. The load on the braces is a compressive load, so the braces should be shaped like columns, but are not. One builder described the 2- by 8-inch planks as "snapping like matchsticks" at a construction site before the walls fell in a windstorm. If four 2- by 8-inch planks had been spiked together to make a hollow column and set diagonally against the wall, with 2- by 8-inch planks laid flat against the walls as beams, the wind load would have been much better resisted with the available material.

Wooden Columns

Most wooden columns are simply smoothed off tree trunks. Large wooden columns, almost always ornamental as well as structural, are hollow. The column consists of curved, usually tongue-and-grooved, sections glued together.

I-Beams versus H-Columns

Whereas steel beams are I-shaped, steel columns are H-shaped, box-shaped, or cylindrical. Beams are shaped like the letter "I" because the depth determines strength. Columns are shaped like the letter "H," and of a dimension that permits a circle to be inscribed through the four points of the "H." A builder may use an available I-shaped steel section as a column but it is wrong to speak of I-beam columns **Figure 2-24**.

Types of Columns

There are three types of columns, which can be differentiated by the manner in which they generally fail. *Piers* are short, squat columns, which fail by crushing. *Long, slender columns* fail by buckling. In buckling, the column normally assumes an S shape. *Intermediate columns* can fail either way.

Understanding the characteristics of long, slender columns yields information and principles applicable to columns and other structures under compressive loads, notably the top chord of trusses.

Euler's Law Columns

Very long, thin columns are known as Euler's Law columns. A straight column, axially loaded, may suddenly collapse. Euler discovered that there is a critical load for a column and that the addition of even a single atom over the critical load can cause sudden buckling and collapse.

Figure 2-24 It is important for fire fighters to use the correct terminology both to demonstrate their professionalism and to ensure safety through common understanding. In this cross-sectional image, the I-beam is located at the top, and the H-column is located on the bottom. Although similar in appearance, the structural properties of these two structural elements differ greatly.

The formula for Euler's Law incorporates the three vital variables in column stability:

$$P_c = \pi^2 \frac{EI}{L^2}$$

in which

P_c is the critical load, the absolute maximum load.

π^2 is 3.1415 squared.

E is the modulus of elasticity of the material in the column. (The modulus of elasticity is a measure of the ability of the material to yield and return to its original shape.)

I is the moment of inertia, a mathematical value for the geometric cross-sectional shape of the column.

L^2 is the length of the column squared.

Once the column starts to yield, there is generally very little reserve strength left and the column is on the verge of total collapse. By contrast, in beam action, there is generally considerable reserve strength beyond initial yielding.

Consider a long, slender column. The load on the column tends to cause it to buckle. If the column is braced rigidly at the midpoint, the effective length of the column is cut in half. Since the square of the length is the divisor in the Euler's Law formula for column loading, cutting the length of the column in two increases the carrying capacity to four times what it was initially. The critical load is four times higher than it was initially. Shortening the effective length of a column by intermediate bracing pays dividends in load-carrying capacity **Figure 2-25**.

The loss of bracing is a cause of column failure. Consider a high scaffold. Several long, narrow columns are securely braced at several points along its height, making several eight-foot segments. The effective length of the column, for this purpose, can be considered to be that of one of the segments. If a bracing connection fails, the effective length of that portion of the column becomes 16 feet (4.9 m). The decrease in load-carrying capacity is geometric **Figure 2-26**. The critical load is reduced to 25%.

Figure 2-25 By bracing a column you can increase the load-carrying capacity.

Column Theory

$$10^2 = 100$$
$$20^2 = 400$$
$$\frac{100}{400} = 1/4$$

100 lb
(45 kg)

400 lb
(181 kg)

20 Feet
(6.1 m)

10 Feet
(3 m)

Figure 2-26 As a column gets longer, its load-carrying capacity decreases.

Temporary Bracing

When a building is under construction, many elements are not in position or are not permanently connected. It is possible that columns might not have the full benefit of the bracing that will be provided by the completed building. Guy or temporary bracing is commonly used in such situations. This bracing may not be adequate to resist high winds or other unexpected loads or it may be vulnerable to fire.

Walls

Walls transmit to the ground the compressive forces applied along the top or received at any point on the wall. A wall resembles a wide slender column. The wall may also be required to resist flexural or bending forces, as does a beam. Wind load is an example of flexural force.

Walls are classified in two main divisions: **load-bearing walls** and **non-load-bearing walls**.

Load-bearing walls carry a load of some part of the structure in addition to the weight of the wall itself. Non-load-bearing walls support only their own weight. **Veneer walls**, **panel walls** (curtain walls), and **partition walls** are some examples of non-load-bearing walls.

A load-bearing wall is more stable than a non-load-bearing wall of identical construction because of the weight of the superimposed load. The load carried by any component in addition to its own dead weight provides stability. Stack up a wall of books or blocks. The wall is easily pushed over. Superimpose your weight on this wall; it is now much more difficult to push over.

The structural consequences of the collapse of a load-bearing wall may be much more serious than those following the collapse of a non-load-bearing wall because of the collapse of building elements supported by the bearing wall. To a fire fighter caught in a collapse, however, it makes little difference whether a masonry wall is load bearing or non-load-bearing.

Within these two broad classes of walls, other descriptors can be applied to walls. A **cross wall** is any wall at right angles to any other wall. The walls should brace one another. Sometimes the bond is so poor, however, that there is no benefit from the cross wall.

A veneer wall is made of a single vertical thickness of masonry; it is designed simply to improve the exterior appearance of the building **Figure 2-27**. Decorative masonry such as stone, brick, or marble may be veneered over common stone masonry, concrete block, reinforced concrete, or steel. By far, however, the greatest use of veneered walls is brick veneer on wood frame.

The veneer wall totally depends on the underlying wall for stability. It should be tied to the wall with metal ties embedded in the mortar. It can be catastrophically unsafe if left to support itself. For example, brick veneer was applied to the visible parapetted portion of a masonry block **fire wall** in an apartment development near Charlotte, North Carolina, to improve its appearance. The brick was applied without ties and was resting on the wood roof. In 1989, the brick collapsed through the roof. Fortunately, no one was in the area, so injuries were avoided.

A composite wall is composed of two or more masonry materials that react together under the load.

To save costly brick used in masonry walls, brick and concrete block (in the past, terra cotta tile) are used to form a composite wall. When walls were all brick, some bricks were **headers**, which were turned on end to cross the wall and tie it together **Figure 2-28**. This design was unsatisfactory when concrete block was used in brick and block composite walls, however. In modern construction, a steel-wire masonry truss is now laid in the wall. As a result, the bricks are all laid lengthwise as **stretchers**. Thus, it is no longer possible to differentiate a brick veneer wall from a brick masonry wall at a glance because both walls now can be laid as all stretchers—a technique also known as a running bond.

The terms panel walls and curtain walls are often used interchangeably to describe non-load-bearing enclosing walls on framed buildings **Figure 2-29**. Technically, panel walls are one story in height and curtain walls are more than one story.

A **party wall** is a load-bearing wall common to two structures. If structural members of both buildings are located in common sockets, fire can extend through the opening.

Fire walls should be able to contain the fire with little or no assistance from firefighting forces **Figure 2-30**. All penetrations of the fire wall must be rated according to the building code in effect, with

Figure 2-27 The use of veneer walls has become commonplace.

Courtesy of the estate of Francis L. Brannigan.

Figure 2-28 Load-bearing brick walls will often have both header and stretcher courses.

Courtesy of the estate of Francis L. Brannigan.

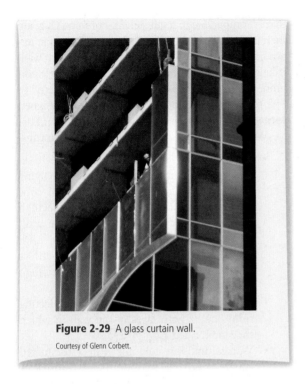

Figure 2-29 A glass curtain wall.

Courtesy of Glenn Corbett.

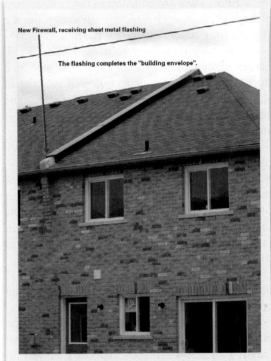

New Firewall, receiving sheet metal flashing

The flashing completes the "building envelope".

Figure 2-30 Fire walls are used to contain fire spread, but penetrations in the wall that do not meet the fire resistance requirements will allow fire to advance through the wall.

Courtesy of Achim Hering.

a rating similar to the fire wall itself. Door openings should be protected with properly rated fire doors. (Fire-rated assemblies are discussed in greater detail in the chapter on the features of fire protection).

Partition walls are non-load-bearing walls that subdivide areas of a floor. They may be required to have some fire or smoke resistance. They also may be required to extend to the underside of the floor above or only to the ceiling line, leaving a void above.

A **knee wall** is a wall typically found in the top floor of a wood-frame home with a peaked roof. This short, 2- to 3-foot-high (61- to 91-cm) wall "squares off" the triangular area at the edge of the room where the sloping roof meets the floor. This creates a void space behind the wall. The resulting triangular spaced area is sometimes used for storage; access to it may be provided through a door—either a full-height closet door or a short access panel or door.

Demising walls are walls bounding a tenant space. A shear wall in a framed building is designed to help resist the force of the wind. It is usually

incorporated in some required enclosure such as an elevator shaft or a stair shaft. In a **rubble masonry wall**, the outer and inner **wythes** (a single vertical thickness of masonry) are laid in courses and finished. The interior consists of stones dumped in at random. Sometimes the stones are loose; in other cases, mortar is dumped in around them.

Cantilever Walls

When a masonry wall is under construction, it acts as a cantilever beam with respect to wind loads received on the face of the wall. When the roof is in place, the wall, with respect to wind loads, becomes a simple beam supported at both ends. High,

freestanding walls are common in the construction of large one-story-high buildings, such as shopping centers, churches, and industrial buildings.

Severe winds may topple a freestanding wall. Earlier, in the section on columns, we discussed the inadequate design of construction wall bracing. In addition, stakes so short that they move when the rain turns the ground to mud are sometimes used—and sometimes fail.

Eccentric loading of the wall by hanging work platforms on one side is another cause of failure. If the wall is braced at one end by another cross wall, the unbraced end will be more vulnerable to collapse. During high winds, there are often gusts that greatly exceed an already high wind speed. In several cases, failure of what had been regarded as well-braced walls has been attributed to gusts.

Precast concrete tilt slab walls (also known as tilt walls or tilt-ups) act as vertical cantilevers when being erected and are braced by tormentors or temporary bracing poles Figure 2-31 . The roof of the building provides the permanent bracing. If the roof fails in a fire, the walls revert to undesigned vertical cantilevers and may collapse.

Wall Bracing

A wall can be compared to a column extended along a line. Most walls have a high height-to-thickness ratio

Figure 2-31 Tilt slab walls being erected. Note the raker shores are holding the walls in place until the roof is completed.

Courtesy of the estate of Francis L. Brannigan.

that is comparable to a tall slender column. They are therefore similarly unstable. President Thomas Jefferson once tried to build a one-brick-thick garden wall. It failed. He reasoned that it was buckling like a column, so he built a buckle into it to brace it internally. The result is the famous Thomas Jefferson-designed **serpentine wall** at the University of Virginia.

Walls can be braced or stiffened by several means:

- **Buttresses** are masonry structures built on the outside surface of the wall. The flying buttresses of Gothic cathedrals, which cope with the outthrust of the roof on the high slender walls, are built away from the wall. A prominent English politician once said, "I am not a pillar of the church, but a buttress since I support it from without rather than from within."
- **Pilasters** are masonry columns projecting from one or both faces of the wall in which they are located Figure 2-32 .
- Wall columns are columns of steel, reinforced concrete, or solid masonry (such as brick or solid block) in a block wall. Concentrated loads such as main girders are applied to the wall directly above the wall column.
- **Cavity or hollow walls** are built of two wythes separated by a space for rain drainage or insulation. Such a wall might be built of two rows of four-inch (102 mm) brick, separated by four inches of space and tied together with steel ties. It is more stable than an eight-inch (203 mm) solid brick wall, just as a column is more stable the farther the material is from the center. Such hollow or cavity walls were originally used to control rain. Rain penetrating the outer wythe drains down the cavity to **weep holes** at the bottom. Currently, the void is often filled with foamed plastic, either rigid board or foamed in place. Sometimes this material is not designed to inhibit the spread of fire. The exact fire problem with these walls is not known, but one can postulate degradation, toxic fumes, and if oxygen becomes available, possibly violent combustion.

Figure 2-32 A set of brick pilasters on the face of an old mill.
Courtesy of Glenn Corbett.

Wall Breaching

If a wall is **homogeneous** (i.e., a good bond exists between bricks, blocks, and mortar), it acts as a unit. A load coming down the wall to a window or doorway does not follow a vertical path, but rather splits and passes on both sides of the opening. Draw a line vertically above an opening, its height at least half the width of the opening. Draw two lines connecting the top of the vertical to the ends of the opening. The masonry in this triangle is the only masonry resting on the lintel.

Roofs

Roofs are very important to the fire service. In some cases, the roof is necessary only to keep the rain out. In other cases, such as with tilt slab concrete buildings, the roof is vital to the stability of the structure, and roof damage can cause wall collapse. Of course, roofs

are the key location where ventilation takes place in many structures during a fire. The danger associated with operating on or under them cannot be overstated.

Roofs may or may not be fire rated. They may be constructed of combustible materials, noncombustible materials, or both. Roofs may be designed to carry large loads or they may be capable of handling only small loads. They may be designed for special purposes such as thermal insulation in a cold storage building or to let in light in an atrium. Roofs may be constructed using a variety of methods and materials. The bottom line: each type of roof is different and presents different challenges. Later chapters in this text discuss specific concerns of roofs in each type of construction.

Conventional wood roofs are discussed in the chapters on wood-frame construction and heavy timber construction. The recent special attention paid to truss roofs should not cause fire fighters to ignore the hazard of the sawn joist roof.

Combustible metal deck roofs are discussed in the chapter that reviews noncombustible construction. Trusses are discussed in detail in the individual chapters dealing with the various types of construction.

While we often think of roofs as "solid" structures, fabric roofs are now more commonly being used in large, permanent structures (unlike the temporary nature of a circus tent). While the advantages of fabric roofs and their related structural frames include quick erection, reflection of the sun's heat, and the ability to bring in daylight, concerns exist over their behavior in strong winds and heavy snow. The Dallas Cowboys' fabric-covered practice facility, for example, collapsed in a strong windstorm in 2009.

A variety of roof styles are shown in **Figure 2-33**. While many of them are specifically associated with wood frame construction, it's important to study each of these and to consider the collapse potential of each type.

Arches

The arch combines the function of the beam and the column. The classic appearance of the Greek temple with its closely spaced columns was dictated by the

Flat Roof

Gable Roof

Mansard Roof

Shed or Single-pitch Roof

Saltbox Roof

Hip Roof

Gambrel Roof

Figure 2-33 Different roof styles.

fact that the Greeks did not know of the arch. The stone beams used in these structures have little tensile strength, so spans must be short. The Romans, however, developed the arch. The arch is under compression along its entire length. If it is properly built, connections are not important to its strength. Many Roman arches used no mortar; the stones were cut and placed perfectly, so that no adhesives were needed.

Because arches tend to push outward at the base, they must be either braced or tied Figure 2-34 . Arches usually are braced by heavy masonry buttresses. Failing buttressed arches often are tied together with steel rods. Some steel arches are tied together across the bottom of the arch, thus eliminating the need for a buttress. In some 19th-century buildings, the brick or concrete floor arch tie rods are visible across the ceiling.

Bowstring-truss roofs are sometimes described as arches or arched trusses, but this terminology is actually incorrect. The truss is a beam, and its thrust is straight down the wall or column. In contrast, the thrust of an arch is outward. When the arch is tied, the thrust is downward, but it lacks the triangles that would make it a truss. This is not the only case where the lines between two construction elements become blurred. Steel arches, as in heavy railroad bridges, may themselves be trusses.

Figure 2-34 Arches tend to push outward at the base.
Courtesy of Glenn Corbett.

The arch is not complete until all of its elements are in place. The keystone is the last stone set in place in an arch, but once in place, it is no more important than other stones (**voussoirs**). An arch is said to spring from its point of support.

The word *arch* brings to mind the graceful Roman segmental arches. A segmental arch is one that describes a portion of the circumference of a circle, but it is not the only arch form. Arches can have one, two, or three hinges—that is, points at which the arch changes direction. Gothic and trefoil arches are found in churches and similar buildings. Flat arches are constructed by tapering masonry units. Hollow-tile flat floor arches are common in 19th-century and early 20th-century buildings.

The availability of laminated wood has made wood arches possible. Many are used in churches. Many exterior masonry arches over windows or doors are false. The exterior wythe is an arch. The interior wythes are carried on wood lintels. This may be an initiator of many masonry wall collapses. If the lintel is steel, elongation may displace the masonry.

Rigid Frames

The **rigid frame** is derived from the arch **Figure 2-35**. Steel rigid frames are widely used for industrial and commercial buildings where clear space is required. Wooden rigid frames are often used for churches. Precast concrete rigid frames also have been used.

Such frames must be tied together at the bottom to resist the characteristic outward thrust of the arch. The ties often are made of steel-reinforcing rods. The rods may connect the bases of the frame or they may be "buttonhooked" into the concrete floor. In this case, the floor is structurally necessary to the building. It is possible that in the future a person cutting into such a floor may consider the tie just another rod and cut it—with surprising, and unhappy, results. In one huge hangar, the ties were cast into concrete beams to prevent this possibility.

In one church it was observed that the ties of a laminated wood rigid frame apparently passed through an ordinary combustible floor-ceiling void. A fire in the basement could cause the ties to fail and perhaps collapse the frame.

Figure 2-35 A rigid frame structure.

Figure 2-36 The exterior of a dome roof in a school.
Courtesy of Glenn Corbett.

Shells and Domes

A shell is a thin plate that is curved. Shells are built of concrete. Wide areas can be spanned with extraordinarily thin shells—shells can be less than 2 inches (51 mm) thick. The shell transmits loads along the curved surface to the supports. An eggshell provides a good example of the strength of a shell in relationship to the weight of material.

The dome is a shell **Figure 2-36**. It can also be considered a three-dimensional arch. There are other forms of shells. These shapes produce roofs of varying architectural designs.

Geodesic domes are formed from a large number of triangles of equal size. They provide structures with very high volume-to-weight ratios. The sheath may be any desired material.

Transmission of Loads

It is important to understand how loads are <u>transmitted</u> from the point of application to the ground. Consider an ordinary brick and wood-joisted building with interior columns. A load is placed on a wood floor. The floor boards are beams and deliver the load to the joists. The joists deliver the load to a masonry wall at one end and to a girder at the other. One end of the girder rests on a masonry wall. The other end rests on a column.

When a load is placed on the floorboards, they transmit the load to the joists on either side of the

load point. The amount of the load transmitted to each joist depends upon the distance of the load point from each of the joists. The nearer joists, of course, receive the greater load.

The load transmitted by the floorboards to the joists is transmitted by the joists to the wall on one end and to the girder on the other. The proportion of load delivered to each support point again depends on the relative distances from the point at which the load is applied to the ends. The load received by the wall is delivered to the foundation, and thus to the ground. The load delivered to the girder is divided among both walls and the column, again in proportion to the distance. The building does it automatically, according to the laws of physics. The structural engineer must calculate the distribution of the loads. This example makes use of a simple masonry and wood-joisted building. Regardless of the size of the building or the construction materials, the principles are the same. Any weight on the roof of a giant high-rise building is transmitted to the ground by the structure of the building.

All loads must be transmitted continuously to the ground from the point at which they are applied. Any failure of continuity will lead to partial or total collapse. Having accurate knowledge of the ground on which a structure is built is vital to its stability.

There is a tendency among those who are concerned about building stability to make light of partial collapse and consider it unimportant. A partial collapse is very important to at least two groups—those under it and those on top of it. If you are caught in a collapse, it is of little importance whether the building collapsed entirely or partially. The 1981 fatal collapse of the suspended walkways (from faulty connections) in the Kansas City Hyatt Regency Hotel, which claimed more than a hundred lives, was only a partial collapse.

Foundations

Ultimately, all loads are delivered to the ground through the foundation. The nature of the ground and the weight of the structure determine the foundation. Foundations can range from simple footings to grade beams or a foundation under the entire wall to

foundations that literally float the building on poor soil. In some cases, wood or concrete piles are driven either to bedrock or until the accumulated friction stops the pile. Almost all foundations today are of concrete.

Connections

Except for the very simplest structures, connections—which transfer the load from one structural element to another—are a vital part of a structure's gravity resistance system. The connection system must be absolutely complete. It is only as strong as its weakest link. A single failure can be disastrous.

Proponents of building materials typically trumpet the virtues of their products, but they may gloss over how those materials relate to other products in real-life conditions. It is important to know about the connections of these materials, however, especially when it comes to firefighting operations. For example, heavy wood beams are slow to ignite, but what is also significant is the connections of wooden girders to cast iron columns, or laminated beams supported on unprotected steel columns.

There are two general types of connections:

- A building is said to be **pinned** when the elements are connected by simple connectors such as bolts, rivets, or welded joints. These are usually not strong enough to reroute forces if a member is removed, so do not assume loads will be redistributed around the missing member.
- In a rigid-framed building, the connections are strong enough to reroute forces if a member is removed.

A **monolithic concrete** structure is one in which the successive poured castings are joined together so that the completed building is like one piece of stone. A monolithic concrete building is rigid-framed. If a column is removed, it is quite possible for the building to redistribute the load to the remaining members. If no member is thereby overloaded, the building will not collapse.

Precast concrete buildings may be pinned or may be made monolithic by the use of **wet joints** in which cast-in-place concrete unites rods that project

from precast sections. Some concrete buildings may have both types of connections. In an ordinary riveted steel-framed building, the collapse of a column will cause the collapse of all elements supported by that column. Some steel buildings have connections that redirect overloads to other sections of the building. This is called **plastic design**.

Failure of Connections

The great majority of buildings stand up. Up to a point a building can cope with undesigned loads, either vertical or lateral. Occasionally, a building not under construction or demolition collapses. A building on fire can be considered a building under demolition. Often structural failure is due to a failure of the connections.

There are numerous ways in which connections can fail:

- Masonry walls shift outward, dropping joists.
- Beams with small bearing areas in the walls on which they rest pull out when the heavy loads are applied to the beam.
- Temporary field bolting of steel gives way in a high wind.
- Steel connectors rust.
- Concrete disintegrates.

In many buildings, and particularly in older buildings, the connections may be adequate as long as the building is axially loaded. However, an eccentric or lateral load, or a shifting wall, floor, or column alignment, may cause collapse even after the building has been standing for many years.

Connections that are capable of carrying a load under normal conditions can drop these same loads quickly under fire attack. Consider the metal joist hanger: instead of a joist resting on top of a girder, the joist hangs on the side of the girder **Figure 2-37**. Heating of this connection in a fire can cause it to fail easily.

Be especially wary of buildings in areas where there was, or is, no building code supervision, such as rural areas that suddenly become urbanized. A Los Angeles City fire fighter died in a roof collapse over what once had been a vacant lot between two brick buildings. To provide a roof over the

Figure 2-37 Metal joist hangers are used in new wood frame buildings. Note that the engineered wood beams do not sit on top of the beam in the center—a point of weakness.

© angelo gilardelli/ShutterStock, Inc.

vacant area, mortar had been removed from between bricks. Pieces of shingle were hammered into the gaps. A **ledger board** nailed to the shingle pieces supported a wood joist roof. The makeshift assembly survived for a number of years until a fire destroyed it.

Connections are concealed in finished areas of buildings. Do not pass up the opportunity to examine basements and attics or buildings under construction or renovation—what you see there is probably typical of the building. Buildings of the same age were probably built by the same methods. Additional problems associated with connections are discussed in other chapters. However, here are examples of some important connection defects.

Sand-lime mortar was used exclusively in masonry work until about 1880. Sand-lime mortar is water soluble. A fire fighter, operating with his unit in the basement of a building, noticed that a hose stream had washed the mortar out from the bricks. He alerted the officer. The building was evacuated and shortly thereafter collapsed.

Gravity connections in buildings simply depend on the weight of the building element to hold them in place. This is especially true for cast-iron columns. A gravity-fit, cast-iron column was one of the causes of the collapse of the Vendome Hotel in Boston, an incident in which nine fire fighters died.

In the past, houses were constructed so that they simply sat on their foundations. Since the 1940s, however, many building codes have required that a structure be anchored to the foundation.

In fact, connections enter into the construction of almost every building. Lightweight wood trusses, for example, are held together with steel gusset plates or **gang nails** (Figure 2-38). Destruction of the fibers holding the gusset plate releases the plate, causing the truss to fail.

Steel is often intermixed with other combustible construction elements. If the building is not required to be fire resistive, the steel is unprotected. Steel heated to 1,000°F (538°C) elongates 9.5 inches (24.1 cm) per 100 feet (30.5 m) of length. If the girder is restrained and cannot elongate, it will overturn and drop its load of wood joists; at higher temperatures, the steel fails.

Unprotected steel connections can often be found in precast reinforced concrete buildings. Unprotected steel rods and cables (which fail at 800°F [427°C]) are often used to tie failing buildings together or to provide some additional resistance to earthquake movement.

It was noted earlier that a load can be suspended on a thin rod as contrasted with the bulky column required to support it in compression. This advantage has not been lost on designers.

Figure 2-38 These gusset plates are used widely in lightweight modern construction within the wood truss system. The plates will fail when introduced to fire, which will cause the truss system to fail.

© Photodisc.

The load cannot stay in tension, however. It must be changed to a compressive load and delivered to the ground. This requires a series of connections. The vulnerable point is the connection most susceptible to fire. It may be located several floors away from the suspended structure. The steel tension rod connection to a wooden beam may be hidden in the cockloft (the space between the top floor ceiling and roof of a building of ordinary construction). The burning of the beam will cause the connection to fail, dropping the rod and its load. Thus, a cockloft fire might cause an interior collapse.

Many codes require that wood joists in masonry walls be **fire cut**. With this technique, the end of the joist is cut off at an angle to permit the joist to fall out of the wall without damaging the wall **Figure 2-39** . The removal of wood lessens the inherent resistance of the joist to fire and can precipitate floor collapse.

Heavy timber buildings were often built with **self-releasing floors** in the 19th and early 20th centuries. Floor girders were set in "anchor boxes" in walls, which allowed the girder to slip out of the wall easily **Figure 2-40** . Wood beams in the building were connected to columns using "post caps." In some cases, wood cleats or steel dog iron similar to a big staple were used to provide minimal stability. Such a floor can be expected to release sooner than if it were tightly connected. Some designers, recognizing this, require tight connections in buildings that they are rehabilitating.

Overhanging and Drop-in Beams

Structural design is often intended to be as economical as possible. The economy may be derived from the choice of material or in the work of erection. Consider a space three bays wide. There are two masonry walls and two lines of columns supporting girders. Sometimes it is more economical to let the two outermost beams overhang the girders by two or three feet. The gap is then closed by a beam dropped in and nailed to the overhanging ends of the outer beams. As

Figure 2-39 A fire cut at the end of a beam.

Figure 2-40 A wall anchor box (bottom) and a post cap (top).

a result, the drop-in beams are connected only by the nailing. They have no support underneath.

Spliced Beams

Long wooden beams are not always available for building needs. Instead, shorter lengths are often spliced together with metal connectors to produce the desired length. The resultant beam will carry its design load, but the connectors may fall out when heated sufficiently, causing collapse. In some buildings, these connectors may have been made to look decorative. Take a second look!

Some years ago, the sports arena in Daytona Beach, Florida, was destroyed by fire. The roof was supported on laminated wood arches. The owners of the building sued the supplier of the foam plastic insulation on the roof, alleging that the flammable plastic had destroyed their sturdy heavy timber building. Pictures of the fire clearly showed that the arches had fallen apart at the connections. When the plaintiffs learned that the defendants had these convincing pictures, they withdrew their suit.

These examples are but a few of the many connection failures that can occur, possibly catastrophically in a fire.

TACTICAL CONSIDERATIONS

1. Gravity exerts great force on buildings 24/7. It never stops. A building's tendency to collapse is increased by fire, while the fire attack crews place additional demands on the structure. When the structure, or part of it, is no longer able to resist the added live and impact loads, collapse occurs. Gravity acts instantly. Once the collapse starts, the results are unpredictable.

2. Company officers should always have an emergency evacuation plan for their crews. Carefully and continuously observe the structure and prepare to evacuate immediately if indicated or ordered. Chief Alan Brunacini once said, "The hardest thing to do is to put a fire fighter into reverse." Instant evacuation from a building, or abandoning the building, should be a drill subject for all firefighters. This sort of thinking does not conform to the image many fire departments have of their operations. From day one of the academy, firefighters are taught to "Hurry up and get in there!" Remember that the proudest ships in the U.S. Navy drill regularly on "abandon ship."

3. A residential structure will usually be designed for a floor load of about 30–40 psf. This includes an adequate built-in safety factor for residences. However, if a building is converted to mercantile or office use without improving the building's design to withstand the increased loads, the loss of structural strength during a fire can cause collapse.

4. Pay attention to buildings that appear to be overloaded or that have loads in excess of the original design. Some clues are HVAC systems on older construction without structural engineering to support the added load, and structures that are internally braced with items such as adjustable posts; these are make-shift attempts at shoring up the increased load. They are not permanent engineered supports and should not be trusted. This should be a red flag and noted on prefire plans.

5. The deployment of hose lines, water, ladders, and fire fighters results in the addition of live loads and impact loads to an existing structure that might already be overloaded. These additional loads can be substantial enough that the increased weight may be more than the building can handle and can raise the potential for collapse. The fire and firefighting efforts will change or increase the loading of a structure. At the same time, the ability of the structure to resist gravity is being attacked by the fire. Unless a building is Type I fire-resistive construction, there is no provision in the design of the structure to prevent failure due to fire. The vast majority of existing structures are not Type I fire resistive. Whatever resistance such structures have to fire-caused structural collapse is purely accidental.

6. When heavy-caliber master streams are used, fire fighters must be conscious of the tremendous weight that they add to the structure under attack. Some of this water will be absorbed into the contents. Paper and fabrics of any type, for example, readily absorb huge quantities of water. Five fire fighters died when piles of scrap from a tissue manufacturer, soaked by water from sprinklers, collapsed on them. Some fire departments have set specific drain times to permit water to drain from structures subjected to heavy master streams before personnel are permitted to enter a building. Paper and fabric storage areas can also absorb substantial amounts of water so that they "swell," increasing their circumference and weight. The swelling can actually push exterior walls away from the roof connections, causing the walls and the roof to collapse. The added weight can increase the concentrated load and cause a floor collapse if the weight increases beyond the designed load limit of the floor. Check the exterior walls for cracks and signs of "pulling away from the roof line" before entering a structure after defensive or transitional attacks.

7. All firefighters must know the rated flow capacities of their nozzles and appliances (gpm) and have the weight of water memorized for quick weight calculations. Water weighs 8.34 pounds per gallon. The math is easy. Simply multiply 8.34 x the gpm

of your water appliance. For example, a 1,000-gpm master stream will potentially add more than 8,000 pounds or 4 tons of weight to a building in just one minute of operation. Modern fireboats can supply well over 20,000 gpm; some fireboat monitors throw between 5,000 and 7,000 gpm per turret. Of course, there will always be water runoff, but it would be inexcusable if the fire department caused a structural collapse that injured or killed fire fighters because crews were oblivious to the amount of weight they added to a structure during master stream operations.

8. Water from rain, snow, or firefighting operations may collect and pond on a flat roof. In the case of such ponding, the fire department may be asked to assist in water removal. Be careful—a flooded roof may be near collapse. Don't add to the weight by introducing a new live load, meaning fire fighters. While operating from ground ladders or an aerial platform, attempt to clear the debris from the drains with a tool. Also, consider placing a dewatering pump on the roof and running the discharge hose off the edge without stepping on the roof. When using master streams on a fire, be aware of overspray onto adjacent flat roof buildings, particularly in freezing weather. The added weight and loading increase the possibility of collapse with such roofs. A roof collapse could set up an air ram that could blow the windows out and collapse the walls. Even staying outside such a building is not always enough. The best tactic is to evacuate the occupants and set up a collapse zone around the building until standing water can be drained from the roof.

9. In these litigious times, incident commanders must be able to predict cause-and-effect actions of the fire department. A structure fire upsets the delicate balance of forces in a building and can result in disastrous changes in loading that the building was not designed to handle. If a catastrophic event is triggered by actions of the fire department, a lawsuit might ensue.

10. Collapse of masonry walls has killed hundreds of fire fighters. Some signs of potential for collapse can be detected during the building inspection or preplanning session before a fire occurs.

11. If you are breaching a wall and find any evidence that the wall is strengthened at that point, stop and start elsewhere. You are likely under a concentrated load.

12. In prefire planning, make note of all concentrated loads, particularly in non-fire-resistive buildings. Old dwellings converted to offices should be examined with special care. Filing cabinets filled with paper are very heavy concentrated loads. The more heavily a structural component is loaded, the sooner it will collapse.

13. The updating of prefire plans with current information can be accomplished from the mobile data computer inside the cab of the apparatus. Updates are made in realtime and immediately distributed throughout the fire department IT computer network. With these systems, prefire information that was once "impractical" to retrieve during an emergency because it took too much time can now be retrieved in realtime.

14. Today's fire officers need to have a solid understanding of fire behavior to forecast events at an emergency. "How big will the fire get?" and "How fast will the fire travel?" are important questions for the incident commander and company officers to consider. Only in recent years has the concept of calculating the fire load in advance gained acceptance, even in fire protection circles. Yet, the absorption of heat by water is the essence of fire suppression by fire departments. Given the widespread prevalence of plastics and synthetic fuels as part of fire loads, it's important to include HRR estimates as part of the criteria in the development of prefire plans. Though extensive data is not available on which to base estimates of HHR, we need to begin somewhere—by using our past experience and trusting our best judgment. Estimates don't have to sound "scientific"; they can be as simple as "Big fire = big water." Using the fire flow formulas can help determine the required gpm that will be required to at least control a fire until it goes out. Calculations of fire load and rate of heat release should be part of a truly adequate prefire plan.

15. Many prefire plans note whether a building is or is not equipped with sprinklers. In contrast, they may not analyze whether the sprinkler system is capable of delivering the amount of water needed to absorb the heat so that the fire can be stopped. At 2:00 a.m., this analysis is irrelevant. If a sprinkler system exists, support it! Certainly, an operating sprinkler system should have some positive effect on a fire even if it cannot completely control it.

Pump operators should automatically hook up to the sprinkler FDC and supplement the system with a supply. Resist the temptation to shut down a sprinkler system too soon. The damage is done. Keep it operating in the fire area until the fire is extinguished. At best, the water spray is cooling the temperature of the fire gases below their flashover levels. The only time supplying a sprinkler system should be stopped is if you're operating on a low-pressure hydrant grid and the sprinkler supply is robbing water pressure needed for attack lines. During preplanning, water supply issues should be noted; in this case, it is better to keep the sprinklers going.

16. The connection that is weakest when subjected to fire is the most significant connection to fire fighters. That connection is the one that will fall on their heads or the one that they will fall through. Suspended structures should be noted and evaluated as part of the prefire plan.

17. The typical theater marquee should always be a red flag—it is a known fire fighter killer. The marquee is actually a suspended beam. If fire destroys the anchoring connection, the beam then becomes an undesigned cantilever.

18. Fire moves easily through hollow columns. In case studies, fire has been transmitted from under porch spaces up to attic spaces through hollow columns, and vice versa. Cast-iron columns were sometimes used as part of the heating system as air ducts in older construction. Fire has extended vertically through cast-iron columns as well.

19. The slightest indication of column failure should cause the building to be cleared immediately. Learn to recognize the signs. Once compression stress limits are reached in a column, there is little reserve strength left and the column is on the verge of total collapse. The failure of a column is likely to occur much more suddenly than the failure of a beam. The failure of a column may precipitate the collapse of the entire portion of the building dependent upon that column.

20. Construction sites where concrete tilt-up slab walls are being erected are vulnerable in high winds until the roof is put on. During high winds, beware of all freestanding walls that are secured by raker shores. These walls can act like a sail in the wind. High winds can exert enough force to push them right over. Establish a collapse zone with an extended area that might be hit by missiles of debris.

21. The proper way to breach a masonry wall is to open a triangular hole. Even before the operation begins, send one member to look for some sort of shoring material. If you cut the top of the hole in a line parallel to the ground, you are depending on the existing mortar above your cut to hold the masonry units in place—certainly an undesigned feature. If a parallel cut is made, it should be supported immediately. If wall elements fall apart easily, indicating that the wall is in poor condition, breaching should cease and you should find another location to force entry.

22. The removal of any part of an arch can cause the collapse of the entire arch. Cutting through tile arch floors, such as for a distributor nozzle or cellar pipe, is dangerous because the removal of one tile can cause the entire arch to fail. A severely unbalanced load or a load outside the center third of an arch may cause it to collapse.

23. When you preplan a building, take a hard look at the structural connections. Do they appear substantial or weak? Are they protected from the heat of a fire? Make notes of this in your preplan.

24. A building with a fire cut or self-releasing floors is designed to collapse inward, without pushing out the walls. It is the duty of a fire officer to ensure that fire fighters are not under the structure when the designed collapse occurs. The only way you will know that such floors are present is from a building inspection. Pay attention to signs, study your prefire plans, and know your buildings.

25. A building on fire should be considered a building under demolition. Take partial collapses seriously. A partial collapse is very important to at least two groups: those under it and those on top of it. If you're caught in a collapse, it is of little importance whether the building collapsed partially or entirely. Call a mayday immediately.

26. Spliced beams make longer beams out of shorter ones and can carry their designed loads. But the connectors may fall out when heated sufficiently, causing collapse. In some buildings, these connectors may have been made to look decorative. Scrutinize them carefully for signs of failure.

27. If a building is not required to be fire resistive, the steel is unprotected. Steel heated to 1,000° F (538°C) elongates 9.5 inches (24.1 cm) per 100 feet (30.5 m) of length. At higher temperatures, the steel fails. Thermal imagers (TI) are a practical way of gauging fire impingement on structural members.

28. Unprotected steel rods and cables, which are often used to tie failing buildings together or to provide some additional resistance to earthquake movement, fail at 800°F (427°C). You need to recognize when temperatures are approaching this point because it is an indicator of potential collapse. TIs are becoming more compact and sophisticated. Directional and pinpoint surface area temperatures will register on TIs, helping company officers determine when the fire is approaching the critical temperature for steel. This is the time to quickly withdraw your crews.

Chapter Summary

- All buildings are subject to one enemy: gravity.
- A load in a building works to destroy the gravity resistance system of the structure.
- Specific terms are used to describe the types of different loads and the way they are placed on a building. It is important to understand and use them correctly. Remember that dictionary definitions are not always identical to the definitions used in the construction field.
- The safety factor represents the ratio of the strength of the material just before failure to the safe working stress.
- At times, two materials are combined to take advantage of the best characteristics of each, making a composite material. In addition, two different materials may be combined in a structural element.
- Buildings are made up of structural elements, such as beams, trusses, columns, walls, roofs, and arches that differ in how they carry the load and transfer it to the next element.
- The manner in which loads are transmitted from the point of application of the ground affects a building's stability in both fire and non-fire conditions. Any failure of continuity will lead to partial or total collapse.
- Connections, which transfer the load from one structural element to another, are a vital part of a structure's gravity resistance system. The connection system must be absolutely complete. It is only as strong as its weakest link.

Key Terms

Arch Combines the function of a beam and a column.

Attitude Vertical or horizontal orientation.

Axial load A load that passes through the centroid of a section under construction and is perpendicular to the plane of the section.

Bar joist Lightweight steel truss joist.

Beam A structural member that transmits forces perpendicular to such forces to the reaction points.

Bent A line of columns in any direction.

Braced frame A structural system that uses diagonal members to provide bracing against lateral wind and earthquake loads.

Bracket Diagonal member that supports what would otherwise be a cantilever.

Brick and block composite wall Consists of an exterior wythe of brick directly mortared or parged to an inner wythe of concrete masonry unit (CMU).

Btu British thermal unit; the quantity of heat required to raise the temperature of 1 pound of water 1°F at the pressure of 1 atmosphere and temperature of 60°F.

Built-up girder Made of steel plates and angles riveted together, as distinguished from one rolled from one piece of steel.

Buttress Mass of masonry built against a wall to strengthen it. Necessary when a vault or an arch places a heavy load or thrust on one part of a wall.

Caloric value Measured in Btu; the amount of heat required to raise 1 pound of water 1°F.

Camber Upward rise.

Cantilever beam A beam supported at one end only, rigidly held in position at that end.

Cantilevered Supported by a cantilever.

Cavity or hollow wall A wall built of two wythes (a single vertical thickness of masonry) separated by a space for rain drainage or insulation.

Centroid The center point at which a body would be stable, or balance, under the influence of gravity.

Chord The outside members (top and bottom) of a truss, as opposed to the inner "webbed members."

Column A structural member that transmits a compressive force along a straight path in the direction of the member.

Composite Built up of different parts, pieces, or materials.

Composite wall A wall composed of two or more masonry materials that react together under load.

Compression Direct pushing force, in line with the axis member; the opposite of tension.

Concentrated load A load acting on a very small area of the structure's surface; the exact opposite of a distributed load.

Continuous beam A beam supported at three or more points. It is considered structurally advantageous because if the span between two supports is overloaded, the rest of the beam assists in carrying the load.

Core construction No external braces involved; bracing is done within the core of the structure.

Cross wall Any wall at right angles to any other wall; the walls should brace one another.

Dead load The weight of a building; the dead load consists of the weight of all materials of construction incorporated into a building, including but not limited to walls, floor, roofs, ceilings, stairways, built-in partitions, finishes, cladding, and other similarly incorporated architectural and structural items, as well as fixed service equipment, including the weight of cranes.

Deflection The deformation or displacement of a structural member as a result of loads acting on it.

Demising wall Wall bounding a tenant space.

Diaphragm floor A floor designed to stiffen a building against wind and other lateral loads such as earthquakes.

Eccentric load A force that is perpendicular to the plane of the section but does not pass through the center of the section.

Fire cut The end of a joist that is cut at an angle to permit the joist to fall out of a wall without damaging the load-bearing wall.

Fire load The potential fuel available for a fire in a building.

Fire resistance The ability of a material to avoid ignition, combustion, and the thermal effects of fire.

Fire wall Wall with a fire-resistive rating and structural stability that separates buildings or subdivides a building to prevent the spread of fire.

Fixed beam Beam supported at two points and rigidly held in position at both points. This rigidity may cause collapse of a wall if the beam collapses and the rigid connection does not yield properly.

Flitch plate girder Made by sandwiching a piece of steel between two wooden beams.

Foundation The lower division of a building that serves to transmit and anchor the loads from the superstructure directly to its earth or rock, usually below ground level.

Gang nail Another name for a gusset plate in a lightweight wood truss.

Girder A beam that supports other beams.

Gravity connection A connection that depends on the weight of the building to hold it in place.

Gravity resistance system All of the structural elements of a building and the connections that support and transfer the loads.

Grillage A series of closely spaced beams designed to carry a particularly heavy load.

Gusset plate (lightweight wood truss) Connecting plate made of a thin sheet of steel used to connect the components of the truss.

Header A masonry unit that overlaps two or more adjoining wythes of masonry to tie them together.

Heat release rate (HRR) The rate at which the potential heat in a fuel is released.

Homogeneous When describing wall construction, a wall that acts as one unit (good bonding exists between bricks blocks, and mortar).

Hurricane bracing Components of a hurricane resistance system that prevent uplift of the components of a structure, including galvanized steel straps used to connect roof trusses to stud walls and anchor bolts used to connect stud walls with sill and sole plates.

Impact load The effect of a moving load upon a stationary structure.

Inverted king post truss A truss incorporating a single compression member; it is inverted because the compression member extends downward.

Joist A beam.

K-bracing An arrangement of braces between columns that resembles the letter "K."

Kilojoule (kJ) Metric unit approximately equivalent to one Btu.

Kilowatt (KW) A unit for measuring the energy release rate of a fire.

KIP 1,000 pounds force.

Knee wall A wall typically found in the top floor of a wood-frame home with a peaked roof. This short wall "squares off" the triangular area at the edge of the room where the sloping roof meets the floor.

Lateral impact load A force that acts on a structure from a horizontal direction, such as wind or seismic forces.

Ledger board A wood board typically attached to a wall's studs that is used to support wood joists.

Lightweight truss A collection of lightweight structural components joined in a triangular unit that can be used to support either floors or roofs.

Lintel The horizontal beam that forms the upper structural member of an opening for a window or door and supports part of the structure above it.

Live load The weight of the building contents.

Load Force or other action that results from the weight of all building materials, occupants and their possessions, environmental effects, differential movement, and restrained dimensional changes.

Load-bearing wall Any wall that carries a load in addition to its own weight.

Megastructure A very large structure.

Megawatt (MW) A unit for measuring the energy release rate of a fire.

Moment The tendency of a force to rotate or twist a structural member.

Moment frame A structural system that utilizes special "moment" connections between columns and beams to resist rotation due to lateral loads such as earthquakes and wind.

Monolithic concrete A construction technique in which all successive poured concrete castings are joined together so that the structure seems to be like one piece of stone.

Needle beam When any change is to be made in the foundation of an existing wall, the wall must be supported. Often holes are cut through the wall, and so-called needle beams are inserted and supported on both sides. They pick up the load of the walls.

Neutral axis The line along which the length of the beam does not change.

Non-load-bearing wall A wall supporting no load other than its own weight.

Overhanging beam A beam that projects beyond its support, but not far enough to be a cantilever.

Panel points The connection points joining ties, struts, and chords in a truss.

Panel wall (curtain wall) Non-load-bearing enclosing wall on framed buildings.

Parallel-chord truss A truss in which both the upper and lower chords are parallel to each other.

Partition wall A non-load-bearing wall that subdivides spaces within any story of a building or room.

Party wall A load-bearing wall that is common to two structures.

Pier A short column of masonry, usually rectangular in horizontal cross section, used to support other structural members.

Pilaster A masonry column projecting from one or both faces of the wall in which it is located.

Pinned A description for structural elements that are connected by simple connectors such as bolts, rivets, or welded joints.

Plastic design Design based on connections that redirect overloads to other sections of the building.

Portal bracing Heavy riveting of girders to columns from the top to the bottom of the frame.

Precast A concrete member that is cast and cured in a place other than its final position in the structure.

Precast concrete tilt slab wall A type of wall that acts as a vertical cantilever when it is being erected and is braced by tormentors or temporary bracing poles.

Q̇ A designation of the HRR; refers to the rate at which a fuel will burn.

Queen post truss A truss with two compression members.

Rakers Diagonal bracing columns.

Reaction The response in structures to the imposed loads, which are generally developed at the supports.

Reinforced concrete In concrete masonry construction, steel reinforcement that is embedded in such a manner that the two materials act together in resisting forces.

Repeated load A load that is applied intermittently.

Rigid frame Structural frame in which all columns and beams are rigidly connected. There are no hinged joints, and the angular relationship between beam and column members is maintained under load.

Rising roof A phenomenon in wood trusses in which differences in moisture levels between the upper and lower wood truss chords cause the truss to bend and create a rise in the roof.

Rubble masonry wall A wall composed of inner and outer wythes of coursed masonry. The space between the wythes is filled with random masonry, sometimes mixed with mortar. Such walls are unstable to a lateral thrust.

Safety factor The ratio of the strength of the material just before failure to the safe working stress.

Sand-lime mortar A water-soluble mixture used in the past as mortar; when water is applied, the mortar can be washed away from the wall.

Self-releasing floor A type of floor in which floor girders are set on anchor boxes in walls and caps attached to columns. A wood cleat or steel dog-iron similar to a big staple is used to provide minimal stability. Often used in heavy-timber construction.

Self-weight Another term for dead load.

Serpentine wall A curving wall.

Shear Forces occurring *within* a building member when opposing forces pull the member in opposite directions.

Shear wall A wall that counteracts the effects of lateral loads such as wind and earthquakes.

Simple beam A beam supported at two points near its ends. In simple beam construction, the load is delivered to the two reaction points and the rest of the structure renders no assistance in an overload.

Spaceframe A three-dimensional pyramid-like truss.

Spandrel girder A girder that ties wall columns together in a framed building.

Steel joist An open web design used for the support of floors and roofs.

Stiffness The capacity of a member or framework to resist imposed loads without excessive deflection.

Strain The actual percentage of elongation (deformation) when a material is stressed.

Stress Force per unit area that produces a deformation.

Stretcher A masonry unit laid horizontally with its length in the direction of the face of the wall.

Structural elements Components of a structure that include beams, trusses, columns, arches, and walls.

Structural frame All members of a structure that are tied together to carry the imposed loads to the substructure, and hence to the ground.

Strut A bracing column.

Suspended beam A simple beam, with one or both ends suspended on a tension member such as a chain, cable, or rod.

Suspended load A hanging load supported from above.

Tension A pulling or stretching force in line with the axis of the body; the opposite of compression, which is pushing, crushing stress.

Tie (truss member) The tensile connecting members of a truss web.

Tie rod A rod in tension; used to hold parts of a structure together.

Torque The measurable turning force applied to a structural member.

Torsion A force tending to twist a structural member.

Transfer beam A beam that typically carries a load around a large opening or over an area in order to avoid intervening columns.

Transmission The manner in which a load is spread from the point of application to the ground.

Triangular truss A roof truss that is triangular in shape; it is used to create a peaked roof.

Truss A type of beam that is a framed structure consisting of a triangle or group of triangles arranged in a single plane in such a manner that loads applied at the points of intersections of the members will cause only direct stresses (tension or compression) in the members.

Tube construction Externally braced structure.

Ultimate strength The highest load that a member or structure can sustain before failure occurs.

Uniformly distributed load A load that is applied evenly over an area.

Veneer wall A wall made up of a single vertical thickness of masonry that is designed to improve the exterior appearance of a building.

Vierendeel truss A rectangular truss with very rigid corner bracing.

Voussoir A wedge-shaped block whose converging sides radiate from a center, forming an element of an arch or vaulted ceiling.

Wall A structural element that transmits to the ground the compressive forces applied along the top or received at any point on the wall.

Wall column A column of steel, reinforced concrete, or solid masonry (such as brick or solid block) in a block wall. Concentrated loads such as main girders are applied to the wall directly above the wall column.

Watt (W) A unit for measuring the energy release rate of a fire.

Web The group of struts, ties, and panel points in a truss.

Weep hole A drainage hole in a masonry wall that allows water trapped inside the wall to escape.

Wet joint Cast-in-place concrete that unites the rods projecting from precast sections.

Wind load The positive or negative force of the wind acting on a structure.

Wythe A single continuous vertical wall of masonry units (one masonry unit in thickness).

Case Study

While conducting a prefire plan at a furniture and appliance store of heavy timber construction, you notice that heavy appliances are stored on the second floor of the structure. You also notice that the roof of the building is being renovated. Answer the following questions about the hazards that might be faced within this building.

1. What type of load is being applied to the floor section of the second floor?

 A. Dead load
 B. Eccentric load
 C. Lateral load
 D. Live load

2. The building being constructed of heavy timber material has a floor assembly that has wood cleats to provide the stability of the floor. What type of floor construction does this signify?

 A. Tie rod
 B. Self-releasing
 C. Rigid framed
 D. Grillage

3. A series of new steel trusses are being added to the roof of the structure. We know from the chapter that steel elongates when heated. At what temperature will steel elongate 9.5 inches (24.1 cm) per 100 feet (30.5 m) of length?

 A. 500 degrees
 B. 1,000 degrees
 C. 1,500 degrees
 D. 2,000 degrees

4. True or False: You note that new gas stoves have been stacked four high on wooden pallets in one particular corner of the second floor. This would be considered a distributed load.

Challenging Questions

1. Which force of nature is always acting to bring a building down?

2. Identify the units of measure when contrasting stress and strain.

3. Identify the portions of a suspension bridge that are in compression and tension.

4. Fire fighters on the roof of a building performing ventilation can be considered what type of load? Could they create more than one type of load?

5. You are a lieutenant conducting a driver training exercise. You notice a crane lifting a new air-conditioning unit onto the roof of a 40-year-old restaurant as you drive by. You park the apparatus and inquire about the building alteration with the restaurant owner. He tells you that the air-conditioning company representative "met with the city" and got its approval. Yet, the restaurant owner cannot produce any paperwork or permits. What do you do? What are your concerns?

Methods and Materials of Construction, Renovation, and Demolition

OBJECTIVES

At the conclusion of this chapter, you will be able to:

- Identify and describe the basic regulations that apply to building construction projects.
- Identify and describe the roles of the typical members of a building design team.
- Identify and describe the roles of the various members of a construction team.
- Recognize why it is important for fire fighters to be able to read building plans and specifications.
- Identify the tasks performed during site preparation prior to construction.
- Point out the unique features of modular construction that influence firefighting operations.
- Identify and describe the hazards to fire fighters during the construction, renovation, and demolition of a building.
- Identify and describe the hazards to fire fighters of the different types of materials used in the construction of buildings.
- Recognize how the shape of a material affects its ability to resist a compressive load or a deflective one.

Case Study

Figure 3-1 The plywood-laden Deutsche Bank building prior to the 2007 fire.

Courtesy of Glenn Corbett.

While buildings undergoing construction and demolition present dangers to fire fighters, buildings simultaneously undergoing decontamination and demolition are even more hazardous. Such was the case in August 2007 when a fire erupted in the former Deutsche Bank building in Manhattan (**Figure 3-1**). The 40-story high-rise, located adjacent to the World Trade Center site, had been heavily damaged in the September 11, 2001, terrorist attacks and remained vacant. As the site was contaminated by asbestos and heavy metals, a plan was developed to decontaminate and demolish the building at the same time. Working from the top down, floors were first enclosed in plywood and decontaminated, followed by demolition. Plywood enclosures were erected on every floor, as well as a special negative-pressure air-handling system for each area (to minimize migration of contaminants) to prepare each floor for decontamination. Unbeknownst to firefighters, contractors had removed the interior stairwells, necessitating the use of an exterior elevator to gain access to each floor. In addition, contractors removed a 42-foot section of the standpipe in the basement, rendering the entire system useless.

When a cigarette-caused fire erupted on the 17th floor, flames spread quickly through the enormous plywood fuel load. Fire fighters, unaware of the disconnected standpipe, were unable to get water on the fire. The fire grew in size, intensified by the specialized air-handling system. Several firefighters became disoriented and called maydays. Some found the exterior "walls" and pushed the plywood panels out of the way, rolling out on to narrow scaffolding many floors above the street. Two fire fighters lost their lives.

1. How are the dangers to fire fighters amplified when decontamination and demolition are conducted at the same time?
2. How dangerous is it to use plywood as a decontamination enclosure in a high-rise under demolition?
3. How important are standpipes and stairwells in a building under demolition?

Introduction

As a fire fighter, it's important for you to understand the inherent dangers of a building under construction, renovation, and demolition. It's also essential for you to understand how the materials used to construct buildings behave under fire conditions. We begin our study through a review of the construction process itself.

Historically, changes in regulations and controls over the construction of buildings (and what goes on inside them) have been incremental, usually as the result of specific fire disasters. During the 19th century, individual laws were enacted in various cities to deal with inadequate egress from buildings (exterior fire escapes came into being), the need for fire walls to prevent fires from spreading from one building to another,

and the need for fire limits that prevented wood-frame buildings from being built in certain areas of a city.

By the 1870s, John S. Damrell, Chief of the Boston Fire Department and veteran of the Great Boston Fire of 1872, was advocating for a "national" set of building regulations. The National Association of Fire Engineers (now the International Association of Fire Chiefs) supported his efforts. It wasn't until the Great Baltimore Fire of 1904, however, that there was a national push for the development of a model **building code** Figure 3-2 . Since the early 20th century, our nation has been building on these early efforts to develop the national model building codes we have today.

It is important for you to remember that the building stock in your community represents a collection of structures built under a variety of regulations—or

Figure 3-2 After the Great Baltimore Fire of 1904, it was decided that a model building code was essential for cities in the United States.

Courtesy of Fire Museum of Maryland.

even under no regulations at all. Most building code regulations are not retroactive, meaning that buildings do not have to comply with the most current regulations in your community. They are "existing, nonconforming" buildings that may contain a multitude of hazards. Still others may be noncompliant because of poor inspection and enforcement effort.

It's also important to note that until only recently, construction regulations rarely considered the safety of fire fighters. For the most part, codes have been concerned with general fire safety (for the community, such as preventing conflagrations) and the life safety of occupants during egress.

Construction Regulations

New structures built today, as well as the construction process itself, have a variety of regulations to meet. Let's take a look at some that have a direct relationship to you as a fire fighter and that apply to buildings:

- Building code: A building code regulates the actual design and construction of new

buildings, providing for legal minimum levels of health and safety. It is the building code that regulates the level and amount of fire protection in a new structure. For example, it is the building code that dictates how and where fire walls and sprinkler systems are to be built and installed. A building code also applies to the renovation, alteration, or demolition of an existing building. A building code is typically not applied retroactively to existing buildings that are not being changed. Building codes are discussed in detail in the chapter that reviews building and fire codes.

- **Fire code**: A fire code regulates the activities that take place in existing buildings, including the maintenance of existing fire protection features such as automatic sprinklers and fire doors, hazardous processes such as spray finishing, the storage of hazardous materials, and general fire safety precautions. A fire code is applicable to existing buildings and hazards. It is the fire code that most fire fighters will have experience using. Fire codes are also discussed in detail in the chapter that reviews building and fire codes.

- **Plumbing, mechanical, and electrical codes**: Plumbing codes apply to the installation of domestic water systems, sanitary systems, and the like. Mechanical codes apply to systems like air-handling systems and smoke control systems. Electrical codes obviously apply to the installation of electrical systems.

- **Zoning regulations**: Zoning laws dictate land usage in specific cities (and are written by the cities themselves). They spell out which types of buildings can be built in specific locations and what they can be used for; for example, zoning codes typically do not allow industrial plants in the middle of an area designated for single family homes. These regulations indirectly affect fire safety in that they tend to push more hazardous occupancies into certain regions of a city.

- **Americans with Disabilities Act (ADA)**: The ADA is a set of regulations passed in 1990 that includes, among other things, requirements for **areas of refuge** for disabled people in multistory buildings. This is important to you as a fire fighter because disabled persons have been instructed to go to these locations and await rescue by you during an emergency.
- **Insurance regulations**: Some insurance companies may have more stringent requirements with regard to fire protection than your locally adopted building and fire codes.
- **Construction safety regulations**: Occupational health and safety regulations—enforced by state and/or federal authorities such as the Occupational Safety and Health Administration (OSHA)—apply to construction sites. They include provisions to prevent falls (through the use of safety barriers) and protection from being hit or crushed. Enforcement of such regulations can benefit fire fighters as well.

⊕ Branniganism

"Try to develop the 'Brannigan Eye,' a term for my practice of looking at everything from the point of view of firefighter safety."

Bolt: © iadams/Shutterstock, Inc.; Texture: © Eky Studio /Shutterstock, Inc; Steel: © Sharpshot/Dreamstime.com

The Building Design and Construction Process

Depending upon the size of a new building, many different professionals may be involved with the design process. Leading the group is the **architect**, who typically is responsible for the overall project and is the individual who actually designs the building in terms of its appearance and layout. It is the architect who decides which materials will be used and how the building will perform.

If the structure is a large building, the design team may call upon a **structural engineer** to design the structural frame of the building. It is this individual who decides the number and size of beams and columns, analyzes the overall strength of the building, the various anticipated dead and live loads, and incorporates **seismic protection** in structures subject to earthquakes.

There are several other engineers involved in the design of a building. These include the **civil engineer**, who designs the layout of the site including parking, drainage, and roadways; the **mechanical engineer**, who designs the heating and air-conditioning systems and the plumbing systems; the **electrical engineer**, who designs the lighting, telecommunications, and emergency power systems; and in some projects, a **fire protection engineer**, who designs the fire protection systems, including sprinkler systems and fire alarm systems.

Once plans have been drawn up, they are submitted to the city buildings department (and, in some jurisdictions, the fire marshall's or fire prevention office) for review and approval (in most jurisdictions). The buildings department typically maintains a close relationship with the fire prevention bureau, although the fire prevention bureau may not have formal broad review/inspection authority over new structures but may have over fire protection systems and other fire-related provisions. It is advisable for fire fighters to work closely with their fire department's fire prevention officers, who can keep them apprised of new buildings in their district.

After plans for a new building have been approved, the building can be built and inspections performed. Once all code regulations have been fulfilled, a certificate of occupancy can be issued by the buildings department.

Numerous individuals and organizations are involved with the actual construction of a new structure. These include the **general contractor**, who typically has ultimate responsibility of building the structure on time and on budget. The general

Figure 3-3 Sprinkler systems are one of many features built by subcontractors.

Courtesy of Scott R. Cherry V.C.C 2000.COM (National Contractor).

contractor, in turn, usually hires a number of **subcontractors**, who build the various systems/components of the building. **Electrical contractors**, **plumbing contractors**, **wallboard contractors**, **fire alarm/security system contractors**, **sprinkler contractors**, and **fireproofing contractors** (who spray a layer of cementitious "fireproofing" onto steel for fire resistance purposes) are some of these subcontractors, to name a few Figure 3-3 .

Building Plans and Specifications

Building plans, often called blueprints, are the drawings used to construct a structure. They are prepared using a **scale**, an instrument with uniformly spaced marks used to measure distance. Scales are like rulers in some ways, but differ from rulers in that the markings on a scale represent measurements that are proportionally smaller than the actual dimension. For example, certain scales will have markings spaced 1/8" (3 mm) apart, each representing a foot in an actual building measurement. Therefore, 3/8" of an inch (10 mm) on a scale represents 3 feet in actuality. The dimensions on a set of blueprints use small units of measure to represent large units of measure so as to be able to fit the drawing on a manageable size piece of paper (or computer screen).

Architects use multisided architect scales that contain scales such as 1" (25 mm), 3/4" (19 mm), 1/4" (6 mm), and 3/16" (5 mm). Because engineers (particularly civil engineers) design roadways and similar things involving long distances, they will often use an engineer's scale, which has scales of 1 inch = 20 feet, 1 inch = 30 feet, and so on.

Building plans provide all of the details needed to obtain a building permit and to construct the building Figure 3-4 . Often, they are broken down by building component. For example, a section of drawings may detail the floor layout including walls, doors, and corridors; a second section of drawings may detail the structural frame; a third section may include plumbing drawings; a fourth section may consist of electrical drawings; and a fifth section may include mechanical system drawings. Fire protection system drawings for sprinkler and fire alarm systems may be included in a set of plans submitted for a building permit; most often, these plans are created and submitted for permit by the fire protection contractors installing the system.

In the past, building drawings were prepared by hand on paper; today, most plans are designed electronically using computers (computer-assisted

Figure 3-4 A page from a set of building plans and a set of specifications.

Courtesy of David Pangburn.

design [CAD] drawings) and are saved as electronic files.

Specifications ("specs") are manuals prepared by building designers that specify which particular type of device or equipment is to be used in the building. For example, if a particular type and brand of sprinkler head, lighting fixture, or water fountain is desired, the specifications manual will name it. Often, the phrase "or equivalent" is used in the specifications manual to provide flexibility to contractors who will be bidding on the construction work. Products will include a technical data sheet ("cut sheet") that can be compared to the specification to ensure compliance.

It is important for fire fighters to be able to read building plans. They should be able to comment on the plans during fire prevention plan reviews and for preplanning purposes. In addition, fire fighters need to be able to read plans during an actual fire to assist in firefighting efforts.

Site Preparation

Preparing a site for construction is necessary for all new buildings. There are some basics that apply to most new structures. Preliminary site work involves testing the soil for quality and load-bearing capacity. For example, soils are classified as to their strength based upon the amounts of gravel, sand, clay, and silt (particles of sedimentary material consisting of very fine particles between sand and clay particles in size; in some cases silt may include small organic particles as well). Once the load capacity of the soil and the structural loading have been established, decisions about the type of **foundation** can be made.

Structures that will be erected on very weak soils, such as those with high silt content, may need to use deep foundations such as **piles** and **caissons**. These are underground "stilts" upon which the building rests, carrying the load of the building to bedrock or a suitable soil region. Piles are made of wood or steel, whereas caissons are typically constructed of concrete.

Below-grade **footings** are shallow foundations used to support smaller buildings and those on

stronger soils. All buildings have three structural components—the **superstructure** (the building itself above grade), the **substructure** (the area below grade to the lowest basement floor slab), and the foundation (from the lowest below-grade slab and the footings/caissons/piles) **Figure 3-5** .

Once the soil has been classified and the structure has been designed, excavation work may commence. Excavating equipment, and sometimes blasting, will bring the excavation down to its desired depth. **Dewatering pumps** are used to remove groundwater as the excavation moves below the water table. Soil walls in the excavation are protected against collapse by the use of braced sheeting using **crosslot bracing**, **rakers**, and **tiebacks**.

The walls of the substructure and foundation are then constructed. Foundations may take many forms. They can simply be a flat slab of concrete (sometimes called a pad) on which the building sits. Other foundations incorporate basements or crawl spaces, often using concrete block as the support for the building above the empty space. Solid concrete walls can be constructed for the basement, poured in place with traditional **formwork** (typical of larger buildings) or constructed of prefabricated interlocking concrete wall

Figure 3-5 The foundation serves as the support to the loads placed upon the structure.

© Justin Kase/Alamy Images.

panels that are bolted together. Still other basements use interlocking insulated concrete forms, often made of polystyrene or polyurethane, in which concrete is poured and reinforcing rods are added, creating a solid insulated concrete basement wall **Figure 3-6**. The interior surface of the insulating foam in the basement is then covered with **gypsum** board. Electrical conduits and plumbing are often placed in the foam—which is therefore a potential location in which an electrical fire may start.

One special type of wall is a **slurry wall** (usually found in large buildings) in which a trench is dug, steel reinforcement is placed in the trench, and a liquid slurry compound is placed in the trench to support the trench walls. Concrete is then poured into the trench, displacing the slurry, which is removed. Once the concrete has **cured**, the interior face of the trench can be removed, exposing the concrete wall.

For many years, fieldstone has been used for foundations and walls, particularly for residential buildings **Figure 3-7**. In some cases, the fieldstone is "dry laid," meaning essentially that the stones are laid without mortar. Whether joined by mortar or not, the stones must be properly positioned ("interlocked") to ensure stability. Be wary of the stability of fieldstone walls, particularly ones in which the stones are laid haphazardly.

Figure 3-7 A fieldstone wall.
Courtesy of Glenn Corbett.

In some areas of the South, wood-frame houses utilized cedar posts (a durable wood) as a foundation. In these so-called pier and beam foundations, floor beams rested upon the array of posts projecting out of the soil. This arrangement creates a crawl space, which poses a danger in the form of hidden fire under the house. In other areas of the South, a "pier and beam" home uses a concrete beam around the perimeter with masonry piers evenly spaced within the perimeter to support the wood floor **framing**.

Modular Construction

While individual small prefabricated buildings have been in existence for more than 150 years, it wasn't until after World War II that **modular construction** really became popular. The use of multiple prefabricated "modules" to speed up construction and provide better quality control for each module unit is very attractive to developers. This construction technique received lots of publicity when two hotels—the Hilton Palacio Del Rio in San Antonio Texas, created for Hemisfair 1968, and the Contemporary Hotel at Walt Disney World, which opened in 1971—were built with fully furnished rooms using this method. (Modular construction must be differentiated from

Figure 3-6 An interlocking, insulated concrete form.
© Christina Richards/ShutterStock, Inc.

manufactured homes, which are factory-built trailers or "mobile homes.")

In the last several years, there has been an increase in the use of the modular construction technique to build wood-frame single-family and multiple-family dwellings Figure 3-8. Individual modules are delivered via tractor trailer, then lifted into place at the building site using a crane. The various modules, each of which is already equipped with wiring and plumbing, are bolted to one another to create the complete home. The builder adds exterior trim, basement stairs, glass fixtures, and the like to finish the home.

One particular completion task is the installation of gypsum board ceilings using a solid foam **plastic** adhesive in some modular homes. The danger of this practice became quite evident when an exterior fire spread to the interior of a two-story home in Acushnet, Massachusetts, in 2008, completely destroying the occupied home. A subsequent investigation by local fire officials revealed that a very large void space existed between the first and second floors—a 22-inch (559 mm) gap between solid 2 by 10-inch (51 mm by 254 mm) wood joists supporting the second floor and 2 by 6-inch solid wood joists supporting the first floor ceiling. Figure 3-9. Flammable solid structural polyurethane adhesive, holding

Figure 3-9 The void space between the first and second floors of the home in Acushnet, Massachusetts, which was consumed by fire.

Courtesy of Kevin Gallagher.

the first-floor gypsum board ceiling in place, lined the entire bottom of this void space. A similar situation existed between the attic and the second-floor ceiling. Both the void space and the attic were quickly consumed and collapsed in the fire.

Another interesting feature of this type of modular construction was the use of a hinged lightweight roof truss. The roof of the Acushnet home was transported flat on the tractor-trailer to the site; once put in place, the roof was "raised" and held in place by small wood members, forming a type of supporting knee wall.

The Hazards of a Building Under Construction

In future chapters, we will discuss details of the actual construction process unique to each of the five types of construction (fire resistive, noncombustible, ordinary, heavy timber, and wood frame). There are, however, some generic hazards to all methods of construction and some particularly dangerous ones that we will discuss here.

Figure 3-8 A modular home under construction.

© Tyrone Turner/National Geographic/Getty Images.

As a fire fighter, you can't pick and choose which calls you will respond to. It is very possible you will be called to a fire in a building under construction. It is important to realize that these buildings are usually much more dangerous than a completed structure. Many of the fire protection features that would be in place when the building is occupied have not yet been completed. Structural elements have not been completed. The danger of collapse is amplified greatly.

It is advisable that you visit buildings while they are under construction to identify the hazards to you and your firefighting operations. You will also get the benefit of seeing how the building is being built—*very important information for your preplans.* Many fire departments actually digitally photograph and videotape the building under construction for training purposes, even years after the building has been finished.

While on site, test the operation of private fire hydrants (to ensure they are turned on and they have the proper threads) and verify the existence of access roads and fire lanes. Make sure that you will have the ability to fight a fire if one should occur during construction.

There are numerous general safety dangers at construction sites. Watch out for open and unsecured openings both outside and inside the building; missing stairwells; hanging, projecting, and moving objects; and open exterior perimeters without the protection of walls.

In terms of fire, there are numerous sources for a fire to start. Certainly, welding and cutting operations are a constant threat and start many fires **Figure 3-10**. Smoking materials often start fires in construction debris. Arson has also been a source of ignition at construction sites.

Construction fires can be devastating. For example, a fire during the early construction framing stage of a wood-frame building (and before the **wallboard** has been installed) will destroy the entire structure within minutes. These fires present a significant conflagration hazard—due to firebrand production and radiant heat—in which the fire can

Figure 3-10 The use of welding equipment poses a vast potential for fire situations.
© AbleStock.

spread to other similar structures as well as occupied buildings.

Falsework or *formwork* used in the construction of concrete buildings is often made of wood **Figure 3-11**. Essentially, the falsework is used to support (and form) poured concrete during the curing process. This presents a tremendous hazard to fire fighters who are exposed to the real danger of collapse should the formwork fail under fire attack. A fire in Key Biscayne, Florida, in 2000 began in the falsework of a 12-story, post-tensioned (post-tensioning is described in the chapter that reviews fire-resistive construction) building. Fire fighters, anticipating the danger of collapse, fought the fire defensively from the outside. A collapse did occur and the building was later demolished.

Although building and fire codes require standpipe systems to be complete up to within a few floors of the top of the buildings under construction (with a limited water supply), do not bet on these systems being present or in operation. Before pressurizing a standpipe system, ensure that all hose valves have been closed, the top of the pipe is capped, and the

Figure 3-11 These supports provide minimal shoring capabilities and are a serious risk to fire fighters who come into contact with them.
© Jason Smith/ShutterStock, Inc.

Figure 3-12 Collapse at a construction site is always a possibility, even without fire.
© arfabita/ShutterStock, Inc.

hose team has been made aware that there may be debris in the system. Pump operators should be advised to bring the pressure up slowly and to be ready to shut down quickly if need be.

Collapse at a construction site is always a possibility, even without fire Figure 3-12. In 2004, a 10-story combination precast/site-cast parking garage collapsed in Atlantic City, New Jersey. Twenty-four workers were trapped in the collapse zone, which extended from the 10th floor down to the 5th floor (the **progressive collapse** was stopped by shear walls at the 5th-floor level). In addition to the Atlantic City Fire Department, a New Jersey State Urban Search and Rescue Team was quickly called. Twenty-one workers survived; three deceased workers had to be extricated from the debris.

Another fire hazard at construction sites is the use of explosives. Used for rock blasting purposes, these materials are sometimes stored on-site (some cities do not allow overnight storage). Explosives present an obvious detonation hazard and can be deadly. In 1988, six Kansas City, Missouri, fire fighters were killed at a highway construction site when unmarked trailers containing explosives caught fire. Explosives should be stored in clearly marked magazines and their whereabouts communicated to adjacent fire companies.

The Hazards of a Building Under Renovation

A building under renovation is also a grave danger. Besides the obvious hazards identified previously with buildings under construction (especially cutting and welding operations), buildings under renovation may be occupied by people and the combustible fire loads of an occupied building. In addition, buildings under renovation may be old and noncompliant, a step away from demolition.

Perhaps the most famous and tragic of all renovation-induced fires was the 1972 Vendome Hotel fire in Boston. The building was being renovated, and after a 3-hour firefighting operation, the building collapsed during overhaul. Renovations made many years earlier had created a ballroom on the second floor, concentrating much of the load from the upper floors onto a single 7-inch (178 mm) wide column. By 1972, new owners were changing the occupancy of the building to a condominium

with new renovations. It is theorized that additional loading (in the form of suspended ductwork hung from beams that the column was supporting), fire fighters, and firefighting water precipitated the collapse. Nine fire fighters lost their lives.

Another problem with renovating older buildings is the prospect of a new occupancy that is inconsistent with the structure. Some buildings were intended for one use and one use only. (Your close relationship with the buildings department is helpful here; make sure that the department is aware of the renovation and is enforcing the building code for the new use.) Removing walls to "open up" a building requires a great deal of engineering to ensure that a collapse hazard is not created.

In any building under renovation, expect to find the fire protection system altered or even shut off—sprinkler valves closed, fire alarm systems shut down. Anticipate these deficiencies and act accordingly.

It is also very possible that egress paths for occupants (and you) have been cut off, blocked with debris, or sealed off. If you are responsible for conducting routine fire inspections of buildings in your district, make sure that you visit all buildings under renovation and notify the appropriate authorities when you find problems.

A common problem with "do-it-yourselfer" renovators is their lack of structural engineering knowledge. Cutting holes in load-bearing walls (for example, between two buildings) introduces the real possibility of collapse. Be aware that such renovations may take place out of sight in a basement, only to come to light when you arrive to find a pile of **bricks** and debris in the street.

The Hazards of a Building Under Demolition

We know that buildings under construction and renovation are particularly dangerous, but we need to add buildings under demolition to the list Figure 3-13 . Except for the life safety of the demolition workers,

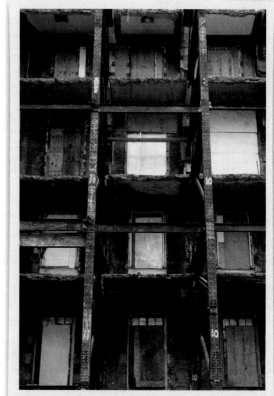

Figure 3-13 You must be aware of how structures are being demolished, particularly if critical supporting members may have been removed. Risk little for these structures.
© John Foxx/Stockbyte/Getty Images.

remember that this is a building that is being thrown away. Risk little for these structures.

The 2007 fire at Deutsche Bank discussed at the beginning of the chapter, which involved a building being both decontaminated and demolished, led to a set of significant regulatory changes in New York City. Among the most important changes were requirements for the use of noncombustible enclosures for hazardous materials abatement (e.g., asbestos) during demolition; however, this change applies only to New York City. In all likelihood, you will find **plywood** abatement enclosures in

most jurisdictions, whether the building is being demolished or not. When such plywood abatement enclosures make their way into a high-rise, the result is an enormous fuel load and a very dangerous situation. Take appropriate precautions.

An earlier incident illustrated just how dangerous a building under demolition can be owing to undermining of the structure. A 2003 partial collapse as workers were demolishing an arena in Grand Rapids, Michigan, trapped workers in an escalator area of the auditorium. The underside of a section of the 5,000-seat auditorium collapsed onto workers standing in the escalator space below. Despite the fact that a large amount of debris filled the escalator opening, all of the workers survived, largely due to the fact that the rest of the structure was still intact, which allowed rescuers and other construction workers to get them out.

Although this incident did not produce any fatalities, it illustrates the hazards inherent in buildings that are being torn down. Systems will have been removed from the building—perhaps even your standpipe system. There will likely be lots of torch-cutting operations, introducing numerous sources of ignition into the inevitable piles of debris.

Several years ago, fire broke out during demolition of a meat-processing facility in San Antonio, Texas. The concrete structure of fire-resistive construction was lined with cork in the refrigerated areas of the plant. Workers with torches ignited a deep-seated fire in the cork lining (very difficult to extinguish—the cork had to be scraped off the walls). Adding to the difficulty in extinguishing this fire was the fact that the workers were cutting ammonia refrigeration lines for the scrap metal. The lines still contained residual ammonia, which is explosive in concentrations of 15–25%. An explosion could easily have occurred.

Dangerous unsupported, freestanding walls are a typical problem at demolition sites. Fractured utilities are another.

In 1998, as small wall sections were removed during the demolition of the Jersey City, New Jersey State Theater, a wall collapsed on an adjacent two-story building. The collapse passed through a second-floor secretarial school (closed at the time) and into the occupied discount store on the first floor. Much of the debris fell into the basement, in which a computer school was located (also closed). Several shoppers, including small children, fell into the basement. Fortunately, they survived. Their rescue taxed the resources of the Jersey City Fire Department.

The Characteristics of Materials

Buildings are constructed of various natural and synthetic materials. Each material—and even different forms of the same material—has certain physical and chemical characteristics that make it more or less desirable for its intended function.

Some materials are strong against compressive forces, others against tensile forces. Some are heat insulators, whereas others absorb sound. Some resist weather; others lack structural strength but provide a good smooth surface that takes a nice coat of paint. Some materials are merely cheap.

Some materials are strong in tension and weak or unsuitable in compression. Manila hemp rope is a good example of this. Other materials are strong in compression and weak in tension, such as natural stone and concrete. The stone arch works because all members of the arch are under compression, and little or no tensile load is placed on a material that is weak in tensile strength.

Some people consider fire solely from the point of view of chemistry. According to this perspective, if a material contributes fuel to the combustion process, it is considered combustible and taken into account for fire formulas. If it does not burn, it is safe. In reality, this perception is untrue. All materials can be damaged by fire even if they do not burn, and the structural damage owing to failure of "noncombustible" materials may be significant.

The fire characteristics of materials are often misunderstood or ignored unless the law directs attention to this vital facet. As each material is covered

in the later chapters, its fire characteristics will be discussed. A brief summary of some important characteristics of common materials that can influence their behavior in fire is provided here.

Wood

Wood used to build structures includes both softwoods (conifers such as pines) and hardwoods (such as oak). Lumber is inspected and graded at the mill for quality purposes. The grade given to the lumber includes such things as span ratings for structural wood panels (which establishes the safe spacing between roof rafters and floor joists that the panel can cover, e.g., a 32/16 panel can cover rafters spaced 32 inches (812 mm) apart or joists spaced 16 inches (406 mm) apart).

Specific terms are used to describe the size of a structural member:

- *Board:* ≤2-inch nominal thickness
- *Dimensional lumber:* 2–4-inch nominal thickness
- *Timber:* ≥5-inch nominal thickness

Nominal thickness is the thickness of the lumber prior to drying in a kiln at the lumber mill. A 2 by 4 inch (51 mm by 102 mm), for example, actually measures 1.5 by 3.5 inches (38 mm by 89 mm).

Woods are sometimes chemically treated for prevention of insect damage. In some cases, fire-retardant salts are pressure-impregnated into wood. Fire retardants do not, however, render wood non-combustible. Their application just makes the wood more difficult to ignite and hinders fire from spreading. Some of the fire retardants used in the past have actually made some wood sheeting lose its strength over time (especially when used in hot attics). A firehouse in Michigan actually found out about this problem when a fire fighter stepped onto the roof and put his leg right through the wood deck.

Wood varies greatly in its compressive strength, depending on the direction in which the load is applied. Wet wood has less strength than dry wood. Plywood is manufactured with the grain of alternate plies laid at right angles to develop approximately equal strength in either direction. **Oriented strand board (OSB)** is another example of a wood panel in which all of the strands are laid at right angles, but the strands are much smaller (3–4 inches [76–102 mm]), which makes OSB cheaper than plywood.

Another group of "manufactured" or "engineered" wood building products—specifically, **lightweight wood trusses** and **wooden I-joists**—deserve special mention because of the inherent danger to fire fighters who fight fires in buildings in which they are used Figure 3-14 . Many fire fighters' lives have been lost due to failures of these two types of members during actual fires. Other engineered products include members that are pressed and glued together: laminated veneer lumber (LVL; thin wood veneers laid parallel to the length of the member), parallel strand lumber (PSL; long veneers laid parallel, but in a more random arrangement), and laminated strand lumber (LSL; similar to OSB but with longer 12-inch (305 mm) flaked wood strands). Lightweight wood engineered members are discussed in greater detail in the chapter that reviews wood frame construction.

Figure 3-14 The lightweight wood I-beams shown here are used in much of today's wood-frame construction. The "longboards" of the past are few and far between.

Courtesy of APA - The Engineered Wood Association.

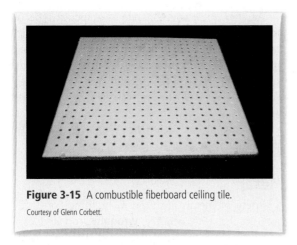

Figure 3-15 A combustible fiberboard ceiling tile.
Courtesy of Glenn Corbett.

Figure 3-16 Spalling effects on concrete from a fire.
Courtesy of Dave Casey.

Keep in mind that wood burns and thus loses structural strength and integrity. The vast majority of structural fires are fought by fire fighters standing on or under wooden structures. Combustible fiberboard, for example, was at one time used for wall paneling, ceiling tiles, and sheathing **Figure 3-15**. It burns persistently. Even when codes require new ceilings to meet flame-spreading requirements, this dangerous old tile is often left above the new ceiling, increasing the dead load. This was a significant factor in the loss of 16 lives in a Tennessee retirement home fire in 1989.

Masonry and Stone

Bricks are smooth-sided pieces of clay that have been cured in an oven. This type of **masonry unit** is quite fire resistant, but can **spall** (lose surface material) when subjected to fire. Like other masonry materials, brick cannot be effectively considered apart from the mortar used with it. Prior to 1880, when **Portland cement mortar** became available, the only mortar used was sand-lime, which is water soluble. Even today, regulations require the use of **sand-lime mortar** in the restoration of old buildings built with soft bricks. Ordinary water leaks or fire hose streams have dissolved sand-lime mortar and caused collapse.

Adobe, which is made of sand, water, clay, and a fibrous material like manure or straw, has been long used to make blocks. Formed and dried in the sun, these adobe blocks have been used in construction for centuries, particularly in the Southwest.

Another type of masonry unit is the hollow-core **concrete masonry unit (CMU)**. This is the typical 8 by 16-inch concrete block, or "cinder block" as it is sometimes called erroneously. (Cinders were used in the past in the manufacture of these blocks.) When used for the construction of walls, CMUs are sometimes reinforced with steel **reinforcement bars (rebar)** and their hollow cores (also known as **cells**) are filled with concrete.

As noted earlier, natural stones are also used for the construction of walls and foundations. These materials include fieldstone and **quarried stone** such as granite, marble, limestone, and sandstone. Natural stone buildings will spall (break off in chunks) when exposed to fire **Figure 3-16**. Granite particularly is subject to spalling. In older buildings, slate and marble were used for stairways, and sometimes for unsupported flooring in library stacks. These structures should be examined for support. If only self-supported, they should be avoided after exposure to heat. Marble can look perfect yet actually may have turned to chalk. In such a case, it will collapse under the weight of a fire fighter.

Concrete

Concrete is a mixture of Portland cement, water, and **aggregates** (coarse aggregate like gravel and fine aggregate like sand). Cement and concrete are not the same thing—cement is a component of concrete. Concrete is cured (hardened to full strength) using formwork (falsework). It is mixed and proportioned to the desired strength; different **admixtures** are added to give the concrete special characteristics such as corrosion resistance if steel-reinforcing rods are to be added to the concrete.

In compression tests, crushing strains are exerted on the material until it buckles or crumbles. Concrete is tested in compression because it has virtually no tensile strength. In tensile tests, the material is pulled apart until failure occurs; for concrete it happens very quickly.

Reinforced concrete is a **composite material**—that is, two elements act together under the load. Concrete provides the compressive strength, while steel provides the tensile strength **Figure 3-17**. Steel is almost equally strong in compression and tension. Usually steel is tested only in tension. If the sample passes, it is assumed the steel also has the required compressive strength.

Figure 3-17 Reinforced concrete frame under construction.
© Daniel Hixon/ShutterStock, Inc.

Aerated autoclave concrete is a special type of lightweight concrete (one-fourth the weight of traditional concrete) that is cured in a factory kiln under high pressure, creating millions of small cells inside the concrete. This type of concrete may be made into slabs, panels, or blocks and can be cut at the construction site. It has performed well in fire resistance testing, but may not have the compressive strength of traditional concrete.

Metal

A variety of metals are used in building construction. In the 19th century, **wrought iron** and cast iron were used in many structures. Wrought iron, an expensive material at the time, is **malleable** (can be shaped) and contains little carbon (0.2%). It has been replaced by steel for many purposes.

Cast iron, in contrast, is a very brittle material with high carbon content (3–4%). If heated in a fire, it can easily fail when struck with the cold water from a hose stream **Figure 3-18**. Cast iron can, in some cases, have good fire characteristics. The casting method determines whether it is good or bad, and this cannot be determined by examination. Poor connections of floor beams to cast-iron columns are probably the chief cause of failure.

Structural steel, in common use today, is an alloy composed of iron and carbon (<2%). It has three negative characteristics in fires:

- It conducts heat.
- When heated, it elongates and may push through any barrier.
- It fails at about 1,000–1,100°F (538–593°C).

Cold-drawn steel, such as cables that are sometimes used to brace failing buildings or as tendons in tensioned concrete, fails at 800°F.

Unlike with cast iron, the application of water to cool steel doesn't cause failure; it simply removes the heat, thereby regaining its strength. Cooling streams should be applied to structural steel members when they have been heated by a fire.

Within the last several years, there has been a dramatic increase in the use of galvanized cold-formed

Figure 3-19 Load-bearing galvanized cold formed steel is used in this multi-story building.

Courtesy of Glenn Corbett.

Figure 3-18 This type of column will easily fail when introduced to heat and then rapidly cooled by a hose stream.

Courtesy of Glenn Corbett.

steel in the form of non-load-bearing "C"-shaped studs as well as load-bearing beams and rafters **Figure 3-19**. Of particular concern are unprotected beams and rafters, which are likely to fail when exposed to fire. While there has not been a significant amount of experience with them, expect failure when these members are subjected to moderate to heavy fire conditions.

__Aluminum__ (known as aluminium in Canada) melts at very typical fire temperatures. This is sometimes beneficial, such as when an aluminum roof melts and vents a fire. Working on an aluminum roof, however, is extremely hazardous.

Aluminum trusses and beams are sometimes used for concrete falsework. They melt and lose strength rapidly when heated. Such a failure occurred in a fire in Fairfax County, Virginia. The glass fiber formwork often used with the aluminum can provide sufficient fuel to create a raging fire. Many other fuels are likely to be present as well. Aluminum ductwork will melt and admit fire to the duct.

Glass

Glass is used in several ways, most commonly in windows and doors. Glass has little resistance to heat or pressures built up by the high temperature of a fire. Often the first sign of a serious fire in a closed structure is the violent pressure failure of the windows, often described by witnesses as an explosion. Some glass windows are double or triple glazed for energy conservation (__thermal pane windows__). They are difficult, if not impossible, for the fire to break. Such windows do not allow for the release of heat from the fire environment, making firefighting difficult and dangerous. Some codes require that some windows in high-rise buildings be made of breakable glass. These are sometimes marked with a Maltese cross for identification purposes.

The concern for terrorism has led to the use of specialized glazing systems for blast resistance. The goal is to prevent the glass from shattering into small pieces and becoming airborne. Glass has killed and injured many people in terrorist bombings; for example, glass was broken in more than 250 buildings in Oklahoma City as a result of the Alfred P. Murrah federal building bombing. It is also important that the entire frame itself be secure enough not to become a projectile. A variety of techniques are used to prevent these problems: special laminates, films (applied retroactively), and special blast curtains (drapes). In addition, special steel frames that absorb the blast energy without failing are used for large glass windows.

Fixed sashes of double-pane glass were a factor in a motel fire in which several persons died. Such windows cannot be readily broken by the items available in a motel room. They should be equipped with emergency opening devices.

Another even more problematic window is the laminated "hurricane-resistant" window, a great advance for protection against flying debris during a hurricane but nearly impossible to break for firefighting ventilation. Carbide-tipped blade saws offer one of the few options of getting through these windows.

Wired glass, when properly installed, has greater resistance to fire than ordinary glass, but it passes radiant heat as readily as regular glass **Figure 3-20**.

Figure 3-20 A wired glass window.
© Alexey Stiop/ShutterStock, Inc.

Glass fiber insulation is a combination of noncombustible glass fibers and combustible binders. It is often installed above metal grid suspended ceilings. If the fire is above the ceiling, this type of insulation will tend to conceal the situation from fire fighters below. If the fire is below the ceiling, the insulation will retain heat in the steel grids. If the glass fiber is not included in the listing of a rated floor or roof and ceiling assembly, the retained heat may cause the ceiling to fail earlier than expected.

Asbestos

Asbestos was long used by itself or in combination with other materials as a fireproofing agent for steel. Inhaled asbestos fibers are a known carcinogen, however, and concern for this hazard has led to the removal of asbestos from buildings. If asbestos was used for fireproofing and removed from a structure, there is the serious question of whether any action was taken to restore the protection to the steel.

Much of downtown New York City is heated by steam supplied at high pressure. A 2007 asbestos-insulated main blew up and heavily contaminated an apartment house. Fire fighters on the scene had no knowledge of the hazard, which was not recognized for several days. A number of pieces of apparatus were out of service until they could be decontaminated.

Synthetic Materials

There are thousands of plastics, including many that are used in construction, which are impossible to discuss in detail. However, the fire characteristics of this material are important to learn. When considering plastics, remember that small-scale laboratory tests of these materials may produce results that are more favorable to the plastic than the plastic's actual performance in real-life situations.

Many plastics can be ignited readily with small flames. Polyurethane imitation wood beams, a commonly used plastic, can be ignited with a match. Plastic imitations of other materials look very realistic and can be quite deceptive. It was reported that

the multifatality disaster at the Six Flags Haunted Castle in New Jersey was started when a guest ignited polyurethane decorations and **interior finish** with a cigarette lighter.

Some plastics emit dense black smoke when burning. The smoke is sooty, sticky, and costly to clean up. It can cause acute distress to those breathing it. Highly toxic gases may be evolved from the burning of plastics.

Thermoplastics (e.g., polystyrene) can produce flaming, dripping plastic, which will produce secondary fires. Such fires can burn down through the wooden floor into the floor void of even 1-hour-rated combustible floor and ceiling assemblies. **Thermosets** (e.g., polyurethane) are plastics that char and burn, but do not flow. An example of this was the cellular foam plastic installed as an interior finish at the Station nightclub in West Warwick, Rhode Island, that burned and killed 100 people in 2003 **Figure 3-21** . The plastic interior finish played a critical role in the rapid development of the fire and smoke.

Foam core panels are plastic panels in which foamed urethane or expanded polystyrene is sand-wiched between two panels of OSB, or OSB on the exterior and gypsum board on the interior surface are being used for roof and wall panels. A home improvement television show recently showed such panels being used on a timber-framed house. The typical panel appeared to be 2 by 8 feet (0.6 m by 2 m). Several panels were shown assembled together to make a large self-supporting roof panel. It appears that this might be a dangerous roof to depend on for ventilation or fire extinguishment. Some plastics melt; others might be consumed by the fire. Some might have a gypsum board thermal barrier; others might not.

Of particular note is the exterior insulation finishing system, more commonly known as EIFS. EIFS is a composite multilayer sheathing panel system; it is composed of a polystyrene foam plastic base covered with a water-resistant coating and fiberglass mesh and an acrylic finish. Some of these panels have been tested as part of a fire-rated wall assembly. These panels are finding their way into many buildings, ranging from tall high-rises to small multiple-family dwellings.

A fire at the Monte Carlo casino in Las Vegas in 2008 involved EIFS wall panels (with an inadequate acrylic finish thickness) and combustible decorative expanded polystyrene three-dimensional "shapes." Most likely triggered by a welder's torch, the fire spread up and across the façade of the upper floors of the building. Subsequent analysis showed that while the non-fire-rated expanded polystyrene decorations served as the primary fuel load, the EIFS also contributed to the fire. Just like any other fire-rated assembly, it is important that EIFS be fabricated and installed exactly as tested **Figure 3-22** .

Resin plastics reinforced with fiberglass are used for corrugated and flat plastic panels and some automobile panels. This type of plastic will burn and leave a matted sheet of glass fibers.

Fire-retardant panels that are made by a similar process, and which have been approved by Factory Mutual for installation without automatic sprinkler protection, are available (where sprinklers would otherwise be required if ordinary panels were used).

Figure 3-21 The synthetic material used on the walls, the overcrowding, and the unapproved pyrotechnics at the Station nightclub in Warwick, Rhode Island, combined to produce one of the most devastating fires in the United States in the last 20 years.

© Robert E. Klein, File/AP Images.

Figure 3-22 EIFS panels being installed.
Courtesy of Glenn Corbett.

Structural members of plastic reinforced by continuous glass fibers oriented in the direction of the load are used for special buildings, such as those where no metal is permitted due to radio frequency interference or where corrosion is a serious problem. Bridges with a clear span of 90 feet (27 m) have been built to be used over tanks emitting corrosive fumes. Even when combustion inhibited, plastics may deform and cause costly damage under fire conditions. The *NFPA Fire Protection Handbook* discusses plastics in detail.

Gypsum

Gypsum, typically used in wallboard (which can be a part of a fire wall or **fire partition**), is an inert material composed of calcium sulfate dihydrate. It is pressed between two sheets of cardboard to create a sheet of gypsum wallboard. Wallboard is part of a rated wall assembly—all components must be tested together in a testing laboratory to get an hourly fire resistance rating.

Gypsum board may contain additional additives to alter its behavior. For several years, polycarbonate plastic backings were added to gypsum board to increase its impact resistance. More recently, gypsum board with a fiberglass mesh core has been developed. In either case, breaching these materials is a challenge. Points of connection at the ends of the sheets are weak points, however, and should be exploited if possible. Cutting through a sheet is difficult and requires a sizable hand tool (e.g., Halligan, axe blade). When you do manage to penetrate fiberglass-reinforced gypsum board, be wary of the entanglement hazard of the remaining fiberglass mesh.

"Alternative" Organic Materials

The move to "green" construction has led to an increase in the use of unusual natural materials. While the conflagration hazard of thatched straw roofs has been recognized for centuries, straw has now found its way inside some modern-day buildings. Some types of "green" homes are now being made of straw bales covered with plaster. Bamboo is taking the place of rebar in some homes. Even mushroom fiber "bricks" are being developed!

It appears that the rate at which these materials are entering the marketplace is outpacing the building codes and the construction classifications. If you encounter them, take the time to establish their propensity for fire and develop a plan of action.

The Effects of Shape

The shape of a material affects its ability to resist a compressive load or a deflective one. Deflection is bending that combines both compression and tension. Shape is not a consideration in tensile loads.

An interesting example concerning shapes and forms of materials involves paper. Set up two stacks of

Figure 3-23 The shape of a material affects its ability to resist a compressive load or a deflective one and can be demonstrated by some simple paper shapes: sheets of paper across a gap between books (a), rolled sheets of paper (b), and pleated paper (c).

books of the same height, about eight inches (203 mm) apart. Lay two sheets of writing paper across the gap. The paper, in the form of thin sheets, is barely able to support itself and, in fact, may sag so badly that it falls Figure 3-23a .

Roll the two sheets of paper lengthwise into tubes about an inch in diameter. Prevent the tubes from unrolling with rubber bands or clips. Place the tubes across the gap. The tubes act as beams that could carry a sizable book without collapsing Figure 3-23b .

Another form of beam can be created from a sheet of paper by folding it into pleats Figure 3-23c . Such a **folded plate** is used at times for concrete roofs. Similarly, **corrugations** give steel a greater ability to span a gap, without unacceptable deflection or bending, than the same steel would have as a flat plate.

Perhaps the most dramatic example of a material's strength also involves paper columns. Roll four paper tubes and stand them on end as columns. Carefully set a large book, such as the *NFPA Fire Protection Handbook*, in place atop the columns. It is possible to stack both of these books on the columns. This heavy weight is supported on less than a half ounce of paper arranged in the ideal shape of a column. Note that the load must be perfectly axial and that any eccentricity in the loading will cause the structure to fail.

Resistance to tensile loads, on the other hand, is directly related to the cross-section makeup of a material (the internal configuration and mass of the material). Regardless of the shape, the paper would tear if tested for tensile strength.

TACTICAL CONSIDERATIONS

1. The (ADA requires new buildings to have *areas of refuge* for disabled people in multistory buildings. This is important for fire fighters and company officers to know because the disabled occupants have been instructed to go to these locations and await rescue by the fire department during an emergency. This regulation is retroactive, so it is an important feature to note on prefire plans. If you discover the building has areas of refuge while investigating the source of a fire alarm, notify the incident commander. Search and rescue teams need to be assigned to check these specific areas on both the fire floor and the floor above for occupants.

2. During high winds, fire fighters should use extreme caution around buildings under construction. Work that needs to be continued on the following day or after the weekend may, therefore, be secured to only a bare minimum standard. There are also numerous subcontractors on site working on a myriad of projects simultaneously who have various levels of safety consciousness. For all these reasons, new construction sites are very dangerous after hours, but especially during severe weather characterized by high winds because material and equipment may or may not be secured. Fire fighters should be wary of flying construction debris at such sites, given that they cannot tell how well the structural elements are secured. At a minimum, crew members should always wear their helmets at construction sites. Helmets, coats, and gloves provide the best protection.

3. Some construction sites can look benign, yet are extremely dangerous. Firefighters must realize that some of these buildings are more dangerous than a completed structure. Many of the fire protection features that would be in place when the building is occupied have not yet been completed. One of the most critical features is the standpipe. If the standpipe installation doesn't keep up with the height of the new construction, you will have to establish your own water supply to upper floors if a

fire occurs there. That will require a long hand hose lay or an elevated platform at a window.

4. Other critical features to watch out for at new construction sites are unfinished stairs and open shafts for future elevator installations. These extreme fall hazards can injure or kill fire fighters when they are the only life hazard at the site.

5. Remember, new construction buildings are in the process of being built. Material is curing, being propped up, or temporarily suspended in place. Structural members are not completely installed. The danger of collapse is greatly amplified because of the lack of stable, permanent structures.

6. Numerous sources of ignition can be found at construction sites; cutting and welding operations and hot tar rooftop operations have started many fires. Don't rule out arson, especially during the wood-framing stage.

7. New construction fires can be devastating. These fires present a significant conflagration hazard owing to firebrand and ember production, as well as radiant heat in which the fire can spread to other similar structures and occupied buildings.

8. A well-documented case study of the effects of radiant heat in an unfinished building is the Santana Row Fire in San Jose, California. On August 19, 2002, an 11-alarm fire in a massive, 9-building, 42-acre, mixed-use, new construction project was believed to have started by hot-work being performed on the roof of a 7-story building. Building 7, the largest in the complex, was the building of origin. Radiant heat caused exposure buildings to ignite. Gusty winds, enhanced by the fire, carried embers the size of 2 by 4 inches (51 mm by 102 mm) up to 0.5 mile (0.8 km) downwind, igniting six buildings of the Moorpark Village and Garden Apartments. In total, 40 apartment units were destroyed, leaving 130 residents homeless, including 34 families. Fortunately, there were no fatalities, but damage was estimated at nearly $130 million. Excellent video of this fire is available at www.youtube.com and www.vimeo.com for study.

9. It is important to visit the sites of structures that are under construction, demolition, and renovation. Familiarize yourself with the hazards present and report any apparent code violations to the fire prevention bureau and buildings department.

10. The application of water to cool steel doesn't cause failure; it simply cools the steel and may even prevent the failure from occurring. This is an excellent tactic to implement in protecting steel exposures that do not have fire-resistant coatings. Remember, the structural steel in common use today is an alloy composed of iron and carbon. It has three negative characteristics in a fire: it conducts heat; it elongates when heated, which may push through any barrier; and it fails at about 1,000–1,100°F (538–593°C). Cold-drawn steel such as cables used for bracing and tensioning fail at 800°F. The thermal imager isn't just a search tool when it comes to steel exposed to fire; it can also be used to monitor flame impingement on structural elements of these materials and risk for their failure. This is one instrument that should not be left on the apparatus during a fire.

11. Some houses are built with aluminum beams; these structures should be expected to collapse early in a fire. Some mobile home trailers have aluminum sides and steel roofs. Because the roof won't melt or vent, there is greater potential for horizontal extension to any adjacent units. Aluminum is also widely used for railings.

12. In one high-rise building, aluminum railings were used to separate the stairway from the smoke shaft. They failed in a multiple alarm fire and presented a serious hazard to fire fighters working in the stairway.

13. Many high-rise buildings contain large glass areas. Units operating inside such buildings should be notified immediately when the failure of such windows is seen from the outside. The sudden inrush of oxygen will accelerate the fire, seriously endangering crews working on or above the fire floor. The failure of windows on upper floors can create a situation known as a *wind-driven fire*. Strong winds at the open window can overpressurize the interior, reversing the direction of the fire— sending it toward fire fighters—as it looks for a new path of least resistance. This event can happen in seconds, creating a blowtorch fire in the hallway. Because fire fighters have little or no time to react and back out, it has led to line of duty deaths. Have a plan to breach a wall or force the door to an adjacent office or apartment for an emergency area of refuge.

14. Falling glass may make outside operations at a high-rise fire dangerous or even impossible. Incident commanders should establish a 200-foot (61 m) safety zone around the building. Falling glass has cut hose lines supplying standpipes, so protect the supply lines from falling glass if possible. Use sheets of plywood or park benches to cover them—anything that would deflect the glass shards. Be attentive to any protected entry ways by which the building can be accessed, such as an underground garage. It may be safer (and necessary) to charge the standpipe by laying supply lines through a protected entry and connecting to the lowest level discharge port.

15. Buildings where asbestos might present a hazard during or after fire suppression should be identified in advance so that technical assistance can be summoned whenever a working fire occurs there. Since you can never be sure of this hazard, the best protection is to have your fire crews wear their entire PPE firefighting ensemble including SCBA, even after the fire is out. There is no reason why we have to breathe in smoke after a fire is extinguished. SCBA bottles can always be exchanged or refilled; your lungs can't.

16. In fires involving plastics, anticipate severe respiratory injuries among the occupants, possibly even mass casualties. Incident commanders should consider setting up a medical branch with the ability to administer large amounts of oxygen therapy to multiple patients simultaneously. Expect heavy toxic smoke conditions and a fast-spreading fire. The toxicity of some plastics makes it imperative that SCBA be worn even through overhauling and outside operations such as a plastic roof fire. Any firefighting on foam core roofs should be done by personnel operating from an aerial apparatus. A building enclosed by foam core panels and fitted with energy-efficient windows might well present flashover and backdraft conditions.

17. After being heated by fire, gypsum starts to break down. This process cannot be stopped.

18. Many fire-resistive partition walls are made of gypsum attached to studs. Sometimes the quickest forcible entry into an occupancy is to bypass a locked door and go right through the gypsum wall.

WRAP-UP

Chapter Summary

- New structures built today, as well as the construction process itself, must comply with a variety of regulations. Inspection is a crucial task to ensure the requirements are being met on the site.
- Depending upon the size of a new building, many different professionals, from architects to construction workers, may be involved with the design and construction process.
- It is important for fire fighters to be able to read building plans and specifications, for use during fire prevention plan reviews, for pre-planning purposes, and to assist in firefighting efforts.
- Preparing a site for construction is necessary for all new buildings. Some basic activities, such as soil testing, apply to most new structures.
- In the last several years, there has been an increase in the use of the modular construction technique to build wood-frame single-family and multiple-family dwellings. These structures have some unique features that may affect firefighting tactics.
- A building under construction is usually much more dangerous than a completed structure. Many of the fire protection features that would be in place when the building is occupied have not yet been completed, and the danger of collapse is amplified greatly.
- A building under renovation is also a grave danger. Besides the obvious hazards, buildings under renovation may be occupied by people and hold the combustible fire loads of an occupied building.
- Buildings under renovation may be old and noncompliant with building codes, as they are only a step away from demolition.

- Except for the life safety of the demolition workers, remember that a building under demolition is a building that is being thrown away. Risk little for these structures.
- Buildings are built of various natural and synthetic materials. Each material—and even different forms of the same material—has certain physical and chemical characteristics that make it more or less desirable for its intended function.
- The shape of a material affects its ability to resist a compressive load or a deflective one.

Key Terms

Admixture A material other than water, aggregate, and cement used as an ingredient in concrete or mortar.

Aggregate Any of a variety of materials, such as sand and gravel, added to a cement mixture to make concrete.

Aluminum A lightweight metal that is both malleable and nonmagnetic. This material has very good conductivity. This noncombustible material has a low melting point and little mass per unit of area, so it disintegrates rapidly in fire.

Americans with Disabilities Act (ADA) A set of regulations passed in 1990 that includes, among other things, regulations requiring areas of refuge for disabled people in multistory buildings.

Architect An individual who is engaged in the design of buildings and who often supervises construction.

Area of refuge An area of a building separated from other spaces by fire-rated smoke barriers in which a tenable environment is maintained for the period of time that such areas may need to be occupied at the time of a fire.

Brick A solid or hollow masonry unit of clay mixed with sand, which is molded into a small rectangular shape while in a plastic state.

Building code Regulates the actual design and construction of new buildings, providing for minimum levels of health and safety. Regulates the level and amount of fire protection in a new structure.

Caisson A shaft of concrete placed under a building column or wall extending down to bedrock.

Cells (concrete) The hollow portions of the core of a concrete block.

Civil engineer An engineer who specializes in the design of parking lots, drainage areas, and roadways.

Cold-drawn steel A type of steel used in the construction of cables that are sometimes used to brace failing buildings or as tendons in tensioned concrete. It fails at 800°F (427°C).

Composite material A material built up of different parts, pieces, and materials intended to act as a single unit.

Concrete masonry unit (CMU) Precast hollow or solid structural block. Sometimes referred to as cinder block.

Construction safety regulations Occupational health and safety regulations (enforced by state and/or federal authorities such as OSHA) that apply to construction sites. These include provisions to prevent falls (through the use of safety barriers) and protection from being hit or crushed.

Corrugation Grooved ridged material, often metal.

Crosslot bracing Internal bracing that transfers the lateral earth pressures between opposing walls through compressive struts.

Cured (concrete) Concrete that is hardened to full strength.

Dewatering pump A pump that removes water from the ground or excavations that hinder construction from an area of the site.

Electrical contractor A subcontractor hired by the contractor to oversee all electrical installation.

Electrical engineer An engineer who specializes in the design of lighting, power telecommunications, and emergency power of a structure.

Falsework Temporary shoring, formwork, beams, or lateral bracing to support the work in the process of construction; also known as formwork.

Fire alarm/security system contractor A subcontractor hired by the contractor to oversee all fire and security alarm installation.

Fire code Regulates the activities that take place in existing buildings, including the maintenance of existing fire protection features such as automatic sprinklers and fire doors, hazardous processes such as spray finishing, the storage of hazardous materials, and general fire safety precautions.

Fire partition A fire-rated assembly that subdivides a building to prevent the spread of fire such as between dwelling units in a multiple dwelling.

Fire protection engineer An engineer who specializes in the design of fire protection systems including sprinkler and standpipe systems, and fire alarm systems of a structure.

Fireproofing contractor A subcontractor hired by the contractor to spray a layer of cementitious material onto steel for fire resistance purposes.

Folded plate A structural element used to strengthen the roof of a structure over large areas.

Footing Thick concrete pads, usually heavily reinforced, which transfer the loads of piers or columns to the ground.

Formwork Temporary shoring, formwork, beams, or lateral bracing to support the work in the process of construction; also known as falsework.

Foundation The lower division of a building that serves to transmit and anchor the loads from the superstructure directly to its earth or rock, usually below ground level.

Framing A system of rough timber structural woodwork that is joined together to support or enclose, such as partitions, flooring, and roofing.

General contractor The prime contractor who oversees and is responsible for the overall work on the site.

Gypsum A natural mineral used in the manufacture of drywall and plaster.

Insurance regulations The rules of an insurance company in regard to coverage. These regulations may require more fire protection for structures than local building and fire codes require.

Interior finish The exposed material on a wall, ceiling, or floor.

Lightweight wood truss A collection of lightweight wood structural components joined in a triangular configuration that can be used to support either floors or roofs.

Malleable The property of a metal that allows it to be shaped by bending, hammering, or extruding without cracking or rupturing.

Masonry unit The type of masonry used on a given structure. Examples include bricks or concrete block.

Mechanical engineer A person trained in a branch of engineering related to mechanical equipment. Usually designs the heating,

ventilation, and air-conditioning systems and plumbing of the building.

Modular construction A construction technique in which a set of prefabricated modules are transported to and assembled at a construction site, creating a building.

Oriented strand board (OSB) A type of board made of layers of strands of wood cut from logs, with a fairly constant width to length ratio.

Pile One of a series of large timbers or steel sections driven into soft ground down to bedrock to provide a solid foundation for the superstructure of a building.

Plastic Synthetic materials (typically hydrocarbon based) that are composed of organic polymers that can be shaped and molded.

Plumbing contractor A subcontractor hired by the contractor to oversee the installation of plumbing within a building.

Plumbing, mechanical, and electrical codes Plumbing codes apply to the installation of domestic water systems, sanitary systems, and the like. Mechanical codes apply to systems likes air-handling systems and smoke control systems. Electrical codes apply to the installation of electrical systems.

Plywood Wood panels manufactured with the grain of alternate plies laid at right angles to develop the approximate equal strength in either direction.

Portland cement mortar The most common type of cement in use today. The powder, when mixed with water, will become a hydrated solid over time.

Progressive collapse Extensive structural failure initiated by local structural damage or a chain reaction of failures following damage to a small portion of a structure.

Quarried stone Natural stones such as granite, marble, limestone, and sandstone. Can be

used in the construction of walls and foundations.

Raker Braced sheeting used in soil walls to protect against collapse.

Reinforcement bar (rebar) Steel bar placed in concrete to increase its overall tensile strength.

Sand-lime mortar Water-soluble mixture; when water is applied to it, the mortar can be washed away from the wall.

Scale (architectural) An instrument with uniformly spaced marks used to measure distance on a set of blueprints.

Seismic protection Stabilization of outer walls, overhangs, floors, and walls to resist destruction by an earthquake.

Slurry wall A foundation wall in an excavation that is heavily reinforced with steel and temporarily filled with a liquid slurry mixture that is subsequently displaced by concrete.

Spall A small fragment or chip dislodged from the face of a stone or masonry unit.

Specifications Manuals prepared by building designers that specify which particular type of device or equipment is to be used in the building.

Sprinkler contractor A subcontractor hired by a contractor to oversee the installation and design of sprinkler systems within a building.

Structural engineer An engineer who is consulted in large buildings to design the structural frame.

Structural steel Steel rolled in a variety of shapes and fabricated for use as load-bearing structural members or elements.

Subcontractor A specialist who undertakes part of a main building contract from a contractor.

Substructure The foundation and footings.

Superstructure Any structure above the substructure.

Thermal pane windows Units composed of two or more glass panes in a hermetically sealed frame.

Thermoplastic Construction material that softens when heated and rehardens when cooled.

Thermoset Plastic that chars and burns, but does not flow.

Tieback Braced sheeting used in soil walls to protect against collapse.

Wallboard A large, rigid gypsum sheet that is fastened to the frame of a building and provides the surface finish.

Wallboard contractor A subcontractor hired by a contractor to oversee the surface finishes of walls that are fastened to the frame of the building.

Wooden I-joist Composite lightweight wood material used in an "I" cross section.

Wrought iron A commercially pure iron of fibrous nature; used for water pipes, rivets, and other ornamental forged work.

Zoning regulations Dictate land usage in specific cities (and are written by the cities themselves). Spell out which types of buildings can be built in specific locations and their permitted uses.

Case Study

You are a lieutenant in an engine company serving a rapidly developing area. One of your responsibilities is to inspect the construction sites within your response area.

1. It is important to remember that the buildings in your area:

 A. Have been built under no regulations
 B. Have all had structural engineer approval
 C. Have been built utilizing a variety of regulations or none at all
 D. Must comply with the latest national codes and regulations

2. The site preparation phase of construction typically involves what specific kind(s) of testing?

 A. Alarm system testing
 B. Testing of the soil for load-bearing capabilities
 C. Testing for trapped methane pockets
 D. All of the above

3. Given your fire protection expertise, you have been asked to assist in the inspection of an old, existing building in the downtown area. While conducting the inspection, you note that the occupancy has changed from a business to an assembly occupancy. What would your next step be?

 A. Contact the contractor
 B. Contact the building inspector
 C. Contact the fire protection engineer
 D. Note the changes on the plans and move forward with the review

4. You note that a "big box" retail store under construction in your district has an EIFS exterior sheathing system. EIFS stands for:

 A. Extruded internal finishing system
 B. Expanded inhibited flashing system
 C. Exterior insulation finishing system
 D. Expanded insulated finishing system

Challenging Questions

1. Identify the differences between fire codes and building codes.

2. What should a fire fighter know about the Americans with Disabilities Act?

3. Identify the role of the architect in the creation of a building.

4. Describe the difference between applying water to a hot steel beam and to a hot cast-iron column.

5. Fill in the nominal thickness dimensions (in inches) for each type of structural wood member:

Board: _____
Dimensional lumber: _____
Timber:_____

6. Assume you are an engine company officer whose district includes a large new wood-frame housing subdivision. What concerns would you have about protecting the complex during construction? What concerns would you have about fire fighter safety?

Building and Fire Codes

OBJECTIVES

At the conclusion of this chapter, you will be able to:

- Explain the history and basis for building and fire codes.
- Explain the differences between fire and building codes.
- Identify other construction codes and standards that have an impact on the fire service.
- Identify important fire protection features of building and fire codes.
- Identify construction-related provisions of fire codes.

Case Study

Figure 4-1 The Great Adventure "Haunted Castle" fire of 1984.

Courtesy of the NFPA.

On a late May afternoon in 1984, fire broke out in the "Haunted Castle" attraction at the Great Adventure Park in Jackson Township, New Jersey (**Figure 4-1**). Eight teenagers became trapped in the haunted house and subsequently died in the blaze. The maze-like "building" was actually a set of 17 interconnected truck trailers; guests walked from one trailer into the next. The dark walk-through haunted house included a variety of fright-provoking scenes, live actors, and a "strobe room." Plastic polyurethane foam was used in decorations in various areas of the attraction. Plywood partitions were also present throughout. The structure did not have a fire alarm system or a fire sprinkler system. There were no specific regulations for "haunted houses" in the building and fire codes of New Jersey or the rest of the nation at the time of the fire.

In the wake of this incident, action on the part of the state of New Jersey was swift. The state's building and fire codes were changed to include fire alarm systems and sprinklers. These regulations were quickly adopted nationally. Today, these "special amusement" buildings require a fire alarm and detection system, full sprinkler protection, a public address system, low-level exit signs, and a Class A rating for all interior finishes.

1. Why does it often take a tragedy to change our building and fire codes?
2. What are some other fires in the past that have led to changes in our fire safety regulations?

A History of Building Codes and Fire Protection

There have been fires and subsequent fire protection regulations almost as long as there have been towns and cities in the United States. Fires struck Jamestown, Virginia, in 1608 and Plymouth, Massachusetts, in 1623. By 1648, New Amsterdam (New York City) had banned the use of wood and plaster chimneys.

Subsequent regulations over the years were adopted sporadically city by city, each of which wrote its own building codes based upon the local fire experiences **Figure 4-2**. It wasn't until 1873 that the National Association of Fire Engineers (NAFE, today known as the International Association of Fire Chiefs) developed a set of eight simple fire protection concerns, dealing with things such as "inflammable materials" as well as limitations on the "excess height of buildings" and fire escapes that were "actually serviceable for invalids, women and children." Although this effort by the fire engineers (chiefs)

Figure 4-2 Fire buckets like these were used during the Colonial era.

Courtesy of New York City Fire Museum.

took a broad look at fire safety, as did a subsequent 1884 report calling for the formulation of a building code, it wouldn't be until many years later that a *national* set of fire safety and building regulations were actually developed.

The real push for national building and fire regulations came in the wake of the Great Baltimore Fire of 1904. Despite the fact that there had been numerous conflagrations before (and after) this fire, the insurance industry took the lead and prepared the *National Building Code* in 1906. The National Board of Fire Underwriters developed this document using the 1884 NAFE report and the New York State Building Code that had just been published.

The National Fire Protection Association (NFPA), organized in 1896, also had its start in building regulations during this period. It issued a sprinkler system standard (eventually becoming NFPA 13: Standard for the Installation of Sprinkler Systems). followed by the *National Electrical Code* in 1897. This document, revised and updated numerous times over the years, is one of the most widely adopted sets of construction regulations around the world.

In later years, other model building and fire codes were developed. **Model codes** are complete regulatory documents that can be adopted by cities and states as they are written. The early model building and fire codes were developed regionally. The *National Building Code,* prepared by the Building Officials and Code Administrators International (organized in 1915) was used in the North and Midwest; the *Uniform Building Code,* which was used in the West and Southwest, was prepared by the International Conference of Building Officials (organized in 1922); and the *Standard Building Code,* prepared by the Southern Building Code Congress International (organized in 1945), was used in the Southeast. In an effort to create a true national building code, these three organizations came together in 1994 to create a single organization and a single set of codes—the International Code Council's (ICC) *International Building Code*, the *International Fire Code,* and several other model codes. The older regionally adopted codes, also known as **legacy codes**, are no

longer produced. The NFPA also currently prepares a model building code, NFPA 5000, *Building Construction and Safety Code*.

Fire Codes and Standards

In addition to **building codes**, the ICC and NFPA develop **fire codes**. In its simplest form, the difference between a building code and a fire code is fairly straightforward: Generally speaking, building codes regulate the construction of a building, whereas fire codes regulate the use of the building after it is built. For example, a nightclub is built with a fire sprinkler system and proper egress under the building code, while overcrowding and maintenance of the fire sprinkler system are covered under the fire code.

Another important distinction between building and fire codes is their application to existing buildings. Generally speaking, unless a building is renovated or modified, building codes are not applied retroactively. A fire code, in contrast, applies to all structures and conditions within them, including existing hazards that were created even before the adoption of the code.

Fire codes and building codes are updated every three years by the NFPA and ICC. Although this keeps the codes up to date, it is up to the cities and states to adopt the codes themselves. If a code is not adopted, it is not enforceable.

Fire protection standards are also important documents. Building codes rely on standards for detailed regulations for installation of building systems, including fire protection systems. Generally, building codes specify *where* systems are to be installed and standards specify *how* the system is to be designed and installed. Standards cover a variety of topics, including the design and installation standards for fire sprinkler and fire alarm systems as well as test standards that cover such things as the ratings of fire doors and fire walls. Examples of design/installation standards include NFPA 13, *Standard for the Installation of Sprinkler Systems,* and NFPA 14, *Standard for the Installation of Standpipes and Hose Systems.*

Test standards are also referenced by building codes. Two of the most important test standards are American Society of Testing and Materials (ASTM) E-119, *Standard Test Method for Fire Tests of Building Construction and Materials,* which is used for testing and assigning hourly ratings to fire walls, columns, and floors, and ASTM E-84, *Standard Test Method for Surface Burning Characteristics of Building Materials,* which tests the flame spread characteristics of interior finishes Figure 4-3 . These tests are discussed in greater detail in the chapter that reviews the features of fire protection.

Fire Protection Basics of the Building Code

It has been said that 75% of a building code deals with fire safety. The following is a review of some of the more important fire protection provisions.

Type of Construction

The five types of construction—fire resistive, noncombustible, ordinary, heavy timber, and wood frame, respectively—are the same five classifications of construction regulated in the model building codes. In the *International Building Code,* they are given numeral designations (Type 1 = fire resistive,

Figure 4-3 The test "tunnel" used to assess interior finish flame spread in ASTM E-84.

ASTM E 84 testing – Image provided by Marcelo M. Hirschler – GBH International – Mill Valley CA.

Type 2 = noncombustible, Type 3 = ordinary, Type 4 = heavy timber, and Type 5 = wood frame). In addition, there are subcategory letters for each of these five classes, referring to the amount of fire resistance provided for each type of construction (e.g., Type IA is fire-resistive construction with a 3-hour fire-rated structural frame).

NFPA 5000 uses the same general numeral designations for types of construction Table 4-1 . However, it uses an additional three-digit numbering scheme to designate the subcategories of each type of construction related to the hourly rating of the structure. The first number refers to the rating of the load-bearing walls; the second number refers to the beams and columns; and the third number refers to the floor construction. For example, a Type III building (also known as ordinary construction) with a "211" rating for its structural elements would have 2-hour fire-rated load-bearing walls, 1-hour fire-rated beams and columns, and 1-hour fire-rated floors Table 4-2 .

An additional type of construction has presented itself over the last several years. These are hybrid buildings, ones that do not fit entirely within one type (classification). Basically, such buildings incorporate materials of more than one type, such as the use of wooden beams and steel columns. Code officials regulate such buildings by classifying the structure by the "least" type of construction present.

⊕ Branniganism

Remodeling operations on older buildings should be studied carefully and frequently for death-trap installations. Don't believe that "All is according to code" is any guarantee of firefighter safety.

Bolt: © iadams/Shutterstock, Inc.; Texture: © Eky Studio /Shutterstock, Inc; Steel: © Sharpshot/Dreamstime.com

The International Building Code even allows buildings with mixed construction types such as a five-story apartment building of Type III construction built on top of a two-story parking garage of Type I construction.

Table 4-1 Fire Resistance Ratings for Types I–V Construction

	Type I		Type II			Type III		Type IV	Type V	
	442	332	222	111	000	211	200	2HH	111	000
Exterior Bearing Walls[a]										
Supporting more than one floor, columns, or other bearing walls	4	3	2	1	0[b]	2	2	2	1	0[b]
Supporting one floor only	4	3	2	1	0[b]	2	2	2	1	0[b]
Supporting a roof only	4	3	1	1	0[b]	2	2	2	1	0[b]
Interior Bearing Walls										
Supporting more than one floor, columns, or other bearing walls	4	3	2	1	0	1	0	2	1	0
Supporting one floor only	3	2	2	1	0	1	0	1	1	0
Supporting roofs only	3	2	1	1	0	1	0	1	1	0
Columns										
Supporting more than one floor, columns, or other bearing walls	4	3	2	1	0	1	0	H	1	0
Supporting one floor only	3	2	2	1	0	1	0	H	1	0
Supporting roofs only	3	2	1	1	0	1	0	H	1	0
Beams, Girders, Trusses, and Arches										
Supporting more than one floor, columns, or other bearing walls	4	3	2	1	0	1	0	H	1	0
Supporting one floor only	2	2	2	1	0	1	0	H	1	0
Supporting roofs only	2	2	1	1	0	1	0	H	1	0
Floor-Ceiling Assemblies	2	2	2	1	0	1	0	H	1	0
Roof-Ceiling Assemblies	2	1.5	1	1	0	1	0	H	1	0
Interior Nonbearing Walls	0	0	0	0	0	0	0	0	0	0
Exterior Nonbearing Walls[c]	0[b]	0[b]	0[b]	0[b]	0[b]	0[b]	0[b]	0[b]	0[b]	0[b]

Reprinted from: NFPA 5000, Table 7.2.1.1 Fire Resistance Ratings for Type I Through Type V Construction (hr).

H: Heavy timber members (*see text for requirements*).

[a]See 7.3.2.1.
[b]See Section 7.3.
[c]See 7.2.3.2.12, 7.2.4.2.3, and 7.2.5.6.8.

Reproduced with permission from NFPA 5000-2012, Building Construction and Safety Code®, Copyright © 2011, National Fire Protection Association. This reprinted material is not the complete and official position of the NFPA on the referenced subject, which is represented only by the standard in its entirety.

Occupancy Classifications

In addition to the five types of construction, building codes classify structures by their **occupancy types**. The building codes are more restrictive on the more "dangerous" occupancy types such as institutional uses (hospitals, jails, and the like, where people are incapable of self-preservation) and hazardous materials storage, as compared to business uses

Table 4-2 Cross Reference of Building Construction Types

Code Source NFPA 5000	I(443)	I(332)	II(222)	II(111)	II(000)	III(211)	III(200)	IV(2HH)	V(111)	V(000)
IBC	—	IA	IB	IIA	IIB	IIIA	IIIB	IV	VA	VB

IBC: International Building Code.

Reprinted from: NFPA 5000, Standard on Types of Building Construction, 2012 edition.

Reprinted with permission from NFPA 220-2012, Standard on Types of Building Construction, Copyright © 2011, National Fire Protection Association, Quincy, MA. This reprinted material is not the complete and official position of the NFPA on the referenced subject, which is represented only by the standard in its entirety.

(such as office buildings). In essence, the occupancy classification relates the hazards of the building and the type of occupants to specific regulations such as allowable area and height.

The *International Building and Fire Codes* uses the following occupancy types; we've also provided some examples. (Note: The following examples are not an exhaustive list—consult the *International Building and Fire Code* for more examples.)

- "A-1" Assembly (e.g., movie theaters, concert halls)
- "A-2" Assembly (e.g., restaurants, nightclubs)
- "A-3" Assembly (e.g., churches, libraries)
- "A-4" Assembly (e.g., arenas, tennis courts)
- "B" Business (e.g., offices, educational occupancies above the 12th grade)
- "E" Educational (e.g., grade school, day care)
- "F-1" Factory Moderate Hazard (e.g., bakeries, laundries, woodworking)
- Factory "F-2" Low Hazard (e.g., foundries, ice plants)
- High Hazard "H-1" (e.g., detonation hazards)
- High Hazard "H-2" (e.g., deflagration hazards)
- High Hazard "H-3" (e.g., physical hazards presented)
- High Hazard "H-4" (e.g., health hazards)
- High Hazard "H-5" (e.g., semiconductor facilities)
- Institutional "I-1" (e.g., halfway houses, group homes)
- Institutional "I-2" (e.g., hospitals, nursing homes)
- Institutional "I-3" (e.g., prisons, jails)

- Mercantile "M" (e.g., department stores, drug stores)
- Residential "R-1" (e.g., hotels, motels)
- Residential "R-2" (e.g., apartment buildings, dormitories)
- Residential "R-3" (e.g., one- and two-family homes)
- Residential "R-4" (e.g., assisted care facilities, residential care)
- Storage "S-1" Moderate Hazard (e.g., books, furniture)
- Storage "S-2" Low Hazard (e.g., cement in bags, fresh fruits, glass)
- Utility and Miscellaneous "U" (e.g., barns, carports, greenhouses)

NFPA 5000 identifies the following occupancy types:

- Assembly Occupancy, >1,000 occupants
- Assembly Occupancy, >300 occupants
- Assembly Occupancy, # 300 occupants
- Assembly Occupancy, Outdoor
- Business Occupancy
- Board and Care Occupancy, Large
- Board and Care Occupancy, Small
- Day-Care Occupancy
- Detention and Correctional Occupancy
- Educational Occupancy
- Health Care Occupancy
- Health Care Occupancy, Ambulatory
- High Hazard Contents
- Industrial Occupancy, Low Hazard

- Industrial Occupancy, Ordinary Hazard
- Industrial Occupancy, Special Purpose
- Mercantile Occupancy
- Residential Occupancy
- Residential Occupancy, One and Two Family
- Storage Occupancy, Low Hazard
- Storage Occupancy, Ordinary Hazard

Area and Height

Building codes regulate the size (the area per floor in square feet before fire walls are required to subdivide the area) and height of the structures (in terms of feet and stories), based upon the construction classification (Types I–V) and the occupancy type. Generally speaking, the more dangerous the occupancy (in terms of life hazard and/or activities taking place inside) and the more combustible the construction type (e.g., wood frame), the smaller and shorter the building that is permissible. For example, the building codes will not allow a wood-frame hospital of 12 stories with 30,000 square feet (2,787 m²) per floor. According to Table 4-3 below, a wood frame hospital of Type V (000) would not even be permitted at all. A wood-frame hospital with 30,000-square foot floors is just too dangerous.

Modifications can be made to the values in the height/area tables. Perhaps the most visible and controversial is that of "automatic fire sprinkler area increases" permitted under the building code. These are one example of a fire sprinkler system trade-off, where the installation of automatic fire sprinkler systems lessens other code requirements for the building, including allowable area and height.

For example, many two-story buildings can have triple the allowable area shown as permitted in Table 4-3. This means that a much larger building can be erected with a more combustible type of construction simply because fire sprinklers are installed. These fire sprinkler trade-offs exist not only for allowable area and height, but also for reductions in egress requirements, structural fire resistance, and increases in quantities of hazardous materials in some cases.

In addition, the allowable area and heights in Table 4-3 have been criticized by the fire service for being very generous. The values in the building code tables have been increased over the years, without much input from the fire service.

What this means to you as a fire fighter is important—the larger the allowable area, the larger the fire area you can expect. The original intent of the area and height limitations was to make a building fire manageable for the fire service to handle, given the type of construction and the occupancy type. The larger, taller, and more combustible the building, the less likely fire control can be achieved. Properly functioning fire suppression systems in a building, such as sprinkler systems, reduce the hazard of increased area and height. However, a disabled sprinkler system in a building that had been permitted to increase its area and height is a very dangerous situation.

Special Hazards

Building and fire codes regulate the special types of hazards present in a building such as atria and high-rise structures. See the chapter that discusses specific occupancy-related construction hazards for a list of specific hazards and some of the building and fire code regulations that are applicable. These special hazards also include underground buildings, hazardous materials use and storage facilities, stages and platforms, and covered malls.

Egress

Building and fire codes set the requirements for emergency egress, including the number of **exit** stairwells, the **travel distance** (the number of feet from any point in a building to a fire-rated stairwell or outside), and the width of a doorway and corridors.

Before we go any further, it is important to understand the three components of means of egress: the **exit access**, the exit, and the **exit discharge**. Typically, most fire fighters and the public use the word "exit" generically, but in "code speak," it is

Table 4-3 NFPA 5000 Allowable Building Height and Area

In each cell the two numbers above the line are height in stories (S = sprinklered / N = nonsprinklered); the number below is the allowable area in square feet. UL = unlimited, NP = not permitted.

Construction Type	Type I		Type II			Type III		Type IV	Type V	
	442	**332**	**222**	**111**	**000**	**211**	**200**	**2HH**	**111**	**000**
	S / N	S / N	S / N	S / N	S / N	S / N	S / N	S / N	S / N	S / N
Maximum building height (ft)	UL / UL	420 / 400	180 / 160	85 / 65	75 / 55	85 / 65	75 / 55	85 / 65	70 / 50	60 / 40
OCCUPANCY										
Assembly >1,000	UL / UL; UL	UL / UL; UL	4 / 4; UL	3 / 2; 15,500	1 / NP; 8,500	3 / 2; 14,000	2 / 1; 8,500	4 / 3; 15,000	3 / 2; 11,500	NP / NP; NP
Assembly >300	UL / UL; UL	UL / UL; UL	4 / 3; UL	4 / 3; 15,500	2 / 1; 8,500	4 / 3; 14,000	2 / 1; 8,500	4 / 3; 15,000	3 / 2; 11,500	2 / 1; 5,500
Assembly ≤300	UL / UL; UL	UL / UL; UL	4 / 3; UL	4 / 3; 15,500	2 / 1; 8,500	4 / 3; 14,000	2 / 1; 8,500	4 / 3; 15,000	3 / 2; 11,500	2 / 1; 5,500
Assembly, outdoor	UL / UL; UL	UL / UL; UL	UL / UL; UL	UL / UL; UL	UL / UL; UL	UL / UL; UL	UL / UL; UL	UL / UL; UL	UL / UL; UL	UL / UL; UL
Business	UL / UL; UL	UL / UL; UL	12 / 11; 60,500	6 / 5; 37,500	5 / 4; 23,000	6 / 5; 28,500	5 / 4; 19,000	6 / 5; 36,000	4 / 3; 18,500	3 / 2; 9,000
Board & Care, large	UL / UL; UL	UL / NP; UL	12 / 7; 55,000	3 / NP; 19,000	2 / NP; 10,000	2 / NP; 16,500	2 / NP; 10,000	5 / 4; 18,000	2 / 1; 10,500	2 / NP; 4,500
Board & Care, small	UL / UL; UL	UL / UL; UL	12 / 11; UL	5 / 4; 24,000	5 / 4; 16,000	5 / 4; 24,000	5 / 4; 16,000	5 / 4; 20,500	4 / 3; 12,000	3 / 2; 7,000
Day Care	UL / UL; UL	UL / 2; UL	12 / 2; UL	6 / 1; 26,500	4 / NP; 13,000	4 / 1; 23,500	2 / NP; 13,000	2 / 1; 25,500	4 / 1; 12,000	2 / 1; 7,000
Detention & Correctional	UL / UL; UL	UL / 7; UL	12 / 7; UL	2 / 2; 15,000	2 / NP; 10,000	2 / 2; 10,500	2 / NP; 7,500	2 / 2; 12,000	2 / 2; 7,500	2 / 2; 5,000
Educational	UL / UL; UL	UL / UL; UL	12 / 5; UL	4 / 3; 26,500	3 / 2; 14,500	4 / 3; 23,500	3 / 2; 14,500	4 / 3; 25,500	2 / 1; 18,500	2 / 1; 9,500
Health Care	UL / NP; UL	UL / NP; UL	12 / NP; UL	3 / NP; 15,000	NP / NP; 12,000	3 / NP; 12,000	NP / NP; NP	3 / NP; 12,000	1 / NP; 9,500	NP / NP; NP

Reprinted from: NFPA 220, Standard on Types of Building Construction, 2012 edition.

Table 4-3 NFPA 5000 Allowable Building Height and Area

Construction Type	Type I 442 S	Type I 442 N	Type I 332 S	Type I 332 N	Type II 222 S	Type II 222 N	Type II 111 S	Type II 111 N	Type II 000 S	Type II 000 N	Type III 211 S	Type III 211 N	Type III 200 S	Type III 200 N	Type IV 2HH S	Type IV 2HH N	Type V 111 S	Type V 111 N	Type V 000 S	Type V 000 N
Maximum building height (ft)	UL	UL	420	400	180	160	85	65	75	55	85	65	75	55	85	65	70	50	60	40
OCCUPANCY																				
Health care, ambulatory	UL	UL	UL	UL	12	11	6	5	5	1	6	5	5	1	6	5	4	3	3	1
Health care, ambulatory (area)	UL		UL		UL		37,500		23,000		28,500		19,000		36,000		18,000		9,000	
Industrial, ord. hazard	UL	UL	UL	UL	12	11	5	4	3	2	4	3	3	2	5	4	3	2	2	1
Industrial, ord. hazard (area)	UL		UL		UL		25,000		15,500		19,000		12,000		33,500		14,000		8,500	
Industrial, low hazard	UL	UL	UL	UL	12	11	6	5	4	3	5	4	4	3	6	5	4	3	3	2
Industrial, low hazard (area)	UL		UL		UL		37,500		23,000		28,500		18,000		50,500		21,000		13,000	
Mercantile	UL	UL	UL	UL	12	11	5	4	5	4	5	4	5	4	5	4	4	3	2	1
Mercantile (area)	UL		UL		UL		21,500		12,500		18,500		12,500		20,500		14,000		9,000	
Residential	UL	UL	UL	UL	12	11	5	4	5	4	5	4	5	4	5	4	4	3	3	2
Residential (area)	UL		UL		UL		24,000		16,000		24,000		16,000		20,500		12,000		7,000	
Residential, 1- & 2-family	UL	UL	UL	UL	12	11	5	4	5	4	5	4	5	4	5	4	4	3	3	2
Residential, 1- & 2-family (area)	UL		UL		UL		UL		UL		UL		UL		UL		UL		UL	
Storage, ord. hazard	UL	UL	UL	UL	12	11	5	4	4	3	5	4	4	3	5	4	4	3	2	1
Storage, ord. hazard (area)	UL		UL		48,000		26,000		17,500		26,000		17,500		25,500		14,000		9,000	
Storage, low hazard	UL	UL	UL	UL	12	11	6	5	5	4	5	4	5	4	6	5	5	4	3	2
Storage, low hazard (area)	UL		UL		79,000		39,000		26,000		39,000		26,000		38,500		21,000		13,500	

HIGH HAZARD
CONTENTS
High Hazard
Contents Not
Requiring Protection
Level 1 through
Protection Level 5

Use values for appropriate occupancy class.
For industrial or storage occupancies, use values for ordinary hazard.

Reprinted from: NFPA 220, Standard on Types of Building Construction, 2012 edition.

(Continues)

Table 4-3 NFPA 5000 Allowable Building Height and Area (Continued)

Construction Type	Type I				Type II						Type III				Type IV		Type V			
	442		332		222		111		000		211		200		2HH		111		000	
	S	N	S	N	S	N	S	N	S	N	S	N	S	N	S	N	S	N	S	N
Maximum building height (ft)	UL	UL	420	400	180	160	85	65	75	55	85	65	75	55	85	65	70	50	60	40
OCCUPANCY																				
Protection Level 1	1	NP	1	NP	1	NP	1	NP	1	NP	1	NP	1	NP	1	NP	1	NP	1	NP
	21,000		21,000		16,500		11,000		7,000		9,500		7,000		10,500		7,500		NP	
Protection Level 2	UL	NP	UL	NP	3	NP	2	NP	1	NP	2	NP	1	NP	2	NP	1	NP	1	NP
	21,000		21,000		16,500		11,000		7,000		9,500		7,000		10,500		7,500		3,000	
Protection Level 3	UL	NP	UL	NP	6	NP	4	NP	2	NP	4	NP	2	NP	4	NP	2	NP	1	NP
	UL		UL		60,000		26,500		14,000		17,500		13,000		25,500		10,000		5,000	
Protection Level 4	UL	NP	UL	NP	8	NP	6	NP	4	NP	6	NP	4	NP	6	NP	4	NP	3	NP
	UL		UL		UL		37,500		17,500		28,500		17,500		36,000		18,000		6,500	
Protection Level 5	4	NP	4	NP	4	NP	3	NP	3	NP	3	NP	3	NP	3	NP	3	NP	2	NP
	UL		UL		UL		37,500		23,000		28,500		19,000		36,000		18,000		9,000	

Reprinted from: NFPA 5000, 2012, Table 7.4.1 Allowable Building Height and Area.

For SI units, 1 ft = 0.3048 m; 1 ft² = 0.093 m².

S: Allowable building height in feet and allowable number of stories above grade in buildings protected with an automatic sprinkler system as specified in 7.4.1.

N: Allowable building height in feet and allowable number of stories above grade in buildings not protected with an automatic sprinkler system as specified in 7.4.1.

UL: Unlimited.

NP: Not permitted.

Note: Within each occupancy category, the top row refers to the allowable number of stories above grade, and the bottom row refers to the allowable area per story.

actually a component of the means of egress. These components are:

- The exit access is the actual path from any point in a building to a fire-rated stairwell (or outdoors).
- The exit is the rated stairwell or exit passageway (essentially a higher rated corridor that connects a stairwell to the exterior of a building when the stairwell is not located on an exterior wall) and the opening (door) to the outside.
- The exit discharge is simply from the outside door to a public way (e.g., the street, public sidewalk). **Figure 4-4**

A complex set of regulations deal with egress. The first and foremost consideration is the calculation of the occupant load, the number of people that the code anticipates can occupy a space given the occupancy type. **Table 4-4** lists various types of occupancies and their occupant load factors. Note that these factors are measured in square feet per person. It is these factors that allow a code official to calculate the total number of people for a space/building and design the means of egress components to accommodate that number of people in a space and building. The number is also posted in places of assembly in the form of an occupancy limit.

Once the occupant load has been determined, the number of exit paths/doors for each room and the entire building must be established. All rooms must have at least one code-compliant means of egress; as the occupant load increases, the number of doors/paths must be increased as well.

With the occupant load and number of exit doors/egress paths determined, designers must then provide a safe path of travel to a public way. This translates into rated corridors (often 1 hour) with rated corridor doors (often 20 minutes), and rated stairwells (1 hour for stairs up to three stories and 2 hours for those four or more stories).

Figure 4-4 Means of egress: The exit discharge (shaded gray), the exit access (shaded yellow), and the exit (shaded blue).

Courtesy of NFPA.

Reprinted with permission from NFPA 101®-2012, Life Safety Code® Handbook, Copyright © 2011, National Fire Protection Association, Quincy, MA. This reprinted material is not the complete and official position of the NFPA on the referenced subject, which is represented only by the standard in its entirety.

Table 4-4 Occupant Load Factor

Use	ft² (per person)[a]	m² lb(per person)[a]
Assembly Use		
Concentrated use, without fixed seating	7 net	0.65 net
Less concentrated use, without fixed seating	15 net	1.4 net
Bench-type seating	1 person/18 linear in.	1 person/45.7 linear cm
Fixed seating	Number of fixed seats	Number of fixed seats
Waiting spaces	See 16.1.6.1.	See 16.1.6.1.
Kitchens	100	9.3
Library stack areas	100	9.3
Library reading rooms	50 net	4.6 net
Swimming pools	50 (of water surface)	4.6 (of water surface)
Swimming pool decks	30	2.8
Exercise rooms with equipment	50	4.6
Exercise rooms without equipment	15	1.4
Stages	15 net	1.4 net
Lighting and access catwalks, galleries, gridirons	100 net	9.3 net
Casinos and similar gaming areas	11	1
Skating rinks	50	4.6
Educational Use		
Classrooms	20 net	1.9 net
Shops, laboratories, vocational rooms	50 net	4.6 net
Day-Care Use	35 net	3.3 net
Health Care Use		
Inpatient treatment departments	240	22.3
Sleeping departments	120	11.1
Ambulatory health care	100	9.3
Detention and Correctional Use	120	11.1
Residential Use		
Hotels and dormitories	200	18.6
Apartment buildings	200	18.6
Board and care, large	200	18.6

Reprinted from: NFPA 5000, Table 11.3.1.2 Occupant Load Factor.

Table 4-4 Occupant Load Factor

Use	ft² (per person)ᵃ	m² lb(per person)ᵃ
Mercantile Use		
Sales area on street floorᵇ·ᶜ	30	2.8
Sales area on two or more street floorsᶜ	40	3.7
Sales area on floor below street floorᶜ	30	2.8
Sales area on floors above street floorᶜ	60	5.6
Floors or portions of floors used only for offices	See business use.	See business use.
Floors or portions of floors used only for storage, receiving, and shipping, and not open to general public	300	27.9
Mall buildings	Per factors applicable to use of spaceᵈ	Per factors applicable to use of spaceᵈ
Business Use	100	9.3
Industrial Use		
General and high hazard industrial	100	9.3
Special purpose industrial	NA	NA
Storage Use		
In storage occupancies	NA	NA
In mercantile occupancies	300	27.9
In other than storage and mercantile occupancies	500	46.5

Reprinted from: NFPA 5000, Table 11.3.1.2 Occupant Load Factor.

NA: Not applicable. The occupant load is the maximum probable number of occupants present at any time.

ᵃAll factors expressed in gross area unless marked "net."

ᵇFor the purpose of determining occupant load in mercantile occupancies where, due to differences in grade of streets on different sides, two or more floors directly accessible from streets (not including alleys or similar backstreets) exist, each such floor is considered a street floor. The occupant load factor is one person for each 40 ft² (3.7 m²) of gross floor area of sales space.

ᶜFor the purposes of determining occupant load in mercantile occupancies with no street floor, but with access directly from the street by stairs or escalators, the floor at the point of entrance to the mercantile occupancy is considered the street floor. (See 3.3.221.2, Street Floor.)

ᵈThe portions of the mall, where considered a pedestrian way and not used as gross leasable area, are not assessed an occupant load based on Table 11.3.1.2. However, means of egress from a mall pedestrian way are provided for an occupant load determined by dividing the gross leasable area of the mall building (not including anchor stores) by the appropriate lowest whole number occupant load factor from Figure 11.3.1.2.

Each individual tenant space has means of egress to the outside or to the mall based on occupant loads figured by using the appropriate occupant load factor from Table 11.3.1.2. Each individual anchor store has means of egress independent of the mall.

Reproduced with permission from NFPA 5000-2012, Building Construction and Safety Code®, Copyright © 2011, National Fire Protection Association. This reprinted material is not the complete and official position of the NFPA on the referenced subject, which is represented only by the standard in its entirety.

An important provision of building and fire codes are limitations on travel distance—that is, how far an occupant must travel to get to a fire-rated stairwell or exit. In practice, this puts limits on how far a person must travel within a room and within a fire-rated corridor, generally a maximum of 200 feet in a building without an automatic fire sprinkler system (buildings with fire sprinklers may have increased travel distance limitations). A **horizontal exit**—a 2-hour fire-rated wall typically running from exterior wall to exterior wall on an individual floor of a building—is used when travel distances

to a fire-rated stairwell or exit discharge are excessive. Horizontal exits (and associated standpipe hose valves) are described in further detail in the chapter that discusses the features of fire protection.

Door hardware is important as well. The type, number, and location of locking devices are limited. Means of egress doors for places of assembly, schools, and hospitals require the use of panic hardware (known as fire exit hardware when used on fire doors), which is characterized by a bar that opens the door with a simple push to avoid a pileup behind the door **Figure 4-5**. When inspecting a facility, make sure that required egress doors are not locked and panic hardware has not been compromised.

Two additional notes on means of egress. Exterior emergency escape and rescue windows are required for all sleeping rooms in residential occupancies, e.g., hotels, dormitories, apartments, one- and two-family dwellings—Group "R" occupancies below the fourth floor for you to rescue trapped occupants. These windows must be a minimum of 5.7 square feet (0.53 m²) in area, with a minimum of 20 inches in width and 24 inches in height (508 × 609 mm).

The use of fixed burglar bars is prohibited by all model codes. These devices are obviously dangerous, with many trapped occupants having been

Figure 4-6 Illegal fixed burglar bars on a window.
© Ryan DeBerardinis/ShutterStock, Inc.

killed by their presence **Figure 4-6**. The authors suggest making their removal a priority for those fire fighters who find them in their response areas.

Fire Protection Systems

Building and fire codes regulate the installation of fire protection systems. They are covered in detail in the chapter that discusses the features of fire protection.

Disabled Occupants

Following the lead of the Americans with Disabilities Act of 1990, the model building and fire codes have included a variety of provisions to increase the safety of disabled persons. These include fire alarm systems with both visual and audible signaling devices as well as ramps for wheelchair-bound individuals.

One of the most important features for fire fighters is the introduction of the "area of refuge." Normally, these are smoke barrier-protected rooms located near stairwells and are identified with the "international symbol of accessibility." **Figure 4-7** These refuge areas may also be expanded stairwell landings. The point is this: Disabled people have been instructed to wait in these protected areas for *you* to come and rescue them. They are provided with a two-way communications system between these areas of refuge and a central control point (usually somewhere in the lobby, but not always). When

Figure 4-5 These fire-rated doors have fire exit hardware to prevent people from piling up behind them.
© Robert Elias/ShutterStock, Inc.

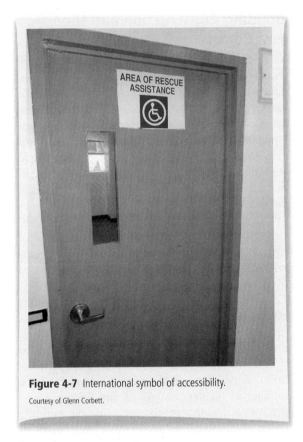

Figure 4-7 International symbol of accessibility.
Courtesy of Glenn Corbett.

preplanning, note the locations of these refuge areas and the communications system.

Elevators

In a multistory building, fire fighters depend on the use of elevators to attack a fire on the upper floors. (Most do use elevators, but some departments are concerned with the safety of elevators and walk up instead.) There is a particular "code" (actually, more of a standard), American Society of Mechanical Engineers (ASME) A17.1, *Safety Code for Elevators and Escalators,* for new elevators and A17.3 for existing elevators, which regulates the design and installation of elevators. The model building codes also contain provisions affecting fire fighter use of elevators as well, including emergency power for selected elevators and smoke/heat vents at the tops of hoistways. For fire fighters, the concern is with the operation of the elevator under emergency conditions. Elevators come in two varieties: electric cable and hydraulic piston.

Today, most elevators have a two-phase operation under fire conditions. Under **Phase I operation**, elevators are recalled to a specific floor; usually, to the first floor lobby of the building by the activation of a smoke detector located in an elevator lobby of a floor. Upon arrival at the first floor, the doors open and allow the occupants to leave. The doors remain open for fire fighters to begin their use of the elevators.

In **Phase II operation**, the fire fighter selects one of the elevators in a bank (a group of elevators) to access the upper floors. Often, emergency generators are capable of running only a single elevator in each bank (make sure your preplans include this limitation). Fire fighters can then control the elevators using a special key and set of procedures (preplans again).

As a result of the World Trade Center disaster on September 11, 2001, the model building codes were recently modified to include two particularly important provisions for elevators in new high-rise buildings: "**occupant evacuation elevators**" and ""**fire service access elevators**." Essentially, these are more "robust" elevators that are expected to work properly under "stressful" emergency situations. Some common features are found in these two types of elevators. For example, both require smoke barrier-enclosed lobbies (smoke barriers are described in greater detail in the chapter that discusses the features of fire protection), direct access to a fire-rated stairwell from these lobbies, specially protected wiring and cables, elevator system monitoring in the fire command center, and standby power for the elevators.

Occupant evacuation elevators are intended for the self-evacuation of occupants in buildings more than 420 feet (128 m) in height as an alternative to an extra mandated stairwell in the building (stairwells required as a result of the September 11, 2001 disaster, known as Phase III operation.) Essentially, they are to be used for occupant evacuation *prior* to Phase I operation of the elevator (automatic recall). Obviously, signage indicating that the elevator may be used for evacuation is required—remember that we have told building occupants for decades *not* to use elevators in

a fire emergency. In addition, during an emergency, a green-colored sign is illuminated, indicating that the elevators may be used for egress. Illuminated red-colored signs indicate that the elevators are not be used and tell occupants to use stairwells. These elevators have special protection against water penetration and a two-way communication system between the lobby and the fire command center. Fire fighters can recall these elevators from the fire command center.

Fire service access elevators are found in new buildings more than 120 feet (36 m) in height. They also have some special features. The hoistways are illuminated (to allow fire fighters to assess conditions in the shaft). The lobbies are a minimum of 150 square feet (13.9 m²) in area and must have a Class I hose outlet in the fire-rated stairwell immediately adjacent to the lobby.

Elevator power controls are usually located on the first floor for hydraulic elevators and on the top floor for electric elevators, where the motors are located. Your preplans should note shutdown procedures and location of the equipment.

A note of caution: For many years, ASME A17.1 has encouraged the use of elevator door restrictors. In essence, elevator door restrictors are metal bars (the type and style varies among manufacturers) used to prevent the occupants of an elevator that is stuck between floors from opening the car and shaft doors and falling down the elevator shaft below. The door restrictor engages when the elevator cab is located 18 inches above or below the "landing zone" (the floor of an elevator lobby). Elevator keys (which come in all shapes and sizes depending upon the elevator manufacturer) typically do not bypass these restrictors.

On September 11, 2001, it is very likely that many New York City fire fighters were killed trying to open elevator doors that were held in place by door restrictors. One company was even found with its hydraulic auto extrication tool, trying to pry the doors open. Although the technology currently exists for a push-button remote release to bypass these door restrictors, they have not become a requirement. Your best bet is to meet with all of the elevator service companies in your district and determine exactly how to open each particular type of restrictor found in your area.

Building Construction and the Fire Codes

While much of the fire code deals with the ongoing maintenance of building code mandated fire-resistive assemblies and fire protection systems, some specific construction type provisions are included as well. These items cover issues such as fire apparatus access roads ("fire lanes"), smoke management fans for high-piled combustible warehouse stock, and flammable liquid storage rooms. While requirements like these are found in the fire code, permits from the buildings department are usually required in order to construct them.

A very important provision is in the model fire codes for fire fighter communications as a result of the difficulties experienced by responders in high-rise and large buildings, such as was experienced by FDNY firefighters on September 11, 2001. Complete emergency responder radio coverage is now required in *all* new buildings (not just high-rises), in the model fire codes. Existing buildings must also meet this requirement; a timetable for compliance to be determined by the local jurisdiction. The importance of communications and the number of incidents in which fire fighters could not communicate with those outside the building or lost communication makes this a truly lifesaving measure.

Code Implications for Firefighting

We have only touched the tip of the code iceberg. As we have described, many factors can affect your firefighting operations, both to your benefit and to your detriment. The authors suggest establishing a close relationship between you, the fire fighter, and your local code officials.

You are encouraged to participate in the code development process, at both the local and national levels. The participation of the fire service in this process over the years has been spotty at best. Your safety and the safety of those you protect lay in your hands. As they say in the lottery: "You've got to be in it to win it."

TACTICAL CONSIDERATIONS

1. Proper fire protection requires both passive and active (dynamic) fire protection features. They work together, providing redundancy in case of individual failures. Beware of changes to the model building and fire codes that have reduced this balance, thus lessening passive fire protection features in favor of sprinkler protection.

2. Watch out for carpeting on walls and ceilings. It is very dangerous, especially when used in day care centers and the like. If you see interior finishes such as carpeting, foam, or other materials on walls and ceilings, notify the fire prevention bureau.

3. It is important to dispatch fire fighters to areas of refuge where disabled people may be found, such as in multistory office buildings. Use the communications system in the lobby to make contact with these individuals. Preplan how you are going to get them out of the building, if necessary.

4. Elevator cabs are often equipped with door restrictors. These devices prevent elevator occupants from opening the cab door and falling down the shaft when the cab stalls between floors. Each manufacturer has a different design. Learn which buildings in your district have the restrictors and how they can be overridden (some are very difficult to bypass).

5. Your knowledge of the building code and the fire code will help you on the fire ground. You will know what to expect, such as where a standpipe hose valve should be located or how a fire wall is expected to behave. You will also be able to more easily spot construction deficiencies, including ones that will have a direct impact on your safety.

Chapter Summary

- There have been fires and subsequent fire protection regulations almost as long as there have been towns and cities in the United States.
- In addition to building codes, the ICC and NFPA prepare fire codes.
- The difference between a building code and a fire code is fairly straightforward: Building codes regulate the construction of a building, whereas fire codes regulate the use of the building after it is built.
- While much of the fire code deals with the ongoing maintenance of building code mandated fire-resistive assemblies and fire protection systems, some specific construction type provisions are included as well.
- For fire fighters, it is important to establish a close relationship with the local code officials.
- Fire fighters are encouraged to participate in the code development process, at both the local and national levels.

Key Terms

Building code Regulates the actual design and construction of new buildings, providing for minimum levels of health and safety. Regulates the level and amount of fire protection in a new structure.

Exit That portion of a means of egress that is separated from all other spaces of a building or structure by construction or equipment as required to provide a protected way of travel to the exit discharge.

Exit access That portion of a means of egress that leads to an exit.

Exit discharge That portion of a means of egress between the termination of an exit and a public way.

Fire code Regulates the activities that take place in buildings. Applicable to existing buildings and hazards.

Fire service access elevators Elevators specifically designed for fire fighter use that are found in new buildings more than 120 feet (36 m) in height.

Horizontal exit A 2-hour fire-rated wall typically running from exterior wall to exterior wall on an individual floor of a building, which is used when travel distances to a fire-rated stairwell or exit discharge are excessive.

Legacy code A source code that is no longer supported or printed.

Model code Complete regulatory documents that can be adopted by cities and states as they are written.

Occupancy type The building code classification for the intended use or purpose of a building or building space.

Occupant evacuation elevators Special elevators intended for the self-evacuation of occupants in office buildings more than 420 feet (128 m) in height.

Phase I operation Elevators are recalled to the first floor lobby of the building by the activation of a smoke detector located in an elevator lobby or on one of the upper floors. Upon arrival at the first floor, the doors open and allow the occupants to leave. The doors remain open for fire fighters to begin their use of the elevators.

Phase II operation A fire fighter selects one of the elevators in a bank to access the upper floors. Fire fighters can then control the elevators using a special key and set of procedures.

Travel distance The number of feet from any point in a building to a rated stairwell or outside.

Case Study

Fire inspectors are tasked with many different responsibilities. One of their major duties is plans review. It is important for fire inspectors to include fire officers in the process of plans review.

The fire marshal within your jurisdiction has asked that three of the fire officers within your department assist with the plan review of a new addition to an existing hospital building. Your battalion chief has asked that you be one of the participants.

1. According to the ICC, a hospital is considered what occupancy type?

 A. Assembly
 B. Business
 C. Educational
 D. Institutional

2. True or False: A fire-rated stairwell is considered an exit.

3. After figuring the occupancy load for a building, what would be the next step?

 A. Calculate the number of exit paths/doors required
 B. Calculate the required flow of the sprinkler system
 C. Determine where standpipe connections are to be placed
 D. Determine whether a fire alarm system is needed

4. You find that rated corridors will have to be established within the building. These are often _____ hour rated.

 A. 0.5
 B. 1
 C. 2
 D. 3

Challenging Questions

1. Describe the difference between a building code and a fire code.

2. Identify and describe the three components of a means of egress.

3. You are a fire captain and are called to look into a late-night overcrowding situation at a one-story department store during the holiday season. Based upon NFPA 5000, what would be the occupant load of the store if the floor area is 100,000 square feet (9,290 m²)?

CHAPTER 5

Fire Behavior and Building Construction

OBJECTIVES

At the conclusion of this chapter, you will be able to:

- Identify fundamentals of fire behavior and key fire phenomena.
- Describe the development of a compartment fire.
- Identify the ways fire is influenced by a building and its characteristics.
- List the ways buildings and their components are affected by a fire.
- Explain how specific building materials behave under fire attack.
- Describe smoke behavior and the effects it can have on the building, occupants, and fire fighters.

Flames: © Dix/Dreamstime.com; Steel texture: © Sharpshot/Dreamstime.com; Chapter opener photo: © Glen E. Ellman.

Case Study

Figure 5-1 The trash compactor in the basement of the Schomburg Plaza complex.

Courtesy of the USFA.

Interpreting and anticipating fire behavior are critical skills for all fire fighters. Such was the case in March 1987, when a fire broke out in the trash chute of a 35-story high-rise in the Schomburg Plaza complex in upper Manhattan (New York City). The fire started in trash jammed between the 27th and 29th floors. Arriving fire fighters were told by building personnel that the fire was in the basement compactor room (**Figure 5-1**); a line was stretched to the compactor and the fire was extinguished. Fire fighters were unaware that burning debris had actually dropped more than 25 floors and ignited the trash in the compactor.

While fire fighters were extinguishing the fire in the basement, fire was spreading vertically from the immediate area of the trash obstruction. The fire spread from the shaft via convection heat currents and radiant heat through "construction openings" in the shaft wall and by conduction into apartments on the 33rd and 34th floors through metal pipe supports and trash chute anchors. In addition, the fire spread to the top floor (the 35th floor) through autoexposure from the 34th floor.

Desperate occupants in the affected apartments made 21 calls to the fire department, in some cases reporting heavy smoke conditions, but were told the fire fighters were putting out the fire and things were under control. Dispatchers did not notify the incident commander (IC) of these calls; the IC, in fact, began returning companies, believing that the fire had been extinguished. By the time it was understood that fire had spread to the upper-floor apartments at the top of the building, seven occupants had been killed, including three who had jumped to their deaths. After the fire, valves in a sprinkler system protecting the trash chute were found to be closed.

1. Are the paths of fire spread always obvious at a fire?
2. How important is it to understand the three methods of heat transfer?
3. What role do dispatchers play during a fire?

The Basics of Fire and Fire Behavior

Although you may have learned the basics of fire and fire behavior in a fire fighter training program, it is a good idea to review this essential information. It is especially important to understand the fundamentals of fire behavior so that you will understand fire's effects on building construction and, in turn, the ways that building construction elements affect fire behavior.

Fire can be defined simply as a rapid oxidation (chemical) reaction that produces heat and visible light. To have such a reaction take place—to have a fire—four essential elements are necessary **Figure 5-2**:

- **Fuel**: a material that will sustain combustion, also known as a "reducing agent"
- **Oxidizing agent**: normally atmospheric oxygen, but may also be chemical compounds known as oxidizers that will release oxygen as they react
- **Heat**: a form of energy that is the source of ignition

- **Uninhibited chemical chain reaction**: a sustained oxidation reaction that produces sufficient excess heat to continue heating unburned fuel, making it available for combustion

An important concept in fire behavior is that of heat transfer. The effects of heat transfer can be seen in the weakening of a steel-frame structure, the rapid spread of fire across a wall surface, or the ignition of an exposure's window frames from a fire across a street. Heat transfer dictates how a fire spreads.

There are three methods of heat transfer:

- **Conduction**: the transfer of heat within an object or between objects through direct contact
- **Convection**: the transfer of heat through circulation within a medium such as a gas or a liquid **Figure 5-3**
- **Radiation**: the transfer of heat through electromagnetic waves **Figure 5-4**

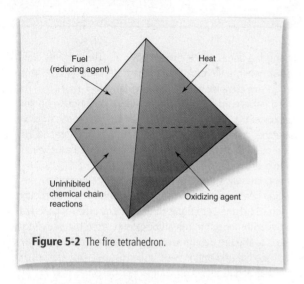

Figure 5-2 The fire tetrahedron.

Figure 5-4 The effects of radiant heat can be seen across the street from the fire as the wooden "monitor" (horizontal light shaft) on top of a mill begins to burn.
Courtesy of Glenn Corbett.

Figure 5-3 The hot gases from a fire left behind this soot and char pattern in a hallway. This effect was not caused by direct flame contact, but rather by convective heat transfer from the hot gases to the wallboard.
Courtesy of Richard Wolfson.

In more practical terms, we can see the effects of heat transfer on the fireground. A bare steel beam above a fire, immersed in the hot gases from the fire below, receives heat energy through the convection (from the hot gas "medium"). As the beam starts to heat up, heat is transferred within the beam through conduction, as well as down the length of the beam. If a combustible material such as a cardboard box is lying against the heated steel beam, it may ignite from conductive heat transfer—that is, from heat moving from one object to another in direct contact.

A large, fully involved wood-frame multiple dwelling in a row of similar structures may ignite adjacent exposures through thermal radiation. Radiation is not visible but its effects can certainly be felt on exposed skin and combustible materials. Large quantities of radiation applied to a combustible material can cause **autoignition**, a phenomenon in which a combustible material ignites spontaneously without the application of a flame or spark. (**Piloted ignition** is ignition of a heated combustible material when a flame or spark is applied.)

The term **autoexposure** has been in use for years within the fire service. You won't find it defined in most fire behavior texts, but it is still important to us. Autoexposure describes the vertical spread of fire on the exterior of a multistory building, from one floor to the floor(s) above. The flames and hot gases exiting a window pass up the side of a building, heating the window and objects in the room directly above, potentially causing them to ignite. In such a case, heat is transferred through a combination of radiation and convection.

Compartment Fires

There is a two-way relationship between building construction and fire behavior—that is, the fire is influenced by the building, and the building reacts to the effects of a fire. Typically, both of these reactions happen at the same time.

Fire scientists use the word *compartment* to describe the enclosure around a fire—an area commonly called a room. Analyzing even a moderate-sized fire in a room is a complicated process, one in which computer models and full-scale testing are helping us to better understand the complex relationship between the fire and the environment around it. We can, however, draw out some basic compartment fire principles.

If permitted to follow its natural course (no extinguishment), a fire (that is not ventilation-controlled) will often pass through these four stages:

1. Incipient: the initial stage of a fire after ignition, usually involving a single or small number of combustibles.
2. Free burning: the secondary stage of a fire in which it spreads to adjacent combustible materials **Figure 5-5**.

3. <u>Flashover</u>: a transition stage in which exposed surfaces within the compartment ignite simultaneously and fire spreads throughout the compartment. Flashover results in full room involvement.
4. Smoldering/decay: a stage in which glowing combustion takes place, without flame. A fire may be forced into the smoldering stage when key components of the fire tetrahedron are in short supply, such as when the oxygen supply is limited.

A variety of terms are used to describe the various elements of a compartment fire. Some of the most important elements are the **<u>fire plume</u>**, which is the column of flames, **<u>smoke</u>**, and heated gases rising above the burning object; the **<u>ceiling layer</u>**, which consists of the hot, buoyant gases that collect at the very top of a room; and the **<u>neutral plane</u>**, which is the interface between the hot ceiling layer and the cooler air flowing into the compartment.

Studying Figure 5-5, you can see how a compartment fire sustains itself and grows. Fresh air is drawn into the fire from the doorway and is drawn into the

Figure 5-5 A compartment fire in the free-burning stage, prior to flashover.

fire itself; this is known as **entrainment**. The hot fire plume consisting of the hot gases and smoke travels upward to the ceiling, where it collects and then moves horizontally. Some of these gases and smoke leave the compartment through any open doors and windows.

A phenomenon known as **flameover** (also known as rollover) may be observed in the ceiling layer. In flameover, hot unburned fuel gases, having gathered in sufficient quantities at the ceiling layer, suddenly ignite. The term *rollover* describes the "rolling" motion of the flames across the bottom of the ceiling layer.

The remaining hot ceiling layer of gases and smoke in the room radiates heat to the objects below. Eventually, overwhelming radiant heat from the ceiling layer drives the room toward flashover.

A fire burning in a room will be influenced by several factors, primary among them the oxygen supply. The location and quantity of oxygen needed to feed the fire will dictate how the fire burns and grows. For example, an open door versus a closed door in a compartment will lead to a different outcome. If the oxygen supply is cut off or greatly reduced, flaming combustion will likely cease and the fire will begin smoldering.

Wind-driven fires, in which substantial winds accelerate a fire and speed up its spread, have come under recent scrutiny. If winds exceed 10–20 mph, their role in driving the fire within a building must be considered from a tactical standpoint. Current protocols call for attacking the fire from upwind and, in some cases, limiting ventilation. To deal with cases where the fire itself has broken a window on the windward side of a multistory building, the New York City Fire Department has developed a "fire blanket" to place over the opening, allowing fire fighters into that unit **Figure 5-6**.

In addition to the oxygen-supplying door and window openings described previously, examples of other construction factors influencing fire behavior include the following:

- Combustible interior finishes or walls, which allow for more rapid flame spread and potentially hasten flashover.

Figure 5-6 The New York City Fire Department deploys a fire blanket.
Courtesy of NIST.

- Vertical openings such as chases and shafts and penetrations, which allow fire to spread vertically.
- The height of the fire room. Rooms with smaller heights may flash over more quickly than taller rooms (the amount of radiant heat at floor level is greater in smaller height rooms).
- The proximity of walls to each other. Narrow corridors with combustible interior finishes, for example, will tend to accelerate flame spread due to the radiant heat from the close, opposing walls.
- Heating, ventilating, and air-conditioning (HVAC) systems, whose operation may provide fresh oxygen to the fire and spread the fire.

Of course, the fire has an effect on the building and its components.as well. The most common example is the charring of wood. As the wood burns and is consumed, it loses strength. Obviously, this is important for load-bearing wood members. There is no way to determine exactly when a charred wood member under load will fail. In the case of light-weight members such as wooden trusses, the connections will likely fail first. It is best to assess the construction technique (traditional wood framing

versus lightweight construction), fire severity, and fire duration to try to anticipate collapse—while understanding that this is not an exact science.

Fires can affect other materials in different ways. Steel, for example, weakens as it is heated to fire temperatures. Aluminum will melt at around 1,200°F (649°C). Some plastics will soften and melt, whereas others will char when exposed to fire. Some types of stone will crack. The bottom line: nearly all construction materials are affected negatively by fire, some more than others. Fire duration and size play significant roles in the degradation of construction materials.

Backdraft

Backdraft is the burning of heated gaseous products of combustion when oxygen is introduced into an environment whose oxygen supply has been depleted due to fire. This burning often occurs with explosive force. Backdrafts, while still fairly rare, have been on the increase, perhaps owing to increased "tightening" of buildings for purposes of energy conservation or security.

Fire Behavior, Fire Fighters, and Buildings

It is imperative to note that collapse is not the only hazard facing the fire fighter in a burning structure. Fire fighters face danger while operating inside structures—for example, being caught up in the rapid development of a fire such as in a flashover or backdraft. Concealed fire that bursts out of a hidden void and the rapid spread of fire over combustible surfaces is equally as hazardous. These hazards may account for as many fire fighter casualties as collapse does. Fire training schools conduct live training in fire-resistive structures where collapse and hidden fire hazards do not exist. Whether intended or not, the message delivered to the firefighting students is simply to "put the wet stuff on the red stuff."

It is not possible to train in the same way for collapse and hidden fire hazards. Live fire training in abandoned buildings has resulted in casualties when the training instructors did not recognize the hazards in the structure or how fire would spread. Special effort must be taken to cover these hazards in the classroom so that it is evident to the student that they are as important as putting out the visible fire.

What can be done to prevent fatalities from such conditions? The ventilation of involved structures is essential in limiting risks to fire fighters operating inside, but the ventilation process can expose fire fighters to the very dangers they are trying to reduce. ICs must recognize conditions that could cause backdraft, flashover, or flameover and could endanger fire fighters operating inside the structure.

It is also possible that these fire development phenomena may not be factored into local fire scene planning and decision making as universally as these special hazards would warrant. A careful and thorough investigation of all abnormal situations should be initiated, whether or not there are deaths and injuries. Encountering unusual heat, smoke, or burning at locations remote from the main fire area could indicate problems. Fire fighters at the fire should use care when dealing with these conditions and, following the fire, attempts should be made to understand why the conditions existed. Additional fire research may be necessary to understand how some materials burn in actual building fires, how some of these unexpected situations occurred, and how fire command and attack procedures might be better designed to reduce this risk.

Fire Growth and Spread

Fire growth and spread are greatly influenced by building construction. Whether or not a compartment goes to flashover, for example, is greatly influenced by the materials lining the walls and ceilings of the space. What follows is a description of these hazards and some of the fire protection methods employed to deal with them.

Flame spread—or the more accurate description, fire growth—is a particularly hazardous fire phenomenon. It was not recognized as such three

generations ago. Unfortunately, today fire growth is still often regarded as a problem for fire prevention staff only, particularly where there is a lack of communication between fire prevention and fire suppression personnel. On November 28, 1942, rapid fire growth was responsible for 492 deaths in the Cocoanut Grove nightclub fire in Boston. At the time, flammable decorations were blamed for the rapid spread of the fire. Inadequate exits and overcrowding were blamed for the monstrous death toll. Pictures from the fire clearly show telltale globs of burned adhesives used to glue highly **combustible acoustical tile** to the ceiling Figure 5-7 .

After other deadly fires with rapid flame spread in a soldier's hostel in Newfoundland and in Mercy Hospital in Iowa, fire officials in the late 1940s began to understand that there was a problem with some building materials. Despite this linkage, early publications on the hazards of certain building materials were often criticized. Today, rapid fire growth is still imperfectly understood, even by some fire suppression forces. In 1982, two Boulder, Colorado, fire fighters died in a training fire in a building lined with combustible fiberboard.

Figure 5-7 The round telltale globs of acoustical tile adhesive at the Cocoanut Grove nightclub fire.

Courtesy of the NFPA.

Examples of Fire Growth and Spread

Serious losses due to rapid fire growth have occurred across the United States. In Los Angeles, California, in 1982, fire claimed the lives of 24 occupants of the Dorothy Mae Apartments. A key factor was flame spread over plywood in the exit hall. In a 1984 Beverly, Massachusetts, fire, 15 people died in the Elliott Chambers Boarding House. Plywood paneling had been used for the interior finish of this structure. Even an activated fire detection system was inadequate in the face of the fast-spreading fire in this incident.

Combustible tile ceilings are often suspended below the floor above, creating a void in which explosive carbon monoxide (CO) gas can be generated and stored. When this gas ignites, there is often an extremely violent explosion. Three of the following five fires illustrate this phenomenon.

A magazine article described a fire in which a fire fighter and three civilians lost their lives in a flashover in a fire-resistive building. There was no mention in the article of what flashed over. A call to the article's author revealed that the fire-resistive building had been "improved" with a suspended combustible-tile ceiling, concealing a void from which the fire could burst out.

In Silver Spring, Maryland, in 1977, a fire broke out in a one-story row of stores. The ceiling was made of suspended combustible tile, which created a combustible void above the tile. The store was well ventilated, and the front windows were out. The surface fire on the tile was extinguished with a big line, but the tile continued to burn within the material and on the back side. Suddenly a huge amount of black smoke under pressure appeared—a signal that a backdraft was imminent. Fortunately the fire fighters were only knocked down and did not suffer serious injury.

Some years ago, several Orlando, Florida, fire fighters escaped from a store with their bunker coats afire. Fire had burst out of the void above a combustible tile ceiling.

Carpeting served to rapidly spread the fire in the Pioneer Hotel in Tucson, Arizona, where 29 people died in 1979. Carpeting also spread the fire in the Harmer House Nursing Home in Marietta, Ohio, where 32 individuals died in 1970. Combustible interior finish has been a factor in other fires leading to fatalities in structures where disabled or elderly people have resided.

Building or Contents Hazard?

Flame spread, or rapid fire growth, can be a problem caused by both the building itself and its contents. The fire growth building problem can be differentiated based on its location:

- Hidden
- Exposed

The fire growth contents problem can result from any of the following elements:

- Furnishings
- Interior finish, including decorations
- Mercantile stock

These categorizations are not meant to be absolute. As is the case with much of this text, it is impossible to make absolute distinctions regarding many factors in building construction. The same material may be, at different times, a hidden building element, an exposed building element, furniture, or merchandise contents in a commercial building. This can be very frustrating to those who want everything to be black or white. Unfortunately, these distinctions are often unclear.

In addition to fire growth, high flame-spread materials may contribute heavily to the fire load and to the generation of smoke and toxic products. These are additional threats to the life safety of both occupants and fire fighters.

The threats posed by particular materials used in building interiors can also be confusing. It is not wise to make any assumptions about them; instead, specific knowledge is needed. For instance, imitation brick observed in one building materials display was made of vermiculite, a noncombustible mineral

material. Imitation stone in the adjacent display was made of glass fiber–reinforced polyester resin plastic; the glass fibers are noncombustible, but the plastic is combustible.

Hidden Building Elements

A wide variety of materials and hidden building elements make major contributions to rapid fire growth. The paper vapor seal on batt-type glass fiber insulation leads to a phenomenal rate of flame spread. **Batt insulation** with paper facing laid in ceilings must be kept free of light fixtures because the heat from the fixture can ignite the paper. In one case, insulation laid on a ceiling was allowed to be in contact with the top of a light fixture. It ignited, and fire spread on the paper covering the insulation. Sprinklers operated below the ceiling, yet had no effect on the fire above the ceiling.

A supermarket with a combustible ceiling was fully sprinklered, including the cockloft. A flammable liquid fire extended to the cockloft and spread rapidly onto the paper backing on the insulated underside of the roof. Twenty sprinklers operated in the cockloft and 90 sprinklers in the store, yet the fire continued to spread. In another case, an electrical fire ignited the paper backing on insulation. The fire spread through the huge undivided roof-ceiling void, which extended over the entire building.

Glass fiber insulation with paper vapor seal was installed around an upstairs bathroom for soundproofing in one of a series of connected (and expensive) dwellings in Lake Benbrook, Texas. The piping was so well insulated from heat that it froze. A torch used to thaw the pipes set fire to the paper vapor seal on the insulation, and 10 houses lost their roofs or upper floors.

Older motels are often built with back-to-back rooms, with each room opening to the outside. A plumbing corridor is often provided between the rooms. For privacy, batt-type fiberglass is sometimes installed between the studs. The paper vapor seal on this insulation is exposed. A plumber's torch could potentially ignite the paper and flash into a

fire extending the length of the corridor and through pipe channels to the corridors above.

Combustible fiberboard is commonly used as insulating sheathing on wood-frame buildings. It is also used as soundproofing. This material can support a fire hidden in the walls.

Foamed-plastic insulation is also used as sheathing, concealed in cavity walls, or glued to the interior surface of masonry wall panels. Foamed plastic applied to walls and ceilings for insulation has been involved in many disastrous fires. When such insulation is installed, it should be protected from exposure to flame by a half-inch gypsum board covering. This covering protects the insulation only against ignition from a small source, however. In a well-developed fire, as the gypsum board fails, the plastic will become involved, possibly suddenly and explosively, as the gypsum falls away.

In what may be one of the costliest fires in history, workers were foaming plastic insulation around wires to prevent cross-contamination between two sections of the Tennessee Valley Authority's Browns Ferry Nuclear Power reactor. Incredibly, workers were using a candle to check for air leaks. A fire started in the foamed plastic and spread over the wiring. The actual cost of wire burned was about $50,000. The overall financial loss to the taxpayers was estimated at $300,000 per day for the year and a half the reactor was out of service.

Foamed plastic may be manufactured so that its flame spread is reduced, but it still can melt. It also lacks dimensional stability. When the plastic is used structurally, this may lead to disaster.

Air-duct insulation commonly installed years ago was usually made of a hair felt with a high flame spread. In contrast, the presently used aluminum-faced foil (not aluminum-faced paper), glass fiber insulation presents little flame spread problem.

Electrical insulation may be self-extinguishing. However, the tests of this material are conducted on wire not under load. When electrical wiring is operated at or above its rated capacity, the heat can break down the insulation and flammable gases can be emitted. The McCormick Place fire in Chicago is thought to have started in the flaming insulation of an overloaded extension cord. Large groups of electrical wires can support self-sustaining ignition.

A spectacular, tremendously smoky, multiple-alarm fire occurred in the cable vault of a telephone company building in New York City in 1975. The fuel for the fire was the insulation in the cables. A number of fire fighters suffered seriously from the smoke created by this fire.

Interior Finishes

At one time, lime plaster was an almost universal finish for ceilings, although some ceilings were also made of embossed steel (tin ceilings) and wooden boards called **matchboarding**. Plaster does not contribute to a fire. In fact, lime plaster absorbs heat, thereby slowing the progress of the fire. Today, many types of interior finishes are used, not all of which retard fire spread.

There are three ways in which interior finishes may increase the fire hazard:

- They may increase fire extension by surface flame spread.
- They may generate smoke and toxic gases.
- They may add fuel to the fire, contributing to flashover.

Commonly used interior finishes include such materials as wallboard, wallpaper, lay-in ceiling tile, vinyl wall covering, and interior floor finish items, such as carpeting.

Low-Density Fiberboard

In the 1930s, **low-density fiberboard** made of wood fibers or sugar cane residue called **bagasse** came into use. Produced in 4- by 8-foot (1.2- by 2.4-m) sheets with a painted surface, fiberboard provided a cheap interior finish that could be installed quickly. Fiberboard sales grew astronomically. This material was used both for initial construction and for rehabilitation. It was substituted for wood sheathing in frame construction. It was also widely used for interior finishes in low-cost construction.

The Celotex Company manufactured many different building materials, including a low-density fiberboard. The name Celotex is widely and inaccurately used to denote fiberboard, regardless of its manufacturer.

When low-density fiberboard is used for sheathing and for soundproofing, it is concealed in walls. A common method of igniting this material is by a plumber's torch. The plumber sees only the gypsum board wall. The fire, however, enters the wall along the copper tubing and ignites the combustible fiber sheathing. Such a fire often goes undetected until it erupts violently. Fires caused by plumbers torches have ignited many buildings under construction; in November of 1990 in Ventura County, California, 350 apartment units under construction were lost in a fire reported to be caused by a plumber's propane torch. A school under construction in New England had fiberboard on wood studs covered with gypsum board. The fire extended to the metal deck roof. The damage in this noncombustible building amounted to millions of dollars.

In a Des Moines, Iowa, shopping center, sprinklers were omitted in one part of the concealed ceiling space. Combustible fiberboard formwork for the concrete roof topping had been left in place. When a fire burned the combustible fiberboard, the estimated loss was $700,000.

An apartment fire in Gainesville, Florida, also exposed the dangers of fiberboard. The walls between units were doubled with two lines of studs, with fiberboard between them. A fire in the laundry room entered the wall void and extended to the roof, then advanced laterally through the attic.

Combustible Acoustical Tile

When punched with holes, fiberboard acquires desirable acoustical properties and becomes combustible acoustical tile. Consider the situation in a building with badly deteriorated plaster ceilings. Replacing the plaster would require removing the old plaster, placing on new plaster, and then painting—three messy jobs. An alternative is for workers to come in over the weekend, nail furring strips to the ceiling and add fiberboard tile, producing a new, attractive ceiling with acoustical

Figure 5-8 Combustible low–density fiberboard acoustical ceiling tiles above a dropped ceiling.
Courtesy of Glenn Corbett.

benefits by Monday morning. Thus, combustible ceiling tiles may be found in addition to the original ceiling, with a void in between the two, offering a space in which fire can spread. In addition, since fiberboard acoustical ceiling tiles today are no longer considered attractive, it has become common practice to install a dropped ceiling below the acoustical tiles **Figure 5-8** .

Industry Opposition to Regulation of Interior Finishes

Years ago, industry vigorously fought attempts at regulating fiberboard. A U.S. Department of Commerce–approved "industry standard" fire test was developed, which amounted to a fire in a thimble full of alcohol. A light flame-retardant coating enabled the fiberboard to pass this test.

In April 1949, a tragic fire occurred in St. Anthony's Hospital in Effingham, Illinois. There were 74 deaths in the fire, mostly infants in the nursery, but also nuns and nurses who would not leave them. The former *NFPA Quarterly*, in two separate issues, reported on the fire and addressed the hazards of combustible tile. Subsequently, manufacturers threatened legal action against the National Fire Protection Association (NFPA), but the association stood its ground.

A major change in industry attitude regarding fiberboard standards came about, not from reasoned explanations by fire protection experts, but as a result of a successful lawsuit against a manufacturer in the wake of the fire. Today, the flame spread characteristics of interior finishes are regulated in building codes (discussed in detail in the chapter that reviews building codes).

Other Hazards Associated with Interior Finishes

Fire hazards may exist as the result of using a wide range of materials for interior finish purposes, including tiles and fiberboard, paper, fabrics, cork, rattan, many types of wood, plastics, and carpeting.

Adhesives

In Hartford, Connecticut, the Hartford Hospital corridor ceilings were made of combustible acoustical tile glued to a gypsum board ceiling. The hazard with this arrangement had been recognized, and the tiles painted with a flame-retardant coating. In 1961, a fire developed in a soiled linen chute, rolled out, and attacked the ceiling; the fire then roared down the corridor. Ultimately, 16 patients died in the incident. Samples of the tile were tested and showed a reduced flame spread not sufficient to support as intense a fire as was experienced. The puzzled investigators removed an entire 25-foot (7.6 m) section of the ceiling and sent it for testing. The flame spread proved to be very high—the difference was due to the adhesive used to attach the tile.

The Clark County, Nevada, Fire Department report on the MGM Grand Hotel fire in 1980 noted that it required 12 tons (10,886 kg) of adhesive to attach tiles to the ceiling of the casino **Figure 5-9**. The flammable adhesive added a large fuel load to the fire.

Adhesives are used in a variety of applications in building construction, including the securing of roof surfaces, the on-site assembly of prefabricated modular units, and the connecting of plastic pipes.

Figure 5-9 The fire effects in the casino area of the Las Vegas MGM Grand.

© Saxon/AP Images.

High-Density Fiberboard

Fiberboard can be manufactured to very high density. Very dense fiberboard was selected for radiation shielding in the Rocky Flats Nuclear Weapons Plant in Colorado. It was regarded as fire safe because it was tested with a blow torch. This is a commonly used but inadequate test. Long continued heat from spontaneously ignited plutonium ignited the fiberboard. The resultant fire and contamination problem was extremely costly.

When punched with holes, high-density fiberboard is called pegboard. At least one manufacturer (Masonite Corporation) produces such hardboard with a very low flame spread rating.

Paper

Some years ago, the former U.S. Atomic Energy Commission (now the U.S. Nuclear Regulatory Commission) was building an extension on Strong Memorial Hospital in Rochester, New York. Crafts-people were working on the excavation. Winter was approaching, so a temporary snow roof was erected to permit work to go on during the winter. The least expensive material available was selected for waterproofing. It proved to be a hemp-reinforced bituminous-impregnated paper of the type used to

protect finished lumber—a material that has a very high flame spread rating. A stove for burning wood scraps was placed in one corner of the snow roof, creating a serious potential for sending a sheet of fire up the face of a functioning hospital. Fortunately, the situation was soon corrected.

Fabrics

The 1944 circus fire in Hartford, Connecticut, in which 168 lives were lost, was fueled by flames spreading rapidly on the gasoline- and paraffin-impregnated canvas that served as the "Big Top" tent **Figure 5-10**. Panicked audience members found it difficult to escape because some exits were blocked by circus equipment. The caged runway for the wild animal show that was erected across the main exit was supposed to be up for only 10 minutes. Remember this disaster when you are asked to overlook the "temporary" blocking of exits.

In another example of a potential fabric hazard, Frank Brannigan was giving a slide-illustrated seminar on combustible decorations in the Northeast. After the seminar, attendees went to a popular local restaurant. The ceiling was draped in burlap. The fire marshal commented that there should be no problem because the place was sprinklered. He had

Figure 5-10 The Hartford circus fire of 1944.
© AP Images.

no idea that fire could race over the burlap much faster than the sprinklers could operate. It is hard to imagine a greater panic-producing situation than fire falling from the ceiling on people, literally igniting hair and clothing.

Cork

A California seafood restaurant has a ceiling decorated with cork floats. It also is sprinklered. The possibility of flames spreading faster than sprinklers can operate is often unevaluated. Cork paneling in an office was a key factor in a five-alarm fire in the Empire State Building.

Rattan

In a south Florida city, a two-story, brick and wood-joisted building had been rehabilitated into a boutique. The ceiling was of rattan. On a visit a few years later, the building was found in ruins. The fire had started on the second floor. In the words of one witness, "It spread so fast it must have been arson!" An equal possibility was that the rattan ceiling itself accelerated the flame spread.

Wood

Wood surfaces are very popular and common. Flame spread over these surfaces deserves special attention.

The unexposed side of plywood or any wood paneling can burn unobserved and protected from fire department streams. For this reason, it is not sufficient to correct such a situation by simply applying a flame-retardant surface treatment to the exposed surface. Although the standard for installation of sprinklers allows for the omission of sprinklers in concealed void spaces with such combustible surfaces in "fully" sprinklered buildings, such treatments are no substitute for exposed plywood surfaces in rooms and other open areas.

When Boston's new John Hancock Building suffered massive losses of glass window panels in the 1970s, plywood was a temporary substitute—so much of it was used that the building was

known as the "plywood palace." At some point, the vertical flame spread hazard was recognized and flame-resistant plywood was used for any further closures.

Plastics

Some materials are easily recognizable. Others are disguised, such as plastic imitation wood sheathing and beams and imitation glass mirrors. **Rigid-foamed polyurethane** has been used for interior finish in many houses. In at least one fatal house fire, the material's ignition propensity and flame spread was a contributing factor. A 1971 fire in a French nightclub, in which 144 persons died, was fueled by foamed plastic used as interior finish.

Forty-nine teenagers died in a fire in the 1981 Stardust disco in Dublin, Ireland. When the Irish government convened a Tribunal of Inquiry to determine the circumstances, it discovered that the ceiling was of mineral tile and, therefore, was not a factor in flame spread. The built-in banquettes and the wall covering, however, were made of plastic. An expensive, unprecedented re-creation of the fire, commissioned by the Irish government, at the British Fire Research Station showed how a small fire in a foamed plastic seat could spread almost explosively, due to the nearby wall covering.

The Irish Tribunal of Inquiry unfortunately has no U.S. counterpart. It is a judicial body with technical expertise, which has subpoena powers and wide authority. Too often in the United States, all the facts of disasters are never revealed due to sealed settlements of civil lawsuits and the narrow scope of criminal investigations.

An interesting feature of Expo '67 in Montréal, Canada, was the United States Pavilion—a geodesic dome covered in clear plastic **Figure 5-11**. One architectural writer marveled at this beautiful, lightweight futuristic building. After the building was turned over to the City of Montréal, a worker accidentally set fire to the plastic with a cutting torch. The clear plastic skin—the equivalent of solidified gasoline—was destroyed in 10 minutes.

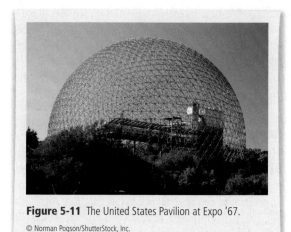

Figure 5-11 The United States Pavilion at Expo '67.
© Norman Pogson/ShutterStock, Inc.

Clear plastic has many inviting uses, but not all designers understand its terrible potential hazard. The plastics supplier for the Expo building was reluctant to provide the material and had impressed on U.S. officials the necessity for strong precautions. The supplier said that when the pavilion was turned over to the City of Montréal, the company wrote to the city explaining the hazard. Some materials are probably too hazardous to use, and administrative controls on their application often are unreliable.

The Isle of Man in the United Kingdom is located in the inhospitable North Atlantic. In an attempt to create a Mediterranean-type resort, a huge entertainment complex called Summerland was built. A major portion of the walls and roof consisted of acrylic plastic, and another large portion was made of asphalt-coated steel, often called **Robertson protected metal**. The interior paneling was of pressed fiberboard. This arrangement created a combustible wall void and set the stage for disaster. In 1973, the potential threat became a reality, when a glass fiber-reinforced plastic stand caught fire outside the building. The fire extended to the asphalt-coated steel, then to the plastic. Fifty people died in the fire.

Even when the problem of combustibility is understood, other fire problems that are possible

with structures incorporating large amounts of lightweight plastics are not apparent, even to architects. The retired dean of architecture at Columbia University once described a proposed pressurized membrane structure that would cover a complex of four to six government office buildings. The structure would encompass 400,000 square feet (37,161 m²). The purpose was to provide an ideal year-round environment independent of outside atmospheric conditions. There was no mention of the deadly atmosphere that would be created within the structure by the release of the toxic combustion products locked within the contents inside the environmental envelope. These toxic products could be released with a single match.

The loss of six fire fighters at the Worcester Cold Storage warehouse in Massachusetts in 1999 was, in part, due to the rapid spread of fire across the interior surfaces of the building Figure 5-12. The heavy timber structure—actually two connected buildings—was built in 1905 and 1912. The building had 18-inch-thick (457 mm) brick walls that were covered with successive layers of 6–18 inches (152–457 mm) of asphalt-impregnated cork (bottom layer), 4 inches (102 mm) of polystyrene or foam glass, and a thin layer of glassboard (the top layer). Additionally, maze-like conditions and limited

Figure 5-12 The Worcester Cold Storage building fire.
© Paul Connors/AP Photos.

numbers of openings were primary factors leading to the loss of the trapped fire fighters who had been looking for a homeless couple.

Carpeting

The use of carpeting has changed. It is now also used on walls and ceilings—not just on floors. It is not always evident to designers or even to fire officials that the location where a material is installed may increase the rate at which a fire can grow. A spectacular 1980 fire in the Las Vegas Hilton Hotel spread from floor to floor outside the building because of flammable carpeting on the walls and ceilings of the elevator lobbies. The fire grew into one terrifying fire front extending many stories in height and claimed eight lives.

It has become commonplace for day care centers to use carpeting on the walls. This practice is very dangerous and must be corrected. Unless the carpeting has achieved the appropriate interior finish fire rating, it must be removed.

Void Spaces

Suspended ceilings of combustible tile form void spaces in which fire can burn undetected until it bursts out furiously. In a 1956 fire, 11 people lost their lives in Anne Arundel County, Maryland, when fire involved the attic above a combustible ceiling in an amusement hall. The raging fire overhead did not appear to be serious.

Three fire fighters lost their lives in a furniture store fire whose spread was aided by the presence of combustible voids. Heavy smoke was reported from one mile away, yet only light smoke was reported from close in at the incident. Close-in positions are not always the best place to see the situation, particularly for a multistory structure.

Fourteen people died in a fire-resistive hotel fire. The third floor, where the fire started, was the only floor paneled in wood. It also had a dropped ceiling, probably made with combustible tile.

Remodeled Ceiling Hazards

A most serious hazard is often created when a building is remodeled. Codes require the installation of a new ceiling that meets flame spread requirements—but they do not require the removal of the old ceiling. When the new fire-rated ceiling is installed below the old ceiling, the dangerous combustible ceiling is left above **Figure 5-13**. Fire can burst down out of that void. Two fire fighters died in Wyoming, Michigan, when fire burst out of the ceiling. Even then, the city did not amend the fire code to prohibit this type of remodeling.

In the Roosevelt Hotel fire in Jacksonville, Florida, in 1963, a number of people died far above the fire floor. The fire was in an old combustible ceiling that had been left in when the new ceiling was installed. The smoke moved through old air ducts and reached areas not affected by the fire itself.

In December 1989 in Johnson City, Tennessee, 16 people died in the John Sevier Center, a fire-resistive unsprinklered high-rise housing elderly citizens. A prime factor in the loss of life was an old combustible tile ceiling glued up to the concrete overhead and left in place when the new suspended grid ceiling was installed. National Institute of Standards and Technology (NIST) researchers noted that CO generation in such a void could be as much as 50 times what is normally generated.

Aircraft Interiors

There have been a number of aircraft tragedies where victims survived the crash but died in the resultant fire. In some cases, the fuel for the fire consisted of the plane's plastic seats and other interior fittings. Effective August 1990, new aircraft have had to meet stricter Federal Aviation Administration (FAA) standards for fire and heat resistance.

Acoustical Treatment

Acoustical or soundproofing treatment is another potential source of trouble. Many acoustical installations have been made without consideration of the flame spread consequences of materials used. The polyurethane foam used in the Station nightclub in West Warwick, Rhode Island, was used for acoustical purposes.

Open-Plan Offices

Flame spread restrictions in codes are typically strictest for corridors and less strict for rooms off these corridors. This has created a potential problem in the modern open-plan office **Figure 5-14**. Where do the corridors begin and end? One fire department solved this dilemma quite simply: if there is no wall, there is no room; it is all a corridor. The strict interpretation of corridors includes the provision that they

Figure 5-13 Fire can spread quickly above this tin ceiling.
Courtesy of the estate of Francis L. Brannigan.

Figure 5-14 An "open-plan" office.
© bhowe/ShutterStock, Inc.

be enclosed, with rated walls and doors. Considering the devastating demonstration of the open office flame spread potential shown in the First Interstate fire in Los Angeles in 1988, the need for corridors to provide both a protected means of egress and a means of compartmentation should be reexamined by code-writing groups.

One little known observation resulting from computer fire simulations of the World Trade Center attacks in 2001 revealed that ceiling temperatures within a corner conference room with full height partitions were substantially lower than the surrounding open office area.

This open office–corridor mindset is prevalent throughout both industry and civilian settings. A proactive fire prevention education program can help to educate the public about these typically unrecognized hazards. The value of sprinklers in regard to flame spread and rapidly changing fire involvement should be stressed.

Alterations

Fire departments should be aware of any building alterations planned in their jurisdictions and ensure that all pertinent codes are obeyed. The authors have

observed a number of instances in which interior finish materials that would not have been permitted under the code during original construction were installed after occupancy. Needless to say, this may be the case in numerous buildings.

A chain of steakhouses decided to change the interior decor of the restaurants to have a rough, plastered look. Combustible paneling resembling plaster was used. It was applied without consulting local code officials. When questions about this material's fire resistance was raised, the management provided certification from the manufacturer stating that the material met local (not-too-rigorous) flame spread standards.

Decorations and Contents

Present-day decorations and building contents present major fire hazards. At one time, a popular decoration that provided a real flame spread or rapid fire growth problem was the Christmas tree. Although there are still tragedies during Christmas season, such as the loss of multiple lives at a San Francisco yacht club a few years ago, the incidence of such fires has been reduced. Regardless of any precautions taken, the number of decorations in places of assembly should be strictly limited to reduce the potential volume of fuel available for a fire.

The San Antonio fire marshal was involved in the removal of several real–cut Christmas trees from hospitals, hotel atria, and art museums. The trees were taken out of the buildings despite a newspaper editorial cartoon showing the "Grinch" with fire marshal emblazoned on his costume dragging out a decorated tree.

Decorations, usually hanging overhead and/or on the walls, and contents such as furniture or stock-in-trade are very difficult to control from a fire prevention perspective. There are few applicable regulations. These regulations, where they do exist, are difficult to write in the manner required by law and are difficult to enforce. Decorations often represent a serious problem because the hazard goes unrecognized even by people who consider themselves fire conscious.

One of history's most serious fires occurred in 1859 in a cathedral in South America. The overhead was decorated with fabric to resemble clouds. When a fire broke out, the fabric accelerated the flame spread, and about 2,000 people were reported to have died.

Closer to home are the fatal fires that have occurred when well-meaning adults create a Halloween haunted house. Often cotton sheeting is used, which has a high flame spread, in proximity to light bulbs (a 100-watt bulb can ignite cotton) or candles. One man lost his life while assembling a haunted house in a school in Virginia. The disaster would probably have been much worse had it occurred when the children were in the house. The presence of the old combustible tile ceiling above the later installed noncombustible ceiling was also a significant factor in the spread of this fire.

A 1984 fire in the Six Flags Haunted Castle in New Jersey, in which eight people died, was reported to have been caused by a visitor touching a lighter to a polyurethane surface. The NFPA report on this fire should be required reading. The extreme hazard of foamed polyurethane often goes unrecognized, yet samples of this foam ignited after a one-second application of flame from a butane lighter. The model fire code regulations that were created following this haunted castle fire have addressed this problem.

Students in a fire-resistive girls' dormitory at Providence College, in Rhode Island, had a contest with cash prizes offered for attractive Christmas decorations. Evergreen trees, made of paper and cotton applied with masking tape, provided continuous fuel along the length of the dormitory corridor. Even fire alarm boxes were covered over. Individual rooms were similarly decorated. Ten students died when a room fire extended through the hall. Dead-end corridors, permitted under the 1938 code, were a serious contributing factor to the severity of the fire. Additional stairways have since been provided.

A water treatment plant under construction had a great deal of glass fiber–reinforced plastic installed for various purposes. There appeared to be little fire hazard; the insurance was cut to the bone. After a fire occurred in the unfinished building, the contractor was paid only a fraction of the loss because he was underinsured. Because of the "glass" in the name, many mistakenly consider this type of plastic to be noncombustible—but it is not.

The Consumer Product Safety Commission created a huge acrylic plastic lobby display to show what the agency was doing for public safety. This pile of "solidified gasoline" was located between the employees' offices and the exits. A young attorney, schooled in fire protection, protested, and it was removed.

A Texas museum was very proud of a walk-through exhibit—a room full of rodeo figures. The larger-than-life-size figures were made of papier-mâché. Visitors in the museum walked on pieces of foamed polyurethane covered in burlap. A potential exit was blocked by bales of hay. Neither the creators of the exhibit nor the visitors had any concept of the hazard. A dropped match or a cigarette lighter could have created a tragic inferno.

A serious 1970 fire occurred in the then Smithsonian Institution National Museum of History and Technology. The heat produced in the fire was unparalleled in the experience of many fire fighters. The source of the heat was not understood until it was learned that huge sheets of clear plastic had hung from the overhead as part of the exhibit. The sheets had been completely consumed by the fire.

Today's Fire Loads

Fire loads and rates of heat release in current homes, stores, and offices are much greater than they were when most of the fire codes were first written. In addition, a further hazard has arisen: the nature of furniture and contents has changed. Solid and foamed plastics have replaced much wood, cotton, wool, and other materials. The hazard presented by these plastics is often unrecognized. Heat release rates—a critical measurement of fire energy output—are covered in detail in the chapter that reviews concepts of construction.

First Interstate Bank Fire

In the First Interstate Bank fire, the floor of origin had wide-open spaces crammed with computers and related equipment. Rapid fire spread over the plastic cabinets and wiring insulation produced a quickly raging fire that was reported from blocks away, while building security staff were busy resetting smoke detectors. The fire growth was thoroughly analyzed by the Fire Research Division of NIST (formerly the National Bureau of Standards [NBS]). This NIST report should be copied and discussed with the management of similar occupancies. It appears that many are still unaware of the hazard of fire loads, while others are fearful of sprinklers, and still others believe that there is plenty of time to deal with a fire.

Hotel Remodeling

A pile of plastic furniture was stored temporarily in a public area of the Dupont Plaza Hotel in San Juan, Puerto Rico. In December 1986, a group of disgruntled employees set fire to it. The fire grew to enormous size almost instantly and resulted in the loss of 97 lives. According to reports, claims settlements from deaths and injuries from this fire have reached $200 million.

Hotels are remodeled regularly. Mattresses and furniture often are removed from rooms and stored in hallways, where they represent a tempting target for arsonists. When carpeting piled in the hallway was ignited in a Fort Worth, Texas, Ramada Inn in 1983, five people died.

Regulations

On a national basis, little has been done to control the hazard of furnishings in buildings. However, the State of California has done a great deal, as has the City of Boston. For its own considerable holdings, the Port Authority of New York and New Jersey has laid down fire load requirements. This action was sparked by a devastating fire in a terminal at John F. Kennedy Airport, which was fueled by plastic seats. Recently developed New York State requirements, however, are concerned with toxicity, not flammability.

Certification of Interior Designers

New York State enacted legislation providing for the certification of interior designers that elevates this vocation to a new level of legitimacy. The New York law does not provide for "grandfathering." Candidates must take two examinations—one on interior design itself, and the other on building codes and fire safety. It would be useful if architecture and interior design writers could be persuaded to take the same instruction.

Residential Fire Tests

In 1980, the NBS conducted fire tests on the contents of typical residential basement recreation rooms. The word "basement" easily could be dropped from this description as immaterial. The room could be a living room or motel room, a fire station recreation room, a doctor's office, a reception area, or an elevator lobby in a hotel. The full report of these tests is a goldmine of information. Do not be deterred by the technical nature of the writing—study it.

The information concerning Test No. 14 is very revealing. The test room was 3.3 by 4.9 meters in size (10.7 by 15.9 feet). The furniture consisted of a sofa, an upholstered love seat, an ottoman, two end tables, a coffee table, and two bookcases. The carpet was ottoman pile with foam rubber backing. The walls were plywood. The ceiling tiles were wood fiber, with a 200-flame spread rating. The fire was started by matches in about a pound of newspaper on the couch. In fewer than 4 minutes, heavy flame was pouring out of the full height of the doorway. Six minutes after ignition, the average gas temperature was reported as 700°C or about 1,300°F.

Available Films

Unfortunately, there are no films or tapes generally available of the NBS tests. However, some years ago, the NFPA produced two movies, *Fire: Countdown to Disaster* and *Fire Power*, which give graphic demonstrations of how fast fire grows in today's interior environments. Although dated, they still

provide a unique glimpse of the power of fire for the uninitiated. In *Fire: Countdown to Disaster*, a sparsely furnished bedroom develops a violent backdraft explosion in fewer than 3 minutes.

Control of Rapid Fire Growth

There are several possible approaches to control the problem of fast fire growth, including eliminating high flame spread surfaces, separating material from the source of combustion, cutting off the extension of the fire, and coating the materials.

Eliminate High Flame Spread Surfaces

Eliminating high flame spread surfaces is the path taken for U.S. merchant ships. Foreign ships may contain combustible trim and veneer. For these ships, reliance is placed on fire suppression from automatic sprinklers whose water supply might fail, particularly during an engine room fire. The rules for U.S. merchant ships severely limit the combustibility of construction materials and surface finishes. The rules for furnishings, however, apparently might not rule out combustible chairs or other contents that can provide the fuel for a huge fire.

Separate the Material from the Source of Combustion

In 1946, Frank Brannigan was investigating a major fire at a naval facility. An engineer who had been working there related that when building an office in the naval warehouse, fluorescent light fixtures were recessed into the combustible tile ceiling. He pointed out to workers at the time that a defective ballast in fluorescent fixtures can reach temperatures of 1,500°F (816°C). His advice to avoid contact with the ceiling was ignored by construction workers. The engineer was happy to help Frank Brannigan locate the fixture that caused the fire.

Cut Off the Extension

A metal-working shop in Richland, Washington, had no fire problems, except one: there was a dip tank in one corner. Fumes from the dip tank rose overhead and coated the wooden roof with oil, creating a high flame spread hazard. A minor fire in the tank spread rapidly under the roof, destroying the plant and closing the business. The tank should have been isolated from the main part of the plant.

Some codes require a metal door sill separating more flammable carpeting from less flammable carpeting. For example, often the carpeting in motel rooms is a higher flame spread hazard than the carpeting in the corridors, so a metal door sill is installed to separate the two. The intent is to prevent the spread of flame from a room fire to the corridor. Frank Brannigan was told by an expert that this would be effective if the door were closed, but would have no effect if the door were open.

Coat the Material

Material can be painted with a fire-retardant coating (remember, there is no such thing as "flameproofing"). One difficulty with surface coatings applied after the hazard is discovered is that they do not protect the hidden surface of the material. Fire, in turn, can burn unimpeded in the void. As mentioned earlier, the surface treatment of the ceiling tile in the Hartford Hospital fire was defeated by the adhesive.

Fire-retardant surface coatings are effective only if applied as specified. It is difficult to get painters to refrain from spreading the material too thin because there is an economic incentive to use as little coating as possible. One fire department requires that the requisite number of can labels be turned over to the inspector. This is not a foolproof method of assurance, but it at least demonstrates the fire department's interest that the job is done right. Another method is for the owner to buy the coating material and to instruct the painter as to how many square feet of surface is to be covered by a gallon.

The architect of the U.S. Capitol is responsible for the design of a number of buildings in

Washington, D.C., in addition to the Capitol itself. All of the building materials used in the original construction or in renovations are either non-combustible or treated with intumescent paints or solutions. Even structures that are constructed outdoors, such as the stand for an inauguration, are treated with intumescent paints. Such precautions proved necessary in 1960, when an electrical fire erupted in the podium at the inauguration of John F. Kennedy. Slowed by the treatment, the fire was quickly extinguished by the D.C. Fire Marshal stationed there for just such an emergency.

Pressure-treated, fire-retardant wood is often used where a wood surface is desired and flame spread requirements must be met. Wooden materials can also be formulated to be flame resistant. One manufacturer, for instance, offers a special hardboard that has a very low flame-spread rating and zero fuel contribution and smoke density. Such a board would be an excellent substitute for the common hardboard and pegboard that have high flame spread characteristics.

Plastics can be manufactured to be more flame resistant. One manufacturer of structural glass fiber–reinforced plastics has produced panels that Factory Mutual (a large industrial insurance company that does a significant amount of its own fire testing in West Gloucester, Rhode Island) will permit to be installed without sprinkler protection.

Products of Combustion

Fires generate a number of products of combustion. This section discusses what the average citizen, and some researchers, call "smoke." In fact, there should be distinctions made among smoke, particles, and fire gases. Many years ago, it was considered helpful to explain to the survivors of fire victims that they did not suffer because they suffocated before they were burned! Over the years there has been a gradual change to recognition that smoke and toxic gases are more significant as fire killers than is thermal exposure.

Smoke

Smoke often provides the first warning of most fires. It can do more damage to property than the fire or water, and it may reduce visibility to zero. Its irritating particles may cause retching and immobility, and those irritants remaining in the respiratory system also may be toxic agents for a surviving fire victim.

By definition, smoke is the generic term used to describe the airborne solid and liquid particulates and gases produced by a fire. It is the solid particulates—typically carbon—that give smoke its opaqueness. The liquid particulates include acrolein and halogen acids. Gases include a myriad of different materials, including CO, hydrogen cyanide (HCN), and carbon dioxide.

Smoke Gases

Smoke gases can cause injury or death when inhaled or, in some cases, absorbed. Some gases can paralyze or slow human ability to function or escape. CO, the most prevalent of the toxic fire gases, has this effect. (HCN is also produced in many of today's fires and is now believed to be as great a danger as CO.)

For most toxic materials, the toxic effect is a product of concentration and exposure time. Haber's rule states that any exposure in which the concentration (in parts per million [ppm]) × minutes exposed equals 33,000 is likely to be dangerous. CO is probably the most common toxic fire gas. According to the *NFPA Fire Protection Handbook*, a 10-minute exposure to 3,500 ppm of CO would be hazardous, possibly incapacitating. Higher exposures with greater concentrations are even more dangerous—12,500 ppm may be fatal after only a few breaths. The toxic levels for other materials can be quite different.

Many other fire gases (such as nitric acid and hydrochloric acid, to name only two) are toxic, and, in sufficient dosage, lethal. In 1929, the Cleveland (Ohio) Clinic, like many other hospitals, used nitrocellulose X-ray film even though safety film was available. A fire in film storage facilities

Figure 5-15 The 1929 Cleveland Clinic fire.
Courtesy of National Board of Fire Underwriters/NFPA.

sent clouds of deadly nitrous oxide gas (one breath can kill) up through the building, and 125 doctors, nurses, and patients died **Figure 5-15**. Some hospitals today use nitrocellulose test tubes for blood tests. There is very likely a substantial storage of these in many supply rooms. A fire in one may tragically duplicate the Cleveland Clinic fire results.

Until the end of World War II, a nitrocellulose base film was used for commercial motion pictures (fire codes still regulate it). Fires involving this film have caused many disasters. There still are places where old, deteriorating nitrate film is stored, such as in libraries and museums, and maybe even in some hospitals.

Many other materials can generate gases in a fire that are hazardous in quite small doses. Many people, including some we might think would know better, are unaware of the hazards. A retired dean of Massachusetts Institute of Technology included nitrocellulose in a list of plastics that might be useful in construction.

Polyvinyl chloride (PVC) is a very effective electrical insulator and has made itself indispensable as a contemporary material. When it burns, however, it emits toxic gases.

Gas Flammability

CO is a flammable gas. When gases ignite in a confined area, a wide range of overpressures can result, ranging from the shattering of glass windows to the shearing of brick masonry walls. Most of the technical studies conducted on CO explosions are made in connection with the design of containers. The authors do not know of any technical investigation of what happens in a typical building fire.

By observation, fire fighters note that, in some cases, CO trapped in a void or pocket simply "lights up"—that is, the gas ignites when sufficient oxygen is available. This effect is sometimes seen as flames appearing in clouds of dark black smoke pouring from windows. This is also a sign that flashover is imminent. When flashover is imminent, fire fighters should be prepared to withdraw.

In other cases, the ignition is more violent, with sufficient energy to break glass windows. This outcome is sometimes observed by neighbors adjacent to a fire in a closed building. Newspaper reports often describe an explosion that called attention to the building. In a few cases, the gas–air mixture proves just right for maximum energy release. When this happens, brick walls may be blown down.

Any or all of these effects have been described as a backdraft. Fire fighters in the vicinity are not in a position to make detailed scientific observations, of course. The many variables involved make a complete scientific study unfeasible; instead, you must understand the subject as best you can with anecdotal information from survivors.

The shockwave of a CO detonation has blown entire buildings apart. CO is not the only gas encountered under explosive conditions, however. For example, gases from burning polyurethane exploded and caused the collapse of a wall on the sound stage of a Hollywood studio in 1974.

Gases can accumulate in any enclosed area. Research at NIST has pointed out that CO can be generated up to 50 times as much in enclosed voids as in the open. Some 47 people died in a 1990 discotheque fire in Zaragoza, Spain. In this incident, a small fire in the large void above the ceiling of

the first floor produced huge quantities of CO and smoke, which poured down the stairs into the basement where the victims died, many of whom were sitting in their chairs.

This tragedy is reminiscent of the 97-fatality fire in Our Lady of Angels School in Chicago in 1958. Some fire professionals share the opinion that the fire had reached the attic overhead and blew down on the classrooms, possibly when a ceiling vent was opened to reduce the temperature. Ventilation of the building does not prevent CO explosions. The gas–air mixture can pocket in a fully or partially enclosed location.

Physical Effects of Smoke

Particulates and gases have different physical effects on people. At one time, filter masks similar to those developed for use in World War I were widely used in the fire service. These masks were dangerous. The filter in that poison gas mask removed solid particles but allowed the odorless CO through. The fire service version contained a chemical that converted small amounts of CO to harmless carbon dioxide (CO_2), but in a number of cases the CO concentration was too high and fatalities resulted.

This same characteristic may be significant in high-rise fires. In such incidents, the smoke particles and CO moving up through the building cool off. Visible smoke often stratifies, with smoke particles becoming deposited on surfaces. In contrast, the colorless, odorless, and tasteless CO can stratify well above the fire, on floors where there is little or no smoke, heat, or air movement. Fire fighters can walk into such death clouds. In 1961, a subbasement fire in a high-rise occurred in New York City. Two fire fighters, who had left SCBA behind, died of CO poisoning on the 20th floor.

Smoke particles can plug up screens. This is a problem in prisons, where screens have been used on windows to prevent passage of contraband. In one case, smoke from a mattress fire plugged the screen and inmates died. This same plugging effect should be anticipated in any location where the outgoing air is filtered, such as where toxic materials are handled.

Smoke Damage

Water damage is often considered the most expensive by-product of fire suppression. It is mentioned in most press accounts and sometimes seems to be regarded as the sole responsibility of the fire department. To varying degrees, fire departments attempt to mitigate this damage. Some departments do extensive salvage work, while others ignore it. Ironically, there is never any mention of water damage at a total loss. In fact, smoke damage may well be the most expensive by-product of a fire, particularly where delicate equipment is involved or health-menacing contamination, such as food spoiled by the smoke, occurs.

Contaminated and Toxic Smoke

It is costly enough to clean up the dirty, oily smoke that comes from materials such as plastics, but if the smoke is contaminated with another health hazard substance, such as radioactive material or **polychlorinated biphenyl (PCB)**, the cost of cleanup can be astronomical. The relationship of health hazards to fire loss is not well understood, though the problem has been around for some time.

The fire problem of radioactive material grows out of the fact that radioactive materials emit energy that can damage living tissue; thus concerns about the safety of personnel might demand abandonment of the property until it is cleaned up. In the Rocky Flats Nuclear Weapons Plant, a fire did about $1 million in conventional damage. The cost of the total cleanup was 25 times greater.

Whether the fire fuel is radioactive or a fire in another fuel disperses the radioactive material, the results can be similar. In the Hanford, Washington, nuclear power plant, a fire spread contamination more efficiently than a nuclear accident. A nuclear experiment went critical (when a nuclear reaction becomes self-sustaining) and spread contamination around the building. It was successfully cleaned up with paper diapers, acetone, and nitric acid. The waste was then packaged to await disposal, but the nuclear cleanup experts had forgotten their chemistry. The sealed

packages ignited spontaneously. The fire spread the same contaminants so effectively that it was not economically feasible to clean up the area. The nuclear power plant was closed off and years later dismantled and buried.

It is easy to control the health hazard of contamination—keep people away. Because the loss of use of a property can be very costly and cleanup can be prohibitive, it is sometimes less expensive to abandon the property.

There have been many serious contamination incidents from PCBs released from electrical transformers during fires. PCBs were used as nonflammable coolants in transformers. There are rules requiring notification of emergency responders of the presence of PCBs, but they are not infallible.

During a relatively small fire in an office building in Binghamton, New York, in 1981, a transformer cracked. PCBs that were used in the transformer coolant heated up and released toxic products into the atmosphere. It took years to clean the building, which finally reopened in 1995. Cleanup costs exceeded the value of the building. The cleanup was slow and costly because of the precautions that had to be taken to protect the workers. Fire fighters should be aware of the extremely toxic nature of PCBs and the fact that these chemicals can be released from any transformer by fire or accident or even in routine maintenance.

The telephone system is the nervous system of the community. Daily life is disrupted when phone lines are out; serious financial losses also can occur when such service is unavailable. Thus, fires in telephone company facilities are not ordinary fires. In addition, the toxicity of the smoke for fire fighters, including the by-products of burning plastic wiring insulation, is a major concern. In 1975, a massive disruption of communications resulted from a fire in a basement cable vault of New York Telephone's main switching center. On May 8, 1988, a serious fire occurred in an Illinois Bell Telephone Central Office in Hinsdale, Illinois. Much of the multimillion dollar loss was due to smoke damage. Fire fighters in both fires suffered the effects of the toxic smoke.

Many of these telephone switching facilities lack sprinkler systems. Some only have detection systems. In another fire, a $30 million loss was suffered when a fire confined to burning PVC-insulated cables and the roof above them generated smoke that corroded precision machinery.

Corrosion

Delicate equipment as well as brickwork and concrete can be damaged or destroyed by corrosive products of combustion. Plastics containing halogens such as chlorine, fluorine, bromine, or iodine form corrosive acids when combined with hydrogen and oxygen or moisture in the air.

Smoke Movement

Generally, fire gases rise due to buoyancy. We are all familiar with the fact that gravity causes things to fall down, but few of us know that gravity pulls down on the surrounding heavier, colder air, causing the lighter, heated smoke of a fire to rise upward. If the smoke is not ventilated, it will cool and remain inside the building, even after the fire is out. Even with PPV and venting in progress, there is still danger from these gases. Cellars and low, unvented areas become collection points. Use of SCBA is critical in the overhaul stage of a fire.

When elevators were first introduced into buildings, designers saw the need to surround the shaft with a cage to prevent people from falling into it. There was no recognition of the fact that the open shaft was a perfect path for the extension of fire upward.

TACTICAL CONSIDERATIONS

1. Like elevators, stairwells, and mail chutes, garbage chutes are one of the truly open shafts in high-rise buildings. The door is a 90° pull-down, self-closing door. The sprinkler is usually above the door opening to the chute. Often it is difficult to see and difficult to reach. If the piping sticks out too much, garbage falling from above can hang up on the sprinkler. There is a variety of scenarios in which the fire in the trash receptacle in the basement, or anywhere else in the chute, can spread to upper floors—vertically, then horizontally. An arsonist who understands fire behavior and sprinkler systems can shut off the main valve or sectional valves so the incendiary fire can spread. Commercially monitored systems make this harder to do without immediate detection. A more likely scenario is when large objects of trash are crammed down the shaft, which acts like a cork in the chute, and wedges the door in the open position. Smoke coming up the shaft can then spread through the wedged open door, filling the small maintenance room with smoke. If that door is open, the smoke can move directly into the hallways.

 Another scenario is when the system is shut down and drained to change a sprinkler. It is difficult for sprinkler technicians to access these sprinklers as well; the time from the previous fire, to when the sprinkler system is recharged and back in service, makes the chute vulnerable to fire.

 The bottom line is this: fire crews need to check each floor, each maintenance room, and open each trash chute door to ensure there is no vertical and horizontal spread from fire. A fire fighter should also check the CO readings using a gas detector during this top-to-bottom investigation.

2. During an actual fire, dispatchers often get calls from residents within a high-rise building reporting an alarm sounding, an odor of smoke, a fire, or all three. If the building has a supervised monitoring alarm system, this confirms that there is a fire and full response is dispatched. Typically, the dispatcher

will then announce to the responding companies, "We've received numerous calls," indicating that callers are verifying a fire. Once the alarm is dispatched, additional calls into the fire alarm center are answered with, "Yes, we're on the way," or "Thanks! We've got it."

3. Though it is important not to tie up 911 lines reporting an alarm where the fire department is already on scene, dispatchers shouldn't be too quick to dismiss callers, especially in high-rise residential buildings. The dispatcher should make note of the caller's room number and relay this information to the IC. For example, if the fire was on floor 4 of a 20-story building, the dispatcher can say, "IC from dispatch, we've received calls from 808, 908, and 1,510 reporting an odor of smoke." If floors 8, 9, and 15 were already checked and reported cleared from smoke, the IC can send crews back up to the actual units to check on the caller. Not only is this good customer service, it will also put the caller at ease. From a tactical standpoint, these apartments are vertically aligned and there may be smoke extension that cannot be determined from the hallway.

4. Unfortunately, fire fighters sometimes judge the danger of smoke based on its density, interpreting light smoke as tenable smoke. Many fire fighters have suffered the effects of CO in light smoke conditions. A fire fighter who is acting "strange" or irrational is suffering the effects of CO. Another common sign is a fire fighter who is alone (without his partner) and going the wrong way—toward the fire. If a fire fighter appears disoriented and is not wearing an SCBA face piece, he needs to be immediately assisted to safety.

5. Firefighting tactics, in one form or another, are always removing one leg of the fire tetrahedron. Even complicated fire incidents are, in their simplest form, removing one leg of the fire tetrahedron. The most common extinguishing agent is water. Water absorbs heat and cools the fuel below its

ignition temperature so it's not giving off fumes. An unorthodox but effective way to use water to remove the fuel leg of the fire tetrahedron is when a magnesium fire is encountered. Magnesium is flammable metal that, when burning, reacts violently with water. Fire fighters are told never to use water on a magnesium fire. However, under certain circumstances (and understanding fire behavior), that is exactly the tactic to use. Left on its own, a large magnesium fire can take a long time to burn itself out. Fire units will have to remain out of service for the duration. Lobbing water from a safe distance using master streams will create a spectacular fire, but it will also burn the magnesium up at an accelerated rate. The faster the fuel is consumed, the faster the fire will go out. The situation is resolved and units can go back in service because the fuel leg of the fire tetrahedron was removed.

6. Heat transfer and fire spread happen through conduction, convection, and radiation. Using a thermal imager, fire fighters can quickly locate concealed fires or determine the location of the fire no matter which medium is used by the fire.

7. During vertical ventilation operations, fire fighters often forget to take a hose line up to the roof. Radiation can cause the uninvolved roof decking or the siding of an adjacent exposure building to autoignite. Hose streams have tremendous reach and can be operated from a safe position; we must keep what we're trying to save wet. If it's wet, it won't burn.

8. Thermal radiation impacting an exposure building may cause ignition of the exposure, including any combustible wall surfaces or combustibles inside the exposure near a window. Hose streams must be applied *directly on the exposure building* to remove the heat. Simply placing a stream between the buildings will not stop the radiant heat; it will pass right through.

9. Many company officers and fire service instructors still teach that fire fighters should not destroy the thermal layer (neutral plane). This has been taught for decades, but things have changed. Fire fighters have hesitated to apply water because they were taught never to destroy the thermal layer. It must be understood that with today's plastics and synthetics fuels, ceiling layer gases must be vented or cooled or they will ignite.

10. Examine the issue of the "thermal balance" and firefighting. First, you have good visibility below the neutral plane. This allows you to search for unconscious victims. Second, as soon as the nozzle opens up into the ceiling, the thermal layer is stirred and visibility is lost. Third, you create steam, which makes the room untenable for victims. Where they may have been saved, the hot, steamy environment can be deadly. This can be avoided if a sprinkler head is activated.

Today's household fuels burn faster and at higher temperatures than the wood, cotton, and paper fuels of the past. The temperature below the neutral plane is much hotter for a victim with exposed skin, wearing household clothing. The survival profile is low. Our modern bunker gear can withstand temperatures above 220°F (104°C)—though it's still uncomfortable. Finally, thermal imagers allow us to see in zero visibility.

Protecting the neutral plane is a plus if you can ventilate. But if you can't, or ventilation is going to be delayed, you must cool the ceiling fire gases below their ignition temperatures with water. It doesn't matter if visibility is good below the neutral plane if the room flashes over.

11. Remember, flameover or rollover may be your last warning sign before flashover.

From a tactical point of view regarding fire behavior, this is the most important concept to understand. Remember that the fire is always looking for the path of least resistance. Learn this. Teach your fire fighters the gas laws of physics. Many tactical considerations are based on the premise. Once you understand this, it will be much easier to predict what the fire will do and where it is going—without surprises.

12. Many fire departments don't have the resources or ability to deploy a fire blanket to cover the entry point of a wind-driven fire. These blankets are heavy because they have to resist the wind without being pushed through the entry point window. There are other tactical options to consider.

- First, trailer-mounted mobile ventilation units (MVU) are becoming more popular for ventilating big-box stores. But depending on the wind speed and the cubic feet per minute (cfm) capacity of the MVU, you might be able

to overpower the wind currents and reverse the effects of a wind-driven fire. If this is not possible, there is no stopping a wind-driven fire until it meets a fire-rated barrier. Keep in mind that everything between the entry point and that barrier will be subject to a blowtorch velocity fire and the survival profile is zero. The fire load will be consumed. You cannot withstand or survive these horizontal-pressurized flames. If the fire barrier (a wall or fire door) holds, the fire will slow down as fuel is consumed or it will burn itself out.

- Another option is attacking the fire from the floor above. A Bresnan distributor nozzle (a 2 ½" appliance that can throw 260–290 gallons per minute [gpm][984–1098 Lpm]) can be inserted if a hole can be cut through the floor. Fog applicators and piercing nozzles can also be deployed, but they may not have enough gpm by themselves. Multiple appliances would have to be deployed. The water-to-steam ratio doesn't apply here because this isn't an oxygen-deficient, confined space. A Bresnan nozzle can also be lashed to an attic ladder and inserted from the floor below if there are no concrete floors. Inserting these nozzles through the walls of an adjacent room can afford a horizontal attack but these adjacent rooms may not be accessible from the hallway. If aerial ladders can reach, the adjacent rooms can be accessed through an exterior window.
- Aerial master streams can be deployed if they can reach. Everything between the wind-driven access point to the first fire barrier is a zero survival profile, so a frontal attack through the window is your only option. The fire needs to be extinguished.
- If the streams can reach, consider deploying portable monitors from the windows of adjacent buildings. This may be a last-ditch effort.
- Finally, protect the fire doors to the stairwell and write off the fire floor. Firefighting efforts will need to be concentrated on the floor above. "Flow the floor" keeping everything wet. Remove all fire load away from windows and exterior walls. Watch for smoke and fire extension through the HVAC system, elevator

shafts, and other poke-through holes. There should be standing water on the floor, especially if it is carpeted.
- Another tactic that may work is attacking the fire from windows on the floor below the fire. Hose streams can be aimed up and out the window directly into the wind. The wind should push the stream back directly into the wind-driven opening and onto the fire floor.

13. Aluminum will melt at approximately 1,200°F (649°C). This is a good number to memorize since most of our ground ladders and some aerials are made of aluminum. But don't forget that aluminum also anneals when exposed to prolonged heat of 350°F (177°C) and higher—that means it starts to lose its rated strength. This is also a number worth memorizing.

14. Hidden fires are just that—hidden! That means fire fighters are not having success uncovering them. The best technology we have today is the thermal imager. Left untended, hidden fires will only get worse and emerge with increased momentum.

15. Understanding fire behavior and the fact the fire will take the path of least resistance, PPV fans can be used to help locate hidden fires. This may seem extreme, but the goal is to locate the fire so it can be extinguished.

 The fans are used to pressurize the space and accelerate the burn by introducing forced, oxygen-rich air. Have the hose lines charged and the teams ready to go with tools to open the walls or ceiling. Keep a fire fighter at the fan the entire time. As soon as fire starts to present, have the fire fighter at the fan shut it off or aim it away from the entry point. Fire fighters now know where the fire is and can dig it out.

16. When fire fighters get trapped in a room and must get out immediately, we teach them to breach the wall and perform a reduced profile, self-extrication into the next room. We need to expand on this concept and apply it to getting patrons out of a public assembly where a fire has occurred and exits are overcrowded by panicked patrons trying to escape.

 Make existing exits bigger by removing doors or tying them open. Don't use wedges because they can be kicked loose. Make windows into doors. Create doors in walls. These extra openings

will draw the fire. Hose lines have to be deployed above the heads of the escaping customers. The goal is to get everyone out as soon as possible.

17. We don't ventilate only for smoke. If there was a fire and smoke conditions were light, fire fighters often settle for natural ventilation. CO is odorless and colorless. Use a PPV fan from the outside or electric fans from the inside. Use the gas monitor to ensure levels are below 35 ppm.

18. A rescue air kit (RAK) is a spare SCBA bottle, face piece, and regulator that is carried in a special bag that is specifically designed for rapid intervention emergencies when a fire fighter is trapped. The RAKs are commercially made, but some fire departments assemble their own versions. It is one of the primary pieces of equipment a rapid intervention team (RIT) carries in for fire fighter rescue. But what do we carry in for civilian search and rescue? It is not a common practice to take in a RAK for civilians, but perhaps we should start. Most fire departments train their personnel to search and rescue the victim by dragging or carrying them out of the building. Then resuscitation efforts begin. In the past, a fire fighter may have given up his face piece to the victim, but this is not a smart thing to do. If the fire fighter is overcome with smoke and the effects of CO, there will be two victims to rescue.

 RITs should be reserved for that purpose alone, but often there are other RAKs sitting unused in the compartment of another apparatus. It may be time to start equipping fire fighters assigned to search and rescue with a RAK for civilian victims.

19. Piercing nozzles are making a comeback. These nozzles are excellent for insulation fires in floors, walls, and ceilings. Insulation fires tend to smolder for hours. Digging it out is a tedious task, especially with "blown-in" insulation. Embers are difficult to find. Some fire departments use Class A foam for these situations but don't rely solely on a foam blanket. The author has experienced a situation where a stubborn (blown-in) insulation fire couldn't be extinguished. There were slight puffs of smoke coming from the insulation but no embers could be found. The decision was made to use Class A foam and apply it heavily. The fire rekindled

5 hours later and a full response was dispatched to the same location for a house fire.

20. Open the walls. Charred wood should be taken down to clean lumber unless the structural stability is in question. In that case, use Class A foam. Consider using piercing nozzles or fog applicators. Create a sprinkler system where none exists. A round nozzle throws a different spray pattern than sprinklers.

21. It is no longer unreasonable to expect that fire fighters should not suffer from smoke inhalation injuries. Unless you're trapped or in a situation where you're going to run out of air, you and your crews should be practicing air management principles. This does not mean to work in the IDLH until your bell starts ringing. Conscientious and safe fire fighters are going to exit the IDLH before their bell starts ringing. Forty-five-minute SCBA bottles provide enough time to make a difference. If things aren't getting better after one bottle, reevaluate the situation. Perhaps you are missing some clues. Make sure your fire fighters are all wearing SCBA in the hot zone.

22. In multistory structures, close-in positions for the command post don't always provide the best position to grasp the scope of the emergency. Consider sending a chief officer up into an adjacent building for an aerial perspective. Even bridges in the immediate area can provide an excellent perspective for a deputy IC to report conditions that cannot be seen from the interior or from the ground. This simple solution is not utilized nearly enough. We are programmed from day one to respond to the address and set up the command post in front of the building. Consider setting up the command post in an upper floor of an adjacent building with a better view of the fireground and leave the Operations Chief at the front of the building. If there was a fire on floor 50 of a high-rise building, it would be difficult to judge fire conditions from the street. However, if the IC went to the 50th floor of the adjacent high-rise building, more information could be gathered to make better strategic decisions.

23. Don't rule out using police or news helicopters to get a close-up aerial view. Technology is now available where video feeds from helicopters can be relayed in real time to computer monitors inside fire apparatus and command vehicles.

24. When we have to enter a below-grade, confined space for a rescue, we check the oxygen percentage before entering the space, often by lowering the gas detector on a tag line to check for CO and oxygen-deficient atmospheres. In an overhead void space, the CO levels must be determined because they can explode under the right conditions. If we had smoke conditions, we know CO was generated. Apply the same tactic, but reverse it. Try attaching a portable gas monitor to a two-pronged rubbish hook or an attic ladder for extended reach.

25. When dealing with a fire in a large space with suspended ceilings, consider requesting additional fire crews equipped with pike poles. Voids need to be checked as quickly as possible. Have a ventilation plan. Remember to bring in roof ladders and attic ladders. Stage them close by. If a suspended ceiling grid collapses on fire crews, the ladders can be shoved underneath the grid to help lift up large sections of ceiling to help free entangled fire fighters.

26. Remember, sprinkler systems do not wait to activate until all occupants are safely out of the building. They fuse at their designed temperature threshold, creating steam, destroying the thermal layer (neutral plane), and reducing visibility. If you look at a sprinkler system in any hotel room, in essence, it is actually a piercing nozzle. The plumbing is inserted through the wall the same way a piercing nozzle would punch through a wall. An activated sprinkler head is actually an indirect offensive attack.

27. If you visit an ARFF fire station at any major airport, you'll notice at least one first-line crash truck is equipped with a preplumbed piercing nozzle. We should follow their lead in ARFF tactics as it relates to attacking fires inside structures that may be difficult to access. ARFF realizes that the plane may be surrounded by flames and aircraft doors may be difficult to breach, especially if the crew and passengers are injured to where they cannot assist opening the doors from the inside of the cabin. The cabin can be quickly filling up with toxic smoke. ARFF doesn't take a defensive posture with passengers on board. The *Snozzle* is ready to deploy and rams the fuselage (with controlled precision) with the piecing nozzle and introduces a sprinkler system into the cabin. They have to stop the production of smoke and ventilate as soon

as possible. Manual efforts are too slow during the initial phases of the attack. The basic strategy is this: introduce a sprinkler system into a confined space.

28. Los Angeles Fire Department (LAFD) fire fighters finally extinguished the First Interstate high-rise fire by what they called "flowing the floor." They removed the fuel load (e.g., furniture, computers, window dressings) away from the windows and exterior walls and toward the center of the room. They monitored the HVAC system and then flowed water on the floor until there were a couple of inches of standing water. If everything is wet, especially the carpeting, it won't burn. Keep exposed fuel wet.

29. Fraternities and sororities are notorious for decorating house/dormitory hallways and bedroom doors with decorations, artwork, notes, and signs. Most of them are made of paper and create an increased fire load in the hallway. Though paper products burn fairly quickly, the flames produced may be high enough to catch the carpeting or other structures on fire that would generate more heat output. The crew going up to investigate a fire alarm should always have a 2.5-gallon (9.5 L) pressurized water extinguisher. Standby crews can hook up to a hydrant and lay a precautionary hose line.

30. Toxic smoke gases are obviously a respiratory health hazard that can be fatal, but many toxins in smoke can also be absorbed through the skin. Make sure your fire fighters are wearing SCBA and proper PPE during overhaul. Limit exposed skin. In hot weather, rotate crews.

31. Many PPV fans are shut off after the smoke clears and occupants are let back into the building before CO readings are taken throughout the occupancy. In one major city, a fire occurred in a restaurant on the first floor of a 12-story building. The restaurant was closed for remodeling and the fire had a head start before fire fighters arrived. After the fire was out and visible smoke was vented from the building, occupants on floors 11 and 12 were complaining of feeling ill. The IC sent a fire crew back up to check the floors with a gas detector. Though the floors were clear from smoke, CO levels were well above 35 ppm. Without taking a CO reading, these floors appeared as though they were unaffected by the fire. CO levels need to be

checked from the top floor down to the bottom. If this is a high-rise building, the IC has to be selective but determine how high CO was able to ascend in the building.

32. Ventilation of the building does not prevent CO explosions. The gas–air mixture can accumulate in a fully or partially enclosed location. This does not mean you shouldn't ventilate; it means there may be enclosed void spaces or rooms that are shielded from the PPV airflow path. Ventilation is the only method by which fire fighters can expel CO from a building. Be systematic and thorough in your ventilation plan.

33. Smoke follows the path of least resistance. PPV is more than just removing smoke from a building.

It introduces fresh air into the environment, creating a more survivable atmosphere for the victims, it cools the environment, it dilutes the products of combustion, it helps lower the temperature of fuels, it can help spread water vapor created from the hose streams, and most important, from a tactical perspective, it changes the direction of the smoke and the fire. Effective ventilation can *make the fire behave*. Remember, especially in vertical ventilation, wherever you make your hole, that's where you're bringing the fire. If you think of ventilation holes as "exit holes," it will help you make the decision on where to cut or break a window when it isn't possible to vent directly over the fire.

Chapter Summary

- Understanding fire behavior—including the effects and hazards posed by heat transfer—is critical for all fire fighters and can influence tactical decision making.
- While compartment fires are complex and dynamic, fundamental aspects of the combustion process can be identified and understood. The most important factor is often the oxygen supply—the location and quantity of oxygen needed to feed the fire will dictate how the fire burns and grows.
- Fire behavior is influenced by building construction; buildings can be negatively affected by fire in a variety of ways, including the materials lining the walls and ceilings of the space.
- Fire fighters are at great risk from rapid fire growth, especially where there is a lack of communication between fire prevention and fire suppression personnel.
- In addition to fire growth, high flame spread materials may contribute heavily to the fire load and to the generation of smoke and toxic products.
- There are several possible approaches to control the problem of fast fire growth, including eliminating high flame spread surfaces, separating material from the source of combustion, cutting off the extension of the fire, and coating the materials.
- Smoke is a generic term used to describe the collective group of solid and liquid particulates and gases.
- Smoke can affect occupants, fire fighters, and buildings in many ways, and may well be the most expensive by-product of a fire.

Key Terms

Autoexposure The vertical spread of fire on the exterior of a multistory building, from one floor to the floor(s) above, via convection and radiation.

Autoignition A phenomenon in which a combustible material ignites spontaneously without the application of a flame or spark.

Backdraft The explosive burning of heated gaseous products of combustion when oxygen is introduced into an environment whose oxygen supply has been depleted due to fire.

Bagasse Low-density fiberboard made of wood fibers or sugar cane residue.

Batt insulation Fiberglass or rock wool insulation with various thicknesses; it may or may not have a paper covering.

Ceiling layer The hot, buoyant gases that collect at the very top of a room.

Combustible acoustical tile Fiberboard in which holes have been punched.

Conduction The transfer of heat within an object or between objects through direct contact.

Convection The transfer of heat through circulation within a medium such as a gas or a liquid.

Entrainment The process of oxygen (air) being drawn into a fire.

Fire plume The column of flames, smoke, and heated gases rising above the burning object.

Flameover A situation in which unburned fuel gases, having gathered in sufficient quantities at the ceiling layer, suddenly ignite; also called rollover.

Flashover A transition stage in a fire in which exposed surfaces within the compartment ignite simultaneously and fire spreads throughout the compartment, resulting in full room involvement.

Fuel A material that will sustain combustion.

Heat A form of energy that is the source of ignition.

Low-density fiberboard Sometimes used as an interior finish, a product made of wood particles such as wood shavings and bound together with a suitable binder.

Matchboarding Ceilings made of embossed steel and wooden boards.

Neutral plane The interface between the hot ceiling layer and the cooler air flowing into the compartment.

Oxidizing agent Normally atmospheric oxygen, but may also be chemical compounds known as oxidizers that release oxygen as they react.

Piloted ignition Ignition of a heated combustible material when a flame or spark is applied.

Polychlorinated biphenyl (PCB) Toxic chemicals used as nonflammable coolants in transformers that may be released during fires.

Polyvinyl chloride (PVC) A commercially produced organic compound used in transformers and capacitors due to its electrical insulator properties and low flammability rating.

Radiation Heat transfer of heat through electromagnetic waves.

Rigid-foamed polyurethane Cellular foam plastic that is used as an interior finish.

Robertson protected metal Asphalt-coated steel.

Smoke The airborne solid and liquid particulates and gases produced by a fire.

Uninhibited chemical chain reaction A sustained oxidation reaction that produces sufficient excess heat to continue heating unburned fuel, making it available for combustion.

Case Study

Fire service leaders have said that the two main subjects fire fighters must know are fire behavior and building construction. Understanding the basics of fire behavior and how buildings react to fire will help you develop an appropriate tactical response to a given situation.

1. The four sides of the fire tetrahedron are fuel, an oxidizing agent, heat, and:

 A. an inhibited chain reaction.
 B. an uninhibited chain reaction.
 C. smoke.
 D. flashover.

2. Which of the following is *not* a method of heat transfer?

 A. Conduction
 B. Backdraft
 C. Convection
 D. Radiation

3. Fill in the blank: _____ is a transition stage in which exposed surfaces within the compartment ignite simultaneously.

4. What is the common name for fiberboard punched with holes?

 A. Batt insulation
 B. Rigid foam insulation
 C. Matchboard
 D. Combustible acoustical tile

Challenging Question

You are a captain who has been temporarily assigned to the fire academy. You are to assist in teaching a basic fire behavior class to new recruits. Using Figure 5-16, fill in each blank to identify important components of a free-burning compartment fire, provide a brief description of each of these components, and generally describe the combustion process in this diagram.

Figure 5-16

Reproduced with permission from NFPA 921-2011, Guide for Fire and Explosion Investigations, Copyright © 2011, National Fire Protection Association. This reprinted material is not the complete and official position of the NFPA on the referenced subject, which is represented only by the standard in its entirety.

Features of Fire Protection

OBJECTIVES

At the conclusion of this chapter, you will be able to:

- Understand the basic concepts of providing fire protection in buildings.
- Define fire resistance and understand how it is determined.
- Identify the primary laboratory fire tests and the characteristics they assess.
- Describe research being conducted to enhance fire protection capabilities.
- Explain the ways in which smoke and fire containment is achieved.
- Describe the various types of fire protection systems.

Case Study

Figure 6-1 The First Interstate Bank fire in Los Angeles.

© Alan Greth/AP Images.

Fire broke out on the 12th floor of the 62-story First Interstate Bank high-rise in Los Angeles, California, on May 4, 1988 (**Figure 6-1**). The building was of Type I construction—a steel frame with spray-applied fireproofing. Automatic sprinkler protection was in the midst of being installed in the building, but the valves from the risers to the sprinkler system on the fire floors were closed.

Fire quickly spread up the side of the building through autoexposure and glass curtain wall panel failure, as well as through the gap between the unbroken exterior glass curtain wall panels and the floor slab. In total, 20 handlines from 4 standpipe risers were used to eventually stop the fire at the 16th floor. Dozens of fire companies were required to bring the fire under control. The danger of a significant structural collapse was a major concern during the firefight.

A postfire investigation revealed that the spray-on fireproofing on the steel was of good quality and exceptionally (and unusually) well applied. The lack of fire blocking at the window curtain wall–floor interface, however, defeated the basic fire protection tenet of compartmentation by allowing the fire to spread through this gap. Active sprinkler protection would likely have kept the fire very small or extinguished it.

1. What is the quality of spray-on fireproofing in buildings across the United States, including your jurisdiction?
2. Given the volume of fire on the affected five floors of the First Interstate Bank, what was the potential for a complete structural collapse?
3. How do fires spread vertically inside and outside of a building, and which features of building construction are intended to stop this progression?

Building Fire Protection: A Short History

From a historical perspective, fire *containment* has always been a primary objective. Fire fighters more than 200 years ago felt a sense of accomplishment if they held a fire to the neighborhood of origin, let alone a particular block. As apparatus and procedures improved, the fires grew smaller in magnitude, reducing down to small groups of structures and individual buildings.

The emergence of building codes in the early 1900s further advanced the elimination of **conflagrations**. The adoption of such codes strived to contain fires not only to a building, but also eventually within a building. Limiting fire spread from building to building is accomplished through several means: restrictions on the area/height of a building, limits on the combustibility of roofs and exterior wall surfaces, minimum separation distances between buildings, limits on openings in exterior walls (doors and windows), and fire-resistive exterior walls. Although all of these features may not be found in all buildings (especially older buildings, which often predate modern code provisions), they form the basis for preventing fires from growing into a conflagration.

Limiting the spread of fire within a building has the added benefit of contributing directly to life safety. Preventing a fire from spreading through a structure protects its inhabitants. This is achieved through the use of **fire-rated** floors, protection of vertical openings such as stairs and elevator shafts, **compartmentation** in the form of **fire walls** and **fire partitions**, fire-resistive construction (e.g., rated columns and beams to prevent collapse and spread of fire), and the use of fire protection systems such as fire sprinkler systems and fire alarm systems.

Life safety for building occupants is, of course, a primary concern in fire protection. Among the most important elements of life safety are proper means of **egress** (exits) and protection against hazards (based on occupancy and the capability of self-preservation). Although life safety for occupants has been addressed for more than 75 years in building codes, incremental improvements are still being made based on tragic, real-world experiences.

Recently, fire fighter safety has become an explicitly stated goal in building and fire codes. A handful of regulations have been based on this goal over the years, including the requirement to place shaftway signs on the exterior of buildings adjacent to elevator and other shafts. This recent change of policy has resulted in a specific chapter in the *International Building Code* dedicated to fire fighter safety as well as general provisions that advocate for fire fighter safety. Very recent code changes to the *International Residential Code* have been made to protect exposed basement wood joists. Recognize, however, that many homes existing today do not have such protection; these code building and residential code improvements do not apply retroactively to existing homes.

Fire Terminology
Confusing Language

Many terms are used to describe materials in advertising or building codes. Some examples are fire rated, fireproof, **flameproof**, self-extinguishing, slow burning, **flammable**, nonflammable, **fire retardant**, nonburning, **fire resistant**, and **noncombustible**. It is uncertain what these terms actually mean and the exact conditions of fire testing required to earn these labels. A sample examined in one fire test may be self-extinguishing, for example, whereas it may rate as combustible if another test is used. Get the full information on the fire tests and the standards used before accepting any claim of superior fire performance.

Although the preferred word is *flammable,* the term *inflammable* may still be encountered. For many years, the NFPA has striven to eliminate this term from fire protection literature because it is confusing—some people

may think it means "not flammable." If it disappears from use, it will soon become archaic, its usage dropped and the confusion avoided.

Noncombustible Buildings

Some buildings are described in codes as noncombustible. This description is not technically accurate. The local fire officer should know exactly what this means. The noncombustibility requirement is limited to certain designated components, so noncombustible buildings can contain significant combustible components, including cornices, wooden interior balconies, and sheathing.

Fire Resistance Distinguished

As a fire fighter, you must distinguish between two uses of the term *fire resistance*. **Fire resistance rating** is a quality ascribed to a wall, floor, or column assembly that has been tested in a standard manner to determine the length of time for which it remains structurally stable (or resists the passage of fire) when attacked by a standard fire. **Inherent fire resistance**, in contrast, is a structural member's resistance to collapse by fire because of the nature of its material or assembly. A heavy wood beam, for example, takes longer to fail from a fire than does a 2 × 4. Likewise, a light steel beam will absorb heat faster and thus fail sooner than will a heavier beam. A masonry wall, itself quite fire resistant, may fail because of the lateral thrust of an elongating steel beam. The inherent fire resistance of a structure has never been formally rated or required by law.

By experience and tradition, the fire service has developed some criteria for determining the inherent fire resistance of a particular type of construction. Unfortunately, these criteria are imprecise. Many times fire fighters risk life and limb without really knowing the degree of risk and without performing any risk/benefit calculation. The generally accepted indications of imminent collapse, such as floors or roofs softening, water flowing through bricks, smoke pushing out of mortar joints, and strange noises, are, at best, grossly inadequate, even for buildings built

many years ago. If these warning signs are solely relied on to alert fire fighters to danger in today's lighter buildings, disasters will be the result.

Some codes include a "protected combustible" classification. Combustible structures are protected with gypsum board and wood assemblies that have passed the standard fire resistance test. When a combustible structure is involved in fire, no code provision, however well written and well meaning, provides real personal safety for the fire fighter. The building is the enemy, and you must know the enemy.

Early Fire Tests

Toward the late 1800s, it became apparent that building structures with "fireproof" or noncombustible materials provided no guarantee against tremendous fire losses. As steel-frame buildings emerged, the necessity for protecting the steel from the heat of fires became understood. A 1921 book titled *Fire Prevention and Fire Protection as Applied to Building Construction* gives fascinating information on the early designs of fire protection for such structures, some of which—such as the Government Printing Office in Washington, D.C.—are still standing. At that time, construction engineers were expected to design buildings "on the job." Destructive fires in these and other unprotected buildings built with innovative but untested techniques helped to bring about the realization that standards and test procedures for fire-resistive construction were necessary.

Wood floors received the earliest attention, as columns were considered to present no problems. One school of thought claimed that if such floors were designed to resist collapse, then they also presented a risk for general collapse of the building. Another school of thought believed that if a wood floor was designed to resist a collapse early in a fire, then the walls would remain standing. The dilemma posed by these opposing viewpoints, it was decided, could be resolved by designing "fireproof" floors.

In 1890, the first fire test of a fireproof floor assembly in the United Sates was conducted for the Denver Equitable Building Company. Hollow-tile floors were tested. The test revealed that porous hollow tiles set in end construction (tile cells at right angles to beams) were superior to dense tiles set in side construction (tiles cells parallel to beams). The floors were subjected to load, shock, fire, and water, and continuous fire tests (24 hours at 1,300°F [704°C]).

Six years later, the New York City Department of Buildings conducted a series of tests on fireproof floors using brick kilns as test furnaces. The central panel of the floor was loaded to 150 pounds per square foot (psf). A wood fire was built on a grate and maintained at 2,000°F (1093°C) during the last four hours. After the fire, a hose stream was applied and the floor was reloaded to 600 psf for 48 hours, with the final load resting on the arch rather than the beam.

Other tests were conducted at Columbia University. Later, gas or oil fires were used in tests for better control. Further, in addition to structural integrity, requirements that the floor must not permit passage of fire were developed.

Between 1896 and 1916, not much fire testing was done on columns. However, the testing that was conducted showed cast iron to be superior to unprotected steel.

Standards for Fire Resistance

New York's Parker Building fire in 1908, the Equitable Building fire in 1912, and the Baltimore conflagration in 1904, which involved many "fireproof" buildings, convinced many people that there was a dire need to develop standards and test procedures to ensure truly fire-resistive construction Figure 6-2 .

The effort to develop standards for fire resistance brought together the National Bureau of Standards (NBS; today known as the National Institute of Standards and Technology [NIST]), Underwriters Laboratories, Inc., the NFPA, and both capital stock and mutual insurance interests (insurance companies have a financial interest in the "successful" outcome of a fire, one in which the claims they need to pay are minimized). Fire resistance tests were conducted during World War I in 1917 and 1918, and early standards were developed that have been preserved in substantially the same form ever since.

Figure 6-2 The Equitable Building fire in 1912.

Courtesy of Irving Underhill/Library of Congress.

Materials and assemblies may be classified based on their fire resistance, or, more accurately, based on their fire endurance—standards for these capabilities are prepared by different organizations. Committees from each organization periodically review and update the standards. The three following standards, while prepared by separate organizations, are very similar in content and structure:

- NFPA 251, *Standard Methods of Fire Tests of Building Construction and Materials* (National Fire Protection Association). This final edition of standard was issued in 2006; the NFPA has decided not to update the standard because, from a technical standpoint, it has fallen behind UL 263 and ASTM E-84, and there is desire to eliminate duplication of standards.
- UL 263, *Fire Tests of Building Construction and Materials* (Underwriters Laboratories, Inc.).

- ASTM E-119, *Methods of Fire Tests of Building Construction and Materials* (American Society for Testing and Materials).

What Does Fire Resistance Mean?

It is important to understand what sort of protection fire resistance is supposed to provide:

- Fire resistance of columns is concerned with resisting collapse.
- Fire resistance of floors is concerned with resistance of passage of fire and collapse.
- Fire resistance of walls is concerned with passage of fire and collapse.
- Fire resistance of fire doors is concerned with passage of fire.

It is equally important to understand what sort of protection fire resistance is *not* supposed to provide:

- Fire resistance, by itself, is *not* specifically directed at ensuring life safety. Many lives have been lost in fire-resistive buildings. Nevertheless, to the extent a building and individual fire-rated components continue to stand under fire attack and allow occupants to evacuate and fire fighters to bring the fire under control, fire resistance can be considered a life safety feature.
- Fire resistance is *not* specifically intended to provide **smoke control**. For example, metal-clad wood fire doors generate carbon monoxide (CO) as they burn. Furthermore, while some earlier fire resistance systems provided smoke containment, it is not part of the three previously mentioned fire resistance standards.
- Fire resistance is *not* concerned with the dollar loss due to fire. There have been huge financial losses in fire-resistive buildings.

Fire Resistance Testing

Essentially, the provisions of the test standards—NFPA 251, UL 263, and ASTM E-119—require a reproducible test fire, a consistent method for conducting tests and classifying results, and specific instruction on the selection and preparation of test

specimens. The objective of these provisions is to ensure uniformity in testing.

The reproducible fire used in the test procedure follows a **standard time–temperature curve** (Figure 6-3). This curve has remained essentially unchanged for more than 80 years. This single curve, created as the result of a small series of test burns in vacant buildings in 1918, represents the temperatures recorded in those tests. The curve does not represent fires of different sizes or fuels, particularly "modern" fires.

In the fire resistance test, samples of typical structural elements are exposed to the standard fire in a test furnace equipped with gas-fed burners. The furnace temperatures are regulated by computer to follow the test curve.

The minimum sizes for test specimens are as follows: columns, 9 feet (2.7 m); beams and girders, 12 feet (3.7 m); partitions and walls, 100 square feet (9.3 m²); and floors and roofs, 180 square feet

(16.7 m²). The load-bearing structural elements are loaded with the prescribed weight for the assembly being tested. The unit passes the test as long as it successfully resists the superimposed load and, in the case of partitions, walls (Figure 6-4), floors (Figure 6-5), and roofs, resists the passage of fire.

Figure 6-4 A wall assembly fire resistance test.

© Javier Larrea/age fotostock.

1000°F (538°C) at 5 minutes
1300°F (704°C) at 10 minutes
1550°F (843°C) at 30 minutes
1700°F (927°C) at 1 hour
1850°F (1010°C) at 2 hours
2000°F (1093°C) at 4 hours
2300°F (1260°C) at 8 hours ot over

Figure 6-3 The standard time–temperature curve as shown in NFPA 251.

Reprinted with permission from NFPA 251-2006, Standard Methods of Tests of *Fire Resistance of Building Construction and Materials*, Copyright © 2005, National Fire Protection Association, Quincy, MA. This reprinted material is not the complete and official position of the NFPA on the referenced subject, which is represented only by the standard in its entirety.

Figure 6-5 A floor fire resistance test.

© Javier Larrea/age fotostock.

Columns have several **thermocouples peened** into the columns. When the average temperature exceeds 1,100°F (593°C) or any of the thermocouples' temperatures exceeds 1,300°F (704°C), the test is ended and the specimen fails.

A floor structure or wall structure must not develop conditions that would ignite cotton waste on the unexposed surface or permit an average temperature rise on the unexposed surface of 250°F (121°C). Under many circumstances, a hose stream is applied to the hot assembly immediately after the fire test to assess its stability when hit by water from a fire hose.

Floor and roof assemblies are often tested under restrained and unrestrained conditions. Essentially, the assemblies are secured on the ends and not permitted to move. This, of course, is a more severe test for these assemblies, given that the expansion of the structural elements as a result of the test furnace heating may cause them to distort and buckle when restrained.

Assemblies that pass the test are rated in units of hours—for example, 1, 2, or 3 hours—depending upon how long the test unit survived the test fire. As appropriate, restrained and unrestrained designations are also indicated. Rated units are listed in references such as the *UL Fire Resistance Directory*, which is published annually. Based upon such **listings** (being "listed" is not the same as being "approved"), local authorities such as the buildings department may "approve" (permit for use as being acceptable to the local jurisdiction) use of these specific materials and assemblies.

Minimum Requirements for Fire Resistance Listing

A listed assembly meets the minimum requirements needed to pass the test. Thus, all components assembled as specified in the listing are required to meet that specific listing. Critics suggest that the NFPA 251, UL 263, and ASTM E-119 standards contain redundant precautions. In reality, nothing could be further from the truth. Owing to the nature of the testing process, the listed assembly is the minimum-quality structure capable of passing the test. Not requiring assemblies to achieve this minimum standard could give a supplier an advantage over its competitors that do meet the standard, as the latter suppliers would inevitably have higher costs.

At times, buildings departments may approve variations from the listing. Such a departure from the listing means that the assembly installed is probably not equivalent to the listed assembly.

The listings are not like a restaurant menu from which one can select one item from column A and another from column B. Put simply, components should not be considered apart from the tested and listed assembly. A very common fallacy is the belief that fire resistance can be achieved simply by nailing up some 5/8" (15.9 mm) "fire-rated" gypsum board. In reality, the fire rating comes from the listing of the *entire* assembly, not the individual components within that assembly.

Also, it is important to recognize that the unit being tested may be seriously damaged or effectively destroyed in passing the test. Indeed, fire resistance standards are often misinterpreted owing to misperceptions about what the testing results mean. A 2-hour fire resistance rating does not mean the building will be relatively undamaged for 1 hour—fire damage can be quite severe even for short exposures. Instead, the fire resistance rating means that *if* the building reacts to the fire in the same manner as the specimens did for that particular type of construction in the test, *and if* the actual fire load (fire impact) does not exceed the fire test load, *and if* the duration of the fire does not exceed the time specified for the components involved, *then* the structure should survive the fire.

However, the time ratings given to fire-resistive assemblies *do not* relate to the real time in an actual fire. Rather, the ratings provide a relative *measure* of comparison, comparing one assembly's duration of endurance to another assembly's duration; a 2-hour fire-rated assembly simply has greater fire endurance when compared with a 1-hour fire-rated assembly.

In the wake of the World Trade Center disaster in 2001, increasing criticism has been leveled at the fire

resistance test, particularly calling into question its ability to represent the "real" fires of today. The bottom line when considering fire resistance ratings is that you need to understand what the test actually represents and what it does *not* represent. Fire resistance—more specifically, fire resistance–rated assemblies—does not necessarily mean noncombustibility. Many listings of assemblies include combustible components, typically wood beams or studs.

The fallacy of arguing that fire resistance–rated assemblies represent a completely safe working platform lies in the fact that the test standard was originally developed for noncombustible floors assaulted by fire *only from below*. In reality, structures that incorporate wood beams or lightweight trusses, for example, might potentially be attacked by fire from both above and below and by fire starting or moving into the combustible void space in the middle of the assembly. For example, an electrical fire starting inside the truss of a 1-hour fire-rated floor assembly is not even considered under the test standard. If a fire starts inside the assembly, forget about the fire rating; no such rating exists. This situation is a major deficiency in the test standard—the standard does not recognize the fact that nearly all such assemblies have electrical wiring running within them.

Fire resistance ratings for building materials must be distinguished from flame spread ratings. It is common (but mistaken) practice to speak of "fire-rated tile." The term is meaningless.

Fire resistance ratings are concerned with the length of time a particular building assembly will continue to perform its structural and/or barrier function in the face of an assault by a standard fire test. Flame spread ratings are concerned with the rate at which fire spreads over the surface of a material, the smoke it develops, and the fuel it contributes to a fire. These two terms are not interchangeable, but in some cases, a single material may be evaluated in terms of either its fire resistance rating or flame spread, depending upon the way it is used. For example, when acoustical ceiling tile is part of a fire-resistant floor and ceiling assembly,

its ability to act as a barrier to the passage of heat is of prime importance. In contrast, when acoustical tile is used solely for its sound-deadening and decorative qualities, such as when tiles are directly adhered to the underside of a concrete floor, then its flame spread rating is of prime importance. As part of the interior finish of a building, its ability to spread flame is a life safety concern. In many cases, both fire resistance and flame spread ratings may be concerns.

The overall fire resistance rating of a building is an accumulation of the scores for hundreds of components. These items are built and installed by persons of all levels of competence. Even though building codes call for certain levels of fire resistance, the entire building as one single unit is not tested. The failure of only one of many elements may be disastrous. Of course, only a real fire is the ultimate test of a building's ability to resist the effects of a fire.

Moreover, one of the most important factors in fire spread within buildings is the interior finishes on walls and ceilings. Fire history is replete with examples of deadly fires in which interior finishes played a deadly role, including the flammable decorations in the Boston Cocoanut Grove fire of 1942, the 1984 haunted castle polyurethane-fed fire at the Great Adventure amusement park in New Jersey, and the 1977 Beverly Hills Supper Club in Kentucky in which carpeting on the walls aided fire spread. The most recent fire where interior finish played a critical role occurred at the Station nightclub in West Warwick, Rhode Island, where 100 people were killed when indoor pyrotechnics ignited polyurethane foam padding on the walls surrounding the performance platform.

Testing and Rating Materials

The first attempts to deal with flame spread by codes failed due to inexact, legally unenforceable language, such as "flame spread no greater than wallpaper." Developing adequate tests is costly and time

consuming. Tests must be consistent; that is, tests performed in the same apparatus on the same material should produce the same results. Tests also should be reproducible; that is, others using similar equipment and procedures should obtain the same results.

Steiner Tunnel Test

The basis for regulation of flame spread today is found in NFPA 255, *Standard Method of Test of Surface Burning Characteristics of Building Materials*, commonly referred to as the **Steiner tunnel test** **Figure 6-6**. The test was developed by the late A.J. Steiner at Underwriters Laboratories, Inc. and is also known as ASTM E-84 and UL 723. Essentially, a test sample 25 feet (7.6 m) long and 2 feet (61 cm) wide forms the top of a tunnel or long box. A gas fire is lighted at one end, and fire progresses along the underside of the top of the box or test panel. Window panels on the side of the tunnel along the length of the tunnel allow for observations of the progression of flame spread.

There are two comparison points in evaluating test results. The flame spread over inorganic reinforced cement board is set at 0. The flame spread over red oak is set at 100. Tests have shown that fire can reach the end of a red oak test panel in 10 minutes. Other materials tested are rated on a scale determined by these points. The formula by which the flame spread index is calculated can be found in NFPA 255.

Flame spread of surface materials is classified as follows:

- Class A: 0–25
- Class B: 26–75
- Class C: 76–200

Most building and fire codes have requirements for ceiling and wall surfaces based on the tunnel test standard. A typical requirement might be for a Class A flame spread rating for exit stairwells in hospitals. Sometimes higher flame spreads are permitted for sprinklered buildings. The model building codes allow for a reduction in the class level when sprinkler protection is available, but in no case can the rating be less than Class C (i.e., A→B, B→C, C).

When the tunnel test is run at Underwriters Laboratories, Inc., **smoke developed** is also measured and indexed. Smoke developed is calculated by measuring the obscuration as the smoke passes a photoelectric cell placed in the stack from the test tunnel. Again, the loss of light from inorganic reinforced cement board and red oak are used as benchmarks. Materials with smoke-developed ratings of 300 or more can be expected to generate substantial amounts of smoke. Only light obscuration is measured in this aspect of the tunnel test—not any of the other effects of smoke or gases, such as toxicity or irritation.

The Steiner tunnel test is widely accepted as a flame spread testing standard. It is almost impossible to read the literature for commercially installed ceiling tiles without encountering a reference to the flame spread rating. This is not wholly the case in the home market. Homeowners generally are unaware of the flame spread testing standard or how it is applied to the tiles they buy.

In some cases, manufacturers have stressed low flame spread ratings, neglecting to point out that the material advertised received a very high smoke-developed rating. All materials listed by nationally recognized testing laboratories are tested for both characteristics.

Figure 6-6 The Steiner tunnel test.
Courtesy of Herbert Blenstein.

Radiant Panel Test

The NBS developed ASTM E-162, *Radiant Panel Flame Spread Test*. Samples for this test measure only 6 by 18 inches (15 by 46 cm) **Figure 6-7**. For some materials, results from this test can be correlated with the Steiner tunnel test results, but no direct relationship should be assumed. The radiant panel test has been used to develop information after serious fires, but would probably not be valid evidence in a prosecution based on noncompliance with a code.

FM Approvals Corner Test

The corner test developed by FM Approvals is designed to simulate an actual fire within the corner of a building **Figure 6-8**. In the test, the walls are up to 25 feet high (7.6 m); the east wall is 50 feet long (15.2 m) and the south wall is more than 37 feet long (11.3 m). This test assesses, among other things, flame spread. It is considered a more realistic test than the Steiner tunnel test because the material is being tested in a more realistic configuration, as compared with the horizontal tunnel test.

Carpet Tests

Floor covering, specifically carpeting, has been an important contributing factor in a number of serious fires. In recent years, many tests and standards have been developed for carpeting. Commercial carpeting presently being manufactured is required to meet a test that measures ignitability of carpeting from a small source such as a dropped match or cigarette. It is a screening test (popularly known as the pill test), and seven out of eight samples must pass the test.

There is also a test to rate carpeting for its ability to spread flame when attacked with a greater ignition source than a cigarette. This test is NFPA 253,

Figure 6-7 The radiant panel test.
Photo courtesy of The Govmark Organization, Inc.

Figure 6-8 The FM Approvals corner test.
© 2007 FM Global, Reprinted with permission. All rights reserved.

Standard Method of Test for Critical Radiant Flux of Floor Covering Systems Using a Radiant Heat Energy Source (also known as ASTM E-648).

The tests cited here are not intended to test carpets for use as wall finish. ASTM E-84 and NFPA 265, *Standard Methods of Fire Tests for Evaluating Room Fire Growth Contribution of Textile Coverings on Full Height Panels and Walls*, are the appropriate tests for this situation.

Critical Radiant Flux of Floor Covering Systems Test

The Critical Radiant Flux Test (NFPA 253) measures a material's ability to resist flame spread. The result derived from the test is the **critical radiant flux (CRF)** of the sample. This is the amount of external radiant heat energy (measured in watts per square centimeter) below which a flame front will cease to propagate. The higher the CRF number, the less flammable the carpeting.

This test uses the radiant panel set at an angle. The carpeting sample is set flat in its normal position. The sample should be tested with the intended underlayment. If the carpeting is flammable, the flame spread will probably increase; that is, the CRF will decrease if an underlayment is present. The underlayment acts as an insulator, keeping the heat from the carpeting from being dissipated to the floor.

Where regulated, the model building and fire codes consider two classes of interior floor finish ratings:

- Class I: CRF minimum ≥0.45 watt/sq cm
- Class II: CRF minimum ≥0.22 watt/sq cm

A reduction in class is permitted in spaces protected by an approved automatic sprinkler system.

CRFs of carpeting involved in several serious fires have been determined. Thus, when regulations are adopted specifying a minimum interior floor finish rating, that may signal a relationship to specific fire danger that is not so apparent in other occupancies.

Don't Be Mousetrapped

Fire departments should be very wary of conducting their own tests. Tests that prove that a particular product is the answer to all problems should be suspect. Testing is a difficult and exacting profession. Leave it to the experts.

It is also difficult and sometimes impossible to determine the fire resistance rating and flame spread characteristics of materials already in place in actual buildings. Both fire resistance–rated and non-fire-resistance–rated assemblies and low and high flame spread materials can have the same appearance.

Field testing can be dangerous. In one case in an eastern city, the fire department was conducting a campaign against flammable decorations. An inspector examined a display for a popular brand of blended whiskey. He decided to test the flammability with his lighter. The display was indeed flammable—the ensuing fire went to multiple alarms. The city paid a heavy judgment and two fire fighters were seriously injured in the fire response.

Research

Research work is under way to bring testing up to today's fire dynamics and replace the standard time–temperature curve. This research will be beneficial to all concerned with fires in buildings.

For example, changes have been made to the U.S. building and fire codes as a result of the World Trade Center collapse on September 11, 2001. Changes were made to improve sprayed–on **fireproofing** in terms of adhesion and cohesion of the substance, mandate wider stairs in some high-rises, provide for a separate high-rise fire fighter stairwell and improved fire fighter elevator access, and require in-building radio signal amplification in high-rises and other structures where communications among fire fighters and other responders might be difficult. In addition, the very first provisions to allow special "hardened" elevators (elevator cabs and shafts that have been fortified against extreme events such as explosions) for civilian egress during an emergency were put in the building code.

Nationally, fire research has taken an interest in building construction and fire fighter safety. In 2004, NIST issued a report on a series of tests in Phoenix, Arizona, in wood-frame buildings. One of the results

of the test was the observation that thermal imagers, while valuable for assessing fire spread, did not offer any help in establishing roof deck temperatures directly under the roof. In the experiments using roofs with both clay tile and asphalt shingles, no observable changes on the camera screen were seen prior to collapse. Instead, water from hose streams cooled the roof and radiation from the fire and smoke plumes "washed out" the fire's signature; as a consequence, the thermal imager did not offer assistance as to the actual temperature conditions of the roof itself. Additional recent testing by NIST has attempted to develop collapse indicator equipment using sonar, laser, and vibration analysis. A prototype vibration analysis tool, "HOBS" (Health of Burning Structure panel), has been developed to monitor vibrations in a structure that might be a precursor to a collapse.

In 2008, Underwriters Laboratories, Inc., with the assistance of the Chicago Fire Department, issued a report entitled "Structural Stability of Engineered Lumber in Fire Conditions," which provided a sobering set of collapse comparisons for "legacy" construction (traditional wood members) versus lightweight structural members (wood I-beams and trusses). In a simulated basement fire involving exposed lightweight wooden I-beams, collapse occurred in a mere 6:03 minutes. The legacy 2- by 10-inch (51 by 254 mm) beam construction outlasted lightweight construction by a 3:1 time margin. Another valuable lesson was also learned: carpet and other floor coverings may conceal the true fire conditions underneath the floor from a thermal imager.

Containment of Fire and Smoke

Generally speaking, there is a hierarchy of fire-resistive assemblies. Building codes specify the level of fire resistance in hours and the number and type of penetrations permitted. Here is a list of fire-resistive assemblies:

1. Fire walls are typically 2- to 4-hour fire resistance rated and are used to create separate buildings. They are defined in the *International Building Code* as "a fire-resistance-rated smoke-tight wall having protected openings, which restricts the spread of fire and extends continuously from the foundation to or through the roof, with sufficient structural stability under fire conditions to allow collapse of construction on either side without collapse of the wall."

2. Fire partitions are typically 1-hour fire resistance rated and are used to create fire-resistive corridors and to separate tenant spaces in covered mall buildings. They are defined in the *International Building Code* as "a vertical assembly of materials…designed to restrict the spread of fire in which openings are protected."

3. **Fire barriers** are typically 1- to 2-hour fire resistance rated and used to enclose shafts, exit stairwells, exit passageways, and **horizontal exits**, and to separate spaces from certain hazardous areas. They are defined in the *International Building Code* as "a fire-resistance-rated vertical or horizontal assembly of materials… designed to restrict the spread of fire in which openings are protected."

Openings in fire-resistive assemblies are required to be fire resistance rated as well. For instance, heating, ventilating, and air-conditioning (HVAC) systems passing through fire-resistive assemblies must be provided with **fire dampers**. Many fire dampers are mechanically operated using a fusible link that drops a metal "curtain" or blade inside, blocking passage of fire Figure 6-9. Other types of fire dampers use pneumatic or electric activation.

Fire dampers typically receive very little attention once they are installed. Shifting or movement of the duct they are attached to can put them out of alignment, preventing the blade or curtain from moving properly. For this reason, fire dampers should be checked periodically. Combined fire and smoke dampers, typically used in **smoke barriers** and smoke partitions, also require periodic maintenance.

Figure 6-9 A fire damper with interior "curtain."

Courtesy of Ruskin Air Management Limited.

Fire Doors

Fire doors can have fire protection ratings ranging from 20 minutes to 3 hours, depending on the type of fire resistance–rated assembly the fire door is located in. One of the most important aspects of a fire door is its ability to close and latch closed. There have been numerous disasters in which the door closer or latch failed (e.g., because it was broken or blocked) and fire spread throughout the building. A New York high-rise fire in 1998 killed four people when the occupant of the apartment in which the fire started jammed a welcome mat under the corridor fire door, allowing fire and smoke to travel up through the building.

A fire door was installed in the American History Museum many years ago over the objections of the Smithsonian Institution staff. It stopped a raging fire from destroying the Star-Spangled Banner. An official who had fought against the fire door later said, "I learned one thing in this investigation. If you are building a building to last 75 years, it will have a major fire, so fire is a design criteria."

Fire losses would be greatly reduced if designers and developers all adopted this point of view. Shopping malls have minimum horizontal barriers, which can allow smoke to pollute an entire mall. One Virginia mall has had two transformer fires that led to multimillion dollar losses owing to its lack of fire barrier features. Fire doors are tested using NFPA 252, *Standard Methods of Fire Tests of Door Assemblies*; UL 10B, *Fire Tests of Door Assemblies*; or the former ASTM E-2074 *Standard Test Method for Fire Tests of Door Assemblies, Including Positive Pressure Testing of Side-Hinged and Pivoted Swinging Door Assemblies*, which was withdrawn in 2007. A furnace is used to test the fire doors.

Fire doors may close by any of three methods:

- Swinging: the most common choice; used in corridors and stairwell openings **Figure 6-10**
- Sliding: typically found across openings in elevator shafts or stairwells in old factories **Figure 6-11**
- Rolling: used in large openings found in modern factories and warehouses **Figure 6-12**

Fire shutters are sometimes used in buildings. They are typically rolling fire doors used to protect small openings such as a ticket booth window in a rated corridor.

There are two types of fire door closure devices: self-closing and automatic. A typical example of the self-closing type is the room or stairway door equipped with an automatic door closer that closes the door after it has been opened **Figure 6-13**. Some stairway door closers are equipped with fusible links, which are designed to fail at a specific temperature and close the door when the heat from a fire is encountered. These links are not sensitive to smoke, however, so the stairway can be totally polluted with smoke before the link operates.

Fire walls with automatic fire doors intended to close only in case of fire are also equipped with fusible links. When the link melts, the overhead rolling door comes down, or the sliding fire door closes horizontally. Sliding fire doors may roll down a track by gravity or be pulled shut by a weight.

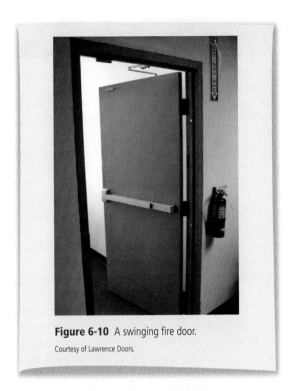

Figure 6-10 A swinging fire door.
Courtesy of Lawrence Doors.

Figure 6-11 A sliding fire door.
Courtesy of Lawrence Doors.

Figure 6-12 A rolling fire door.
Courtesy of Lawrence Doors.

Figure 6-13 A self-closing device.
Courtesy of the estate of Francis L. Brannigan.

In either case, they should be checked by lifting up on the weight that holds the door open. The door should close properly without any assistance. Many existing fire doors are not in a state of readiness, and a variety of problems occur: broken closers, broken latches, mechanical blockages preventing them from completely closing (such as a broken track or a damaged drum on an overhead rolling fire door in a warehouse), or the proverbial door chock. Inspections and vigilance are necessary to ensure that fire doors remain in a good state of repair. Inspections should include the following items:

- Operating fire doors and shutters.
- Raising the counterweights of automatic sliding and counterbalanced doors to be sure they close.
- Inspecting doors for damage.

- Inspecting all hardware including latches, guides, and thresholds for proper function.
- Checking the fusible links of automatic closing devices to ensure they are free of paint and other foreign matter that might impair their operation. (Note that the fusible links of sprinklers in one paper plant did not operate properly because they were covered with oily paper dust.)
- Checking the proper operation of "door coordinators" that ensure that a set of opened fire doors ("leafs") close in the proper order, ensuring the gap between the doors is properly sealed **Figure 6-14**.

Figure 6-14 Originally, this pair of fire doors was to be self-closing with each door leaf shutting in a specific order, using a "coordinator." The door on the left is supposed to close first, followed by the door on the right. With door coordinator removed, the doors close in the wrong order. Note the extended metal plate on the right door's edge, which is supposed to seal the opening when both leafs are closed properly.

Courtesy of Glenn Corbett.

As with many devices or materials, such closure devices may be completely inadequate to control smoke movement. Unfortunately, it is a common practice to block open self-closing doors.

Many fire reports note that the spread of fire and smoke in such occupancies was due to doors left open. The fact that stairway and corridor smoke and fire barrier doors are often blocked open, however, should not prevent fire departments from advocating for self-closing doors in apartment and hotel occupancies.

Non-fire-resistive buildings and older fire-resistive buildings were usually built without any consideration of the control of fire and smoke movement. As buildings are altered, there may be code requirements to upgrade the smoke and fire control. Some such alterations are more effective than others.

Using Smoke Detectors for Release

The development of the smoke detector has provided equipment that is sensitive to smoke and that may help to eliminate the problem of blocked open doors. In one method designed to prevent this practice, the doors are held open by electromagnets. The door latch system can be triggered automatically by the smoke detectors on either side of the door by fire, smoke alarm, or sprinkler water flow alarm. These doors should be designed to permit manual release as well. When an alarm cuts the current to the magnets, the door releases and becomes a typical self-closing door. Such a system is easy to test if the door contacts the magnetic latch when pushed to the wall.

Some doors are too heavy for magnetic latches, but may be held by mechanical latches that are released in the same way. Two-leaf doors should be checked to see that the doors close in the designed sequence as discussed previously. The roller that accomplishes this task is frequently damaged from constant use.

In other cases, the smoke-sensitive doors are controlled individually. One or two detectors (one on each side) trip a door as smoke is detected. In some cases, the detector is integral with the door closer. If the door is open and the detector senses smoke, the door is released. Such doors should not have latches to hold them open.

Stairwells

When it was learned from such fires as the 1908 Parker Building fire in New York that fire could extend up open stairways in "fireproof buildings," the concept of compartmentation—that is, creating fire areas no larger than one floor—was developed. With this strategy, all connections between floors were designed to stop the spread of fires. For example, stairways in new buildings were enclosed and provided with self-closing doors with a fire protection rating. Outside metal fire escapes or true fire towers with "atmospheric breaks" (open to the outside air) were installed.

Many times, however, open interior "access" or "convenience" stairs are installed. These create a single fire area consisting of two or more floors, as was the case in the One Meridian Plaza fire in Philadelphia. The use of these open stairs allows smoke and fire to travel easily from one floor to the floor above.

Self-closing doors are considered a nuisance to many occupants who must move from floor to floor. For this reason, they may block stairway doors open with wooden wedges or other items. These makeshift devices are sarcastically called "four-hour fusible links" by inspectors. A few years ago, the United Kingdom mounted a campaign to enclose stairways and install fire extinguishers. To their chagrin, inspectors found that in some cases the extinguishers were used to block open the stairway doors.

In the United States, the stairway doors at the office of a well-known consumer safety advocate were blocked open with bundles of safety pamphlets. To counter this blocking problem, fusible links or door checks with integral 165°F (74°C) fusible links have been used. The closing of stairways does not often come naturally or easily. Indeed, the open grand staircase is often a crowning architectural feature of a monumental building. One of the most famous examples is found in the Paris Opera, where the volume of the stairway appears to be greater than that of the auditorium. An architect who worked in a city hall admitted that the open stairways could cause a loss of life, but felt that closing up the stairways would destroy the beauty of the building.

After several multiple-fatality hotel fires in the 1940s, an effort was made to enclose open stairways. In many cases, the corrective measures failed in that often the stairway was left open at the lower floors. In some building restorations, a "monumental" stairway was left unenclosed. The operative, yet strange, argument is that sufficient exits are already provided by protected stairways—a position that ignores the open stairway's danger as a transmitter of smoke and heat.

Horizontal Exits and Smoke Barriers

In past years, the concept of each floor being a single fire area was adequate. Today, however, floor areas in buildings have increased tremendously. This has necessitated horizontal exits. Horizontal exits (typically a 2-hour–rated wall with rated fire doors) are used to reduce travel distances and to subdivide (compartmentalize) floors.

From a firefighting perspective, this means that stairwells—and the standpipe hose outlets they often contain—are farther apart in multistory buildings. For that reason, building codes require hose outlets at the openings of a horizontal exit; this is to ensure that a sufficient number of hose outlets are available throughout each floor, avoiding excessively long handline hose stretches **Figure 6-15**.

Smoke barriers (typically 1-hour fire rated and with smoke-protected openings, including "smoke" dampers and door sweeps at the bottom of fire

Figure 6-15 A noncompliant horizontal exit, missing the hose outlet.

Courtesy of Glenn Corbett.

Figure 6-16 This smoke barrier allows safe egress of patients in a health care facility.

Photo courtesy of Won-Door Corporation.

doors) are used in health care facilities because they can be used to move nonambulatory occupants to a smoke-free area—it is difficult to move such individuals vertically **Figure 6-16**. Smoke barriers are also used in certain underground buildings. The horizontal barriers and smoke barriers are often resisted

by building developers because closing down the opening to door size can damage architectural concepts aimed at highlighting a building's "openness."

Penetrations in Fire-Resistant Walls and Floors

It is a certainty that holes and penetrations will be created in fire-resistant assemblies when a building is being built. It is also a certainty that holes will be created during the life of the building. The reasons for the holes are numerous—pipes, conduit, wires, cables, and so on. The individuals creating the holes have no idea they have compromised the "fire integrity" of the wall or floor in question.

Holes must be properly sealed with listed through-penetration firestop systems. These fire-tested "systems" actually comprise a set of components, including special caulking, insulation, and supports. The entire system must be installed as designed and listed, otherwise, the system will not properly seal the opening in an actual fire. The firestop system must match the hourly fire resistance rating of the wall or floor it is protecting.

Escalators

Escalators are widely used in department stores and public occupancies. There is strong resistance to enclosing escalators. A number of methods of protecting the openings with various types of sprinkler or spray systems have been adopted in various codes.

One method involves the use of water spray nozzles directed downward through the opening. The fire department should be aware of this type of system because it can cause considerable water damage. Blame for this damage may be put on the fire department.

In some systems, a line of sprinklers is located around the escalators. They are typically used in conjunction with a glass partition around the opening that is 3 or 4 feet (91 or 122 cm) deep. The sprinklers are shielded from one another to prevent one sprinkler from cooling the other. Although these sprinklers may extinguish a nearby fire, they would have no effect on smoke or gases moving from a fire not in the immediate

vicinity. This code-accepted method of sprinkler protection is an inadequate method to protect against vertical smoke migration. In short, life safety in such occupancies depends on the full sprinkler protection of the building and not partial protection at floor openings.

Thoughts on Smoke and Compartmentation

Fires produce huge volumes of heat, smoke, and gases. Starving the fire is one of the reasons sometimes advanced as a justification for compartmentation. In some cases, a fire has been cut off completely by compartmentation, but such cases are rare. The successes tend to be offset by tragedies that occurred when openings were made into otherwise unventilated compartments. In 1932, for example, eight fire fighters died when they opened the door on a smoldering fire in an unventilated storage room in the subbasement of New York's Ritz Tower Hotel. There have been a number of similar disasters.

Venting a fire, even at the risk of increasing its volume, is often necessary to make the conditions inside the fire building somewhat habitable for occupants and fire fighters. This necessity was recognized more than 100 years ago in the case of multistory fires in tenements in New York City. Packed with hundreds of immigrants, these buildings were deathtraps. A single wooden stairway, with wood and glass doors to tenant apartments, rose through the center of a building. Fire fighters learned early on that opening the roof over the stairway relieved the pressure on the upper floors, where potential victims were often at the windows. In 1903, the New Law Tenement House Act required, among many improvements, that all stairways be masonry enclosed, noncombustible, and have a skylight, so that fire fighters could easily vent the stairway.

In 1930, Assistant Chief Thomas Dougherty—a man far ahead of his time—wrote the book *Fire*. In it, he not only explained the necessity of ventilation, but also offered a design for automatic vents tripped by fusible links to be installed atop stairways. Some readers may object: "Automatic vents violate the cardinal rule, 'Don't vent until you have water.'" As with all rules, however, there are exceptions to this one.

Theaters where live actors perform on stage with scenery were the first buildings ever designed so that a fire could occur and the occupants would be protected from the combustion products. Automatic vents are provided above the stage to reduce the pressure on the proscenium fire curtain. The stage may burn out, but the audience has a greatly improved chance of escape. This concept evolved from the hundreds of deaths that occurred in theater fires. However, in Chicago's 1903 Iroquois Theatre fire, in which 603 people died, the stage vents were fixed in a closed position, and a gas light fixture blocked the curtain from closing. The program had boasted that the structure was "absolutely fireproof."

In 1919, the Holland Tunnel was being planned to connect New York and New Jersey under the Hudson River. It was designed to provide enough airflow through the tunnel to reduce the maximum amount of CO that could be generated by automobile exhausts to an acceptable level. The permissible level for CO was determined by having volunteers drive cars, until affected, around a "racetrack" cut into the coal mine at the Bureau of Mines Experiment Station near Bruceton, Pennsylvania. This was the basis for the permissible level for 8-hour exposure to CO. In the 1950s, the permissible level was cut in half. Tests showed that the airflow required to produce this level would be about 50 miles per hour (mph) (80.5 kph). Ole Singstadt, the design engineer, objected. He claimed that if a fire occurred in the Holland Tunnel, the occupants of cars in the tunnel would be incinerated. He argued for a triple tunnel arrangement, consisting of an air supply tunnel below the car tunnel and an exhaust tunnel above it, so that toxic gases would be vented vertically, directly above the fire. The Port Authority management balked, citing the cost. Singstadt, a tough, world-class engineer, pointed out that he was the only person in the world who could build the tunnel, and it would be built his way or no way. In the end, the Holland Tunnel was built his way.

In 1951, a truckload of carbon disulfide burned in the tunnel. Fire showed out the New Jersey-side vent tower, and it was necessary to use hose streams to cool the fans **Figure 6-17** . Everyone in the tunnel escaped, although unfortunately one fire officer died.

Figure 6-17 Two of the extraction fans in the Holland Tunnel failed to operate due to the tremendous heat of the 1951 fire.

© stf/AP Images.

The Eurotunnel (the "Chunnel") under the English Channel from England to France was completed in 1994. In this tunnel, cars and trucks are carried on trains. There are two separate train tunnels. A service tunnel between the two train tunnels is kept at a higher pressure than the train tunnels to exclude smoke. Access is provided through "smokeproof" doors. Reports estimated that 1,500 passengers could be evacuated from a burning train in 25 minutes. In November 1996, a fire in the Eurotunnel involving a train carrying cars and trucks did substantial damage, but no lives were lost.

Fire department ventilation has traditionally consisted of opening and breaking windows and cutting holes in the roof to allow smoke and heat to escape by gravity. A number of years ago, smoke ejectors were developed to increase the volume of airflow. Later, large fans were used to exhaust smoke, generally during overhaul. When this technique is used, it is necessary to prevent outside air from entering the air stream, thus diluting the flow for combustion products. Positive-pressure ventilation (PPV) has now taken hold across the United States.

Fire Protection Systems

In addition to the use of fire barriers as a means of fire containment and life safety, fire suppression and detection systems are another critical means of providing fire protection. What follows is a brief review of these systems, how they impact your firefighting efforts, and how you can use/support them.

Automatic Sprinkler Systems

For many years, automatic sprinklers were familiar only to fire protection personnel. Most citizens, including those who reported on fires, were completely unfamiliar with them. Times have changed, of course. One of the first questions many reporters now ask at serious fires is, "Is there a sprinkler system in the building?"

Automatic sprinklers are fast becoming a fact of life in buildings. They are being installed in new buildings in compliance with codes to reap the substantial benefits granted by such protection, and in recognition of the potentially catastrophic financial consequences, particularly liability, presented by serious fires. Slowly the need to retrofit sprinklers into buildings that should have been sprinklered when they were built also is being recognized.

Sprinklers were once almost exclusively installed in factory and mercantile buildings. For decades, the only sprinklered high-rise office building in the country was the headquarters of the National Board of Fire Underwriters. They are now included in every occupancy and are gaining greater acceptance in one- and two-family dwellings.

Sprinkler systems are installed under one of three different standards: NFPA 13, *Standard for the Installation of Sprinkler Systems*; NFPA 13D, *Standard for the Installation of Sprinkler Systems in One- and Two-Family Dwellings and Manufactured Homes*; and NFPA 13R, *Standard for the Installation of Sprinkler Systems in Low-Rise Residential Occupancies* ("low-rise" in this case means residential occupancies four stories or fewer and less than 60 feet in height).

Types of Sprinkler Systems

There are four major classifications of automatic sprinkler systems. Each type of system includes piping for carrying water from a source of supply to the sprinklers in the area under protection.

- *Wet pipe systems:* These systems employ automatic sprinklers attached to a piping system containing water under pressure at all times. When a fire occurs, individual sprinklers are actuated by the heat, and water flows from the sprinklers immediately.
- *Dry pipe systems:* These systems have automatic sprinklers attached to piping that contains air or nitrogen under pressure. When a sprinkler is opened by heat from a fire, the pressure is reduced to the point where water pressure on the supply side of the dry pipe valve can force open the valve **Figure 6-18**. Then water flows into the system and out any opened sprinklers. Dry pipe systems are typically installed in areas of the building subject to freezing.
- *Preaction systems:* These are systems in which there is air in the piping that may or may not be under pressure. When a fire occurs, a supplementary fire-detecting device in the protected area is actuated. This opens a water control valve, which permits water to flow into the piping system before a sprinkler is activated. When sprinklers are subsequently opened by the heat of the fire, water flows through the sprinklers immediately, the same as in a wet pipe system. The advantage of these systems is that water typically will not flow if a sprinkler or pipe is broken, a desirable feature for areas that have easily damaged contents.
- *Deluge systems:* These systems have all sprinklers open at all times. When heat from a fire actuates the fire-detecting device, the deluge valve opens and water flows to, and is discharged from, all sprinklers on the piping system, thus deluging the protected areas, usually flammable liquid operations **Figure 6-19**.

Figure 6-18 A dry pipe sprinkler valve at a fire training facility. The gauge on the top measures air pressure while the bottom gauge measures water pressure.

Courtesy of Glenn Corbett.

Figure 6-19 Water flows from all of the sprinklers in a deluge system as soon as the system is activated.

Courtesy of Tyco Fire and Building Products.

Design and Operation of Sprinkler Systems

Today, sprinkler systems are hydraulically designed, meaning that the pipe sizes selected are based upon the flow and pressure necessary to supply the sprinklers. With the exception of deluge systems, sprinkler systems are designed with only a certain number of sprinklers operating, not all sprinklers in the system. (The design would anticipate anywhere from 4 sprinklers on some residential systems to upwards of 50 sprinklers on some commercial building systems.) Each sprinkler is designed to flow a specific amount of water at a specific pressure to achieve a **density** of water (gallons per minute [gpm] per square foot [ft²]) over a specific area of coverage (sprinkler operating area). Systems are designed to have an adequate water supply to meet this **demand area** (the anticipated overall size of the fire with all sprinklers operating).

The demand is based on the type of hazard presented. NFPA 13 has developed a set of occupancy classifications: light hazard, ordinary hazard group 1, ordinary hazard group 2, extra hazard group 1, and extra hazard group 2. See Appendix A for a list of examples. In general, the more significant the fire threat (e.g., museum versus flammable liquid spraying), the higher the density (i.e., the higher the flow and amount of water discharge).

It must be noted that although a system could have had an adequate water supply at the time of design and installation, the water supply may have deteriorated over time or the contents could have changed. As an area grows and develops, the demands on the public water supply could leave a sprinkler system with an inadequate flow or pressure. The change in building contents can cause the sprinkler system to be inadequate for the contents being protected.

Prewetting of exposed contents is an essential part of ordinary sprinkler operation. Sprinkler systems first slow, and then stop, the advancement of the fire through prewetting. Thus, the ability of the water to penetrate the material is key to sprinkler system effectiveness. Many palletized materials, however, are shrink-wrapped in an envelope of tight plastic. Pallets that have shrink-wrap on all sides including the top are considered fully encapsulated and are problematic to protect because sprinklers cannot prewet the material. In some cases, this has resulted in a failed sprinkler attack.

Sprinkler Installation Incentives

The two model building and fire codes provide different incentives for the installation of sprinklers. It has been estimated that there are over 200 incentives (reductions of the requirements) in the *International Building Code*.

Some who are opposed to these incentives or concessions argue that "they are giving away the store" to get sprinklers installed. Indeed, some fire protection professionals believe that a lot of fire protection eggs are being put into the one basket. The question is, "Who, if anyone, is watching the basket?" This question is discussed in detail later in this chapter. First, you will learn about the code-related, site development, and tax or insurance incentives that make sprinkler installation attractive.

Detailed information and provisions of specific codes are available from the National Fire Sprinkler Association. Typical code incentives or concessions generally are available in the following areas:

- Heights and areas
- Construction of corridors and tenant separations
- Interior wall, ceiling, and floor finishes
- Travel distances to exits
- Exit widths
- Standpipe requirements, hose station, and water flow
- Fire detection systems
- Draftstopping in attic spaces

Site development incentives:

- Fewer fire hydrants with greater spacing
- Reduced fire flow, small supply pipe
- Increased allowable distance from public access way
- Street width reduction
- Cul-de-sac allowances

Tax or insurance incentives:

- Elimination of value of sprinkler system from assessed valuation
- Property tax rebates
- Elimination of water department fees
- Insurance premium reductions

Opposition to Sprinklers

Not everyone is enthusiastic about the spread of requirements for automatic sprinklers. The concrete and gypsum industries and the manufacturers of spray-on fireproofing or other forms of protective encasement of steel beams object to trading off passive fire protection for sprinklers, citing the importance of protecting the structure of the building, the imperfection of mechanical systems, and the possibility of explosion, arson, or earthquake. In their view, sprinklers are a secondary defense. Another chief argument involves the loss of compartmentation. Compartmentation alone is not the solution to proper fire protection, however.

Furthermore, there is still much opposition to any requirement for retroactive installation of sprinklers. Although much of the opposition is financial, the specious argument that such requirements are unconstitutional has found some favor. Despite this, in the wake of the collapse of the World Trade Center in 2001, New York City passed legislation requiring sprinklers in existing unsprinklered high-rise office buildings. Sprinklering existing buildings can also help to bring existing buildings into compliance with current building codes.

Many opponents claim that sprinklers are ugly. However, there is no evidence that anyone ever refused to enter a building because it was unattractively sprinklered. Much of the cost, particularly of a retroactive installation, is caused by hiding the sprinkler system. If the argument of overall sprinkler cost is an issue, the fire department should be prepared with the cost of a bare bones system, and point out that aesthetic costs are the option of the owner, not a fire protection requirement.

The probability of a serious fire in any given office building or other building with many occupants is extremely low. It is also a fact, however, that in the typical unsprinklered glass-enclosed office building with interior stairways and a substantial fire load, the consequences of a serious fire during working hours could be very severe, with multiple fatalities.

Those who do not wish to provide the sprinkler protection often argue from "good experiences." In effect, they "need" fatalities to be convinced. They are unconvinced by rational arguments. Unfortunately, this is the argument often accepted by many public officials.

The charge is sometimes made that there is little experience with sprinklers in high-rise buildings. After the terrible Triangle Shirtwaist fire in New York City in 1911, the law was changed to require that all factories more than six stories in height be sprinklered. In the greatest migration of an industry in history, the entire garment industry moved into high-rise sprinklered loft buildings (rented factory space) in midtown Manhattan in the 1920s and 1930s. From these buildings, a large body of experience on successful sprinkler operations has been accumulated.

After the spectacular First Interstate Bank fire in Los Angeles in 1988, passage of a sprinkler law for all new and existing high-rise buildings in California seemed assured. However, this was not the case. All buildings built after 1974 and over 75 feet high (22.8 m) throughout the state of California must be sprinklered. In Los Angeles, all pre-1974 high-rises must be sprinklered. A law that would have required a statewide retrofit failed, however, according to one source, the rejection of this measure was due to the large number of state-owned buildings that would have required sprinklers.

Architects who know the codes can often circumvent sprinkler requirements. Because of the loss of fire fighters' lives in cellar fires, some cities have required sprinklers for basements in excess of a certain size, usually 2,500 square feet (232 m²). One of the solutions is to cut the basement into sections with fire walls, so that no section is larger than the legal limit. It is likely that the fire doors will not function in such buildings, so the result can be a much more difficult and dangerous fire.

Another way to circumvent the regulations is to keep the building barely below the height at which sprinklers are required. Although the building was sprinklered, one of the authors of this text (Corbett)

encountered a nine-story building that did not meet the definition of a high-rise because it was 74.5 feet (22.7 m) in height!

Much opposition to sprinklers develops from a lack of understanding of the threat of fire and the operation of sprinklers. In Evanston, Illinois, the Byer Museum was destroyed by fire in 1985. The unsprinklered Dallas Biblical Arts Museum was destroyed in a 2005 fire.

The National Gallery of Art in Washington, D.C., had a policy of not lending its art to sprinklered museums. The Smithsonian Institution was actually asked to return some works on permanent loan when National Gallery officials learned that the Smithsonian was retrofitting its facilities with sprinklers. The Chief of the Fire Protection Division of the Smithsonian and a museum director met with National Gallery curators. As a result, the National Gallery now will lend art to sprinklered museums. There are still a few museums, however, that will not.

In 1990, in Chicago, a $75 million loss occurred in two buildings being renovated that housed nine art galleries. The loss of artwork amounted to $50 million. Fire doors had been removed from connecting openings. The building in which the fire started had been sprinklered, but the system was out of service pending the installation of a new system.

The valuable exhibits in one Canadian museum are protected by sprinklers, but for some reason the lounge and gift shop are not. Fire in the contents of the latter areas could generate enough smoke to seriously damage the exhibit area. The heat from a fire in the lounge and gift shop might set off sprinklers, yet the water would not be hitting any fire. Such "economy" or "selective placement" of sprinklers is "penny wise and pound foolish."

Perhaps the most visible opposition to sprinklers has been over the recent national model code requirement for sprinklers in one- and two-family homes. While sprinklers have been installed in many cities in the western United States for decades, the new regulations adopted in the rest of the country have been met with stiff opposition from home builders at the local and state levels of government. In some cases, home builders have been able to block sprinkler legislation or have enacted legislation repealed. The builders' primary complaint? The systems cost too much and will put home affordability out of the reach of many prospective homeowners.

Popular Misconceptions About Sprinklers

Fears arise out of some of the many misconceptions about sprinklers.

The Sprinkler System Will Discharge on Even a Trifling Fire

This is not true. In fact, it takes a sizable fire to activate a conventional sprinkler. In some cases, the sprinklers should be supplemented with other fire protection measures. If the fuel for the triggering fire happens to be valuable artwork, a serious loss may take place in the first few minutes. For example, valuable original prints of birds by John James Audubon are displayed in a sprinklered wooden house in Key West, Florida—but the amount of fire needed to trip the sprinklers already would have caused extensive damage to the collection by the time the sprinklers came on.

The Entire Building Will Be Drowned When the Sprinklers Go Off

In ordinary systems, the sprinklers go off one at a time, not all at once. Only in deluge systems do all the sprinklers discharge at once. Television perpetuates this myth. Only sprinklers on which heat is impinging will activate.

Water Does More Damage Than Fire

This is not true. There is no "water damage" at a total loss. Wet materials ranging from books to computers have been successfully salvaged. In contrast, heat-damaged or burned material is destroyed. In 1960, the fleet aircraft carrier USS *Constellation* suffered a devastating fire. Many computers that had been soaked with salt water and subjected to freezing for months were recovered; burned and heat-damaged computers were not. The major damage in many of today's fires is caused by the greasy, often corrosive smoke from burning plastic.

It is often recommended that laboratories be equipped with sprinklers. Some laboratory scientists have a great fear of water damage and endorse CO extinguishing systems as a panacea. Opposition to sprinklers can be overcome by pointing out that sprinklers provide academic freedom—the opportunity to attempt something dangerous without causing a disaster.

A prominent fire protection engineering firm designed a sprinkler system for a nuclear laboratory with 1/4" (6.4 mm) sprinklers (strainers had to be provided in the lines) to satisfy the fears of scientists about water damage. The water discharge was calculated to be just enough for the existing fire load. No controls existed to prevent the typical buildup of fire load, however. Such a system could be characterized as a "tinkler" system, designed only to annoy the fire.

This is not the case. At times the efficiency of sprinklers is deceiving. Consider the operation of one sprinkler that extinguishes a fire. The amount of burned material is relatively small. If the sprinkler flowed 15 minutes—a reasonable amount of time—about 500 gallons (1893 L) of water would be dumped. The observer concerned about water damage has no idea of what might have happened had the fire continued to burn.

The Pipes Might Leak

This is unlikely. Sprinkler piping is hydrostatically tested after installation for 2 hours at 200 psi (pounds per square inch)—much more extensive testing than domestic piping is subjected to. Sprinkler leakage losses are minor and are often due to careless use of lift trucks or freezing. If sprinklers are undesirable due to fear of possible leakage, then logically *all* plumbing should be eliminated. A properly installed sprinkler system is supervised, and an alarm will sound and the fire department will respond when water flows. No such alarm is provided on domestic piping.

Smoke Is the Big Killer So Smoke Detectors Are Better Than Sprinklers

A smoke detector does nothing to suppress the fire that is generating the toxic smoke; instead, the sprinklers suppress the fire and the production of smoke. However, sprinklers are slow to operate on a smoldering fire that is generating toxic gas. Smoke detectors are necessary for life safety and provide early warning, especially in sleeping occupancies, even though the sprinklers will control the fire as soon as it breaks out.

There have been tests of smoldering fires in which temperatures generated by the fire were too low for sprinkler activation, yet lethal concentrations of CO gas were produced. The concentrations developed over several hours, but potential victims in such situations may be disabled or disoriented before a fatal dose is received.

The NFPA has no record of a multiple-death fire (i.e., a fire that kills three or more people) in a completely sprinklered building where the system was properly operating. The only exception to this record exists in an explosion or flash fire or where fire fighters or plant employees were killed during fire suppression operations.

We Have Smoke Detectors and the Fire Department Is Right Down the Block

The fire department may be working at another incident. Patrick E. Phillips, chairperson of more than one NFPA fire alarm committee, has repeatedly warned that fire protection and fire detection are not synonymous. Only if the fire alarm is received where there is someone ready, willing, and able to control the fire could detection be considered to be protection. Many fires spread so fast that even immediate detection and response are ineffective. The myth that there are 15 minutes available to control a fire should be laid to rest. Consider this sad repudiation of this myth: a full–box alarm assignment of the Boston Fire Department was only a short distance away from the Cocoanut Grove Night Club where 492 people died in 1942.

Sprinklers Cause Damage to Libraries

Some librarians are opposed to sprinklers, citing a number of arguments against their use: the water will wash the classification labels off the backs of the books, water will damage the books, books are hard to burn, or we've never had a fire before.

For firefighting purposes, the major concern is not the local branch library, but rather the main library with its extensive and often irreplaceable collections.

The 1986 Los Angeles City Library fire is a case in point Figure 6-20 . This massive concrete structure was built in 1926. A concrete fort enclosed book stacks and racks for book storage. Upward ventilation was provided throughout the stacks to prevent mildew. It would be hard to imagine a better built-to-burn design. Not only did the design extend the fire, but the open construction permitted water to flow down to books below. The arsonist ignited the fire at the top of the stacks. The manual firefighting effort required 1,250,000 gallons of water. The cost of the fire loss was about $20 million. By contrast, the initial fire could have been controlled or extinguished by operating no more than two or three sprinklers.

The solution is full automatic sprinkler protection. Unfortunately, library stacks are perfectly arranged for spreading fire. They are simply racks

Figure 6-20 The Los Angeles City Library fire of 1986.

© Ben Martin/Time & Life Pictures/Getty Images.

storing combustible material without fire separation between levels. Ventilation is almost impossible—spaces are very constricted and passages are tight and tortuous. The heat produced in such a fire is extreme.

A number of municipal libraries are now sprinklered. An architect planning a new library asked the then Assistant Director of the Salt Lake City Library (Frank Brannigan's daughter Eileen), "What do you like most about your library?" Her reply shocked him, "It's sprinklered." She went on to explain that libraries are public buildings, making libraries vulnerable to arson. With the sprinklers in place, a disturbed person with a grievance can't destroy the library. As much as 85% of recent library fires were due to arson.

Smoke Detectors Set Off All the Sprinklers

A television show showed a cigarette setting off a smoke alarm, which in turn set off all the sprinklers in a building. The network has reportedly defended such misrepresentations by claiming "creative license." If someone advances this as a fact, it might be useful to mention that it was probably seen on a fictional TV show and is therefore fiction.

In addition, all fire department personnel should be alerted to the potentially serious damage caused by comments to the media made by uninformed occupants or careless remarks by fire fighters. The fireground commander should make it a point to explain to reporters the success of the sprinkler operation, if that is the case. Providing accurate information about sprinklers should go a long way toward dispelling the myths surrounding them and enhancing their acceptance.

Fire Service Activities to Correct Erroneous Opinions About Sprinklers

In years past, the sprinkler industry did little or nothing to overcome negative public perceptions. Promotion of the use of sprinklers was left to the public fire protection forces and the insurance industry. The Home Fire Sprinkler Coalition has worked to increase the knowledge to all sectors related to residential sprinkler systems.

Sprinkler Demonstrations

Some simple demonstrations have been useful in correcting erroneous opinions. Many mobile sprinkler demonstrators have been built to show the lifesaving efficiency of quick-response sprinklers. Conversely, some technicians in laboratory and computer occupancies are concerned that sprinklers will go off in a minor fire. To demonstrate the amount of fire it takes for a standard sprinkler to operate, consider the following demonstrations.

Attach a sprinkler to a length of steel pipe fed by a garden hose. Build a sizeable bonfire of fast-burning fuel. Hold the sprinkler over the fire. The sprinkler will be extremely slow to activate. Many who feared that the sprinkler would operate on a trifling fire will be amazed and converted.

Sprinklers and Flammable Liquids

Many people believe that sprinklers will be totally ineffective on flammable liquids. At the Norfolk Naval Base, a demonstration was used to show that sprinklers would control the fire and make it possible for employees to get close enough for a fire extinguisher attack.

Fire Service Misconceptions About Sprinklers

The Building Is Sprinklered; There Is No Problem

Sprinklers are not a universal remedy for all fire problems. Inspectors have noted the presence of ceilings of burlap and other high-flame spread materials—a hazard that should not be dismissed lightly because "it's sprinklered." It is very doubtful that the sprinklers would operate in time to control the flame spread. Put simply, if a life hazard is so severe that it cannot be controlled by sprinklers, then the hazard should not be permitted to exist or other protection measures must be in place.

Supplying the Fire Department Connection Is a Secondary Operation

The sprinkler system should be backed up by fire fighters as soon as possible after the responders' arrival.

The more water a sprinkler discharges, the more heat that is absorbed. The supply pressure should be the maximum safe pressure because the increased density of the water pattern may be crucial. Some sprinkler systems have a higher working pressure than others. Preplans should include a statement as to the pressure to be supplied and maintained. The 1991 fire in the Meridian Building in Philadelphia was stopped by seven sprinklers on the 30th floor. It was estimated that the discharge from these sprinklers was three to four times the normal amount due to the pressure supplied by the fire department. It is possible that the sprinklers could have stopped the fierce fire attack if they were operating at only normal pressure.

Sprinklers Should Be Shut Down As Soon As Possible to Prevent Excessive Water Damage or to Clear the Air

Factory Mutual statistics show 23 fires in 10 years with losses of $43 million (1986 dollars) in which premature closing of valves by fire fighters was a major factor in exacerbating the damage. When a sprinkler system is controlling a fire, cooled gases are often driven downward and obscure vision. Sprinklers should not be shut off as long as hot water is falling down. All visible fire should be extinguished. The fire fighter who shuts a valve should be in full personal protective clothing with a radio and should remain at the valve so it can be opened instantly if needed.

Residential Sprinkler Systems Are the Same as Other Sprinkler Systems

Residential sprinklers, as defined by NFPA 13R, *Standard for the Installation of Sprinkler Systems in Low-Rise Residential Occupancies*, and NFPA 13D, *Standard for the Installation of Sprinkler Systems in One- and Two-Family Dwellings and Manufactured Homes*, are not the same as sprinkler systems used in other structures. Residential sprinkler systems are intended to limit flashover and to hold a contents fire until the occupants can escape. There have been a significant number of successes with these systems **Figure 6-21**.

Residential structures, however, are only partially sprinklered under these standards. To make the

Figure 6-21 A residential sprinkler.
© Justin Kral/ShutterStock, Inc.

systems affordable while maintaining life safety, sprinklers are omitted from certain areas in residences. This is particularly significant when a residential sprinkler system is used in a multiple-family dwelling with truss floors or in a building with interconnected voids such as in rehabilitated buildings with lowered ceilings.

Some object to the term *partial sprinkler system*. They argue that this term should be applied to systems where sprinklers are omitted from areas vulnerable to fire. NFPA 13R and 13D systems are based on data that indicates the omitted areas are not vulnerable to fire. Unsprinklered areas include small closets and bathrooms as well as unused attics.

It is possible that data relevant to the combustible truss void that exists in truss floor buildings was not available at the time when NFPA 13R and NFPA 13D were developed; this lack of evidence in the database might explain the omission of this concern from the standards. It is a fact that fires originate in void spaces that would be unprotected by sprinklers. Electrical fires and failures of metal fireplace flues are but two possible causes of fires. Years ago, a number of the fires investigated at the NBS (today known as NIST) started in or very early extended into void spaces, which would be unsprinklered under NFPA Standards 13D and 13R.

The following few cases are representative of other fires that started or developed in areas that were unsprinklered or would be unsprinklered—and thus unaffected by sprinklers—under the standards cited:

- *Williamsburg, Virginia:* A "sprinklered" renovated dormitory at the College of William and Mary was destroyed by a void fire.
- *Syracuse, New York:* Although not a "void" fire, four fire fighters lost their lives in a fire in a fraternity house that had been equipped with a corridor sprinkler system for life safety.
- *De Kalb County, Georgia:* Fire resulting from a plumber's propane torch developed in the truss void of a nearly completed three-story apartment house with no contents load. The sprinkler system was operational, but the sprinklers didn't fuse. A fire fighter fell through the second floor into the fire when the truss collapsed. Fortunately, his partner used a hose line to protect him from the fire until he could be pulled out.
- *Orlando, Florida:* Fire started in a defective ceiling light fixture in a first-floor apartment of an unsprinklered three-story, brick-veneered, plywood-sheathed, truss-floor apartment house. The fire extended to upper floors on the outside of the wood sheathing, concealed by the brick veneer. The truss floor of the third floor sagged more than 1 foot.
- *San Antonio, Texas:* A spilled flammable liquid ran down into the truss void of an unsprinklered apartment house and was accidentally ignited. Fire fighters had barely complied with orders to get out when the floors collapsed.
- *Lake Benbrook, Texas:* An outside rubbish fire extended through the exterior wall into the truss void of an unsprinklered townhouse.

In plain language, the sprinkler systems in many buildings are partial sprinkler systems. This labeling is not meant to denigrate the usefulness of sprinklers in saving the lives of occupants and protecting property. As stated, there have been many successful sprinkler operations. Balancing the potential structural failures with the positive life safety benefits, Frank Brannigan was strongly in favor of the partial systems—in fact, his house was partially sprinklered

in 1962. However, fire departments should not expect to receive the level of structural protection performance from residential sprinkler systems that they are accustomed to receiving from standard sprinkler systems.

The Fire Department and Sprinklers

In 1898, the International Association of Fire Chiefs resolved to have nothing to do with the practice of pumping into the newfangled invention called the automatic sprinkler. Today, the situation is entirely different: fire chiefs are enthusiastic supporters of automatic sprinklers. Some fire departments, however, seem to act as if "sprinklers are none of their business."

In the past, some insurance companies exhibited exactly the reverse attitude. That is, until high-rack warehouses presented a new problem, insurance companies, relying heavily on sprinkler protection, seemed to operate as if the fire department did not exist. For instance, Norman Thompson (then Director of Research for Factory Mutual) wrote a very important book, *Automatic Sprinklers*, published in 1951 by the NFPA, which led to the adoption of the then-new spray-type "standard sprinkler." The book makes no reference to the fire department supporting the sprinkler system. In a list of 20 things to do when fire occurs in the plant, the last item was "call the fire department."

There is no doubt that a properly designed, adequately supplied and properly inspected, tested, and maintained sprinkler system can almost guarantee that a fire will not destroy a building and will save lives. To "sell" sprinkler protection, however, concessions may be made with respect to other fire protection features, such as exit travel distances, flammability of surface finishes, size of fire areas, protection of void spaces, and other time-proven fire protection and life safety features.

It was noted earlier that almost all fire protection emphasis today is being placed on sprinklers, with the responsibility for them left with fire departments. The buildings department remains out of the picture after the Certificate of Occupancy is issued. The fire department or fire prevention bureau then assumes the responsibility to ensure the system is operating properly after the building is built. The insurance company often compensates for its risks by raising the premiums or refusing to renew the risk.

Because sprinkler protection is so important, the fire department's responsibility cannot be permitted to end when the sprinklers are installed and approved. Fire fighters making fire suppression, pre-planning, or fire prevention inspections should be acutely aware of the conditions that can cause sprinkler failure.

The fire department should take action when notified of a sprinkler system that is out of service or has other impairments. As part of monitoring its coverage area, it should keep up-to-date on systems being installed or that are temporarily out of service. When a sprinkler system is pronounced "out of service," the degree of impairment should be given. For example:

- The system can be supplied through fire department connections (FDC).
- Mechanics on scene can restore the system in 10 minutes.
- The system is completely unavailable.
- The second floor is out of service.

In Cartaret, New Jersey, a fire in the S&A Plastics plant severely exposed a Unocal storage facility containing hundreds of drums of flammable solvents. This building, in turn, exposed a number of tanks. Fire Superintendent Richard Greenburg had reviewed the plans for a sprinkler system being installed in the drum storage building. He had the system supplied through the FDC and, assisted by a Unocal employee, opened the valves. Fourteen sprinklers checked the spread of the fire. Five sprinklers had opened before the water was turned on, demonstrating that the potential for a disastrous extension was real.

Why Were the Sprinklers Installed?

As you have seen, there are several reasons for installing sprinklers. The nature of the reason may help to determine fire department action concerning impairments. The model building and fire codes

permit a number of trade-offs (often resulting in cost reductions) when sprinklers are installed. For instance, there can be a greater travel distance to exits and narrower exit components than what is required in an unsprinklered building. This provides substantial savings in the building of a shopping mall, for example.

What action should the fire department take when it learns that the sprinkler system is out of service in a department store, a shopping mall, or any structure with a population of thousands? The life safety of all the occupants depends almost totally on adequate sprinkler protection. Use the fire code provisions that allow for the implementation of a fire watch.

There have been several fires in unsprinklered department stores with a heavy loss of life. Over 100 Christmas shoppers and employees died in the Taiyo Department Store in Kumamoto, Japan, in 1973. Sprinklers were being installed in the new addition of the store that was under construction, but the law did not apply to the existing structure.

The fire department action should be the same as if a large proportion of the exits were blocked. In such a case, the fire department would have no hesitation in shutting down the store until the condition was corrected. This doesn't mean that many stores would have to be shut down. When the alternatives are properly explained to management before the crisis arises, the work can be done while the store is closed.

A fire department that does not accept the scenario, or doesn't plan to take any action when sprinklers installed for life safety are shut down, would be well advised to send the following inquiry to the city attorney:

Dear Counselor:

Consider a building in which building code modifications, particularly reduction in egress requirements, were permitted because the building was sprinklered. What is our legal position if we were notified that the sprinklers were turned off while the building was occupied, no special action was taken in such a case, a loss of life occurred during a fire, or a key factor in loss of life was that the victims could not get out in time?

In addition to legal action against the city, is it reasonable to expect that senior officers could be sued successfully?

Sincerely,

Fire Chief

The legal situation may vary depending upon why the system was installed, so the fire department should identify the reason why each sprinkler system was installed. If the sprinkler system was installed for life safety or in accordance with another code provision, there still may be a legal responsibility to take action if the sprinklers are out of service. Even if there were no legal requirement for the installation of sprinklers, the loss of the sprinklers may create such a serious situation that the fire department can act under general legal provisions authorizing the abatement of serious hazards.

All too often, an old building of ordinary construction with sprinklers shut down or removed could produce a disastrous fire that destroys or threatens to destroy adjacent structures. For example, in 1983, a spectacular fire in Donaldson's Department Store did $90 million damage to the adjacent fire-resistive high-rise of the Northwestern State Bank in Minneapolis, Minnesota.

Fire Department Policy

The fire department should have a formal policy on the subject of sprinkler systems out of service. The policy should be reviewed and approved by the governing authority, making it a legal requirement. The policy should cover items such as the following:

- *Fire department notification:* It may be necessary to seek the cooperation of other licensing authorities to place the burden of notifying the fire department directly on the sprinkler contractor

or building owner. This does not exclude required notification by the owner or tenant.

- *Formal legal action:* Situations may arise where the fire department should take legal action, such as shutting down a department store or requiring a fire watch.
- *Informal action*: Situations may arise where informal actions are taken, such as modifying prefire plans, providing emergency water supply, working with management to establish fire watches, or shutting down hazardous processes.
- Authority to modify requirements, accepting alternative methods and/or materials that are equal or superior in performance.
- Formal personnel instruction.

Fire Department Instruction

A program of fire department education regarding impairment of sprinkler systems should include training on the following topics:

- A basic knowledge of sprinkler protection, including types of systems and the adequacy and reliability of the water supply
- An understanding of the different reasons why sprinklers are installed
- Fire department policy on procedure and practices involving sprinkler systems
- Detailed knowledge of situations that decrease or destroy the efficiency of sprinklers

Company-Level Inspections

Sprinkler impairment can be divided into water supply and water distribution problems. The design and adequacy of the water supply system is beyond the scope of this chapter. It is assumed here that the designed supply is adequate. At times, however, this assumption is not valid. An old wooden hotel in Alaska had been sprinklered, down to a sprinklered shower stall. The unconnected sprinkler riser brought the revelation that the contractor never finished the job. In addition, the Siamese connection was located so that a supply line would make it impossible to open the exit door.

Many sprinkler systems in low-rise buildings depend solely on the city water main pressure to provide adequate sprinkler flow. In many areas of the country, continuing drought has forced water supply utilities to reduce pressure to conserve water. The pressure at upper stories of low-rise buildings may be inadequate for sprinkler discharge. In this case, the owner should be required to install a pump to provide adequate pressure.

The fundamental purpose of a sprinkler system is to hit the incipient fire with enough water to control it. There are too many ways for failure to occur to cover here, but a few examples should get fire fighters looking more critically at sprinklered buildings and make routine inspections more productive.

Making the following determinations is well within the capability of adequately trained fire suppression personnel performing an inspection. Some can be determined visually; others require answers from management or maintenance personnel. Any questions should be referred to the designated fire department authority for resolution. All deficiencies should be formally documented, but the system should provide for immediate verbal reporting of serious problems. Fire department policy should direct the action to be taken in various circumstances, including notification of the fire prevention bureau.

Outside the Building
Check for several items:

- Are flow tests performed regularly to assure that the designed supply is actually available?
- Are static supply elements, such as overhead or surface tanks, or pond intake, in good condition and protected from freezing where necessary? If the system depends on pumps, is there a check on maintenance and test procedures? Is the power supply vulnerable to a fire? Is the pump on and in the automatic position?
- If the power company cuts power to the building, will the pumps fail? Are emergency

power supplies adequate? Is the emergency system tested regularly? Where a pump is installed, a manifold for testing the fire pump is usually provided. Each 2 ½" (64 mm) outlet indicates 250 gpm (946 L) of pump capacity.

- Are fire department Siamese connections available **Figure 6-22**? In earlier years, they were optional. In many industrial plants, fire department connections were omitted from individual buildings. As a result, when a fire main is placed out of service, the sprinklers served by that main are out of service. If there is no FDC, it is not possible to use hose lines to supply the system. All systems should have FDC connections whether or not required by code. If there is a possibility of pumping contaminated water into the sprinkler system, the check valves that prevent backflow into the domestic system should be checked regularly.
- Are FDC properly protected from damage? All openings should be capped. When caps are missing, an open pipe is a target for debris that might cut off the flow of water. The heavy brass caps popular years ago frequently were removed. Use cast iron or plastic caps to protect today's FDC.
- Are all FDC labeled to indicate the type of service and area covered? The piping from the FDC to the system is dry and thus unsupervised.

Nothing is apparent if it is broken. In modern construction, connections are often quite distant from the building, under driveways and parking lots and subject to traffic pressure and corrosion. Outside valves are most often post indicator valves (PIV), which indicate open or shut **Figure 6-23**. At a minimum, the valves should be locked as a precaution. The best system includes an electrical central tampering alarm. Some PIVs are located horizontally in the wall of the building. Sometimes the valve is located in a curb box on the sidewalk or in the street. Often this box is difficult to locate. It may even be paved over or fire apparatus may be parked on it. A sign should be posted on the building giving distances and directions to help locate the valve box.

Inside the Building

Sprinkler valves are made to indicate their position. If the stem is protruding, the valve is open; hence, this type of device is called an OS&Y (outside, screw, and yoke) valve **Figure 6-24**. Are the valves open? In the best system, electrical tamper switches transmit a signal if the valve is turned. At the very least, valves should be

Figure 6-23 Post indicator valves are usually found outside of a building and control an underground valve assembly.
© Jones and Bartlett Learning. Courtesy of MIEMSS.

Figure 6-22 Fire department connection.
© Jones and Bartlett Learning. Photographed by Glen E. Ellman.

Figure 6-24 Two OS and Y valves with tamper switches. The valve on the left is closed (stem concealed) and the valve on the right is open (stem is fully out).

Courtesy of Ralph G. Johnson (nyail.com/fsd).

chained and locked to prevent tampering with them. The sprinkler valves are often hidden in locked closets. In some malls, the sprinkler valve for one store is located in another store. Signs should indicate the location, and it should be entered on the preplan.

Items to note on the inside of the building include the following:

- Is there a system for reporting any valves closed for maintenance? The system should assure that all closed valves are reopened. In a huge department store, 29 sprinkler valves were closed and 28 were open. A fire fighter fatality resulted when a fire occurred in the area covered by the closed valves.
- The details of the system for controlling shut valves and the special precautions taken when valves are shut are very much the fire department's business. This is especially the case if the sprinklers were installed by law or if exemptions were granted with respect to other code requirements.
- Is there a potential for freezing of all or portions of the sprinkler system? Are adequate precautions taken?

- Are old sprinkler systems or inadequately designed systems in use?

Dry pipe systems that have "gone wet," either for fires or by accident, are often found to be plugged by rust particles/scale carried by the inrushing water. A record should be kept of these occurrences and they should be reported to a designated authority for investigation. An accidentally discharged dry pipe system can be detected by the presence of high pressure, indicated by a pinned needle on the "air" pressure gauge on the dry pipe valve.

Often, industrial buildings and warehouses are built on speculation; the builder sometimes puts in a minimum sprinkler system to make the building attractive to tenants. The fire load introduced by the tenant may be far beyond the capacity of the system. Although specific analysis is beyond the company inspection level, some fire loads, such as boxed records, foamed plastics, aerosols, or rubber tires, are easily recognizable as being beyond the capacity of the typical sprinkler system.

One of this book's authors (Corbett) experienced the problem of undersized speculative warehouse building system designs. Any such problem must be corrected before a tenant moves in. The solution to this problem is to document the method of storage and the materials stored through the use of a "commodity letter," developed by the San Antonio Fire Department. The letter documents what is being stored in the building, how it being stored (racks, solid piles, shelves), aisle widths, and other pertinent information.

Many shopping malls are only partially sprinklered. Often a structure was built as an open mall and the stores were unsprinklered. Later, when the mall was enclosed, sprinklers were provided in the open area but not in the individual stores. In other cases, major department stores that anchor the mall are sprinklered while other stores are not. In one case, a fire in an unsprinklered central mall area broke windows in a department store and supermarket. More than 200 sprinklers were triggered in those two stores, which had to be closed for 2 months following the incident. Damage was $7 million.

Mall investors often want to keep their costs down and sometimes only provide the sprinkler main, leaving each store owner to install sprinklers. A serious hazard exists when a store under construction is left unsprinklered. All vacant stores in a mall should be required to be sprinklered because they are often rented temporarily to other occupants for storage.

The construction of a high-rise building represents a tremendous investment. The owner wants to start using or renting available space as soon as possible. This leads to a dangerous situation of construction work, with all its hazards, being carried on in an occupied building. Even if the building is sprinklered, often the most hazardous floors—those under construction—are not sprinklered because the sprinklers go in after the ceiling is in place.

The buildings department should be alerted to this hazard, and insist that the sprinklers be in service throughout the building before any partial occupancy is permitted. If the buildings department is complacent, the fire department should take every means possible to mitigate the hazard. Prospective tenants, who are relying on the sprinkler system, should be put on notice as to the hazard.

There seems to be great interest in preserving historic buildings. Often these buildings are just minimally restored. Sometimes the buildings department requires sprinklers in these structures. Often the sprinklers just cover the accessible areas and do not cover the voids. This may be a deliberate cost-saving decision. The sprinkler system is needed for life safety and property protection, and the sprinklering of the voids would make the cost prohibitive. If this is the case and the fire starts in or penetrates the void, the building may be destroyed. As noted earlier, a "fully sprinklered" renovated dormitory at the College of William and Mary was destroyed because of a void fire.

In some historic buildings, sprinklers are installed in the nonpublic areas and omitted in the public areas in deference to "historic authenticity." This is called selective placement of sprinklers. While doing restoration work, an architect recently recommended sprinklers be installed in the attic and the basement of the Old State House in Boston—but not on the intervening floors because the sprinklers were not "architecturally compatible." However, providing adequate fire protection without compromising historic values has often been accomplished.

The historic 187-year-old John C. Calhoun Mansion, called Fort Hill, is maintained on the campus of Clemson University in South Carolina. A museum of valuable art, furnishings, and historic memorabilia from the 1700s and 1800s, it has been sprinklered since 1966. In May 1988, an arsonist set a fire inside Fort Hill, but the sprinklers extinguished it with moderate damage, and the historic structure still stands. Two weeks before the fire, a visiting historic preservation conservator referred to the visible, nonauthentic sprinkler system as an embarrassment. He recommended a fixed Halon deluge system. The university wisely disregarded his recommendation.

McGill University in Montréal housed two important research units in an old mansion called the Lady Meredith House. There were no sprinklers in this structure. A basement fire spread up through the unfirestopped walls and open staircases, necessitating five alarms. Many research papers were lost. Fire officers trying to convince a director of such an enterprise to spend scarce research money on fire protection might point out that computers often are a target of thieves, and fires are often set to conceal the theft.

Are all areas of the building "seen" by sprinklers? Unsprinklered voids often exist. An old office building in Memphis, Tennessee, has been extensively modified over the years, including the installation of a light-hazard sprinkler system. The sprinklers were obstructed by air ducts, however, and the system was not properly maintained. When a fire originated in a void space, it took scores of fire fighters 3 hours to literally "dig" the fire out of the walls, cocklofts, and other concealed spaces.

The sprinkler protection of a dinner theater was impaired by the construction of a mezzanine without sprinklers underneath it. A fire occurred underneath the mezzanine and gained headway until ceiling sprinklers operated and sounded the alarm. The loss was much greater than it would have been had proper sprinkler protection been installed.

In some cases, unsprinklered voids are created by a building within the building. Offices on factory floors, special storage rooms, and mezzanines are typical of the structures that can provide a place for the fire to grow unchecked, burst out, and overpower the sprinkler system. These areas should be sprinklered or their ceilings should be of lightweight plastic panels approved for such use, which melt out when exposed to fire. Plywood will not burn through in time.

In some cases, large ceiling expanses of department stores have been fitted with panels to conceal the sprinklers. This might be dangerous. Melting and falling panels might be taken by occupants to mean the structure is failing and cause panic.

A $2 million loss occurred in a Midwest shopping mall when fire extended through a combustible void above the ceiling of the sprinklered building; the roof was a conventional metal deck. Another major loss was suffered at Tinker Air Force Base in Oklahoma, when a metal deck roof fire, started by roofers, could not be controlled by sprinklers installed only below the ceiling line.

Are sprinklers uncompromised? Sprinklers should remain unpainted and untainted by other materials. Painted sprinklers should be replaced. Spray fireproofing has been applied to sprinklers. To counter the belief that sprinklers are ugly, concealed sprinklers have been developed. A flat plate, usually enameled to match the ceiling, conceals a sprinkler. It will drop out when heated. Concealed sprinkler covers should not be painted when the ceiling is painted. Paint might interfere with the cover dropping.

Exposure protection is sometimes provided by sprinklers outside the building. In cold climates, these are subject to freezing. Protecting them from freezing should not require shutting down the entire system.

In other cases, open outside sprinklers all discharge when a valve is opened manually. Such systems are very useful and can release firefighting units for other tasks. The fire department should know the location of the valve and how to get to it when the building is closed. This information should be incorporated into the prefire plan.

Sprinkler Fraud

In 1984, the fire protection community was shocked and dismayed to learn of a major case of sprinkler fraud. As part of an investigation of an arson fire at a sprinkler installation company, a sprinkler contractor in California was discovered to have installed unconnected sprinklers that were simply glued to the ceiling. The authors were told there were a number of other similar situations. Although many new trends start in California and work eastward, let us hope this does not become a trend with sprinkler fraud. In any case, those inspecting and checking sprinkler systems should look beyond the obvious.

Management of Sprinkler Systems

Management is receiving lots of attention in the fire service today. Unfortunately, much fire service management effort is concerned with managing the fire department, rather than managing the fire problem. The integrity of sprinkler protection is a big fire management problem, and fire departments would do well to get started on managing it!

It's also a good idea to make sure that you have established sprinkler system protocols in your standard operating procedures. What company will supply the fire department connection (first or second due engine)? When is it permissible to shut down the system? What about system impairments? A good place to start is NFPA 13E, *Recommended Practice for Fire Department Operations in Properties Protected by Sprinkler and Standpipe Systems.*

Special Situations Involving Flammable Liquids

Flammable liquid fires can be controlled with sprinklers, but they present special problems:

- Most flammable liquids float on water. Thus, there is the potential for flowing flaming liquid, which can spread fire and injure fire fighters.
- Flammable liquids have a high rate of heat release.

- Flammable liquid containers can result in a BLEVE (boiling liquid/expanding vapor explosion).
- Aerosol cans may act as flaming rockets.
- Runoff water may create a significant contamination problem.

The best method for controlling flammable liquid fires is to keep the containers from overheating and control the flowing fire on the floor. The containers can be kept from overheating if the sprinklers can wet every container exposed to heat. The use of conventional sprinklers on flammable liquid fires creates the problem of controlling runoff to prevent environmental problems. It appears that foam water sprinklers would substantially reduce the problem. Isolation in a flammable liquids storage room is typically required for flammable liquids. Isolation in specially designed rooms (preventing rockets from leaving the room) for large quantities of aerosols is also required. A floor fire can be aided by the use of sills and trench floor drains that are designed to draw off spills of liquids and water to designated areas.

Protection of Glass Fire Barriers With Sprinklers

The use of glass, on both the exterior and interior of buildings, for separations that may be designed to resist fire or that may accidentally serve this purpose, is increasing. In some cases, lines of sprinklers have been used to protect the glass. It is important that all the glass be wet. In the case of hot fires, the wetting must occur early to avoid thermal shock to hot glass. Window treatments, such as draperies, should not be installed between the glass and the sprinklers.

Early Suppression/Fast Response Sprinklers

Sprinklers have a number of opportunities to fail to perform properly in high-rack storage warehouses. In addition, the supply piping for in-rack sprinklers limiting the movement of the racks themselves has been a problem.

Early suppression/fast response (ESFR) sprinklers were developed not just to control the fire,

but to suppress it. This requires early discharge of a larger quantity of water—as much as 100 gpm per sprinkler in some cases. In fact, ESFR sprinklers are designed to approach the old fire fighters' slogan, "Put the wet stuff on the red stuff." Conventional sprinklers cool the ceiling and prewet surrounding storage. ESFR sprinklers dump most of the water directly on the fire.

The orifice of an ESFR sprinkler is 3/4" (19 mm) in size; conventional sprinklers have 1/2" (13 mm) openings **Figure 6-25**. ESFR system installations are new—in most cases, the existing piping of traditional systems is not adequate in size. Tests have indicated that ESFR sprinklers can protect rack storage of high-density plastics up to 25 feet (7.6 meters) in height without the use of in-rack sprinklers.

The use of ESFR sprinklers permits greater flexibility in warehouse layout because the need for in-rack sprinklers is eliminated for many commodities up to a specified height.

Preplanning Considerations

Here are some preplanwning considerations:

- A sketch of the system showing fire pump location, risers type (wet pipe, dry pipe, etc.), and location(s) (typically, a single riser can cover an

Figure 6-25 Compare the conventional sprinkler's 1/2" orifice (left) with the 3/4" orifice of an ESFR sprinkler (right).

Courtesy of Glenn Corbett.

area up to 52,000 square feet (4831 m²) for most occupancies with a limitation of 40,000 square feet (3715 m²) for high-piled stock warehouses and extra hazard occupancies), main control valve, and significant building details such as the location of fire barriers, stairwells, and corridors

- The total water supply demand of the system (flow in gpm and pressure in psi), which is printed on a sticker on the riser or on a metal card attached to the riser
- Any connections to the fire alarm or other systems (e.g., a **water flow switch**, which detects the movement of water in the systems and transmits a signal to the alarm system)
- Fire department connection location and the areas it serves
- Types of sprinklers, including any unusual characteristics
- Location of manual release for preaction and deluge systems

Some basic firefighting considerations for operating sprinkler systems are summarized here:

- Supply the FDC.
- Verify the main riser control is open (usually an OS&Y valve).
- Note the pressure on the main riser gauges (supply side and discharge side of the alarm valve/dry pipe valve/deluge valve). Is the discharge gauge moving, possibly indicating water flow? Are they appropriate readings (e.g., does it read zero on the supply and discharge gauges, indicating a closed valve in the street)?
- Ensure the fire pump (if present) is working and read gauges; manually start the fire pump if necessary.
- Do not close the valve until you are sure the fire is out; only close the valve after the fire has definitely been controlled. Leave a fire fighter at the valve with a radio to reopen the valve quickly if necessary.
- Do not replace sprinklers or reset dry pipe valves. Let the contractor handle it; post a fire watch if necessary until the system is restored.

Standpipe Systems

Standpipe systems—fixed networks of piping and hose valves (and sometimes hose)—are installed in tall and/or large buildings to provide quick fire attack. They have been in existence since at least 1870, perhaps even earlier. Today, there are three classes of systems, based upon their user and hose size. According to NFPA 14, *Standard for the Installation of Standpipe and Hose Systems*, these classes are:

- *Class I system:* A Class I standpipe system provides 2 ½" (64-mm) hose connections to supply water for use by fire departments and those trained in handling heavy fire streams.
- *Class II system:* A Class II standpipe system provides 1 ½" (38-mm) hose stations to supply water for use primarily by the building occupants or by the fire department during initial response.
- *Class III system:* A Class III standpipe system provides 1 ½" (38-mm) hose stations to supply water for use by building occupants and 2 ½" (64-mm) hose connections to supply a larger volume of water for use by fire departments and those trained in handling heavy fire streams.

These three classes also dictate the amount of water flow and pressure from a standpipe, which has varied over the years. Generally speaking, today Class I and III systems provide (or are capable of providing) 250 gpm (946 Lpm) at 100 psi at each hose valve. Years ago, the pressure requirement was only 65 psi, based upon the use of 2 ½" (64 mm) hose. The increase to 100 psi recognized the use of automatic and fog nozzles, although even this is not enough for nozzles that operate at 100 psi, given the friction loss in the hose. Class II standpipes supply 100 gpm (379 Lpm) at 65 psi. These are often called **houselines** and are intended for use by building occupants.

Standpipes can also be classified by the water supply provided for the systems:

- **Automatic-wet**: An automatic-wet standpipe system is a wet standpipe system that has a water supply that is capable of supplying the system demand automatically.

- **Semiautomatic-dry**: A semiautomatic-dry standpipe system is a dry standpipe system that is arranged through the use of a device, such as a deluge valve, to admit water into the system piping upon activation of a remote control device located at a hose connection.
- **Manual-dry**: A manual-dry standpipe system is a dry standpipe system that does not have a permanent water supply attached to the system.
- **Manual-wet**: A manual-wet standpipe system is a wet standpipe system connected to a small water supply for the purpose of maintaining water within the system or sharing a water supply with an automatic sprinkler system but not having a water supply capable of delivering the system demand attached to the system.

Standpipe Operations

The use of Class I and III standpipes by the fire service has had many years of success, particularly through the 1980s. These success stories are based upon the "classic" 2 ½" (64 mm) handline with smooth bore nozzle. The fire service then began to change, but the standpipe systems did not. The proliferation of 1 ¾" (44 mm) hose and lower staffing levels pushed fire departments to use the 1 ¾" (44 mm) attack line with automatic/fog nozzles, even for high-rise fires, while the standpipe pressures remained at 65 psi.

This all came to a head in 1991. At the tragic One Meridian Plaza high-rise fire in Philadelphia, improperly set pressure-reducing valves (PRVs) cut the available hose valve pressure to very low levels— the 1 ¾" (44 mm) hose with 100-psi nozzles the Philadelphia Fire Department were using were practically useless. The fire burned vertically unabated through the building until it was stopped by ten sprinklers at various locations that had been retrofitted. Three fire fighters were killed at the fire.

It is the opinion of the authors that high-rise fires must be fought with 2 ½" (64 mm) hose lines, not 1 ¾" (44 mm) or 1 ½" (38 mm) hose lines. In addition, smooth bores are the nozzle of choice because of their reach, penetration, and (importantly) low operating pressure.

Pressure-reducing valves, which are used to cut the high pressures found on the lower floors of tall high-rise standpipe risers, are a source of major concern for fire fighters. This is often the case when a few tall **zones** are used. (Zones are the individual riser segments, one on top of the other. In tall structures, there is an economic incentive to have fewer zones to avoid having to install more pumps.) It is important to ensure that the valves are tested frequently to ensure that they are set at the proper pressure. Waiting until a fire occurs to learn about this problem is too late— some of these valves are not even field adjustable!

Here are some standpipe preplanning considerations:

- A sketch of the system showing fire pump location, risers and hose valve locations, main control valve and riser isolation valves, as well as significant building details such as the location of fire barriers, stairwells, and corridors
- The type of water supply and flow/pressure characteristics
- Types of hose valves, including any pressure-reducing valves and their mode of operation (and how to adjust them)
- FDC and the areas/zones it serves (it is common to have more than one standpipe FDC serving different floors of a building) as well as the proper pressure needed to supply it
- Connections to a fire alarm system or other fire protection system

Basic firefighting considerations for operating a standpipe system include the following:

- Supply the FDC. If it is blocked/clogged, you may pump through the second-floor hose valve in a stairwell as a backup if the hose valve is not a pressure-reducing valve.
- Verify that the main control valve and riser isolation valve are open.
- Note the pressure on the main riser gauge at the ground/basement level to establish that the system is operating properly. (There should be a pressure gauge at the top of each riser as well.)
- Ensure the fire pump (if present) is working and read gauges; manually start the pump if necessary.

- Be wary of dry standpipe systems—they will take longer to fill/pressurize and may have been vandalized. (Watch out for hose valves that have been opened by vandals before the fire.)

Fire Alarm, Detection, and Communications Systems

Fire detection and alarm systems have been around for many years, dating back to the 1800s. Today, a myriad of detection devices and other hardware are available to alert occupants and (in some cases) the fire department that a fire has been detected. The design and installation standard is NFPA 72, *National Fire Alarm Code*.

Initiating devices (which transmit a "change of condition" signal) include the following:

- Manual fire alarm boxes (pull stations)
- Spot-type smoke detectors, which include photoelectric and ionization smoke detectors
- Line-type smoke detectors, which cover an area such as the projected beam detectors found in atria
- Duct smoke detectors
- Spot-type heat detectors (including nonrestorable fusible element detectors, fixed temperature detectors, restorable bimetallic strip and disc detectors, combination rate of rise/fixed temperature detectors, rate compensation detectors, etc.)
- Line-type heat detectors (including heat-sensitive wires for industrial applications such as cable trays)
- Gas detectors
- Flame detectors (ultraviolet or infrared)
- Water flow switches
- Water pressure switches
- Supervisory switches such as tamper switches on OS&Y sprinkler valves

The following indicating devices signal the "change of condition" to occupants:

- Strobes
- Horns
- Chimes, buzzers, and sirens
- Speakers
- Lamps

Despite the fact that there is a standard for fire alarm systems, no single standard has been established for the actual design of panels (although fire departments can request such within their jurisdiction according to NFPA 72). This fact has been the bane of many a fire fighter who is forced to try to decipher a fire alarm message at 3:00 in the morning. Although some improvements have been made with textual displays, fire alarm system activations can be problematic—all the more reason for preplanning!

Depending upon the age of the system, alarm and trouble signals may simply be indicated by small red and yellow lights, respectively, or may actually be printed out on a computer monitor screen in a modern system. The key is to know what the system is telling you—that is, the type of activation (e.g., elevator lobby smoke, water flow switch) and the location of the activation (e.g., second floor, fourth floor, west wing).

Often, the initiating devices are grouped by zones. These zones, in turn, may be grouped either by areas of a building or by type of device, especially in small buildings (e.g., smoke detector zone, manual pull station zone).

In addition, some fire alarm systems are co-located with other fire protection and life safety system controls in a fire command center in high-rise buildings. This arrangement allows fire fighters with portable handsets to talk with the lobby command post from telephone jacks positioned in various locations, such as stairwell landings, elevator lobbies and cabs, and so on. In the opinion of the authors, such systems are not very useful because they limit the mobility of fire fighters. High-rises and covered malls also may have public address systems, allowing officers to make announcements to the building occupants.

Fire fighters should take the following aspects of alarm and detection systems into consideration while doing their preplanning:

- Area(s) protected by the system.
- Types of detection and other initiating devices.

- Location of alarm panel and any **remote annunciators**. These small panels indicate the fire detector activation but do not provide all of the main alarm panel features; they are often placed at entry points to the building Figure 6-26.
- Type of panel and type of zoning, if present. If it is an older system with simple red, yellow, and green lamps, it is critical that you have a diagram/sheet indicating which zone corresponds to which area/type of device in the building! Otherwise, all you will know is "zone three" has been activated—but what does that mean?

Generally speaking, water flow switches are more reliable indicators of the actual location of a fire than smoke detectors, which can be "fooled" by migrating smoke. In a large or tall building with multiple detector activations, it is usually best to use the flow switches for the fire floor/area.

Do not reset the system until the device of origin has been established. Modern smoke detectors typically use a solid red light on the individual detection device to indicate activation. Use the **silence switch** on the panel if you have established

Figure 6-26 A remote annunciator.
Courtesy of Glenn Corbett.

that the activation is actually a false alarm. Many modern panels can store a "history" of alarm activations—scroll through them to see the sequence of activations.

Public Education and Alarm Systems

Smoke detectors have saved many lives in residential fires. Unfortunately, all too often these detectors are found to be missing or out of service. The educational effort stressing the importance of continuously operative smoke detectors should be intensified, particularly among low-income and non-English-speaking residents.

In some cases, smoke detectors are relied on to the exclusion of automatic sprinklers. As noted earlier, however, fire protection and fire detection are not synonymous. Only if the fire alarm is received where there is someone ready, willing, and able to respond and control the fire can detection begin to be considered protection. In some cases, the fire fighters are not available; in others, the fuel is of such a nature or the fire is so inaccessible that the fire is out of control when the fire suppression forces arrive. In the case of a high-rise building, a realistic estimate of the time between the alarm and the start of extinguishment is at least 20 minutes.

Unwarranted and Delayed Alarms

An *unwanted alarm* is defined by NFPA 72, *National Fire Alarm and Signaling Code*, as "*any alarm that occurs that is not the result of a potentially hazardous condition.*" Some communities have instituted penalties for repeated, unwanted alarms. However, such programs should have effective means to prohibit building managers from failing to transmit alarms. In several large-loss fires, such as those occurring at Los Angeles's First Interstate Bank, Philadelphia's One Meridian Place, and Houston's Westchase Hilton (a multiple-fatality incident), delay or failure to transmit the alarm received from the fire alarm control panel detectors was a factor in elevation of the fire's severity. The City of New York sued a building management company

to recover the costs of fighting a fire, including subsequent medical costs, that sent 20 fire fighters to the hospital; the city charged that the building management had delayed transmitting the alarm.

Smoke Management Systems

Perhaps the most mystical of systems is the smoke management system. These systems are intended to do just what their name implies: manage smoke. Most fire fighters have limited or no experience with them.

Smoke management systems can take different forms, depending upon the building code and the design intent of the systems. These types can be categorized as follows:

- Smoke control: Usually in the form of pressurization to prevent movement into protected areas, such as **stairwell pressurization systems**
- **Purge**: Venting of smoke
- **Zoned smoke control**: A combination of pressurization and venting; typical in high-rises where the fire floor is vented and the surrounding floors, above and below, are pressurized to prevent migration
- **Air flow**: The use of high air velocity to stop smoke movement

The key to understanding the operation of a smoke management system is to look at the original design and testing criteria (including the type/size of fire it is protecting against). If designed and tested properly, you will understand what the system is supposed to do. More importantly, it will tell you what you as a fire fighter should or should *not* do with the system!

Different design requirements and guidelines exist. The *International Building Code* includes design criteria. In addition, NFPA 92, *Standard for Smoke Control Systems* contains design criteria and has detailed information that should be consulted.

These systems run the gamut from simple to complex. They can be found in atria, stadiums, covered malls, underground buildings, and older high-rises. (The requirement for "general" smoke control in high-rises was removed from the model building codes years ago; only stairwell pressurization system requirements remain.) For existing systems, you must do your homework and figure out the design criteria for such systems—don't wait until a fire occurs.

Most smoke control systems will have some type of control panel. This panel may include diagrams of the area protected. Most use toggle switches to turn on/off various parts of the system, including fans and dampers. The wrong flip of a switch can pressurize the fire floor and exhaust areas that you want to keep smoke out of—watch out!

Determine whether the system is automatic (the usual case), manual, or both. In the case of an automatic system, determine what activates it. Spot-type smoke detectors are notoriously bad initiating devices for system activation because they can be remote from the fire's actual location. Most systems use water flow switches and, in the case of atria, projected beam detectors.

In all cases, try to be present when the system is tested. You will learn a lot.

Some things to consider when preplanning a smoke management system:

- General description of the system and the type of smoke management (smoke control, zoned smoke control, purge, air flow).
- Location and extent of system.
- System design criteria. (What is it trying to achieve?)
- Is it automatic, manual, or both?
- What types of initiating devices activate the system?
- A step-by-step sequence of system operation: what turns on, what turns off, what opens and what closes, etc.
- Location and description of control panel.

Firefighting considerations for smoke management systems include the following concerns:

- Assess the system operation: Is it working properly? If not, should you shut it down?
- When activating a system manually, let all fire fighters know so they will not be endangered.

- Leave a fire fighter with a radio at the system controls to ensure quick activation/shutdown if necessary.

Other Fire Protection Systems

Fire fighters will encounter other types of protection systems, protecting specific types of hazards. These include **dry chemical systems**, **wet chemical systems** (often used for cooking hoods), **foam systems** (used for flammable liquid tank protection), **carbon dioxide systems** (sometimes used to protect materials that can be damaged by water, such as record storage vaults and for printing presses), **clean agent systems** (which are used to protect sensitive equipment and have been replacing older **Halon systems**), **water mist systems** (also used to protect sensitive equipment), and **water spray systems** (which can protect exposed LPG tanks and electrical transformers).

Each of these systems is installed to protect against a specific hazard. When preplanning these systems, understand what they are trying to do: Are they intended to put the fire out or just cool the exposed hazard? How hazardous is the agent— is it relatively harmless like water mist or deadly like a carbon dioxide discharge? Is the system **total flooding** (i.e., the agent is applied throughout the entire compartment/area) or local (i.e., the agent is applied to just the piece of equipment/hazard)?

TACTICAL CONSIDERATIONS

1. It usually falls to the fire service to maintain a memory of unusual fire hazards. There is very little "out of date" experience in the fire service. Many buildings standing in your district have been around for decades and have outlasted the careers of numerous fire fighters. Hopefully before they left, they shared their knowledge of those building hazards that have the potential to kill fire fighters. There is no reason not to know the dangerous hazards of buildings that have been around your city for any length of time. Get out and inspect them. Update preincident plans and make sure that everyone knows the hazards that may have existed for decades. Knowledge is the best survival tool we have. Know your buildings.

2. Begin with prefire planning. Prefire planning starts with a scenario of the fire potential in a building with all its ramifications. The rate at which a fire will grow is one of them. It makes a tremendous difference to know that the fire-resistive building is filled with fast-burning materials. It is likely that the usual, initial small line tactics will be ineffective and waste time that is needed to bring heavy-caliber streams into play.

3. Any location where nitrocellulose or any other especially toxic material is stored demands intensive preplanning with emphasis on defensive operations. In any given situation, one gas might be more toxic than another. SCBA may be required for these defensive operations.

4. Company officers must be well informed as to flame spread potential. Preplanning will forewarn personnel of possible flashover or heavy fire situations. Personnel should be well trained in the possibility of hidden fire. Learning by experience alone may be fatal. Tactics and equipment should be developed to deliver the water necessary to control fast-spreading, high-heat-output fires.

5. Insulation is often blown into an attic space with a large flex hose. It's quick and economical. Unlike batt insulation, which is wrapped and unrolled between the ceiling joists in long single strips and easy to pull out in sections to check for hidden fires, blown insulation represents one giant blanket of insulation. Many fires have rekindled because of undetected hot spots embedded in the insulation. During overhaul in an attic, Class A foam is recommended for final extinguishment. But even with Class A foam, hot sparks may still be burning somewhere underneath the insulation. Often, even a thermal imager may not register the heat source. When it is impractical to remove all the insulation from an attic space, it's best to post a fire watch and rotate crews for the next 24 hours to prevent a rekindle. This author speaks from experience.

6. An estimation should be made of the potential for rapid flame spread over interior finishes. For instance, many churches with exposed wood plank surfaces present the potential for extreme flame spread and, thus, a fully developed fire. Fire can develop with unimaginable speed.

 i. In a Florida church fire, photographs showed only light smoke as units arrived. Within a minute, heavy fire was rolling out of every window and door. A fire department that has not preplanned properly for this type of fire may well waste time using handlines that will be completely useless. This fire was suppressed in 14 minutes by three heavy-caliber streams positioned so as to sweep all surfaces.

7. A carpeting fire will fill a corridor with flame. It should be fought with a solid stream, directed through the fire to wet the carpet beyond the fire, thus stopping extension. A fog stream may push the fire.

8. When fighting a fire in any corridor, there could be more than two fire fighters on the hose line. Putting three or more fire fighters in reverse to quickly back out of a corridor in untenable conditions can be confusing, clumsy, and slow. Always pay attention to other doorways and rooms within the corridor that can be forced open for an area of refuge. If you're on

the nozzle and therefore the last one out on a hose line, it may be quicker to force a door into a tenable space. Then you can find a window or breach walls to make a safe exit. Have an escape plan to survive the event.

9. The traditional *below-grade*, interior attack may often result in smoke damage, which could be reduced by changing the access point where the attack is initiated. In some cases, fires in basements and laundry rooms must be attacked from the interior stairway, resulting in smoke damage to the floors above. Smoke damage can be reduced with an indirect attack such as through a basement window or by selecting an alternative entry point like a rear basement door. Of course, an exposure line should be laid to protect the door at the top of the stairway. Utilize the reach of the stream. Crews can protect the stairwell from a safe location on the floor above without standing right over a weakened floor from a basement fire.

10. Smoke can do more damage to property than the fire and water. Fire fighters should place a high priority on controlling smoke damage to the extent possible. In malls, this would include closing any available doors, particularly the sliding doors installed on the major flagship department stores.

11. A realistic estimate between alarm time and the start of extinguishment in high-rise buildings is 20 minutes.

12. The fire department should educate the management and occupants of buildings on the importance of keeping stairway doors closed. Signs should be provided, employees and occupants should be educated, and citations should be issued when necessary. It is foolish to have a law requiring enclosed stairways and then take no steps to ensure they are closed.

13. Recognize that interior stairways can emit toxic smoke and gases throughout the building. This requires an early, painstaking search of all floors above the fire. At a fire, command must designate which stairways are to be used for fire suppression and which one for evacuation. Label stairway doors with tape or marker for incoming crews

14. Fire fighters should carry sufficient wedges to hold electronically released fire doors open when operations make this necessary. Because many fire doors do not operate properly, units should be assigned to check all fire doors and close those that should be closed. Fire fighters should never advance through a doorway protected by a fusible link-tripped fire door without blocking it to prevent its closing or dropping behind them. The force of the door can kink, pinch, or sever a charged hose line, leaving the attack crew unprotected on the opposite side of the fire door.

15. In some fire departments, checking that fire doors operate properly is a routine ladder company responsibility. In any case, this responsibility should be preassigned and not left to chance. Fire doors in fire walls and both horizontal and overhead roll-up doors are triggered by fusible links, sometimes located high up in the overhead. A burst of heat might trip the fusible link. It may be impossible to raise an overhead fire door and it can be very difficult to open a sliding fire door.

16. Fire departments should keep up with current literature and training seminars on PPV. Critiques of PPV operations often do not understand the physics and science behind it. They should be made to articulate the practicality of their alternative ventilation plan and should be made to develop a body of recorded experience with PPV techniques. Vertical ventilation is effective and at times necessary, but with lightweight truss construction or built-up commercial roofs, it can be dangerous, time-consuming, and labor-intensive. Fire departments that do not have the resources or staffing should consider becoming experts at using PPV. This ventilation technique is simple, fast, and effective if used correctly with knowledge and experience. The use of PPV on initial attack should be carefully monitored.

17. Many fully enclosed air-conditioned buildings are under positive pressure. Opening a single window on the leeward side of the building in the fire area will permit the excess pressure to vent, taking much of the smoke with it. Opening several windows can weaken and dilute the effect.

18. The fundamental purpose of a sprinkler system is to hit the incipient fire with enough water to control it. An experienced apparatus driver doesn't need to be told to charge and support the sprinkler system. This simple but essential task is often overlooked at fires.

19. The sprinkler system should not be shut off as long as hot water is falling down. This is an indication of the high heat atmosphere above you.

20. If responding to a library fire, be aware that 85% of recent library fires were due to arson. Note also that older libraries often used marble for the flooring. If unsupported, as in the Library of Congress, such floors can look good after heat exposure, but fail under a fire fighter's weight.

21. Some FDC for sprinkler systems cover only certain portions (floors) of a building.

22. Fire departments should know where sectional valves are located for partial sprinkler systems. In addition, the piping of some sprinkler systems in high-rise buildings may be limited in terms of its working pressure, providing only as little as 200 psi. When supplying a fire department connection, know the pressure limitations and know which parts of the building you are supplying.

23. The fire department should not manipulate any valves or any other equipment during an inspection or test. Such action would leave the department exposed to being accused of error or negligence if a system failure occurs. Permanent files should be kept of such notification because it may be years before a problem arises.

24. A critical point for standpipes: know the pressures and flow capabilities of all standpipe systems. A standard hydraulic formula adds 25 psi for the standpipe and 5 psi per floor on top of the required friction loss calculations. Your current "high-rise pack" may be incompatible with some of the systems in your city. In addition, be wary of PRVs—know how they work and if they can be overridden. Enlist the assistance of fire prevention personnel or fire protection engineers to develop a list of all the systems in your community.

25. Company officers should drill their crews on how to set up an emergency standpipe system and a backup standpipe. In the rare event of a standpipe failure, damaged threads on numerous standpipe valves, or threads on the high-rise pack that do not match those of the standpipe valves, be prepared to recover by establishing a water supply in the stairway. This can be done by laying LDH or 2 ½" (64 mm) hose up the stairwell or by weaving a 2 ½" (64 mm) hose vertically through the windows of the stairway landings. Weaving helps support the weight of a charged vertical 2 ½" (64 mm) supply line. Attach a 2 ½" (64 mm) gated wye to the end of the hose. At a minimum, it must reach the landing of the floor below the fire floor.

26. Class I standpipes are vertical hollow tubes. If for some reason the standpipe cannot be charged at the Siamese FDC on the exterior of the building, 2 ½" (64 mm) hose lines can be laid to the first floor (or lowest level) discharge valve of the standpipe. Using the proper couplings and a Siamese connection, two 2 ½" (64 mm) lines can be attached to the discharge valve for supplying the standpipe.

27. An apparatus driver who encounters a Siamese FDC (standpipe or sprinkler connection) with frozen or rusted female swivels that cannot be freed can still make the connection by inserting two 2 ½" (64 mm) double male couplings, then attaching two 2 ½" (64 mm) double female couplings. In essence, you've created a new Siamese FDC. Figure out a way to brace or prop up this connection to support the added weight.

28. Another way to connect to a Siamese FDC with frozen or rusted female swivels is to twist the 2 ½" (64 mm) supply hose counterclockwise 4–5 turns. Insert the male coupling into the female swivel and untwist the hose clockwise to make the connection.

29. When a 2 ½" (64 mm) gated wye is attached to a standpipe discharge valve with multiple hose lines, consider supporting the weight on the valve with a hose strap or some other method. Another method is to have a tested short 10-foot (3 m) "pony" section of 2 ½" (64 mm) hose. The pony section is attached to the standpipe. All other connections are made on the hose at floor level. This keeps all the extra weight from hose lines and appliances off the standpipe valve.

30. Seasonal public assemblies, such as haunted houses, have had their share of disasters with flammable contents and decorations that have led to multiple fatalities. Many decorations in haunted houses are black to create a dark, scary atmosphere. Quick and effective ventilation efforts are critical in such public assemblies. Removing smoke and heat will introduce an increased survivable atmosphere as well as improve visibility to allow able patrons a chance to escape and search teams to quickly cover the area and rescue those who can't.

WRAP-UP

Chapter Summary

- Fire containment has always been a primary objective of fire fighters.
- Some buildings are described in codes as noncombustible. This description is not technically accurate. The local fire officer should know exactly what this means.
- Rated fire resistance is a quality ascribed to a wall, floor, or column assembly that has been tested in a standard manner to determine the length of time for which it remains structurally stable (or resists the passage of fire) when attacked by a standard fire.
- When it was learned that fire could extend up open stairways in "fireproof buildings," the concept of compartmentation—or creating fire areas no larger than one floor—was developed. Fire-resistive assemblies—fire walls, fire partitions, and fire barriers—are key elements in that approach to fire protection.
- Fire suppression and detection systems are another critical means of providing fire protection. They include automatic sprinkler systems; standpipe systems; fire alarm, detection, and communications systems; and smoke management systems.

Key Terms

Air flow The use of high air velocity to stop smoke movement.

Automatic-wet standpipe system Standpipe system attached to a water supply capable of supplying the system demand at all times; it requires no action other than opening a hose valve to provide water at hose connections.

Carbon dioxide system A fire protection system intended to protect materials that can be damaged by water; it uses carbon dioxide to suppress the fire.

Clean agent system Gaseous fire extinguishing agent that does not leave a residue when it dissipates.

Compartmentation Subdivision of a building into small areas so that fire or smoke is confined to the room or section in which it originates.

Conflagration Rapid involvement of a fire situation that goes beyond the normal attack situations that fire departments encounter.

Critical radiant flux (CRF) The amount of external radiant heat energy (measured in watts per square centimeter) below which a flame front will cease to propagate.

Demand area The "expected" area of sprinklers that will activate as anticipated by the sprinkler design standard; essentially the projected size of the fire.

Density The unit rate of water application to an area or surface; expressed in gpm/ft^2 $[(L/min)/m^2]$.

Dry chemical system An automatic fire extinguishing system that discharges a dry chemical agent.

Egress Adequate exits within a building.

Fire barriers Assemblies that are typically 1- to 2-hour rated and used to enclose shafts, exit stairwells, exit passageways, and horizontal exits, and to separate spaces from certain hazardous areas.

Fire damper A mechanical device installed in an HVAC system at the point at which it passes through a fire-resistive assembly so as to block the spread of fire through the fire-rated assembly.

Fire partitions Assemblies that are typically 1-hour rated and are used to create fire-resistive corridors and to separate tenant spaces in covered mall buildings.

Fireproofing Material applied to structural elements or systems that provides increased fire resistance; usually serves no structural function.

Fire-rated The classification indicating in time (hours) the ability of a structure or component to withstand a standardized fire test. Does not necessarily reflect performance in an actual fire.

Fire resistance rating Quality ascribed to a wall, floor, or column assembly that has been tested in a standard manner to determine the length of time it remains structurally stable (or resists the passage of fire) when attacked by a test fire.

Fire resistant Construction designed to provide reasonable protection against fire.

Fire retardant Substance that helps delay or prevent combustion.

Fire walls Assemblies that are typically 2- to 4-hour rated and are used to create "separate buildings" within an overall structure.

Flameproof Resistant to fire.

Flammable Combustible that is capable of easily being ignited and rapidly consumed by fire. Flammables may be solid, liquid, or gas.

Foam system A firefighting system that provides for the delivery of a proportioned foam and water mixture for use in fire extinguishment.

Halon system A fire protection system that provides for the transfer of halogenated agents between fire extinguishers, supply containers, and recharge and recovery containers so that none of the halogenated agents escape into the atmosphere.

Horizontal exit A fire-rated wall typically running from one exterior wall to another, used when stairwells in a multistory building (or exterior exits in a one-story building) are too far away for egress purposes. The wall effectively creates two areas of safety on a given floor.

Houseline Class II standpipe system that supplies 100 gpm at 65 psi. Intended for use by building occupants.

Inherent fire resistance Flame resistance that is derived from an essential characteristic of the structural member.

Initiating device A system component that originates transmission of a change-of-state condition, such as in a smoke detector, manual fire alarm box, or supervisory switch.

Listing A compilation of fire-rated assemblies that have been tested by a testing laboratory and found to meet the minimum requirements of the test standard.

Manual-dry standpipe system Dry standpipe system that does not have a permanent water supply attached to the system.

Manual-wet standpipe system Wet standpipe system connected to a small water supply for the purpose of maintaining water within the system or sharing a water supply with an automatic sprinkler system, but not having a water supply capable of delivering the system demand attached to the system.

Noncombustible A material that, in the form in which it is used and under the conditions anticipated, will not aid combustion or add appreciable heat to an ambient fire.

Peened Forcible attachment of a thermocouple to a steel column or beam.

Purge Flow of air or an inert medium at a rate that will effectively remove any gaseous or suspended combustibles and replace them with air.

Remote annunciator Device indicating an off-standard or abnormal condition by both visual and audible signals.

Semiautomatic-dry standpipe system Standpipe system that is attached to a water supply capable of supplying the system demand at all times and that requires activation of a control device to provide water at hose connections.

Silence switch When flipped, the fire alarm is silenced but still active.

Smoke barrier A 1-hour fire-rated assembly that has also been designed and tested to minimize the migration of smoke.

Smoke control System that utilizes fans to produce pressure differences so as to manage smoke movement.

Smoke developed In the Steiner tunnel test, a calculation of smoke spread made by measuring the obscuration as the smoke passes a photoelectric cell placed in the stack from the test tunnel.

Stairwell pressurization system System that introduces positive pressure into a stairwell to provide a tenable environment within the stair tower in the event of a building fire.

Standard time–temperature curve The profile of the furnace temperatures that are dictated by the test standards NFPA 251, UL 263, and ASTM E-119.

Standpipe system An arrangement of piping, valves, hose connections, and allied equipment installed in a building or structure, with the hose connections located in such a manner that water can be discharged in streams or spray patterns through attached hose and nozzles, for the purpose of extinguishing a fire, thereby protecting a building or structure and its contents in addition to protecting the occupants.

Steiner tunnel test A common term that refers to NFPA 255, *Standard Method of Test of Surface Burning Characteristics of Building Materials*.

Thermocouple An electrical temperature-measuring device used in laboratory fire testing.

Total flooding Act and manner of discharging an agent for the purpose of achieving a specified minimum agent concentration throughout a compartment.

Water flow switch A switch that detects the movement of water in a sprinkler system and transmits a signal to the alarm system.

Water mist system Distribution system connected to a water supply or water and atomizing media supplies that is equipped with one or more nozzles capable of delivering water mist intended to control, suppress, or extinguish fires and that has been demonstrated to meet the performance requirements of its listing.

Water spray system Special fixed pipe system connected to a reliable fire protection water supply and equipped with water spray nozzles for specific water discharge and distribution over the surface or area to be protected.

Wet chemical system An automatic fire extinguishing system that discharges a wet chemical agent.

Zone Defined area within the protected premises.

Zoned smoke control Smoke control system that includes smoke exhaust for the smoke zone and pressurization for all contiguous smoke control zones.

Case Study

You and your engine company are dispatched to a water flow alarm in a three-story apartment building of lightweight wood-frame construction on a cold winter night. Your prefire plan shows that this building is equipped with an NFPA 13R automatic sprinkler system. Upon arrival, you can hear an electric bell ringing. You are met by a security guard who advises that there is no fire and he is unable to find an activated sprinkler head.

1. Which of the following areas are provided with sprinklers according to NFPA 13R?

 A. Small bathrooms
 B. Small closets

C. Kitchens

D. Unused attic

2. True or False: Sprinklers will likely be found inside the floor trusses of this building.

3. If the building was built of Type II noncombustible construction and was 8 stories, which standard would have been followed for the installation of a sprinkler system?

A. NFPA 13D

B. NFPA 13

C. ASTM E-119

D. ASTM E-84

4. True or False: The water flow switch in this system will become activated only as the result of a sprinkler activation.

Challenging Questions

1. Define a smoke barrier and describe how it is useful in a hospital. How does a smoke barrier differ from a horizontal exit?

2. Why is it important to know that fire resistance testing for floor assemblies using the ASTM E-119/UL 263 test standards only test the floors on their underside?

3. During a fire inspection, you note carpeting on the interior walls of a day care center. What should you do to deal with this situation?

4. You are a captain of a truck company in a suburban district. You are made aware that the new covered mall in your area will be equipped with a smoke management system. What should you do to prepare utilization of this system during an actual fire?

Wood-Frame Construction

OBJECTIVES

At the conclusion of this chapter, you will be able to:

- Understand and correctly use the terminology associated with wood construction.
- Identify and describe six types of wood-frame structures classified as Type V construction.
- Identify and describe the specific fire protection differences between balloon-frame and platform-frame construction.
- Describe the construction of a platform-frame building.
- Understand how a truss is constructed and how it performs from a fire protection perspective.
- Explain the difference between firestopping and draftstopping.
- Describe the behavior of engineered and manufactured wood products under fire attack.
- Describe the different types of sheathing, siding, and roofing materials.

Case Study

Figure 7-1 The Green Bay, Wisconsin, floor collapse, August 2006.
Courtesy of Green Bay Metro Fire Department.

In August 2006, Green Bay, Wisconsin, fire fighters responded to a typical house fire—a house fire like many others that occur across the United States every day (**Figure 7-1**). The two-story, single-family home was built in 1999, using lightweight wood trusses. A fire had erupted in an unfinished area of the basement and then spread into a finished area.

Responding fire fighters, believing a person was trapped, began search operations on the first floor. One team of two fire fighters went to the right, another pair to the left. As the team on the left made its way through a foyer to a bedroom, the lightweight truss floor collapsed, sending the two fire fighters into the basement. Both fire fighters were trapped; one fire fighter, a lieutenant, was killed. A subsequent investigation determined that the floor collapsed over a basement hallway between the finished and unfinished basement areas. The collapsed truss floor was a "heated" floor with embedded heating coils in concrete covered in ceramic tiles, which added weight to the floor along the tiles.

1. How quickly do lightweight wood floor trusses collapse in a typical basement fire?
2. How common are lightweight wood truss floors with built-in heating systems?
3. Would heat from a fire be hidden from a thermal imager in a floor system like this?

Introduction

In the United States and Canada, most fires are fought by fire fighters standing on, in, or under wooden structures. These are some basic facts about **wood**:

- Wood is combustible. As it burns, its structural strength is lost. Unless the fire is checked, at some point gravity will act and a collapse will result.
- Wood construction creates combustible void spaces in which fire can hide, grow, and burst out.
- Wood in thin sections can have a very rapid flame spread.

Definitions

Some basic definitions are necessary for a better understanding of the elements of wood construction. Building terms are often local in origin, and often the construction trades use different words for components than do engineers or laypersons. The term **wood framed** is typical of the imprecise terminology common in the building field. A framed building is usually understood to be a building with a skeleton of beams and columns, in which the walls are just **curtain walls**. (These types of walls are non-load-bearing; they carry no weight other than their own weight and are installed only to keep out the weather. The frame itself carries the entire weight of the structure as well as live loads.) The opposite term is **wall bearing**, indicating a structure where the entire structural load is carried on the walls. The **wooden-walled building**, which is usually referred to as a frame building, is actually a wall-bearing building that carries the load of the structure and contents. As this brief summary suggests, the various types of structures can all be rather confusing.

Wood-Frame Terminology

A variety of terms apply to wood-frame construction. It is important that you know the proper word to describe a particular building component, as in all

types of construction. In particular, you should be familiar with these terms:

- **Chamfer**: To cut off the corners of a timber to retard ignition.
- **End matched**: **Lumber** with tongues and grooves at the ends.
- **Engineered wood**: In the construction trade, this typically refers to laminated timbers. In this text, it refers to wood modified from its natural state or lumber combined to make elements.
- **Glued laminated timbers**: Planks glued together to form a solid timber **Figure 7-2**.
- **Heavy timber**: Lumber that is 8 inches or more in its smallest dimension. (Note: some building code definitions of heavy timber specify smaller dimensions.)
- **Joists**: Floor beams.
- **Legacy construction**: In terms of wood-frame construction, the use of solid wood members as opposed to lightweight wood trusses and wooden I-joists.
- Lumber: Wood that has been sawn and planed.
- **Matched lumber**: Tongue and grooved lumber (usually lengthwise).

Figure 7-2 Glued laminated timbers are used to form large cross sections, longer lengths, curved shapes, or any combination of these.
© Ms Deborah Waters/ShutterStock, Inc.

- **Oriented strand board (OSB)**: Board made of layers of strands of wood cut from logs, with a fairly constant width-to-length ratio. OSB is said to have less expansion from moisture. It is available in sizes larger than 4 by 8 feet (1.2 by 2.4 m). An **oriented flakeboard** is also available.
- **Plywood**: Layers of wood veneer laid in different directions, glued together under pressure.
- **Rough lumber**: Lumber that is left as sawn on all four sides.
- **Splines**: Wooden strips that fit into grooves in two adjacent planks to make a tight floor.
- **Studs**: Columns in frame buildings, usually nominal 2 by 4 inches (2×4) or 2 by 6 inches (2×6).
- Wood: A hard fibrous material forming the major part of trees. It is usually milled or otherwise processed for use in construction.
- **Wood lath**: Narrow, rough strips of wood nailed to studs. Plaster is spread on wood laths. Rarely used in new construction, wood lath is present in many existing buildings **Figure 7-3**.

Uses of Wood in Buildings

Wood is used in many ways. Its significance relating to fire varies with its uses—structural, non-load-bearing, or decorative:

- Wood is used structurally to carry the building loads.
- Non-load-bearing uses of wood include wood lath for carrying plaster and wood wainscoting, which is a thin grooved board used for ceilings and wall panels.
- Wood is commonly used for roofs of buildings that may appear to be of noncombustible construction.
- Wood is used for interior trim or within older partition walls of otherwise fire-resistive buildings.

Figure 7-3 Wood lath is found in older types of construction. It can pose a problem for fire fighters as they open up walls and ceilings when checking for fire extension.

© DigitalVues/Alamy Images.

Types of Wood-Frame Buildings

Wood is used to carry the major structural loads in many forms of Type V construction:

- Log cabin
- Post and frame
- Balloon frame (sometimes called eastern framing)
- Platform frame (sometimes called western framing)
- Plank and beam
- Truss frame

Log Cabin

The log cabin was not invented by Daniel Boone or Davy Crockett, but rather came to the East Coast of the United States from Sweden and to the West Coast from Russia. Prefabricated log cabins were sold in Moscow in the 1600s. It was easier for early settlers to use whole trees or slightly dressed lumber for home building than to saw trees into boards. Later, the humble log cabin would be concealed under siding or even brick veneer.

Many older buildings, even multistory buildings, are concealed log cabins. The walls of such a building carry unexpectedly heavy loads and therefore have potential for serious collapse. Know your buildings by investigating them, especially those that appear to have been masked by a veneer or another wall structure.

Log cabins are coming back into fashion. A number of companies offer dressed logs and sometimes these are prepared for assembly. The interior surfaces of many of these cabins are usually wood, often either boards or plywood, which would increase the intensity of a fire, as contrasted to the typical residential gypsum board interior. Fire departments in areas where such buildings exist should be aware of this hazard Figure 7-4 .

Figure 7-4 A typical log cabin.

© Fanfo/ShutterStock, Inc.

Post and Frame

Post and frame construction came to America from England, Germany, and France. Such a building has an identifiable frame or skeleton of timber fitted together. Joints are constructed by **mortise and tenon** (socket and tongue), fitted together to transfer loads properly Figure 7-5 . They are pinned with wooden pegs called **trunnels** (a New England term for a tree nail).

Post and frame buildings are the ancestors of the high-rise because they are framed, not wall-bearing, buildings. In other words, the walls are not structural or load bearing. Original post and frame buildings are part of our heritage, and every effort should be made to preserve them. Recently, new post and frame buildings have been constructed as homes and commercial buildings. Like log cabins, the interior finish often consists of wood. Many new post and frame buildings are enclosed with foam core panels. Foam core roofs present the possibility of failure.

Deceptive Imitation Post and Frame

In the 1920s, many balloon-frame houses were built with the exterior finished to resemble post and frame construction. They were called **English Tudor** style homes Figure 7-6 . Recently, this style has had a renaissance. Buildings of several other types of construction also have been decorated to resemble the post and frame style. Some ordinary brick wall-bearing buildings have been dressed up to give the appearance of post and frame. In one case, the fake post and frame trim was removed two decades later to modernize the building. At Walt Disney World and other amusement parks, buildings of steel construction are decorated to have a post and frame appearance.

Balloon Frame

Balloon-frame construction started in 1833. Augustine Taylor, a Chicago carpenter erecting St. Mary's Church at Fort Dearborn, is credited with hitting

Figure 7-5 This mortise and tenon joint between a post and beam is beginning to come apart in this 100-year-old barn.

Courtesy of Glenn Corbett.

Figure 7-6 An English Tudor style home.

© Steve Holderfield/ShutterStock, Inc.

upon the idea of fabricating a wall of ordinary studs and nailing it together rather than using a post and frame assembly method. Inexpensive machine-made nails had become available by that time, which made it possible to erect the entire wall at once without the skilled labor necessary to do the cutting and framing required for post and frame construction. This type of construction was derided by many, who said it was so light that it was like a balloon. Unfortunately, as will be seen, it can be compared to a balloon in another way. Balloon-frame construction became an almost universal construction method for multistory wooden buildings until the middle of the 1900s.

In a balloon-frame building, the studs run two or more stories high from the foundation to the eave line Figure 7-7. At the floor line, a horizontal board called a ribbon board is nailed to the studs. The joists rest on the ribbon board. The channels between the studs may be open from the cellar to the attic, and the joist channels (space between the joists) are open to the stud channels. Thus, fire can spread through all the interconnected spaces from cellar to attic and across the ceiling. Present-day methods may use firestopping (discussed later in this chapter), but earlier construction methods, when many of these buildings were built, did not.

The interconnected voids can be considered to be one big balloon. This effect was once observed at a fire in an old balloon-frame house, where there was a heavy fire in the porch ceiling. Each time it was hit with a fog stream, the fire ballooned out at the rear of the building.

When balloon-frame buildings are remodeled, firestopping may or may not be installed. Installing firestopping in an old building is very costly. In one case, a historic balloon-frame college building was fully rehabilitated. Even though sprinklers were installed, fire could have raced through the voids if they were not firestopped. The final cost of the renovation was about two-thirds of the cost of a new building.

Until recently, it was reasonable to assume that any wooden building built before 1940 and more than one-story high was probably of

Figure 7-7 An illustration of a balloon-frame house; note the continuous studs that run from the basement to the attic, allowing a basement fire to rapidly extend.

balloon-frame construction. Later buildings used platform-frame construction. Recently, however, the Victorian-style construction has come back into favor, and expensive multistory platform-frame suburban homes are being built to look like 100-year-old farmhouses.

Some construction texts recommend balloon-frame construction as superior to platform frame for masonry-veneer houses (see the discussion of brick veneer later in this chapter). Shrinkage of the wooden building, however, could damage the brick veneer. The smaller vertical shrinkage potential associated with a balloon-framed structure contrasts with the greater horizontal shrinkage potential of platform-frame construction.

Some years ago, two six-story brick veneer apartment houses were built in an Ohio suburb using balloon-frame construction. Studs were spliced to achieve the required height. The apartment houses had many deficiencies and presented a major fire hazard. The fire chief fought successfully to have them torn down before they were ever occupied.

Fires in Balloon-Frame Buildings

The fire hazards found in balloon-frame buildings are well known. As an example, a chief officer, who was a student of Frank Brannigan in a fire science course, described a fire that the first-due company reported as "outside rubbish." The officer ordered some of the siding removed and found fire in the walls. Shortly thereafter, another officer reported that an upstairs bedroom on the opposite side of the house was untenable, even though no fire was showing. The fire had quickly crossed over through the joist channels and had literally surrounded the bedroom. A good fight saved the building, but the damage was extensive.

Another California fire demonstrated that sometimes even sprinklers cannot protect balloon-frame structures. An incendiary trash fire, which started against an exterior wall of a Victorian building, penetrated into void spaces. Responding fire fighters checked for extension but found nothing. Fire broke out 4 hours later. Much structural damage occurred

before the sprinklers activated on the fire that broke out from the voids.

Plumbers (or amateur plumbers) are often the culprits for starting fires in the walls of balloon-frame buildings. Whether a plumbing torch was used to braze a copper tube or to thaw out a pipe, many a balloon-frame building (and even non-balloon-frame buildings) has succumbed to these sources of ignition.

The basement of a balloon-frame building is usually the worst place for a fire to start. It has unimpeded access to the entire structure through the unfirestopped walls. The fire will move quickly up to the top of the structure and into the attic **Figure 7-8**. It is critical that fire fighters anticipate this spread, sending both truck companies and engine companies with hose lines to each of the upper floors and opening up walls to hit the spreading fire. It is also critical to get the attic/roof opened up because the fire will quickly emerge in that area. Consider designating an "extension sector" and assigning a fire officer to observe the building on all sides, watching for signs of fire spread such as blistered paint, smoke from the eaves, a sudden increase in the volume of smoke, and so on, below the operating forces or other signs that the building "built to burn" is succeeding and it is time to evacuate.

The use of **vermiculite** inside wall stud channels of a balloon-frame home, perhaps for insulation, was observed in one fire incident. The sand-like material acted as a firestop, slowing down a significant fire that had erupted in an overstuffed basement.

Platform Frame

In platform-frame construction, the first floor is built as a platform **Figure 7-9**. This means the **subflooring** is laid on the joists, and the frame for the first-floor walls is erected on the first floor. The second-story joists are then placed, the second-story subfloor is laid, and the second-floor walls are erected on the second-story subfloor. The third floor, if any, is built in the same way. Three stories are typically the limit for platform-frame construction.

Figure 7-8 An illustration of a rundown, overcrowded, old Victorian balloon-frame house. Note all of the potential problems and danger signs that may be found in such buildings. These problems include easy upward fire spread from basement to attic when doors are open, blocked and broken interior stairs, window air conditioning units blocking access, as well as individual locks on interior doors, and very importantly, unimpeded fire spread from the basement to the attic through the exterior wall void space.

Wood Light Frame Construction

Ridge Beam (ridge pole)

Joist

Rafter

Sheathing

Top Plate

Sill

Stud

Header

Foundation

Figure 7-9 An illustration of a platform-frame house.

In a balloon-frame building, there is structural continuity from top to bottom. In contrast, in the platform-frame building, there is no continuity from top to bottom. If construction methods are as described, there are inherent barriers (the load-bearing and non-load-bearing partition walls throughout the building) to limit the spread of fire through the walls. There is a good possibility of confining a fire to one portion of the structure, even if the fire gains access to a concealed space.

However, there is one common vertical means of fire spread in both balloon-frame and platform-frame buildings: the open stairwell. In single-family homes, this is the quickest way for smoke and heat to rise to the top of the home. It is essential that this vertical "chimney" be protected by a hose line, both to prevent extension and to provide a means of egress.

Another common construction feature provides a bypass for fire, and from a fire protection point of view, converts a platform-frame building into a balloon-frame building. **Soffits** (false spaces above built-in cabinets, usually in a kitchen or in the undersides of stairways and projecting eaves) provide a connection, generally without a firestop, between wall and joist spaces. Furthermore, only a thin sheet of wood or composition board makes up the top of the built-in kitchen cabinets. A kitchen fire that extends into such a cabinet will enter the soffit space quickly. Fire can then extend to the joist spaces in a multi-story building or to the attic spaces in a one-story building. Such soffits are often built back to back in multiple-family dwellings so that fire may extend immediately to the adjacent occupancy. Interconnection of the soffit void with pipe openings may cause fire to spread vertically throughout the building.

Constructing a Platform-Frame Building

Once a foundation and basement have been built, the **rough carpentry** of the platform frame may begin. The first-floor **sill** (usually rot and insect decay resistant) is laid on the foundation wall. The floor joists follow, usually spaced 16 or 24 inches (41 or 61 cm) on center (oc). Subflooring, in the form of plywood or OSB, is laid on top of the joists.

Bridging (or cross-bracing) is placed between the joists to keep them erect and to help share the loading placed on the floor. Floor openings are created for stairwells and the like through the use of **headers** (along the end of the opening) and **trimmers** (along the length of the opening). Essentially, the floor joists are cut short to create the opening, and the headers and trimmers are attached to the joists to carry the load around the opening.

The wood walls, consisting of studs spaced 16 inches oc with a **bottom plate** and a set of **top plates**, are next constructed on the floor itself. Once one wall has been completed, it is tilted up into place and temporarily braced to hold it in place. The other walls are then constructed, one at a time, and tilted up into place. Once all of the walls have been erected and tied together, the bracing is removed, and sheathing panels (sometimes insulating) are placed over the studs. A successive floor is built on these walls, and the process is replicated.

The top set of walls receives the roof framing. Consisting of **rafters** and a **ridge board** (sometimes called a ridge pole), the pitched frame forms a peaked roof. (Truss roofs, described later in this chapter, are an exception to this description.) **Hips** and **valleys** give the roof a variety of sloped shapes, usually for aesthetic purposes.

Stairways between floors are framed using **stringers**, placed on both sides of the steps. Stringers support the risers (vertical stair components between treads) and treads (horizontal stair components on which people place their feet). The taller the riser and smaller the tread, the steeper the set of stairs.

Plank and Beam

Architects and interior designers, looking for more attractive, economical, or at least different construction methods, have adapted industrial plank and beam construction techniques to residential, assembly, and commercial structures **Figure 7-10**. Instead of using multiple floor beams placed 16 inches (41 cm) apart, heavier beams are used and are spaced much farther apart. Instead of thin, rough subflooring or plywood, thick, finished tongue and groove

Figure 7-10 A plank and beam building.

planks are used. The planks are thick enough to span the wider gap between the beams without deflecting. The finished plank surface is the ceiling.

From a firefighting point of view, this type of construction has the positive benefit of reducing the volume of concealed space in which fire can burn shielded from hose streams. The interior finishes used, however, often have high flame spread and smoke-developing characteristics. Surface damage may be serious, even in an easily controlled fire in which collapse or hidden fire are not factors.

In some residences, to provide an open effect, partitions are only head high, except in bedrooms and bathrooms. This increases the life hazard for nighttime fires because the kitchen, dining, living, and family rooms are essentially a single open area in which fire can easily build up.

Many churches and similar buildings with high open spaces are built of rigid laminated wood frames or arches and heavy plank roofs. Any fire that reaches the surface of the wood will spread rapidly and develop into a huge volume of fire. Any intermediate structure, such as a sacristy or robing room, should be sprinklered or of built of noncombustible construction because of their particular hazard. Eliminating the intermediate fire load greatly reduces the potential for ignition of the roof.

Lightweight Trusses and Other Wooden Members

The current practice of designing buildings of lighter and lighter materials and substituting geometry for mass has led to significant strides in the architectural field. It has also created significant problems in the field of firefighting. Although we will also discuss trusses in the context of other types of construction (including lightweight wood trusses in ordinary construction, described in the chapter that discusses heavy timber construction), it is important that we have a conceptual understanding of the

deadly problems that trusses (and other lightweight members) present in wood-frame construction.

Knowledge of the hazard of trusses has developed gradually in the fire service. In 1948, Frank Brannigan wrote the first article to appear in the fire service press on the then new concept of prefire planning. The article described a Navy drill hall that was converted into a warehouse. The pictures clearly show 100-foot timber trusses, but the preplan does not mention them as a hazard. Any fire department preplanning such a structure today would give serious attention to such trusses.

When the first settlers arrived in this country, huge virgin forests yielded inexpensive wooden beams and columns. Structures were massively built of wood and masonry, without regard to their weight. Times have changed, however, and today the emphasis is on lighter-weight structures. For example, lightweight gypsum panels are advertised as suitable replacements for heavier masonry shaft enclosures. The steel industry seeks lighter fireproofing for steel to replace the concrete that was once commonly used.

The weight of materials and the weight of the structure required to support heavy materials are now important cost considerations. Suppliers of materials lighter than those previously used proudly brag about weight savings. For example, the Empire State Building weighs 23 pounds (10 kg) per cubic foot, whereas modern high-rises weigh as little as 8 pounds (3.6 kg) per cubic foot.

The largest practical open areas and the maximum amount of offsite prefabrication are other important considerations in today's construction. Many of the techniques for making buildings lighter and providing large open areas without columns have been available for centuries **Figure 7-11**. The hammerbeam truss, for instance, used in churches and other monumental Gothic structures, was invented in the 1200s and used for Westminster Hall (not Westminster Abbey), which still stands in London. In recent years, increased material and labor costs have forced builders to maximize the use of more economical construction techniques.

Figure 7-11 The medieval church of San Miniato al Monte relies on wooden trusses to support the roof.

© Renata Sedmakova/ShutterStock, Inc.

Trusses are most often used as beams, but there are many trussed columns, especially in huge buildings and special-purpose structures, such as high-energy transmission lines and tall radio towers.

The truss satisfies many building requirements:

- It is lighter in weight than solid construction, thus reducing the weight of supporting walls or columns.
- It provides long clear spans, thus giving maximum flexibility in the use of space.
- Many trusses, such as lightweight wood trusses and steel bar joists, can be delivered prefabricated. Huge steel trusses can be delivered partially prefabricated, with the final connections made in the field.

Lightweight Wood Trusses

The lightweight wood truss has been in general use since the 1960s; one of the first articles on the firefighting implications of this type of construction was written by Frank Brannigan in 1971. Composed of 2- by 4-inch (51 by 102 mm) pieces of wood, the lightweight wood truss is pieced together on a factory table and subsequently joined together with light-gauge galvanized steel gusset plates (also known as "gang nails") **Figure 7-12**. These plates contain a series of small 3/8" (9.5 mm) "teeth" that are pressed into the wood surface; it is these small teeth that hold the whole truss together.

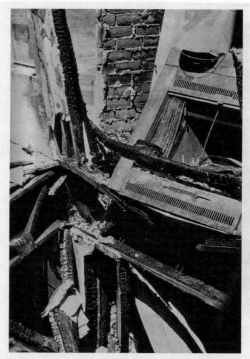

Figure 7-12 The remains of a lightweight wood floor truss in a wood frame apartment building. An accidental fire involving a spilled flammable liquid ignited inside the void space of this truss, ultimately leading to the involvement and collapse of the second floor, the third floor, and the roof.

Courtesy of Glenn Corbett.

Figure 7-13 A finger joint in a lightweight wood truss.
Courtesy of American Wood Council, Leesburg, VA.

Another type of lightweight wood truss uses a **finger joint** and adhesive to join the wood truss together (Figure 7-13). The fingers at the end of the pieces of wood forming the web of the truss are fit into slits in the top and bottom chords of the truss, using phenol and resorcinol formaldehydes as an adhesive. In this case, it is friction and glue that hold the truss together. Some open-web trusses with finger joints manufactured in Oregon between 1974 and 1980 reportedly came apart at the joint in normal use.

Dangers of Lightweight Wood Truss Floors

Lightweight wood truss floors are a known hazard to fire fighters. They are built of a small cross section of wood, with the high surface-to-mass ratio characteristic of kindling. Metal gusset plates dig into the wood approximately 3/8" (9.5 mm). Heating of the gusset plate will decompose tensioned wood fibers holding the teeth. Not only is there the hazard of early collapse, but the void is also a reservoir for explosive carbon monoxide (CO) gas waiting for a fire fighter to pull the ceiling. Firefighting tactics based on sawn joist floors will kill fire fighters if used on buildings with truss floors.

The truss floor building ordinarily gives no outward indication of its presence. The only solution is for the fire department to preplan, record, and retrieve the construction information on the fireground. During a fire, the presence of truss floors may be disclosed by smoke or fire pushing through the wall at the floor line (this is less likely in a brick veneered building). The exterior sheathing plays a critical role in whether this effect is visible.

Dangers of Lightweight Wood Truss Roofs

Wood truss roofs are just as hazardous to fire fighters as lightweight truss floors. In addition to the likelihood of collapse under fire conditions, they present another hazard: they lack the ridge pole that is present in a legacy construction, meaning that the fire department's roof ladder hooks will hang on the roof deck sheathing (such as plywood) rather than a beam. During construction, the roof trusses must be temporarily held erect by the use of wood slats to tie them together; it is the roof itself that holds the roof trusses upright (Figure 7-14). Be wary of exposed lightweight wood roof trusses in a windstorm—many of them have fallen over like dominoes at construction sites.

A lightweight wood truss roof involved in fire is a very deadly place to be. Any attempt at ventilation must be performed using an aerial device such as a tower ladder. Expect the collapse of lightweight wood truss roofs when they are involved in fire.

Figure 7-14 A lightweight wood truss roof under construction.

© Portlandia/ShutterStock, Inc.

The Truss Void Itself

The wood truss floor introduces a catastrophic new dimension into fire suppression in combustible buildings—a **truss void** or trussloft. The hazards of the cockloft or void between the top-floor ceiling and roof, which is not high enough to be called an attic, have long been familiar. This floor–ceiling truss void can properly be called a trussloft.

All comparative strength controversy aside, the joisted floor found in legacy construction has one tremendous advantage over the truss: each joist acts as a fire stop. The area of the void is limited. The truss void, in contrast, represents a large area in which explosive CO can accumulate. The use of trusses provides a huge increase in the volume of concealed voids in the structure. Voids are almost invariably interconnected horizontally and vertically by utility pathways. The generation and accumulation of CO in void spaces is not as well recognized by the fire service as it should be. It is both toxic and explosive. The hazard of hidden fire is as important to the fire service as collapse.

The National Fire Protection Association's (NFPA) *Fire Protection Handbook* provides good information on CO. A rule of thumb is that any exposure in which the product of concentration (in parts per million [ppm]) × duration (minutes) = 33,000 is likely to be dangerous.

A 10-minute exposure to 3,500 ppm of CO would be hazardous and possibly incapacitating. Higher ppm values are even more dangerous—12,500 ppm can be fatal after a few breaths. The potential for loss of life or incapacitation of occupants from CO generated by a fire in the void even before the alarm is sounded cannot be dismissed. The flammability range of CO is from 12.5–74%; its ignition temperature is 1,128°F (609°C). When an ideal mixture of CO in air is present, a detonation can result that is sufficient to blow a building apart. Generally, a deflagration—a huge sudden burst of flame—appears.

Catastrophe Potential

A most serious threat to life safety by lateral extension exists when the trusses are extended out, either as cantilevers or as continuous beams, to support the balcony that is the only exit for occupants and the only access point for the fire department. Where the truss passes over the exterior wall, firestopping or draftstopping in the form of a sheet of gypsum board is "buttered into place" with cement. This "firestopping" is penetrated for the lighting circuits of the balcony. In any event, the use of such an unreliable barrier to the extension of fire from the void is inexcusable. In some cases, the ceiling of such a balcony is plywood, permitting early access to the void for fire originating on the balcony or coming out the window.

It is a credible scenario that the stairway exit can be involved in fire and collapse before the occupants have escaped. In many suburban areas, a typical response might consist of two engine companies and a truck, each with three or four fire fighters. It is impossible for this response to remove, via ground ladders, all the occupants of a three- or four-story truss-floored multiple-unit dwelling. Even with a greater response, it is unlikely that all the occupants could be located and removed before interior collapse forced withdrawal. The potential for a disaster should force a code requirement that the exit facilities be totally separated from the truss void.

Automatic Sprinklers and Truss Voids

Some truss-floored multiple-unit dwellings have been protected with residential sprinkler systems, meeting the requirements of NFPA 13R, *Standard for the Installation of Sprinkler Systems in Low-Rise Residential Occupancies*. The purpose of these systems is to prevent flashover, save lives, and permit the occupants to escape. There have been a significant number of successful operations. However, these are partial systems and should not be expected to provide the fire suppression or control that full sprinkler systems have provided. The sprinkler systems cover selected portions of occupied spaces. It is very likely that the sprinkler system will control any fire originating in the protected areas of an apartment. However, if the fire originates in or penetrates the void, the sprinklers will not be in a position to control the fire.

A fire in a relatively new, NFPA 13R–protected, three-story wood-frame condominium in New Jersey began on a top-floor unsprinklered balcony in a plastic bucket for discarded cigarettes. The fire spread from the bucket, up the wood balcony walls, and into the eave overhang. The unsprinklered attic ignited, and fire traveled across the entire top of the building, unimpeded due to the lack of required draftstopping. The top floor was destroyed, and water severely damaged the lower floors.

In an older 13R-sprinklered, combustible dwelling in Texas, the floor trusses were **cantilevered** out to form the exit balcony. The balcony was one entire nonfirestopped void. The firestopping at the front wall included pieces of gypsum board buttered into place with cement. Very likely it was penetrated in one or more places. A failure of one of these barriers would admit fire to the entire balcony truss void. It is not hard to envision a serious loss of life in such a structure. Unfortunately, this outcome might misguidedly deal a fatal blow to the concept of residential sprinkler protection.

In DeKalb County, Georgia, fire resulting from a plumber's propane torch developed in the truss void of an almost completed three-story apartment house. There was no contents load. The sprinkler system was operational, but the sprinklers did not fuse. A fire fighter fell through the second floor into the fire when the truss collapsed. Fortunately, his partner used a hose line that protected him from the fire until he could be rescued.

Lightweight Wooden I-Beams

The wooden beam sawn out of a tree trunk is inefficient in terms of weight and cost. Recently, the legacy construction **sawn beam** has been replaced by wooden I-beams (composite wood joists).

Look at the end of a steel I-beam. The steel is extruded through a die, so the designer can choose the most efficient shape. Both the top flange and the bottom flange are wide to cope with compressive and tensile loads. The web, which separates the top from the bottom, is thin, just sufficient to keep the top and bottom apart. The newer I-shaped beams use 2- by 4-inch boards for the upper and lower flanges and OSB for the web; in the past, these wood I-joists used plywood webs. The present-day versions are glued together using phenol and resorcinol formaldehydes, and manufactured in lengths up to 60 feet (18.3 m).

Unlike the case with the sawn beam of legacy construction, the "fat" or surplus (structurally unnecessary wood along the sides) is no longer present in these lightweight wood I-beams. A sawn wooden beam can be thought of as "containing" an I-beam with "surplus" wood along the sides. The surplus wood is what makes it possible for fire fighters to stand and operate on a burning structure. This "fat" has been the basis of interior firefighting tactics. As long as only the "fat" or surplus wood is burning, a fire fighter is relatively safe. Until it burns away, the beam is structurally sound. The gradual sagging of sawn beams due to the loss of the exterior wood often gives warning of impending collapse.

The web of the I-beam is often penetrated by utilities, and in some cases sizable holes are cut into it so

that any fire gets a grip on both sides of the I-beams at the same time, guaranteeing early failures. Often, cut-off ends of I-beams are used as "firestopping." Be alert to this possibility. A Fairfax County, Virginia, fire fighter went through a wooden I-beam floor. The house was under construction, and the floor was the only fuel for the fire. In another case, a residence in Clayton County, Georgia, suffered a severe loss when wooden I-beam floors collapsed.

Recent tests of wood I-beams, as discussed in the chapter that reviews the features of fire protection, have shown failure in just over 6 minutes in a simulated basement fire at Underwriters Laboratories, Inc. (UL). In an actual 2009 fire in Clark County, Nevada, a very small fire involving a pile of textiles in a commercial building led to near-complete destruction of several wood I-beams above the fire. Despite the fact that there was no direct flame impingement on the beams, the convection heat was sufficient to destroy the joist webs and compromise their points of support.

Wood I-joists, along with lightweight wood trusses, can be considered the most dangerous of all structural members to fire fighters today; their proliferation ensures that all fire fighters across the United States will encounter them in an actual fire. They fail incredibly fast, often soon after the arrival of fire fighters on the scene. All fire fighters, including incident commanders and company officers, must be extremely vigilant and wary when it comes to them—it is critical that the location and extent of a fire be firmly established and that the potential for collapse be assessed. Given the fire service's long list of supreme sacrifices due to buildings using these lightweight structural members, the decision to "write the building off" should be at the top of the incident commanders' strategic alternatives.

Truss Frame

The U.S. Forest Products Laboratory has developed a method of construction called truss-framed construction. This is different from conventional construction in which truss roofs, and perhaps truss floors, are substituted for sawn beams. However, the structure is still held together by nails **Figure 7-15**.

Truss frame is engineered construction in which the roof and floor trusses and studs are tied into a unitized frame. The studs are an integral part of both the roof and floor trusses. There must be meticulous attention to detail in the manufacture and erection of these structures. A chief advantage of this type of construction is speed. It is sufficient to note here, however, that the small-dimension lumber (two by fours or less) will burn faster than larger solid lumber. In addition, unsupported spans in trussed structures are subject to total collapse, and the loss of a stud due to fire could precipitate the collapse of the integral roof or floor truss. The floor truss void (trussloft) in which hidden fire can spread provides a hazard to fire fighters that affects firefighting tactics. From the firefighting point of view, these are truly disposable buildings.

Truss-frame structures are plane structural components. Their design assumes that every truss member will remain in its assigned position under load. Permanent bracing must provide adequate support to hold every truss in its design position and to resist lateral forces.

Firestopping and Draftstopping

One of the major problems of combustible construction is fire spread through hidden voids. Firestopping is often required by code to be installed to prevent the spread of fire. To the best of the authors' knowledge, there are no tested standards for wood firestopping or draftstopping.

There is a difference between firestopping and draftstopping. Firestopping and draftstopping limit the spread of fire by preventing the movement of flame, hot gases, and smoke to other areas of the building. Firestops typically limit *vertical movement* through relatively small, concealed passages such as under stairs and inside walls. Firestopping material may consist of at least 2-inch (51 mm) nominal lumber, two thicknesses of 1-inch (25 mm) nominal

Figure 7-15 Note the points of connection in this sketch of a two-story truss-frame house.

lumber with broken up lap joints, or 23/32" (18 mm) plywood or other approved materials.

Draftstops, in contrast, limit *horizontal movement* through large concealed passages such as open-web floor trusses or attics. Draftstopping may consist of at least 1/2-inch (13 mm) gypsum board, 3/8-inch (9.5 mm) plywood, sheet metal, or other approved materials, usually applied parallel to the main framing members.

A lack of draftstopping or compromised draftstopping in the attic of multiple-unit dwellings has led to their destruction. For you to have a chance at stopping a fire in an attic, the draftstopping must be properly installed and maintained. Proper draftstopping allows you to get ahead of the fire by opening up ceilings adjacent to the involved area and "surrounding" the fire.

Types of Firestopping

This text defines two types of firestopping. *Inherent firestopping* comes as a result of the normal building construction. This kind of firestopping is incidental to necessary structural purposes. An example is the floor-to-wall seal when a masonry panel wall is built directly on a concrete floor. It is reasonable to assume that such inherent firestopping is in place and effective.

Another type of firestopping, which we call *legal firestopping*, is installed with no other purpose than to meet the requirements of a code. Such firestopping may provide a barrier to the spread of fire in the interior voids of the building. However, this type of firestopping is often ineffective, and questions regarding legal firestopping may always remain: Was the firestopping properly and completely installed? Was the building inspector competent and vigilant? Was it tampered with penetrations after installation?

All firestopping must be in place to be effective. The lack of firestopping in one stud channel (the space formed by the sides of two studs and the exterior and interior walls) is sufficient to transmit fire all

the way from the cellar to the attic. In older houses, both the exterior sheathing and the lathing on the interior walls are made of wood, so all four sides of the chimney-like stud channels are combustible.

Lack of firestopping is particularly critical in balloon-frame buildings. It is important to note that firestopping was unheard of by many builders when many of these structures were being built more than 60 years ago.

If the temperature rises and the volume remains the same, the pressure rises. Building owners do not know that leaving an "inconsequential" opening in the firestopping will, in effect, create a nozzle. Nor do they realize that if firestopping is not perfect, it might as well be omitted.

Wood firestopping is often made from the cut-off ends of joists. Unbelievably, the cut-off ends of wooden I-beams have been used as firestopping. There is, of course, no seal because the wood "cut out" to make an I-beam creates a space.

There are those who believe building practices are different on the West and East coasts. Regarding firestopping, however, the authors have observed uniformly unsatisfactory practices. Recent firestopping or draftstopping practices consist of a sheet of gypsum board "buttered" into place, or a piece of thin plywood or flake board poorly fitted (and often penetrated by utilities and wiring). These techniques are about as effective as a posted notice forbidding fire to pass.

Firestopping, once installed, often is removed or penetrated for the installation of such items as heat ducts, electrical cables, sprinkler systems, and central vacuum cleaner systems. In one case, it was removed to install the fire main requested by the fire department and was not replaced. Because the firestopping is often not necessary to an installer making an improvement or fixing an existing item, it is unlikely that it will be replaced.

Trusses create a new hazard, creating a truss void or trussloft in each floor and in the attic. When this hazard is discussed, truss proponents sometimes argue that firestopping will mitigate the problem. If draftstopping or firestopping is installed by someone who understands that its purpose is to stop a gas under pressure, and not to satisfy an inspector, it may limit the spread of fire. Note, however, that even if the draftstopping installed around the perimeter of the affected space is successful, this will not prevent the collapse of the affected area—at least 500 square feet in a single-family residential building and 1,000 square feet in other buildings. This is a large enough area to develop a severe backdraft explosion or to provide a significant collapse. In multiple-unit dwellings, the recommended location for firestopping is along the tenant separations so that the entire floor–ceiling area, above and/or below the unit on fire, is a collapse area.

Protecting Wood from Ignition

It is sobering to realize that most fires are fought by fire fighters standing on, in, or under a combustible structure. Eliminating this basic undesirable combustibility characteristic of wood as a building material has long been a goal of those interested in fire protection. Early fire prevention books extolled the virtues of whitewash in fireproofing wood. In the early 1800s, ads for products like "Blake's Patent Fireproof Paint" were common.

Attempts to protect wood by encasing it in cement-like products were found to be dangerous because the wood, not being exposed to the air, tends to decay. In a San Antonio, Texas, building, one of the authors saw wood columns that had been "fireproofed" with sheet metal cladding. Serious wood decay can occur unobserved with this practice.

Impregnation

Wood cannot be made fireproof or noncombustible. It can be made fire retardant by impregnation with mineral salts, which slow its rate of burning. Some codes accept impregnated wood use in noncombustible spaces. Impregnation is accomplished by placing the wood in a vacuum chamber,

drawing out moisture from its cells, and forcing mineral salts into the wood. Sometimes the wood is pricked to increase penetration. Mineral salts do not penetrate deeply, and the removal of surface wood may destroy fire retardant treatment. Such wood is sometimes called pressure treated, but should not be confused with wood that has been pressure treated to resist decay.

Pressure treatment can significantly reduce wood's flame spread, as measured in the American Society for Testing and Materials (ASTM) E84 Steiner tunnel test. UL subjects pressure-treated wood to a test for 30 minutes rather than the usual 10 minutes. Pressure treatment can significantly reduce the hazard of wood construction. Given enough exposure to fire, treated wood will burn, although at a slower rate.

UL provides various classifications of treated lumber. For detailed information consult "Lumber, Treated (BPW)" in the latest issue of the *UL Building Materials Directory*. Such wood often is called fire retardant treated (FRT) wood.

Some chemicals previously used to impregnate the wood have leached out and corroded metal connectors. Some advertising literature for impregnated wood claims materials now used are noncorrosive.

Some years ago, to avoid parapetting fire walls through the roof, builders adopted a practice of placing FRT plywood, one sheet wide, on both sides of a firewall that did not penetrate the roof. Potential deficiencies of this method of attempting to stop the extension of the fire over the top of the firewall occur because the plywood delaminates (layers separate) and fire passes over the top of the wall. Recently, it also has been found that plywood treated with certain chemicals decays from heat and is subject to failure if walked on.

The problem is compounded if treated plywood is used not only along the line of the fire wall, but also randomly throughout many roof areas. In some cases, the entire roof is treated plywood. In an attempt to provide some sort of barrier to the spread of fire over an unparapetted fire wall, some

jurisdictions require a sheet of gypsum board to be attached to the underside of the roof, on each side of the fire wall. Presently, there is no evaluation of the adequacy of this concept.

Surface Coating

The fire hazard or flame spread of wood can also be reduced by the application of intumescent coatings that swell up when heated. As a substitute for pressure impregnation, one manufacturer suggests that all surfaces be painted with the proper coating at the job site and touched up as necessary. This method eliminates the objection by carpenters that treated wood dulls cutting tools.

One of the major problems with a surface coating is the tendency for it to be spread thinner than its recommended thickness. A gallon of surface coating must not be applied to more than the area specified for it. One inspector requires that surface coating container labels be delivered to show that the required amount of coating was purchased. It is always recommended that the owner purchase the material and hire a painter to apply it, in an attempt to achieve the recommended coverage. It also needs to be reapplied at certain intervals to retain its proper function.

Applying surface coatings to existing combustible installations such as plywood does not completely solve the problem because the unexposed surface of the paneling remains untreated. However, in one fire, intumescent-coated plywood wainscoting did not burn significantly, while an untreated chair rail molding burned completely.

Dangerous Treated Wood

Plywood and lumber treated to resist insects and moisture (pressure treated) have been used for basement walls. These materials are also widely used for exterior structures such as decks. They are, of course, combustible and might cause a collapse. The fumes from such pressure-treated wood are toxic, so scraps of this wood should not be used for

kindling. A fire investigator became ill and required hospitalization after spending some time investigating the ashes of a fire that had involved pressure-treated (for retarding decay) wood.

Plywood and Other Wood Building Materials

About 100 years ago, it occurred to someone that one of wood's limitations—the lack of shear strength along the grain—could be overcome by slicing the wood into thin layers, placing the layers at right angles to one another, and gluing the entire mass together. We know this product today as plywood, which is just about equally strong in all directions. This was the first of many ways in which natural wood was engineered into a useful product.

A basic problem of plywood exposed to fire is that it delaminates, which increases the surface area and the rate of heat release. Plywood can be impregnated to render it fire retardant. The problems associated with some fire-retardant plywood roofs were noted earlier in this chapter. Plywood can be used as an interior finish, as a building sheathing without structural value, and as a structural material in floors, roofs, or walls. Such construction is often described as stressed skin or diaphragm, in which the plywood provides some of the structural strength of the building, particularly in providing resistance to shear stresses.

Spliced and Laminated Timbers

The shortage of big trees from which solid timbers could be sawn led to the development of **spliced timbers**. Various metal connections are used to transfer loads so that spliced timber acts like a single timber.

In a fire, the heated metal connections can destroy the wood and the timber may fail. More recently, laminated timbers have been developed.

Plank-like sections of nominal 2-inch (51 mm) (or thinner) boards are glued together under pressure to produce large arches, beams, girders, and columns. Such timbers are also known as Glulam, a trade name. Sometimes bolts are used to supplement the glue. When highly finished, these timbers are very attractive. When combined with wood planks, they can provide the necessary structural strength together with an aesthetically pleasing interior finish. Laminated timbers apparently burn like solid heavy timbers and do not delaminate like plywood.

Do not mistake laminated wooden beams (which do not delaminate) with plywood as used in wooden I-beams (which does delaminate). Laminated wood sections were spliced together to form arches for a Daytona Beach, Florida, sports arena. In a fire, the wood was only charred but the arches fell apart at the metal connections.

We usually think of arches as having a characteristic segmental arch shape. Two-hinge arches of laminated wood are available, combining in one member both column and girder. They provide a straight-walled structure with a flat roof and a clear floor area. The Forest Products Laboratory in Madison, Wisconsin, is constructed of arches that provide a floor area the equivalent of five stories in height with a 60-foot (18.3 m) span.

Paper Wrapping

Laminated timbers and other finely finished wood are shipped in a protective paper wrapper. This cover is kept on as long as possible during the construction period. This paper is hemp-reinforced and coated with a bituminous moisture repellant. It ignites readily, has a high flame spread, and could contribute to a severe loss in a building under construction. In one fire, this paper was responsible for the extension of a grass fire to piles of packaged lumber.

In another case, an addition to a hospital was under construction. A wooden snow roof was erected over the excavation. Hemp-reinforced treated paper was selected to waterproof the roof. A potbellied

stove in a change room spewed out sparks. A fire could have easily sent a sheet of flame up the face of the hospital.

Planks

Tongue and groove roof planks, for plank and beam construction, were fashioned in the past by wasteful methods. Such planks are now often fabricated without waste by gluing three boards together with the center board protruding on one side and indented on the other. During a fire, such planks may separate like plywood, causing the boards to fall from the overhead. It might be wise for a fire department to run its own test on materials used locally.

Chipboard

Often wood chips are glued together to make flat sheets. These **chipboards** (also known as particleboard) are sometimes used for the floors of mobile homes. Some chipboard is water soluble and has dissolved in fires.

On a television show many years ago, large plywood or chipboard roof panels with a gypsum board interior surface were sandwiched around a thick plastic foam core to create panels. The panels were then installed between widely spaced roof beams to provide a roof that was otherwise unsupported. It could be dangerous to vent such a roof. Even away from the fire area, heat from the fire might melt some foamed plastics and the roof may fail. If the foam is uninhibited, a fierce fire with dripping plastic would be possible.

Flitch Plate Girders

<u>Flitch plate girders</u> are a composite of a steel plate sandwiched between two solid sawn wood joists, or in some cases, plywood **Figure 7-16**. These materials present a problem to fire fighters, as failure of the connection between the wood and steel or burnout of the plywood could cause failure.

Figure 7-16 A flitch plate girder.

Sheathing

Sheathing is the covering that is applied to the studs or framing of a structure. The exterior surface covers the sheathing. Tongue and groove boards laid diagonally to provide shear strength were used in older houses. Many old houses were built without sheathing, and fire spreading though walls can come out through joints in the siding. Today, plywood is commonly used for this purpose.

In recent years, low-density black fiberboard that has been moisture- and vermin-proofed has been used in residential construction because it can be installed quickly and has relatively high insulation value. (Celotex is often misused as a generic name for any low-density fiberboard. In fact, the Celotex Company manufactures many building products, and many other companies manufacture low-density fiberboard.)

Low-density fiberboard material carries a warning: "Combustible. May burn or smolder if ignited." A common method of ignition of this material is the plumber's torch. A similar material is often used for sound conditioning. In one school, sound conditioning sheathing was applied directly over wood studs and covered with a gypsum board. Later, a plumber used a propane torch to sweat a copper pipe. The flame from the torch ran along the pipe and found the fiberboard. This caused a fire that extended vertically to the metal deck roof.

The brand-new school was destroyed. It was uninsured because it was believed to be fireproof.

As noted previously, plywood is also used for sheathing. In some cases, a building is totally sheathed in plywood. In other cases, plywood is used at the corners to provide shear strength.

Gypsum sheathing is found in some localities, particularly where combustible sheathing is not permitted. It is similar to the gypsum wallboard used in interiors. Combined with brick veneer surfacing and gypsum interior surface, it is used to provide rated fire-resistive exterior walls.

Foamed plastic is now widely used for sheathing. It may or may not be flame inhibited. In any case, if exposed to fire, it may degrade and give off noxious fumes.

Siding

The siding on a building is the outer weather surface installed over the sheathing. Many different sidings are used on frame buildings. All combustible siding readily extends fire vertically, and is easily ignited by an exterior fire such as trash or burning foliage.

Wood siding is usually **novelty siding**, laid on horizontally. Novelty siding is often called clapboard, but true clapboards, which are cut from a log in a wedge shape, are rare.

Board and batten siding consists of boards laid on vertically, butt to butt, with strips nailed over the joints. Sometimes plywood is used, and **battens** are nailed on 1-foot (30.5 cm) centers to simulate board and batten. Battens have been used on painted concrete block walls to simulate such siding.

Plywood siding (sometimes referred to as "T1-11" plywood, a trademark of the APA-Engineered Wood Association) is delivered in 4-foot-wide (1.2 m) sheets, side matched. It gives the appearance of 4-inch-wide (102 mm) strips about a half inch apart. Shingles and shakes (longer, thicker shingles are called shakes) are also used for siding. As siding, they do not present the extreme conflagration hazard of wood-shingled roofs. However, they can become ignited when burning shingles or brands fly about and set fire to rubbish or dry foliage.

Asbestos cement shingles have been used both in new construction and as a replacement over deteriorated wood siding. They are noncombustible, but the presence of wood trim, and often the old wood siding, renders a building with this type of siding as vulnerable to a grass or rubbish fire as a wooden building. Asbestos cement shingles can explode when heated, and flying particles can cause eye injuries and possible inhalation hazards.

Asphalt felt siding, also known as gasoline siding, is usually made to look like brick or stone **Figure 7-17**. It is often used as replacement siding over wood. It burns readily and produces dense black smoke. Asphalt siding is very similar to asphalt roofing that meets certain roofing standards. This has led people to believe that it has some fire resistance value. It has none.

Vinyl siding is made to look like wood siding. It is a thermoplastic that will deform (droop), burn, and drip under attack from a fire **Figure 7-18**. It is often used to cover up old wood siding.

Metal siding, when used on residences, is usually made up to resemble some other material. Currently, aluminum siding is made to look like clapboards.

Figure 7-17 Asphalt felt siding made to look like bricks.
Courtesy of Glenn Corbett.

Figure 7-18 Vinyl siding melted from fire venting from an opening. Note the charred original wood siding underneath.

Courtesy of Glenn Corbett.

Embossed sheet metal has been made to look like imitation stone.

Corrugated metal siding is used on industrial buildings of wood or steel framing. Metal siding can present severe electrical hazards, both from stray electrical currents and from lightning. Thorough bonding and grounding of the siding and metal roof will do much to eliminate such hazards. This is rarely done, however, and the possibility that metal siding is energized should be a factor in firefighting.

Stucco is a thin concrete surface that can be used on any structure, such as brick, block, hollow tile, or wood. It is not correct to describe a structure as a stucco building. The correct description is of stucco over something else. In the case of wood, stucco is laid on metal lath. The lath makes it difficult to make openings in the wall. Stucco is used both in initial construction and in rehabilitation of old buildings. Some assert that the stucco makes the building safer from fire. This is untrue as far as life safety is concerned. However, the use of stucco does improve exposure protection over that of comparable wood-surfaced buildings.

In some cases, a coat of gray stucco is applied, and then a coat of red stucco is laid on over the gray. Red stucco is then removed in lines, leaving the gray stucco visible. The effect is that of brick masonry.

Brick Veneer

Brick structures have so-called good press. Brick veneer siding is popular for wood-frame residences, garden apartments, and smaller commercial buildings in areas where brick is economical. This brick is not structural; this is, it carries no load except itself. Although initially more costly than wood siding, the savings on painting over the years may make brick veneer economically attractive.

Brick veneer walls are laid up from the foundation in one wythe (one thickness of brick in a wall). Such a wall, if not attached to something, is very unstable because it is thin. In a veneered wall, galvanized steel anchors are nailed to the studs (nailing anchors to the sheathing is a poor work practice). The anchors are bent at a right angle and embedded in the mortar between two brick courses (or layers). If there is a fire in the wall, the nails may detach from the studs due to **pyrolytic decomposition**. The resultant free-standing wall is unstable. It is not stabilized by the compressive load of the structure as is a bearing wall. On the other hand, its collapse will not affect the stability of the structure. The collapse of a veneered wall of an upper story may be an impact load to a first-floor extension or an adjacent structure.

Until a few years ago, it was easy to identify a brick veneered wall compared to a solid masonry wall. All the bricks were laid as stretchers (laid lengthwise). There were no headers (laid across the wall with the ends showing). In a bearing wall, headers can be seen, usually every seventh row. More recently, the use of metal truss masonry reinforcement has made it possible to omit the headers from a solid masonry wall. If the brick wall is all stretchers, you should suspect that it is brick veneer, but you cannot be certain this is the case.

Many buildings combine brick veneer and solid brick masonry. Bearing walls may be solid masonry, while nonbearing walls may be veneer. Lower floors may be solid and upper floors veneer. This cannot be

detected after the building is finished and points to the necessity for recording information about buildings as they are built.

An old wooden house in New York had been rehabilitated by the addition of a brick veneer wall. It was heavily damaged by fire. During overhaul, an interior collapse knocked down the brick wall. A lieutenant was killed while pushing an injured probationary fire fighter out of harm's way.

Stone Veneer

Natural or artificial stone and cast concrete are also used as veneer. Permastone is one trade name. Stone can be used in the same manner as brick.

Today, natural stones are very popular Figure 7-19. Wire lath and mortar are applied to the sheathing; the stones are stacked and mortared to the wire lath base.

Wood Shingle Roofing

Wood shingles or shakes are split pieces of wood used for roofing or siding. Shakes are larger than shingles. Some of the greatest fire disasters in history have been due to the spread of fire by wood shingle roofs Figure 7-20.

If consideration of a fire is limited to the problem of the building in which the fire originated, the wood shingle roof might be a more desirable

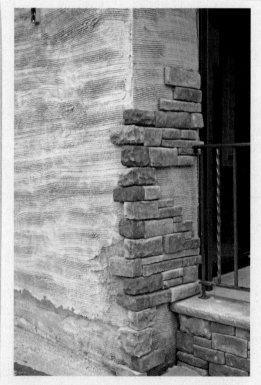

Figure 7-19 A stone veneer wall under construction.

Courtesy of Glenn Corbett.

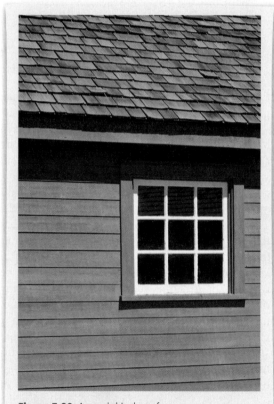

Figure 7-20 A wood shingle roof.

© Design Pics/Thinkstock.

material because by burning through, it vents the fire. However, the conflagration hazard presented by wood shingles should be the dominant consideration.

In recent years, wood shingles and shakes have made a strong comeback. It seems as if each generation must learn the lessons about these materials anew. In 1959, the NFPA warned that a major conflagration risk was present where large areas of wood shingle roofs existed. The Los Angeles conflagration of 1961 proved the prediction to be accurate. There are many areas now where the majority of houses have wood-shingled roofs. In some areas, they are permitted wherever frame buildings are permitted.

Fast fire department response, one-story buildings, wider spacing between buildings than in bygone years, and the fact that we have no extensive numbers of shingles on closely spaced 2½-story buildings that existed 50 years ago have combined to keep the conflagration rate low. But given the coincidence of hot dry weather, brush fires that engage a large portion of available firefighting forces, high winds, and a hot fire in a wood-shingled structure, the threat of conflagration is still very great.

On the afternoon of July 31, 1979, the Houston City Council was considering an ordinance with minor restrictions on the use of wood shingles. The ordinance was tabled. The day was hot and dry. At about the same time, the fire department responded to an alarm for a fire at the Woodway Square Apartments. By that evening, 30 apartment buildings were destroyed, hundreds of people were left homeless, and an estimated $44 million in damage was incurred. The fire spread because of the wood shingle roofs. Ironically, the owner of the apartments was in the process of replacing the shingle roofs. The next day, the shingle ordinance passed.

In the early evening of March 14, 1988, residents of apartments in Davis, California, also were the victims of a fire spread by wood shingles. Nine of 18 buildings in the complex were damaged. Over 100 University of California students lost personal property. Damage was in excess of $1.5 million.

Wood shingle proponents stress that brush fires are more responsible for conflagrations than shingles.

The Davis and Woodway Square fires, and many other fires, however, did not involve a brush fire. They were purely structural fires. Yet, wood shingles do spread fires that begin in brush, as seen in the San Diego County firestorm of 2003. The San Diego City Council, which had rejected tougher restrictions on wood shingles, was forced to address the issue after the wildfires killed 16 people and destroyed 2,427 homes.

Testing laboratories rate wood shingles in accordance with NFPA 256, *Standard Methods of Fire Tests of Roof Coverings.* Elements considered include flame exposure, spread of flame, and resistance to burning brands or flying pieces of burning wood. Roofing materials are classified as *A*, *B*, or *C*. The least fire resistant is C rated. In addition, for wood shingles, a flying brand test is required to determine if the roofing will produce brands. Wood shingles that meet Classes A, B, or C standards are available.

Know the building code provisions governing wood shingles in your area. If rated shingles are required, bear in mind that only bundles of shingles are labeled, not individual shingles. You cannot tell treated from untreated shingles by looking at them. Note also that treated shingles exposed to weather effects will lose their effectiveness.

Tile Roofs

Some homes may feature a heavy clay tile roof instead of wood shingles. Clay tile was particularly popular in the 1920s and 1930s, although it has been in use for a very long time and is still popular today **Figure 7-21**.

The Apple Valley, California, Fire Department had a narrow escape when thousands of pounds of tile collapsed while fire fighters were making a fog attack on an attic fire. As is typical, the failing roof showed no visible signs of weakness. The interior residence was clear. The fire fighters were not accustomed to thinking of roof collapse.

This incident points to the need for better exchange of information among fire departments. In January 1989, two Orange County, Florida, fire fighters died when the truss-supported tile roof of a store, in which

Figure 7-21 An older home with a clay tile roof.
Courtesy of Glenn Corbett.

the attic was involved, caved in on them. In a now famous video, three members of the Phoenix, Arizona, Fire Department fell through a tile roof supported on light trusses after ventilating the roof. Review of the fire on videotape revealed that the acceleration of the fire, immediately after ventilation, seemed to provide the final weakening of the roof supports.

Imitation Timber

Styles in interior decor seem to run in cycles. In contrast to the modern interiors that were popular a few years back, many current styles feature wood. Very often wood is made to appear structural when it is not. Often it is not even wood. Few if any of these practices are effectively regulated by code. Some of these deceptions should be noted:

- Unprotected steel beams or columns boxed in wood to look like wooden structural members.
- Unprotected steel encased in plaster surfaced to look like wood.
- False wood beams that are actually hollow wooden boxes. (The support system should be examined. Decorative beams have been found suspended from aluminum wire. In one case, they were just glued to the ceiling.

The temperature up at the ceiling can be hundreds of degrees higher than it is at the floor level.)
- Polyurethane imitation wood beams and fittings such as brackets. (These are readily ignitable and burn furiously. Such imitations are often found in conjunction with plywood veneers. The hazards include massive fire load overhead and the potential for collapse.)

It is becoming fashionable to make a masonry building appear to be of wood construction. The Teton Lodge, in Teton National Park, Wyoming, is made of concrete. The wood formwork marks were left evident and the entire building stained brown. Much of the interior finish is plywood.

Brick or block buildings have been covered with a wood veneer that is mounted on furring strips, making it easy for the wood to burn on both sides. In one case, a concrete block store was covered with shingles stapled to plywood sheets and mounted on vertical wood trusses. The shingles were said to be treated, but apparently the plywood and wood trusses were not.

Making Wood Construction Safe

Wood is a uniquely renewable resource. It has served humanity for ages and will continue to do so. Unfortunately, it is combustible. The problem of its combustibility can be dealt with only by complete automatic fire sprinkler protection, being properly designed with adequate water supplies, and being competently maintained. Structures with unsprinklered floor or attic voids do not have complete sprinkler protection.

Given this protection, a wooden structure is superior from a life safety point of view to many unprotected fire-resistive structures. A case in point: visitors to Teton and Yellowstone Parks will see two hotels. One is of concrete but it was lined with plywood when last seen. There were no sprinklers. Another is an old wooden hotel that is fully sprinklered. The authors would much rather sleep in the fully sprinklered wooden hotel.

TACTICAL CONSIDERATIONS

1. Fire fighters operating on, in, or under burning wooden structures are in a hazardous situation. Wooden structures under attack from fire are on their way to collapse. Always have an exit plan and be prepared for immediate evacuation. There should be a clear signal when an emergency evacuation is ordered. Some fire departments use a series of short air-horn blasts, supplemented by an emergency radio message. Today's fire communication centers have the ability to transmit an "evacuate the building" tone over all the portable radios on the fireground, but voice transmissions over the radio alone are insufficient. All fire fighters should be trained to evacuate the building immediately upon hearing the signal. Leave the equipment behind. Don't try and retrieve it—no length of hose or tool is worth a fire fighter's life. However, if you have a tool in hand, don't drop it. You may need it in case you or your crew become trapped.

2. Fire departments should drill on "Evacuate the building!" Consider ending your hands-on training drills with this evolution. Fire fighters need to practice a quick exit in realtime. They may have only a second or two to survive the event, so make them hustle. In one training evolution, a fire fighter was using a hose strap (webbing) to operate a 2½" (64 mm) hose line. The rookie fire fighter had the strap over his head and across the shoulder. The training officer saw the potential entanglement hazard and yelled, "Look out! The wall is collapsing!" The fire fighter quickly shut down the nozzle, dropped the hose line, and ran. Not realizing the strap was over his head, it yanked him to the ground like a dog on a short leash. The training officer came up to the rookie to inform him that he did not survive the wall collapse, then proceeded to show him how to properly use the hose strap without putting it over his head.

3. Most fires are fought by fire fighters standing on, in, or under wooden structures. Thin wooded sections can have a very rapid flame spread. Remember, in newer construction, floors are stronger than roofs.

4. If fire fighters are not successful in getting ahead of the fire, confining it, and extinguishing it, building collapse is usually the final event in any structure fire.

5. Wood construction creates combustible void spaces in which fire can hide and burst out. These spaces can also collect superheated smoke and fire gases that can lead to a backdraft or smoke explosion.

6. The key word in glued laminated timbers ("Glulams") is *glue*. Glue accelerates the flame spread and burn speed, weakening what would otherwise be a large timber cross section.

7. OSB or oriented flakeboard quickly succumbs to fire—but it can also lose its structural strength and fail when it becomes saturated with water.

8. Plenty of balloon-frame construction can be found throughout the United States, and it will be here for decades to come. The key point to remember about balloon-frame structures is that they have channels between the studs that are open from the cellar to the attic. In other words, there are no firestops. Teach your crew to identify these structures. One indicator is tall, narrow windows in older construction homes.

9. The basement of a balloon-frame building is usually the worst place for a fire to start. The path of least resistance for heat is up. Without firestopping, the fire can travel throughout the entire structure unchecked. Fire will move quickly up to the attic. You need to immediately send a company to check the attic with a thermal imager.

10. When a fire has entered the inner structure of a balloon-frame building, it can quickly spread to every part of the building in all directions. Investigate all parts of the building immediately. Be aware of the potential for intense fire buildup in void spaces. Don't wait for the smoke, heat, and fire to make conditions untenable before you decide to check the attic. ICs should make use of every available thermal imager on the fireground.

11. If a fire starts in the basement, you need to quickly check the attic. If a fire starts on Floor 1, Floor 2, or above, you need to check the basement because the

wall spaces are open channels where burning embers and debris can fall from the upper stories down to the foundation. This can start a new fire from below. As fire travels up, the two fires can join, trapping fire fighters above.

12. In a balloon-frame building, the problem with an interior attack is that fire moves up through multiple stud channels. As the first channel is opened up, visibility becomes obscured with smoke and steam. If a thermal imager is not available, it's easy to miss a channel and the fire may get past you.

13. Many fire departments today lack the bent swivel tip that was used on the old controlling nozzle. With the swivel tip, the fire fighter could quickly give it a dash up and down the stud channel. This is not easily done with today's equipment. Many old Navy fog applicators can still be found, however; they are an excellent tool for extinguishing these types of fires. Another effective nozzle is a piercing nozzle. The spray produced by this nozzle converts to steam and travels through the same channel as the fire.

14. The safest and most effective way to attack a fire in a balloon-frame structure is by removing the exterior siding. Fire fighters working from ground ladders can quickly rip the siding, exposing the channels and the wall studs. The exterior sheathing is cosmetic, while the interior sheathing is often part of the interior wall. The wall studs don't always carry the weight of the wall. The 1- by 6-inch (2.5- by 15.2-cm) wood sheathing or lath and plaster carry a lot of the weight of interior walls. That's why it is best to use a piercing nozzle or fog applicator for interior attacks.

15. In a wall fire, the attic is considered an extension of the wall. Resist the urge to cut a ventilation hole in the roof. This will "complete the chimney" and draw the fire into the attic and the structural members of the roof. If the roof stays intact, the atmosphere, though charged with heat and smoke, is an oxygen-deficient atmosphere and will not sustain a fire. This allows crews to extinguish the wall fire from the exterior fire.

16. Soffits or void spaces can exist in built-in cabinets in a kitchen, the underside of stairways, bathrooms, or projecting eaves. In balloon-frame construction, these void spaces, which generally lack firestops, can be a direct path for fire to enter the wall and joist spaces and can allow the fire to quickly spread vertically and horizontally. Whenever you respond

to a kitchen fire or food on the stove that is "out by occupant" (extinguished by the occupant), don't take the person's word for it. Civilians usually don't recognize that fire can extend into the walls. The cabinets above the stove must be checked with a thermal imager. If heat registers, you must remove the cabinets and open up the walls. Remove all charred wood until you reach natural wood. Then check the attic.

17. The best and fastest tactics for protecting exposures should be studied and implemented before the fire gains a serious foothold. A heavy-caliber stream from a master stream appliance directed at a specific spot on the exposure might fan out to provide effective coverage. Remember, a water curtain doesn't protect an exposure—radiant heat can travel right through the spray. You must cover the exposure with water. If water is flowing over a combustible surface, that exposure will not burn. In another case, it might be necessary to use the slower procedure of getting lines to each floor. If fire extension protection is planned for, the defense will more likely be successful. Additional resources should be called for and staged immediately. Traffic control also can become a serious problem at such spectacular fires, so call for police.

18. Fire burning in balloon-frame walls destroys the structural integrity of the building. Collapse is a serious threat—especially in buildings that have experienced previous fires. Fires in balloon-frame buildings should be observed from the exterior by a fire officer specifically assigned to that function and positioned far enough away to see the entire building. This fire officer should watch for signs of fire spread, such as heavy volumes of smoke pushing out from voids in the building, as well as intense heat buildup and structural failure, which may not be apparent from close range.

19. Don't be fooled by a simple trash bin or dumpster fire on the exterior of a balloon-frame structure. Fire may have extended into the walls or into a soffit under projecting eaves. If the walls are clear, continue to inspect the soffit void space and the attic. In one case study from Dallas, Texas, fire in a soffit void space extended up into the attic. A previous fire had weakened the structural integrity of that roof. When a fire lieutenant (not knowing the previous history) went to the roof with his crew to cut a hole for vertical ventilation, the roof collapsed. The lieutenant, who

was not wearing his SCBA face piece, fell through the roof, became trapped in the burning attic, and was killed. Fire fighters must wear their SCBA face pieces during vertical ventilation operations. There is little or no reaction time when a roof collapses—and certainly not enough time to don a face piece.

20. Traditional church fires can be quite spectacular. Where a plank beam ceiling is already or about to become involved in fire, prepare for a heavy stream attack. Solid stream tips should be used to reach the surface of the wood with maximum water flow. Fog streams, which lack penetration capabilities, are ineffective and would be eaten up by the heavy fire. When the surface is wet, the fire is extinguished. In preplanning, select locations for master stream placement so that all areas are covered. This sort of solid stream attack worked successfully at a fire in a Florida church.

21. The compression load of a truss is carried on the top chord, typically a light wood member such as a 2 by 4 inches (5 by 10 cm) or 2 by 6 inches (5 by 15.2 cm). Loss of wood can precipitate a crushing failure in compression. The entire tensile load is carried on the bottom chord, so just one break can precipitate failure. The lightweight wood truss uses metal gang nails or gusset plates as the connectors in the web. These nails (or teeth) are only 3/8-inch (9.5 mm) deep—that's all that penetrates the wood. Keep this in mind during your risk–benefit analysis. Time is of the essence. Always cut a vent hole in a lightweight truss roof from an elevated platform or an aerial ladder, a skill that needs to be practiced because apparatus placement isn't always ideal. Remember, loss of wood or loss of a gusset plate can cause a collapse.

22. NFPA 13R deals with residential sprinklers for other than one- and two-family dwellings. One of the purposes of installing sprinklers is to prevent flashover. A sprinkler system will most likely control any fire originating in the protected area. However, if the fire originates or penetrates a void space, the sprinklers will not be in a position to control the fire. Some fire fighters are reluctant to introduce a fog applicator into an attic void space; they ask, "Isn't this technique essentially introducing a sprinkler system into that space?" In doing so, you would be creating a situation no different than if the space was already sprinklered. Sprinklers are fused by heat, not necessarily by flame contact. Consider using this tactic before opening the roof.

23. Many communities where wood construction is popular are developing farther away from urban centers, which means fire departments may have to travel greater distances with fewer members responding. Dry weather and high winds can accelerate these fires. If first-in units are understaffed, consider a "blitz attack"— a master stream tactic utilizing the booster tank. (Most booster tanks on apparatus now carry between 500 and 1,000 gallons (1893 and 3785 L) of water.) This attack needs to be backed up immediately with a supply line. When a fire is gaining momentum, you have to make the heaviest hit possible. The British thermal units (BTUs) generated by a growing fire have to be overwhelmed by gallons per minute (gpms; big fire, big water); otherwise, the fire will not go out. If a blitz attack doesn't extinguish the fire, at least it can knock the fire down to a handline-size fire (fewer gallons per minute flow). This slows the fire down, buying you some time to regroup so you can get handlines in place. As far as life hazard is concerned, if a fire is large enough to require a blitz attack, the rescue profile is probably zero.

24. In an apartment fire in Seattle, Washington, the first-in engine company was in a position where a blitz attack would have been the ideal initial tactic. When fire fighters saw people trapped on an adjacent balcony, they decided to throw a ladder and went into rescue mode instead of laying hose lines. Though no lives were lost, the delay in applying first water to slow the fire down allowed the fire gain enough headway that it ended up destroying the entire apartment complex. Remember—sometimes the best tactic to save lives and property is to put the fire out or at least to slow it down until additional units arrive.

25. In areas where many buildings have wood shingle roofs, prepare for a blitz attack to knock down the original fire quickly. Call for additional units immediately. Don't wait for confirmed extension.

26. Whether the wall is a veneer or a structural wall is immaterial to the fire fighter struck with falling bricks. Some fire fighters place too much reliance on newer helmets. No matter how strong the helmet, it is worn on a human body with limited impact resistance. The wearer of the super-helmet may find that the frontpiece becomes a belt buckle! Remember, through the helmet may be impact rated, your neck isn't.

Chapter Summary

- Type V construction includes six types of wood-frame structures—log cabins, post and frame, balloon frame, platform frame, plank and beam, and truss frame.
- In a balloon-frame building, fire can spread through all the interconnected spaces from cellar to attic and across the ceiling.
- The current practice of designing buildings of lighter and lighter materials and substituting geometry for mass has led to significant strides in the architectural field, but it has also created significant problems in the field of firefighting. Trusses are one example of such a development.
- One of the major problems of combustible construction is fire spread through hidden voids.
- Firestopping is often required by code to be installed to prevent the spread of fire. There are no tested standards for wood firestopping or draftstopping.
- Wood cannot be made either fireproof or noncombustible.
- Wood is a uniquely renewable resource, but it is always combustible. The problem of its combustibility can be dealt with only by complete automatic sprinkler protection, being properly designed with adequate water supplies, and being competently maintained.
- Structures with unsprinklered floor or attic voids do not have complete sprinkler protection.

Key Terms

Asbestos cement shingle A "fireproof" roofing shingle that is composed of cement reinforced with asbestos fibers.

Asphalt felt siding A type of siding manufactured by saturating a dry felt with asphalt, then coating it with a fine glass fiber.

Balloon frame A wooden structure in which all vertical studs in the exterior bearing walls extend the full height of the frame from sill to roof. No firestops are present within the walls.

Batten A type of siding that uses vertical strips of wood to cover joints.

Bottom plate A structural member laid on the subfloor where the vertical studs are installed.

Bridging Braces placed between parallel-frame members to prevent movement from their vertical axes. Also known as cross-bracing.

Cantilevered A type of construction in which an overhang is supported from only one end where one floor extends beyond and over a foundation wall.

Chamfer To cut off the corners of a timber to retard ignition.

Chipboard Wood chips that are glued together to make flat sheets. Often used in the floor construction of mobile homes.

Corrugated metal siding Grooved panels utilized on industrial buildings of wood or steel framing; it poses an electrical hazard from stray electrical wires or lightning.

Curtain wall A non-load-bearing wall that carries no weight other than its own. It is installed only to keep out the weather.

End matched A descriptor for lumber with tongues and grooves at the ends.

Engineered wood In the construction trade, laminated timbers. In this text, wood modified from its natural state.

English Tudor Balloon-frame structure finished to resemble post and frame construction.

Finger joints Interlocking wood pieces that are glued together to create a longer piece of dimensional lumber.

Flitch plate girder Composite of a steel plate or plywood sandwiched between two beams.

Glued laminated timber Planks glued together to form a solid timber.

Header A joist that parallels floor/roof beams and is used to create an opening.

Heavy timber Buildings constructed with non-combustible or limited-combustible exterior walls and floors made of large-dimension combustible materials. Also known as Type IV construction.

Hip The rafter at the angle where two sloping roofs or sides of a roof meet.

Joist Wooden 2 × 8's, 2 × 10's, or 2 × 12's that run parallel to one another and support a floor or ceiling and are supported in turn by larger beams, girders, or bearing walls.

Legacy construction In terms of wood-frame construction, the use of solid wood members as opposed to lightweight wood trusses and wooden I-joists.

Log cabin Structure constructed of entire tree trunks, as opposed to boards sawn from trees.

Lumber Wood that has been sawn and planed.

Matched lumber Tongue and grooved lumber (usually lengthwise).

Metal siding A type of siding that is often made to look like another type of siding—for example, stone or clapboards.

Mortise and tenon Joints that are fitted together to transfer loads properly.

Novelty siding Combustible siding laid over the sheathing of a building. This type of siding comes in many different shapes and is laid horizontally.

Oriented flakeboard Another name for OSB.

Oriented strand board (OSB) A type of board made of layers of strands of wood cut from logs, with a fairly constant width-to-length ratio.

Plank and beam Used in modern construction, a wood-frame structure built with heavier beams that are spaced much farther apart than the traditional 16 inches.

Platform frame A structure in which subflooring is laid on the joists, and the frame for the first-floor walls is erected on the first floor.

Plywood Wood panels manufactured with the grain of alternate plies laid at right angles to develop the approximate equal strength in either direction.

Plywood siding Four-foot-wide sheets that give the appearance of four-inch-wide strips about a half inch apart.

Post and frame A wood-frame structure with an identifiable frame or skeleton of timber fitted together. Joints are constructed by mortise and tenon, fitted together to transfer loads properly.

Pyrolytic decomposition The chemical decomposition of a compound into one or more other substances by heat alone; pyrolysis often proceeds combustion.

Rafters Wood members used to support the roof sheeting and loads.

Ridge board A structural member placed on the ridge of the roof onto which the upper ends of rafters are fastened.

Rough carpentry All of the wood framing members and sheathing of a building. Also known as framing.

Rough lumber Lumber that is left as sawn on all four sides.

Sawn beam Wooden beam sawn out of a tree trunk. To some extent, these beams have been replaced by the wooden I-beam due to the weight and cost of the true wooden beam.

Sill Component found on the bottom of the frame of a wood structure. It rests on and is bolted to the foundation.

Soffit False space above built-in cabinets, usually in a kitchen, or in the undersides of stairways and projecting eaves.

Spliced timber Due to the shortage of solid timbers, these members are joined together

by various metal connectors to transfer loads so that the spliced timber acts as a single member.

Spline Wooden strips that fit into grooves in two adjacent planks to make a tight floor.

Stringers A component of a set of stairs used to support risers (vertical stair components between treads) and treads (horizontal stair components on which people place their feet).

Stucco An exterior plaster finish made of Portland cement.

Stud A column in a frame building, usually nominal 2 by 4 inches (2 × 4) or 2 by 6 inches (2 × 6).

Subflooring Laid on top of the joists, the structural member that serves as the base for the finished floor.

Top plate Top horizontal member of a wood-frame wall that supports the ceiling joists.

Trimmer A piece of lumber that supports a header and can be found in floor openings such as stairwells (runs perpendicular to floor joists).

Trunnel Wooden peg used to pin together mortise and tenon joints.

Truss frame Developed by the U.S. Forest Products Laboratory, a type of wood frame that substitutes sawn beams for truss members used in roofs and floors. It is held together with nails.

Truss void Found within a truss roof system, a void space located between the top floor ceiling and the roof. Also known as a cockloft.

Valley The lower slope formed by the connection of two inclined planes of a roof.

Vermiculite Mineral used as bulk insulation and as an aggregate in insulating.

Vinyl siding A thermoplastic that will deform and drip when introduced to a fire situation. Often made to look like wood siding.

Wall bearing A descriptor indicating a structure in which the entire structural load is carried on the walls; the opposite of a curtain wall.

Wood A hard fibrous material forming the major part of trees. It is usually milled or otherwise processed for use in construction.

Wood framed Building with exterior walls, interior walls, floors, and roofs made of combustible wood material.

Wood lath Narrow, rough strips of wood nailed to studs. Plaster is spread on wood laths. Generally no longer used, wood lath is present in many existing buildings.

Wooden-walled building A wall-bearing building that carries the load of the structure and the contents.

Case Study

Your last company officer often reinforced the message that the two most important things to learn in the fire service are fire behavior and building construction. You have just been promoted and have been assigned to a ladder company. You decide to hold a training drill in which you will discuss wood-frame construction.

1. While discussing different terms to be used, you mention that not all wood buildings are considered to be "wood frame." This leads to some discussion. What would be your explanation?

2. You discuss the different types of wood-frame buildings. Which type is not one that you would discuss?

 A. Plank and beam
 B. Bowstring
 C. Balloon frame
 D. Log cabin

3. While discussing lightweight wood trusses, you explain that some lightweight wood trusses are joined with metal gusset plates and others are joined with adhesive and a _____.

A. trunnel
B. flitch plate
C. finger joint
D. moment connection

4. As you move on and start to discuss the hazards associated with lightweight wood I-joists, you make certain to mention that in a simulated basement fire, they failed in just over _____ minutes.

A. 10
B. 6
C. 12
D. 15

Challenging Questions

1. Identify six examples of wood-frame buildings.

2. Describe the similarities and differences between firestopping and draftstopping.

3. Explain the possibilities of fire spread in a balloon-frame house.

4. You are a fire fighter assigned to protect an alleyway between two old wood-frame apartment buildings, both sided with asphalt shingles. What are your concerns?

5. You are an engine company officer assigned to attack a fire on the second floor of a three-story wood-frame apartment building with wood truss floors and a wood truss roof. As an engine company officer, what are your concerns for the safety of your company while operating in this building?

Flames: © Dix/Dreamstime.com; Steel texture: © Sharpshot/Dreamstime.com; Chapter opener photo: © Jones and Bartlett Learning. Photographed by Christine McKeen.

CHAPTER 8

Heavy Timber and Mill Construction

OBJECTIVES

At the conclusion of this chapter, you will be able to:

- Identify and describe the characteristics of heavy timber buildings and specific characteristics of mill construction (Type IV construction).
- Identify the dimensions of heavy timber members.
- Describe why the misnomer "slow burning" is inappropriate for heavy timber buildings.
- Identify the specific hazards of these conflagration breeders.
- Identify specific problems with mill construction.
- Identify the dangers of concealed spaces in renovated mill buildings.
- Identify the differences between new heavy timber buildings and those built in the past.
- Describe the hazards of vacant heavy timber buildings and structures under demolition.

Case Study

Figure 8-1 The Passaic, New Jersey, conflagration of Labor Day 1985.

Courtesy of Glenn Corbett.

While not as common today as they were in the 1800s and early 1900s, conflagrations still erupt in cities today. At the heart of some of these conflagrations are the heavy timber mills of yesteryear.

On Labor Day of 1985, an arson fire began in a stack of vats of naphthalene pushed up against an old mill in Passaic, New Jersey. The fire spread rapidly to the interior of the adjoining multistory factory. Although the complex was very old, it had become the home to dozens of small businesses.

Upon their arrival, fire fighters stretched supply lines to a yard hydrant and began applying water to the fire. Unbeknownst to them, the water supply to the yard hydrant system had been shut down years earlier. They quickly ran out of water and were forced to pull out.

Fire spread throughout all of the buildings in the complex (**Figure 8-1**). It grew so massive that attempts at containment within the complex were unsuccessful. Immense quantities of radiant heat spread fire to numerous exposed three-story homes and another mill across the street. A lack of available water (the city hydrants were used to capacity) prevented the exposed mill from being extinguished with handlines at the ready. Brands landed across the river that bounded the complex, starting additional fires. Ultimately, 1 fire fighter died of a heart attack, 60 businesses were lost, and dozens of homes were consumed.

1. What threat do old mill complexes present to a community?
2. How important is the connection between fire prevention personnel and firefighting forces?
3. How is water best used in a developing conflagration?

A History of Heavy Timber and Mill Construction

Heavy timber construction, and its subcategory of **mill** construction, had its origins in medieval England. "Half timber" homes utilized exposed large-dimension, hand-hewn structural members connected with mortise and tenon joints; exterior walls featured the exposed wood members, with brick or plaster filling the gaps between them. (The aesthetic appeal of this look was revived in the English Tudor–style buildings of the 1920s and 1930s discussed in the chapter on *Wood Frame Construction*, although they were not actually constructed as half timber structures.)

Half timber homes were constructed in early American settlements. Later, post and frame construction (discussed in the *Wood-Frame Construction* chapter) used basic elements of the structural framework of half

Figure 8-2 The Barker Tavern in Scituate, Massachusetts, is one of the oldest remaining post and beam buildings in the United States, dating from 1624.

Courtesy of Glenn Corbett.

timber construction **Figure 8-2**. Some of these buildings (such as those constructed from the Dutch sandstone

of New York and New Jersey) used stone walls, whereas others used wooden walls. By the 1800s, smaller dimension wood members had replaced large timbers in residential buildings.

The use of large wooden members was not completely abandoned, however. Large structures, such as barns and churches, continued to use these timbers. From the early 1700s, churches also incorporated an important feature—masonry walls.

Also of note were the mills—that is, the factories that emerged during the Industrial Revolution. In the 1790s, the first mills established in the United States, including those in Pawtucket, Rhode Island **Figure 8-3**, and Paterson, New Jersey, used large timbers for the structural frame, but maintained wooden exterior walls.

It became readily apparent that while the massive beams and columns were difficult to ignite, the exterior wooden walls of such structures were particularly vulnerable to a fire, particularly exposure fires. The construction of mills in the United States quickly shifted to masonry exterior walls (as churches had done many years earlier) during the first half of the 1800s. This new form of factory became known as "mill construction."

The specific features of mill construction were a matter of local preference and insurance regulations. It was not until the late 1800s, as a result of a movement led by insurance executive Edward Atkinson of the Boston Company, that a standardized form of mill construction emerged. Beyond his advocacy for substantial wood members (an obvious position for an insurance executive), Atkinson pushed for the elimination of concealed spaces as well as vertical elevator and stair shafts *outside the main structure*.

Mill construction fell out of favor around World War I with the arrival of reinforced concrete **Figure 8-4**. While the technique of mill construction is no longer being used for new construction, many mills are still in existence in the United States.

Characteristics of mill construction include the following features:

- Exterior load-bearing and non-load-bearing walls are solid masonry, usually either brick or stone, such as granite or sandstone **Figure 8-5**.
- Columns and beams are of heavy timber with cast-iron connectors used to cover joints where the fire might obtain a hold. **Chamfered columns** have beveled corners for the same reason.

Figure 8-3 The Slater Mill in Pawtucket, Rhode Island, built in 1793.

Courtesy of Joseph E. B. Elliot/Library of Congress Prints and Photographs Division.

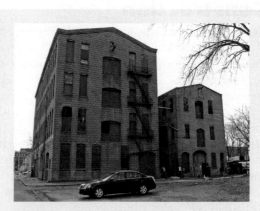

Figure 8-4 A reinforced concrete factory mimicking the look of heavy timber.

Courtesy of Glenn Corbett.

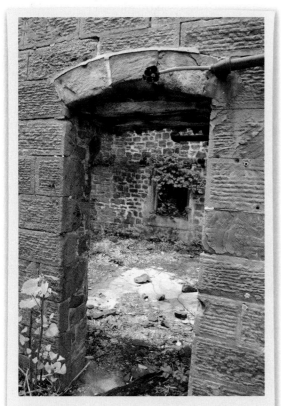

Figure 8-5 The remains of the 1836 Colt Gun Mill, Paterson, New Jersey. Note the use of sandstone for the exterior walls.

Courtesy of Glenn Corbett.

- Floors are of thick grooved, splined, or laminated planks Figure 8-6.
- Roofs of thick splined or laminated planks are supported by beams or timber arches and trusses Figure 8-7.
- Openings between floors are enclosed by adequate fire barriers.
- The ends of girders are **fire cut** to release in the event of a collapse without bringing the wall down. Sometimes a **cast-iron box** is built into the wall to receive the end of the girder.

Sometimes the beams are set on a **corbelled** brick shelf Figure 8-8.

- There are **scuppers** (drains in the wall) to drain off water Figure 8-9. The combination of waterproof floors and scuppers is provided to reduce water damage on lower floors. If the scuppers do not drain, there may be a serious increase in the floor load due to retained water. The scuppers should be checked for debris. A broom handle will hold the scupper open and increase the flow. In removing the water that flowed from a ruptured sprinkler riser in a huge Navy warehouse at Norfolk, Virginia, fire fighters discovered that the concrete floors had been graded uphill to the scuppers, making them ineffective. Note that scuppers were provided in sprinklered buildings to remove water, but not in unsprinklered buildings. What did the designers think the fire department was going to use to extinguish the fire? (Fire protection is illogical at times.)
- Concealed spaces are eliminated. The finish is "open" without voids, such as can be created above dropped ceilings or behind wall sheathing. If a paneled office or display room has been constructed within the open floor area of a mill building, it may provide a destructive hiding place for fire. Such spaces can prevent detection and extinguishment.
- A large number of windows are used to provide maximum lighting to the interior of the mill. Monitors found on some peaked mill roofs provided ventilation and light to the top floor Figure 8-10.
- One-story "sawtooth" mill structures became popular around the turn the early 1900s Figure 8-11. They featured a set of windows at the vertical portion of each "tooth."
- The protection of vertical openings and the division of the building into sections by fire walls are vital. If fire walls are to have any meaning in limiting the loss to the area in which the fire starts, openings in them must

Roof Planking

Flush type roof used where parapet wall protection is not required.

Roof beams anchored to girders, roof girders to top story columns and to exterior walls

Roof Beam

Roof Beam

Roof Beam

Paper

Roof Girder

Wall plate bolted to wall.

Finished Floor

Top story columns anchored to floor girders.

Scupper

Roof Girder

Floor Planking

Floor Beam

Floor Girder

Wood Bolster

Floor Beam

Floor Beam

Floor Girder

Anchor Bolt

Floor Beam

Floor Girder

Column

Malleable iron post cap (see details for other types of connections)

Corners Chamfered

Figure 8-6 Construction of a mill floor.

Roof Planking

Flush type roof used where parapet wall protection is not required.

Purlin

Purlin

Purlin

Wall plate bolted to wall.

Purlin anchored to trusses

Pitched roof truss with two-piece spaced members.

Bolts and timber connectors at truss joints and splices.

Anchor Strap

Figure 8-7 Construction of a mill roof.

be protected by operable, self-closing, or automatically closing fire doors suitable to the intensity of fire to be anticipated.

- Most important is an automatic sprinkler system with a water flow alarm and an alarm connection to the fire department. A dependable system of supervisory locks, tamper alarms, and visual checks is necessary to make sure that the sprinklers are continually in service. If repairs must be made, special fire prevention and fire suppression precautions must be taken.

- Special hazards—particularly hazardous industrial processes or hazardous materials—should be located in detached buildings.

Figure 8-8 A

© Justin Kase zsixz/A

Scuppers in an exterior wall, in line with

Corbett.

Figure 8-10 A heavy timber building with a covered-over roof monitor.

Courtesy of Glenn Corbett.

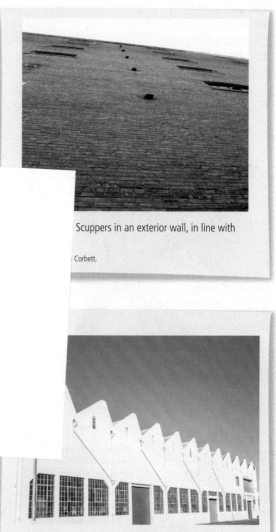

Figure 8-11 A mill with a sawtooth roof.

© Brians/ShutterStock, Inc.

Properly built and maintained, a mill building can be a structure in which fires can be brought under control before the building is involved in the fire. If one or more of the vital characteristics is absent, either by original design or because of the way the building was modified, the building can become involved in the fire. Once this happens to any substantial degree, the heavy timber characteristic

becomes a liability, not an asset. The so-called **slow-burning** characteristic is of no value once the fire department must fall back to defensive tactics, and in truth the structure becomes long burning. It is hard to understand how a living tree in the forest that has not been rained on in 3 months presents "serious fire danger," whereas a tree trunk used in a building that hasn't been rained on in over 100 years is "slow burning."

⊕ Branniganism

"Slow burning" is a misnomer when it comes to heavy timber buildings.

Bolt: © iadams/Shutterstock, Inc.; Texture: © Eky Studio/Shutterstock, Inc.; Steel: © Sharpshot/Dreamstime.com

Slow Burning?

Proponents of heavy timber construction often advance the term *slow burning* as an unqualified advantage because of collapse resistance. In fact, the slow-burning characteristic is an advantage only as long as the fire department can maintain interior offensive operations. Once the fire department must operate defensively, slow burning delays extinguishment and prolongs air pollution and exposure to toxins for fire fighters and civilians.

Once beyond an interior offensive attack, be prepared for a massive fire, perhaps the biggest of your career. A fully involved, multistory, heavy timber building is a **conflagration breeder** for a variety of reasons:

- Tremendous amount of radiant heat from each flaming window opening, which can ignite exposures hundreds of feet away.
- Production of numerous large fire brands (including wood planks), carried by the tremendous updraft of the fire erupting from the interior of the building, propelling them for hundreds or thousands of yards downwind.
- Large collapse zones for tall walls, which drop onto other buildings and start secondary fires.

- Fire spread to wood-frame homes that often surround these buildings.
- Shut-off sprinklers, the lack of or turned-off standpipe systems, open equipment shafts, and large open uncompartmented areas in old factories. The lack of sprinklers in most churches and synagogues.
- Improperly maintained fire doors and lack of compartmentation.
- Sealed-up window openings, particularly in old mills, including sawtoothed roof mills **Figure 8-12**. Ventilation will be extremely difficult or impossible.

When fighting an uncontrollable fire in a heavy timber building, expect the fight of your life. You will need a large water supply, numerous heavy-caliber streams, a large collapse zone, an extensive brand patrol downwind, and a lot of luck to prevent a conflagration (and hopefully a few empty parking lots surrounding the burning building).

A safety note: it has been observed that stair shafts often remain intact in multistory heavy timber buildings after a fire. They can potentially be an area of refuge for evacuating fire fighters, even if the rest of the building collapses.

Figure 8-12 This old mill had it's basement windows completely sealed with masonry and aboveground windows covered with metal grates.

Courtesy of Glenn Corbett.

Conversions, Modifications, and Preservation of Old Heavy Timber Buildings

Many of these sturdy old buildings have been around for a long, long time, and often efforts made to remodel them have detracted from their original fire characteristics. For example, steel trusses or girders may have been inserted, making it possible to remove some columns. The new trusses, by their nature, do not have the inherent resistance of the massive wooden columns and beams they replaced. If the new trusses were of steel, 1-hour fire protection would be required for them in many codes; however, there has been many a slip between the enforcement of the code and the construction of the building.

Many such buildings are no longer used for their original purpose **Figure 8-13**. They are used for tenant factories, schools, storage, and discount stores, and, in some cases, are being subdivided into apartments. In many instances, the basic principles set forth earlier are seriously compromised.

Fire loads are often beyond the capacity of the installed sprinklers, the sprinkler system is turned off, or the sprinkler system was removed years ago under the provisions of an older building code. Unsprinklered areas are created. The result can be and has been disaster.

The substitution of unprotected steel for heavy timber construction or the use of steel to reinforce aging or overloaded timber structures should be noted **Figure 8-14**. Such steel can fail early and trap fire fighters who believe they are operating in a heavy timber building. Fire departments should study alteration permits and investigate when building materials are delivered to the building being renovated.

Beams are often penetrated for electrical conduit, and in older buildings for gas pipes. Long bolts and nuts are used to attach timbers to other members. Any of these metal components can provide a path for heat to reach the interior of the building. Destruction of wood by pyrolytic decomposition may cause failure. Long solid timbers are hard to obtain, so watch for repairs consisting of spliced timbers with overlapping joints and metal connectors. These would be subject to early failure at the joint.

Figure 8-13 A self-storage facility in an old mill.
Courtesy of Glenn Corbett.

Figure 8-14 This mill was partially consumed in a large fire. Rather than tear the building down, the owner replaced the missing portion with noncombustible construction.
Courtesy of Glenn Corbett.

The Methodist Church of Gibson City, Illinois, is a building of masonry exterior walls and interior timber framing. It became necessary to remove a timber column carrying an 18-ton (16,329-kg) load. A one-story-high steel truss of steel tubes and solid steel tie bars was inserted into the building to permit the removal of the column. A description of this engineering feat was silent as to any fire protection provided to the steel. Unless the steel were protected, the truss could fail in a fire before the wooden column failed.

Full fire sprinkler protection, adequately maintained, is the only fire protection measure that can reasonably be expected to prevent a disaster in a heavy timber building. Problems arise when buildings are abandoned, are converted to multiple-unit occupancy, or are being prepared for rehabilitation into apartment or office occupancies. The fire department should anticipate these problems and plan accordingly. The political authorities are often inclined to "go easy" on a developer who promises to provide employment or housing or whatever is perceived as a good use of an aging building. In many of these conversions, assistance is provided by local government to better local economic conditions. Great pressure may be put on fire authorities to waive red tape and unrealistic regulations in a "slow-burning" building. The cost of sprinklers may be perceived as a stumbling block to the project, so this type of fire protection may be given short shrift in the renovation. In addition, there can be a question of who is responsible for the sprinkler system, particularly in a multitenant building.

Of particular note is the use of ceilings in newly renovated heavy timber buildings, particularly in the case of residential units and office spaces in mills **Figure 8-15**. The creation of concealed spaces (e.g., ceiling enclosures, equipment enclosures) is common in these conversions, despite the fact that current building codes do not allow such conditions. Fire fighters should point out this code violation to local building and fire code officials, who may not be aware of such an issue.

Figure 8-15 Note the sprinkler pipe "drop" in this mill conversion—is this for a future ceiling that has yet to be installed?

Courtesy of Glenn Corbett.

The fire safety of a heavy timber building depends on careful maintenance of the features listed earlier. In some cases, components such as old fire-resistive stairwell doors or elevator doors with fusible link closers may need to be upgraded and replaced.

Realistic prefire scenarios, forecasting the severe economic impacts of such a disaster, presented by top fire department offices to plant management and city officials might get attention, particularly if the loss of jobs and tax revenue is emphasized.

New Buildings of Heavy Timber Construction

While new mills are not being constructed, heavy timber construction continues to be desirable, particularly for churches and synagogues. Today, heavy timber construction is classified as **Type IV** construction. Unfortunately, a building conforming to the code definition of heavy timber construction probably lacks at least one or more of the features that the mill construction designers had learned were vital for fire safety. Typical departures from the true mill construction (and current regulations for heavy

"Dump" Buildings

"Dumps" (old, outdated, and poorly maintained buildings) are fire fighter killers. What should be done with them if your local elected officials want to save them? The obvious answer is that the building should be sprinklered. As a matter of fact, considering the hazard of fighting basement fires, some codes require sprinklers in mercantile basements or basements over a certain size.

The owners and occupants of such premises often fight the requirements for sprinklers vigorously if they are not required by code. In many cases, the building is of such low value that sprinklers cannot be paid out of "insurance savings." It costs as much to sprinkler a dump as a first-class building. But if a building is not worth sprinklering, is it worth a life? Why not abandon the outmoded concept that every dump building is worth saving? Why not identify the buildings that present an unwarranted life hazard and deal constructively with the problem created by their existence? Why not be the first fire department to write a letter like the following?

XYZ Fire Department

To the owner and occupants of the premises on 123 Blank Street:

A recent survey of your building by this department shows it to be of heavy timber construction without adequate separation into fire areas. In the event of a serious fire, the property is subject to total loss. During fires in buildings of this type, this department and other fire departments across the country have suffered unnecessary injuries and deaths. To prevent your building from becoming a death trap to our fire fighters, we recommend that you install a complete wet-pipe automatic fire sprinkler system. When these improvements have been completed and if they are maintained, it will be most unlikely that there will ever be a serious fire in your building.

We recognize that you may find these fire defense measures to be economically difficult, if not impossible. However, we have a duty to protect all of the taxpayers from the cost of injuries and to prevent the deaths of our fire fighters. Until you can comply with our recommendations, in the event of a serious fire in your building, our fire department will not permit our members to enter the building except to rescue human beings. Firefighting efforts will be confined to protecting nearby properties and extinguishing fire in your property from safe points of vantage.

In view of this determination on our part that a serious fire in your building represents an unacceptable risk of unwarranted expense to the city, and a risk to fire fighters, it would certainly seem prudent on your part to try to ensure that any fire that may occur in your building will be minor. The following recommendations are pertinent:

1. Be sure that all your employees understand that the fire department is to be called immediately in the event of fire or the smell of smoke. This is to be done before any search is undertaken or any attempt is made to fight the fire. We only have a few minutes to save your building—do not waste them. You can call us at this number: 555–1234.
2. Other pertinent recommendations [add as necessary].

Sincerely,

Fire Chief I.M. Withit

timber buildings) include cast-iron or unprotected steel columns, steel or part-steel trusses, unsprinklered void spaces (again, not permitted under building codes but found in actual buildings), highly combustible contents, unprotected vertical openings (also not typically permitted in building codes but found in actual buildings), and inadequate or no sprinkler protection.

Heavy timber structures are not being built in record numbers today, but they still find a niche for certain types of occupancies. Perhaps the most common is the modern heavy timber church, which substitutes laminated timbers (Glulam) for solid timbers (due to the inability to secure such large timbers) **Figure 8-16**. These churches also may use heavy timber trusses with masonry walls, including concrete block. Large places of assembly (such as indoor arenas and transit stations) have also used heavy timber construction recently **Figure 8-17**.

Like buildings of ordinary construction, heavy timber structures use masonry load-bearing walls and wood structural members. However, by definition, buildings of heavy timber construction have massive timbers as structural elements **Table 8-1**.

Be wary of architects and contractors who claim to be erecting a heavy timber structure. Many do not know the actual dimensions required. Such was the case when an architect claimed that an old downtown department store in San Antonio—which was to be incorporated into a new covered mall building—was a heavy timber structure. In fact, floor beams were nowhere near the 6- by 10-inch (15.2 by 25.4 cm) minimum dimension required by code. Construction plans and calculations had to be redone.

As theory goes, heavy timbers are difficult to ignite. They will char on the surface only. Structural stability is theoretically assured (this is the reason that testing laboratories such as Underwriters Laboratories, Inc., do not test heavy timber members—they have *inherent* fire resistance). Although this is generally true, the floors of most heavy timber structures that have been used as factories or warehouses have been soaked with a variety of flammable and combustible fluids over the years. This obviously alters the ignition and combustion characteristics of the floor planks and beams.

It is important to note that the fact that a particular material may burn slowly is much less important than how the structure is connected. All loads must be delivered to the ground; the entire path must be examined. Study the connections.

Vacant Buildings

Baltimore, Philadelphia, Lynn (Massachusetts), Minneapolis, Indianapolis, and Montréal are

Figure 8-16 A laminated wood truss in a modern church.
Courtesy of Glenn Corbett.

Figure 8-17 Large glued laminated timbers are used to form the roof and walls of this arena.
Courtesy of APA - The Engineered Wood Association.

Table 8-1 Minimum Dimensions of Heavy Timber Structural Members

Wood columns, sawn or glued laminated, supporting floor loads, no less than 8 inches in any dimension

Wood columns, sawn or glued laminated, supporting only roof loads, no less than 6 inches in width and 8 inches in depth

Wood beams and girders, sawn or glued laminated, supporting floor loads only, no less than 6 inches in width and 10 inches in depth

Wood beams and girders, sawn or glued laminated, supporting roof loads only, no less than 4 inches in width and 6 inches in depth

Framed or glued laminated arches (grade to floor line) and timber trusses that support floor loads, no less than 8 inches in width or depth

Framed or glued laminated arches (grade to floor line) that do not support floor loads, no less than 6 inches in width and 8 inches in depth for lower half of member height and not less than 6 inches in depth for the upper half of the member height

Framed and glued laminated arches for roofs that spring from top of walls/abutments and timber trusses that do not support floor loads, no less than 4 inches in width and 6 inches in depth

Source: NFPA® 5000™, Building Construction and Safety Code™, 2013 Edition Chapter 7, Section 7.2.5.5.1., 7.2.5.5.2., 7.2.5.5.3., page 86.

Reproduced with permission from NFPA 5000-2012, Building Construction and Safety Code®, Copyright © 2011, National Fire Protection Association. This reprinted material is not the complete and official position of the NFPA on the referenced subject, which is represented only by the standard in its entirety.

some of the cities that have suffered massive downtown fires in old combustible interior masonry buildings, some of which were heavy timber, that were being remodeled. In these types of fires, the buildings were usually wide open, with fire barriers having been removed or rendered inoperable. Sprinkler systems had been removed or disabled, and the fire had possession of the building upon fire department arrival. The only fuel was the timber interior structure of the building. Multimillion-dollar damages also were sustained by nearby structures. Many valuable assets and plans for downtown rehabilitation have gone up in smoke. In the Minneapolis fire, floors 6 through 16 of an adjacent fire-resistive high-rise office building were gutted when exposed to a fire in a department store that was being demolished. The loss was almost $100 million.

Similar problems are presented by huge structures not planned for rehabilitation but simply vacant because of economic changes. Sprinklers are often turned off to prevent freezing of the water in the pipes, and the structures, often occupied by vagrants who kindle cooking and warming fires, are an open invitation to arson. Three Detroit fire fighters died in

a spectacular fire in a vacant building. Detroit had appropriated over $300,000 only days before the fire to tear down the building.

The tragic mill fire in the Worcester Cold Storage building, which was consumed on December 3, 1999, saw many of these problems lead to disaster. While this landmark fire was discussed at the very beginning of this text, it is important to reemphasize some of the important issues—an abandoned heavy timber building with homeless people starting fires, the lack of working sprinkler protection, a combustible interior finish, and almost no windows.

Full automatic sprinkler protection is absolutely necessary to prevent the heavy timber building from becoming a spectacular fire. Some such buildings have been recycled for other uses and the sprinklers eliminated. Spectacular total loss fires are a credible risk.

The reality, unfortunately, is that sprinkler protection is often turned off in these vacant buildings. That also includes "yard" hydrants within the surrounding complex **Figure 8-18**. (Yard hydrants— hydrants with 2½" (6.4 cm) outlets and no pumper connection—were intended for use by industrial fire

brigades, which, it was envisioned, would attach handlines to them. They were never intended to supply pumpers; they should not be used unless it is clear that doing so will not steal water from the sprinkler system.) Although the sprinkler system may be turned off, it may still be possible to feed the system through the fire department connection if the system is still intact. Attempt this only if the fire is still confined to a limited area of the building and *it is safe to do so.* Preplanning is critical.

Figure 8-18 A yard hydrant in an old mill complex.

Courtesy of Glenn Corbett.

TACTICAL CONSIDERATIONS

1. Prefire plans should call for heavy-caliber, smooth-bore master streams, which provide maximum reach to cover the interior surfaces of heavy timber structures. The objective is to wet all the wooden surfaces. Due to the high British thermal unit (BTU) output and radiant heat associated with these fires, fog streams lack the velocity and penetration to be effective in these circumstances and evaporate before reaching the interior surfaces. Ineffective streams should be shut down; they waste water and rob pressure that should be used to operate master streams. Hose placement and monitor vantage points for applying large-caliber streams should be determined in advance. Dry-run drills should be undertaken. A wet drill that is effective in determining the reach of your hose streams and appliances is set up at the corner of the longest exterior wall of the building. Flow water just long enough to determine the reach and the capacity of the existing water system to effectively supply these lines.

2. Heavy timber structural elements may provide a fair measure of structural stability and a wider safety margin to operate during a fire—if the connectors and any supporting structural members are equal in rating to the largest timbers in fire resistance. If not (as is usually the case), structural collapse will be determined by failure of the weakest components. This is particularly true of timber trusses.

3. Concealed spaces (concealed floors and roof) are not permitted in heavy timber construction under the building code. Nevertheless, be wary of spaces that may have been created when an old heavy timber mill was converted to commercial work lofts or for multifamily residential use.

4. The best defense against a raging fire in an old mill of heavy timber construction is a functioning fire sprinkler system. Reliance on offensive firefighting is inadequate; access problems, radiant heat, and other unforeseen factors may inhibit manual firefighting efforts. Sprinklers are on duty 24/7 and can fuse and operate—applying water on the fire, regardless of the location and interior conditions. Apparatus operators should know the pump pressure required to support a sprinkler system. The initial pump pressure should be 100 pounds per square inch (psi). If it is a confirmed fire with fused sprinklers the maximum pump pressure should be 150 psi. Hooking up to the sprinkler Siamese connection should be an automatic task. A good driver doesn't have to be ordered by the incident commander to support an operating sprinkler system.

5. Protection against radiant heat from a fully involved mill requires the application of water directly onto the surface of the exposure buildings. The term *set up a water curtain* is often misused when protecting exposures. A water curtain is just that; it consists of water lobbed or "raining down" between the fire building and the exposure building, providing little or no exposure protection. Radiant heat waves pass right through a water curtain and can ignite the exposure building. The exposed walls and roof need to be wet from direct water application to prevent autoignition. If it's wet, it won't burn. Concentrate on exposure protection; the main body of fire in the original building is most often beyond your control.

6. Many fire departments back off from checking private fire protection systems, especially the female swivels of Siamese standpipe and sprinkler connections. They often are stuffed with trash and debris or frozen and rusted tight. This attitude can be summarized as "It's their responsibility, not ours." Not so! If you get called to a fire at this location at 2:00 in the morning, it's now your problem. Notify the fire prevention bureau of such situations. A good driver can quickly assemble the right couplings to correct the problem of frozen swivels or knows how to "back-twist" the hose counterclockwise to make the connection.

7. Expect a massive fire in a heavy timber structure in the following circumstances: if there are concealed spaces that cannot be reached with hose streams, if there

are other obstructions to interior firefighting efforts, if the contents' fire fuel has a fast, high-heat release potential, or if the building is unsprinklered. Given this massive fire, heavy water supplies will be required. A key point of the prefire plan is to determine how the water is to be applied in sufficient quantity to stop the production of heat and protect exposures.

8. Vacant heavy timber buildings that are slated for remodel or demolition may be far more hazardous than a leaking gasoline tank truck on a downtown street. Many U.S. cities undergoing urban renewal and rehabilitation have suffered massive downtown fires in these types of buildings, where the buildings were left unsecured, open to vagrants (who can start cooking and warming fires), and where sprinkler systems had been removed or disabled. Three Detroit fire fighters died in a spectacular fire in a vacant building. If the fire had a significant head start and has possession of the building upon fire department arrival, the incident commander needs to ask himself or herself, "At what point does this building no longer have any value?" Don't wear out or take risks injuring your fire fighters on loser buildings!

9. Protecting property requires a risk–benefit analysis, but it is still a core value of the fire department. Incident commanders should take calculated but limited risks in trying to save property from fire, especially in mill construction and heavy timber buildings that are occupied by businesses. Losing an operational business to a fire in a heavy timber building can have a severe economic impact on a community measured in terms of lost jobs, production, and tax revenue for the city. Fire departments should place the goal of maintaining the city's revenue and employment base high on their priority list. Speed in laying hose lines, establishing a water supply, and applying water to the fire is essential for success.

10. If the fire flow capability of available resources exceeds the required fire flow for the building, an offensive attack can be initiated. Conversely, if the fire flow requirements exceed the fire flow capability of available resources, a defensive operation is required. In other words, if you cannot deliver the required gallons per minute to extinguish this fire, you have to go defensive!

Chapter Summary

- Historically, heavy timber construction, including mill construction, has been used for large structures that have been uniquely vulnerable to fire.
- Many of these sturdy old buildings have been around for a long, long time, and often efforts made to remodel them have detracted from their original fire characteristics. For example, steel trusses or girders may have been inserted, making it possible to remove some columns. The new trusses, by their nature, do not have the inherent resistance of the massive wooden columns and beams they replaced.
- Today, heavy timber construction is classified as Type IV construction.
- Proponents of heavy timber construction often advance the term *slow burning* as an unqualified advantage because of collapse resistance.
- In fact, the slow-burning characteristic is an advantage only as long as the fire department can maintain interior offensive operations.
- Once the fire department must operate defensively, slow-burning delays extinguishment and prolongs air pollution.
- A building conforming to the code definition of heavy timber construction probably lacks at least one or more of the features that the mill construction designers had learned were vital for fire safety. Of particular note are ceilings placed directly under beams, creating dangerous concealed spaces for fire spread.
- Full fire sprinkler protection, adequately maintained, is the only fire protection measure that can reasonably be expected to prevent a disaster in a heavy timber building.
- Problems may arise when heavy timber buildings are abandoned, are converted into a multiple-unit occupancy, or are prepared for rehabilitation into apartment or office occupancies. The fire department should anticipate these problems and plan accordingly.
- Today, heavy timber structures are not being built in record numbers, but they still find a niche for certain types of occupancies such as churches and synagogues.
- Sprinkler protection is often turned off in vacant buildings made of heavy timber construction.

Key Terms

Cast-iron box An iron box built into a wall to receive the end of a girder.

Chamfered column A wooden heavy timber column that has been cut at an angle (beveled) on each of the corners to make it more difficult for fire to ignite the column at that location.

Conflagration breeder A structure that presents severe exposure problems that are capable of initiating a conflagration—a large, multiple-building fire that is not easily contained.

Corbelled A series of projections, each one stepped progressively outward from the vertical face of the wall as it rises up to support a cornice or overhanging member above.

Fire cut The end of a joist cut at an angle to permit the joist to fall out of a wall without acting like a lever and pushing the wall above it out of place.

Heavy timber Buildings constructed with noncombustible or limited-combustible exterior walls and floors made of large dimension combustible materials. Also known as Type IV construction.

Mill The earliest form of heavy timber construction; a type of factory that emerged during the Industrial Revolution.

Scupper An outlet in a wall of a building for drainage of water from a floor or a flat roof.

Slow burning A characteristic of a building that should allow a fire in that building to be brought under control before the building itself becomes involved.

Type IV Heavy timber construction.

Case Study

You are a captain assigned to an engine company that is in the process of completing a preplan for an old textile mill that is currently being converted into a new apartment building. The building is 4 stories high, 150 feet (45.7 m) in width, and 450 feet (137.2 m) in length. You take a walk through of the building, noting issues of concern for your firefighting efforts.

1. While walking across the second floor, you note wooden columns that have been cut at an angle on each corner. Which term describes this feature?

 A. Corbelled
 B. Laminated
 C. Champhered
 D. Dissected

2. On your preplan, you note that heavy timber construction is classified as which of the following?

 A. Type III construction
 B. Type I construction
 C. Type V construction
 D. Type IV construction

3. You notice that gypsum board ceilings are being installed in the apartments. Is that permitted? Yes or no?

4. You see that the old beams in the mill have fire cuts on their ends. Will they act like a lever on the exterior wall of the mill when they fall away during a fire? Yes or no?

Challenging Questions

1. What are some features of mill construction that may not be found in heavy timber construction?

2. You are the incident commander for a fire involving a 6-story, 100-year-old heavy timber warehouse in a heavily populated, urban neighborhood with numerous wood-frame apartments. Fire is quickly extending up through the building from the first floor. What are your concerns about a conflagration? How do you deal with the problem?

CHAPTER 9

Ordinary Construction

OBJECTIVES

At the conclusion of this chapter, you will be able to:

- Understand the details of ordinary construction, including features found in modern structures of ordinary construction.
- Understand how the structural stability of a masonry wall is compromised.
- Identify specific wall and wall component problems.
- Recognize collapse indicators.
- Identify the problems associated with interior structural elements.
- Identify fire hazards associated with roofs.
- Understand fire spread through void spaces of ordinary construction.
- Explain how masonry walls and fire doors act as fire barriers.

Case Study

Figure 9-1 The deteriorated bowstring truss roof of this abandoned laundry fell and killed two fire fighters in Chicago.

Courtesy of NIOSH.

Understanding the potential for collapse of a structure during a fire is at the core of studying building construction. This potential can be multiplied by the addition of problems such as age, renovation, abandonment, and subsequent deterioration. Such was the case in December 2010 when the bowstring truss roof of a 1920s-era abandoned laundry of ordinary construction in Chicago collapsed on several fire fighters, tragically killing two of them (**Figure 9-1**). An investigation revealed that a rubbish fire in a room at the rear of the laundry had extended to the wood truss roof; the extent of this extension was not readily apparent to fire fighters. Some fire fighters reported that at floor level, there was very little visible smoke immediately prior to the collapse, which occurred soon after the fire was declared under control. It is likely that water leakage had rotted the trusses when the building was abandoned.

1. What is the potential for collapse of abandoned, unmaintained buildings?
2. Is it possible for a significant fire in a bowstring truss roof to remain hidden from fire fighters immediately below the truss?
3. How does the potential for collapse differ between abandoned and inhabited structures?
4. Which type of abandoned building marking system is used in your community?

Introduction

There have been some comments that this chapter is concerned with hazards that exist primarily "back East." It is true that many of the examples cited here occurred in older eastern cities, but many of the best photos that illustrate these examples were taken west of the continental divide. Those who built the structures of the East migrated westward and exercised their talents from Seattle to San Diego. The fact that a fire department has only a few of the buildings described in this chapter in its area is one reason to learn from the bad experiences of others, so that it may avoid these problems itself.

In addition, lateral transfer is becoming more common among the highest positions in the fire service. An ambitious junior fire officer cannot afford to neglect any aspect of this subject simply because that problem doesn't exist in his or her current department.

This chapter should be studied in conjunction with the chapters on every other type of construction.

A wide range of construction characteristics—void spaces, balloon frames, tile arch floors, marble stairs, timbers, unprotected steel, plain and reinforced concrete—may be found in ordinary and other types of construction. Some elements may masquerade as being fire resistive.

Classifying Ordinary Construction

The term **ordinary construction** describes an almost infinite variety of buildings. In simpler days, it was generally known as brick and wood-joisted construction. Today, this construction uses a variety of walls and interiors. The chief common characteristic of ordinary construction is that the exterior walls are made of masonry. However, a few cast-iron-front buildings still stand, particularly in the Soho neighborhood of Manhattan Figure 9-2.

The incorrect spelling or pronunciation of words such as masonry (not masonARY), lintel (not

LENtil), and spalling (not spalDING) reflects on the competence of fire professionals and underscores the need for accuracy in usage and pronunciation. Proper use of technical words or trade jargon helps to bolster the perception of technical knowledge and authority. For example, fire fighters should avoid using the term *brick building*. Buildings of different types of construction use bricks, both real and imitation. If you mean a building with brick load-bearing walls and wood joists, say *ordinary construction*.

Ordinary construction is classified as **Type III construction**. It can be described as "Main Street, USA." When Disneyland was created back in 1955, the organization wanted to build its version of Main Street, so it chose the appearance of typical brick-front commercial buildings from the 1800s. The actual structures in the Disney parks are imitation brick on steel frame and are fully sprinklered.

Although we often think of ordinary construction in terms of old "Main Street," there are many new structures being built of ordinary construction. One of the more common examples is the one-story **strip mall** with lightweight wood roof trusses and concrete block walls.

Figure 9-2 Built in the 1850s as commercial buildings, these buildings in lower Manhattan have decorative cast-iron-front walls. Even though these walls are not masonry, they are classified as ordinary construction.

© Robert Crum/ShutterStock,Inc.

In February 1989, two Orange County, Florida, fire fighters died in the collapse of a lightweight truss roof of a store in a strip mall. The trusses spanned 50 feet (15.2 m) front to back. The roof collapsed in less than 12 minutes. The fire was above the ceiling and did not appear to be threatening from the inside. The first-due engine was the only unit on the scene when the collapse occurred.

Masonry walls in ordinary construction may consist of brick, stone, concrete block, terra-cotta tile, adobe, precast, or **cast-in-place concrete**. The wall may be all of one material, different materials may be used in different areas, or different materials may be combined to form composite construction.

Today, load-bearing wall assemblies sometimes use open-cell polystyrene panels, in which rebar and concrete are inserted into the cells for structural strength. Wood joists are hung from these panels. Even though they are eventually covered by veneers, these plastic structural members add another layer of fire safety problems for fire fighters.

Most building codes used to have a provision for so-called **fire limits**. Within the fire limits, a structure could not be built unless the outer walls were constructed of masonry to limit fire extension. Wood-frame buildings were banned inside the fire limits. These regulations were established in many older cities only after major conflagrations occurred.

As we have seen, codes and standards divide types of buildings into five broad classes. Unfortunately, many of the buildings a fire fighter must cope with were built by people who used building materials that seemed best suited to their purpose or were readily available. Often, little or no thought was given to distinctions between types of construction as classified by building codes, the National Fire Protection Association (NFPA), or insurance underwriters.

Characteristics of Ordinary Construction

The simplest ordinary construction building consists of masonry bearing walls, with wood joists used as

simple beams spanning from wall to wall. The joists usually are parallel to the street frontage of the building (or the smallest building dimension.) The roof may be similar to the floor in construction or it may be peaked by using rafters or trusses. In most cases, a **cockloft** (void space) separates the top floor ceiling from the roof. Ventilators for the cockloft are sometimes seen in the side walls **Figure 9-3**.

Bearing and nonbearing walls use similar construction materials and are often identical in appearance. In the typical downtown business or commercial building, the side walls are the bearing walls, whereas the front and back walls are nonbearing **Figure 9-4**.

The simple wood beam floor is satisfactory for buildings up to a practical limit of about 25 feet (7.6 m) in width. For a wider building, or one with an irregular floor plan, interior masonry walls or a column, girder, and beam system must be provided. There is no limit to the ingenuity of builders. Many possible combinations of building materials are used. Columns may be wood, brick, stone, concrete block, steel, or cast iron. Different materials may be used for columns in the same building. Balloon-frame stud walls may provide intermediate support. Girders may be wood or unprotected steel. The connection

Figure 9-4 It is critical that responding fire fighters know the location of load-bearing walls within a structure, which is impossible to tell from the exterior of this building.

© Kanwarjit Singh Boparai/ShutterStock, Inc.

Figure 9-3 Although not common, some multiple dwellings of ordinary construction have cockloft vents between the top floor ceiling and roof deck; in this case, they are covered with metal weather shields.

Courtesy of Glenn Corbett.

systems that attach the beams to the girders and the girders to the columns are of almost infinite variety. It is in the weakness of these connections that the principal interior collapse potential is often found.

As in wood construction, void spaces are an inherent part of ordinary construction. Some fire protection measures, such as embossed metal or tin ceilings, which were intended to prevent the extension of fire from the usable space to the void space, also prove to be barriers to the fire department's efforts to reach the fire once the fire penetrates the void space.

As a general rule, there is no effective fire separation within an ordinary construction building, either from floor to floor or within floors. Even where fire separations exist up through the regular floors of the building, they often are imperfect or nonexistent in attic spaces.

There is an inherent limit to the height of masonry buildings due to the necessity for increasing the thickness of the wall as the height of the building increases. The tallest old-style masonry bearing wall building in the United States is the Monadnock Building in Chicago **Figure 9-5**. It is 15 stories high,

Figure 9-5 Under today's standards, ordinary construction buildings are usually limited to no more than six stories, unlike the Monadnock Building shown here.

Photo courtesy of David K. Staub

and the masonry walls at the ground level are several feet thick. About the time it was built, steel-frame construction was developed, and masonry soon was being used only for fire-resistant panel walls, supported on the steel frame, and was not load bearing.

In recent years, high-rise brick or concrete block buildings with no wall thicker than 12 inches (30.9 cm), and medium-rise brick buildings with no wall thicker than 8 inches, have been developed, supplanting the traditional practice of ever-increasing wall thickness. These reinforced masonry structures depend in great measure on integration with reinforced concrete.

Masonry Construction Terms

Some of the principal terms used in masonry construction are defined here:

- **Adobe**: Bricks made of clay, water, and straw, dried in the sun. After the bricks are laid, the building is then parged (parging is defined later in this list) with a slurry of the same material.

- **Ashlar masonry**: Stone cut in rectangular units.
- **Cantilever wall**: A freestanding wall that is unsecured at the top. This wall acts like a cantilever beam with respect to lateral loads, such as wind or a hose stream. Precast concrete walls are very dependent on the roof for stability. If the roof is affected by a fire, these walls are likely to fall.
- **Cavity walls**: Hollow walls in which wythes are tied together with steel ties or masonry trusses.
- **Composite wall**: Two different masonry materials, such as brick and concrete block, used in a wall and designed to react as one under load Figure 9-6.
- **Concrete masonry unit**: A precast hollow or solid structural block made of cement, water, and aggregates (CMU).
- **Coping**: The masonry cap on top of a wall Figure 9-7.
- **Cornice**: A projecting decorative ledge at the top of a masonry wall.
- **Course**: A horizontal line of masonry.
- **Cross wall**: Any bracing wall set at a right angle to the wall in question.
- **Flying buttress**: A masonry pier at a distance from the wall and connected to it. Such buttresses resist the outward thrust of the roof. They are used mostly in Gothic architecture.

Figure 9-6 Used together, the components of this wall will support the load.

© Photos.com

Figure 9-9 A commercial building with fieldstone walls.

Courtesy of Glenn Corbett.

Figure 9-7 Rust-colored coping stones on the top of a stucco covered parapet wall.

Courtesy of Glenn Corbett.

Figure 9-8 A header course of bricks (middle row).

Courtesy of Glenn Corbett.

called piers, buttresses, pilasters, or columns. These components may be built inside or outside the building. Where visible, they indicate where the wall is strongest, often where the concentrated loads are applied, and where *not* to attempt to breach the wall.

- **Parging (pargetting)**: The process of covering a masonry wall with a thin coat of concrete.
- **Rubble masonry**: Masonry composed of random stones Figure 9-9 .
- **Rubble masonry wall**: A wall composed of an inner and outer wythe of coursed masonry. The space between is filled with random masonry sometimes mixed with mortar. Such walls are unstable to a lateral thrust.
- **Solid masonry walls**: Masonry units (either solid or hollow) laid contiguously with the joints filled with mortar.
- **Stretcher course**: Bricks laid so that the long side is visible Figure 9-10 .
- **Terra-cotta tiles**: Made of clay and fine sand, terra-cotta is fired in a kiln. Terra-cotta is both structural and decorative and is used in ornamental facings Figure 9-11 . Structural terra-cotta has been replaced to a large extent by concrete block.

- **Header or bond course**: Bricks laid so that the end is visible Figure 9-8 .
- **Hollow masonry walls**: Two wythes of masonry with an air space between. The wythes are tied together or bonded with masonry.
- **Masonry columns**: Masonry bracing incorporated into unstable masonry walls; often

Figure 9-10 Stretcher courses of bricks.
Courtesy of Glenn Corbett.

Figure 9-11 The Filene's building in downtown Boston under extensive renovation. Note the clay terra-cotta tiles, which were used to enhance fire resistance over 100 years ago, underneath the exposed second floor.
Courtesy of Glenn Corbett.

- **Unreinforced masonry**: Ordinary masonry walls are not reinforced, so they have no resistance to lateral movement. As a consequence, they are very vulnerable to earthquake collapse.
- **Wythe**: A vertical section of a wall, one brick or masonry unit thick.

Renovation and Restoration of Ordinary Construction

Over the years, most old buildings have undergone extensive modifications. Usually, such modifications have had a detrimental effect on the structure from a fire suppression and protection point of view, creating collapse potential or interconnected voids from which fierce fire can burst out on the unwary.

This is not a new problem. The site of an 1890 fire in which 13 Indianapolis fire fighters were killed, the Bowen-Merrill Building, is now home to a restaurant frequented by many fire instructors attending the Fire Department Instructors Conference (FDIC). The building was essentially a front and back with the stores on the other two sides providing the side walls. Over the years, some interior walls on the first floor had been removed and iron poles placed above the joists to support the upper floors. Additionally an arch in the basement had been removed, further weakening the structure. A fire brought the building down on the fire fighters.

"Old news!" you say. In fact, within the past several years we have seen these conditions and many others in existing buildings. It is also important to note that alterations over the years, including reconstruction from fires that may have occurred many years ago, may cause one part of a building to be quite different from another. What you see on first observation may not be representative of the entire building. Furthermore, interior alterations and finish may make it difficult or impossible to determine the true nature of the building **Figure 9-12**.

A case in point is the dining hall at the National Fire Academy, which was used during the Civil War as a temporary hospital for victims of the Battle of Gettysburg. When the building was first taken over from St. Joseph's College, it was not too difficult to trace the growth of the building over the years by examining the basement and the attic. The more recent interior improvements have made it much more difficult to do so. Cast-iron columns mark the boundaries of the original structure.

Figure 9-12 It is important for fire fighters to understand the history of the structures within their responding districts. What changes have been made over the years?

© Stephen Aaron Rees/ShutterStock, Inc.

Preservation

Old buildings may be preserved for a number of reasons. They may have undeniably historic value, such as Independence Hall in Philadelphia, or they may be a part of our architectural heritage. They may be preserved to rehabilitate a specific neighborhood or a developer may fix up an old building to create an attractive apartment, shop, or office rental space.

In such projects, it sometimes seems as if the fire safety of the occupants has a lower priority than the preservation of the original structure. In any case, it is important to remember that fire fighter safety is left up to the fire department—in short, you are on your own.

The call for preservation of older multiple-unit dwellings from the 1800s in some urban areas has

Branniganism

There is no magic formula for preventing or combating a fire in a historic vacant building. First, you should have an aggressive code enforcement division. Its goal should be to prevent a fire and work with preservationists to ensure that as many precautions as necessary are taken to reduce the threat.

Bolt: © Iadams/Shutterstock, Inc.; Texture: © Eky Studio/Shutterstock, Inc.; Steel: © Sharpshot/Dreamstime.com

led to an unusual situation. The walls of these structures, typically brick, are retained while the entire core of the building, including floors, is gutted. The wood joist floors are replaced with lightweight steel joists in some areas, while others are retrofitted with lightweight wood trusses and lightweight wooden I-beams. It is critical that fire fighters take note of such changes and plan accordingly.

The Owner's Rights

The right to own a building and to maintain it as one pleases is a fundamental right anchored in the U.S. legal system. Some laws have whittled away at this right, albeit rarely without legal challenge. Often only in the presence of clear evidence of existing public danger will the courts order an owner to repair or demolish a property.

It is certainly not a commonly held belief that the potential collapse of a building in the event of a fire is adequate grounds to infringe on the owner's rights. The fire department is left, then, with the duty of examining buildings for possible collapse and adjusting fire suppression tactics accordingly.

In addition to neglect of maintenance, alterations—with or without building permits—may prove just as deadly for fire fighters. A 2005 Bronx, New York, fire in an apartment building led to the deaths of two fire fighters and severely injured four other fire fighters when they all had to leap out of windows 50 feet (15.2 m) above the ground. Occupants on the top floor of the multiple-unit dwelling had illegally installed partitions in the unit, blocking the fire fighters' escape when the apartment flashed over. A New York State Supreme Court justice overturned a criminal conviction of the owners—the justice ruled that the prosecutors did not prove that the owners knew the partitions had been installed and, therefore, the owners could not be held accountable for the fire fighters' deaths.

Fire Resistance

The line between ordinary construction and early **fireproof** construction is not clear-cut because building development is evolutionary. Many of the

hazards of early fireproof buildings can be found in ordinary construction. Unsupported marble stairs are a good example.

Portions of an ordinary construction building may have some degree of fire resistance, either initially or as a result of legal action. Some examples include installing a properly enclosed, fire-rated stairway in an old school or providing a fire-rated barrier around a special hazard such as a boiler room. Rarely does this piecemeal provision of fire-resistive features alter the fundamental nature of the building, however.

In some codes, **concrete topping** was required over first-floor wood floors for fire resistance or to provide a sanitary floor. This layer represents additional dead weight and may confuse fire fighters as to the true nature of the floor. It may also conceal the heat below. Concrete topping on the floor was a factor in at least two building collapses that led to multiple fire fighter fatalities.

Some additions or modifications may be truly fire resistive, particularly in public buildings. It is a principle of most building codes that the fire-resistive and non-fire-resistive portions of a building should be adequately separated. Unfortunately, there are a number of cases where this principle was not carried out.

Recent Construction

Many buildings constructed in recent years have departed from simple ordinary construction, though their development does not necessarily represent an improvement. In modern construction, a non-combustible void can accumulate explosive carbon monoxide gas as readily as a combustible void. A noncombustible void can contain combustible wiring and thermal insulation.

In addition, the use of solid sawn wood joists in earlier buildings of ordinary construction has been replaced by the use of lightweight wood trusses and wooden I-beams in new buildings of ordinary construction. This presents not only the problem of fire spread in the void spaces of these new buildings, but also the possibility of larger accumulations of explosive carbon monoxide. Many new strip malls

and small business occupancies are constructed of lightweight wooden members.

Buildings of ordinary construction often use large amounts of wood "flourishes." In early 1991, a Los Angeles County, California, fire fighter suffered fatal injuries when a wooden façade collapsed. The use of wood truss roofs, wooden cornices, canopies, "colonial" belfries, combustible interior wall and ceiling finishes, and even wood veneer over masonry leaves few truly noncombustible commercial or institutional buildings.

The desire for wider spans and the availability of construction cranes have led to the widespread use of unprotected steel for roof framing. One office building is a case in point. The floors are of bar-joist construction with left-in-place corrugated metal forms and concrete topping. The ceilings are of lay-in tile. A casual inspection might indicate that the building is of fire-resistive construction, using floor and ceiling bar-joist assemblies. In fact, the building is of ordinary construction. It did not need to be fire resistive because the code does not require a fire-resistive building. The steel framing of the roof is unprotected steel. The roof itself is plywood. The ridge beam runs the length of the units. Given a fire in the wooden roof, enough heat would be generated to elongate the beam and push out the brick gable ends of the building.

In another building, steel beams run end to end through two openings in the fire walls, on which they are supported. The roof is plywood on wood trusses. The elongating steel would push down the end gables and might pull down the fire walls.

General Problems of Ordinary Construction

The problems presented to the fire department by ordinary construction can be divided into the following areas:

- The structural stability of the masonry wall
- The stability of the interior column, girder, and beam system
- Void spaces
- The masonry wall as a barrier to fire extension

These are not hard-and-fast divisions—they are all interrelated. An exterior collapse *will* usually cause an interior collapse. Conversely, the elongation of steel beams or an interior collapse may cause the walls to come down. A carbon monoxide explosion in a void space can demolish a building. A minor failure of an exterior wall may permit fire to extend to another building.

Discovery of Hazard

Many fire texts cite indications of building failure that may be observed on the fireground. Typical indicators include smoke or water flowing through walls, soft floors, a small partial collapse, walls out of plumb, and time since arrival on the scene. Although these are all good indicators, sometimes they are insufficient or, tragically, appear too late.

Early in Frank Brannigan's study of this subject, two books were noted as key texts: *Construction Failures* and *Building Failures*. Both of these excellent texts cover similar ground. The authors were consulting engineers who were often retained after a collapse to determine the cause of the failure. If an investigation after the fact could determine that the clues to impending disaster were evident before the disaster, perhaps the fire service could examine buildings beforehand and determine whether a particular building was likely to fail in a fire. These authors helped to move the consideration of fire-caused building collapse from anecdotal experience to analysis.

Many of the ordinary constructed buildings in a typical city were built before present members of fire departments were even born. There is still ample opportunity to study such buildings ahead of the time of fire and establish preplanned tactics to minimize the risk from collapse. Unfortunately, too many fire reports indicate that the fire fighters were completely unaware of situations that presented serious fire suppression and collapse problems.

In this text, many clues to potentials for disaster are given. The list of such signs could, of course, be almost endless. Use these examples as a guide to help you estimate hazards in buildings where your department must fight fire.

Some of the clues are evident from the street; others require detailed examination, which may not always be possible. In any event, be aware that an ordinary construction building, however sturdy and well maintained, was built without any thought as to what would happen to the building in the event of fire. By code definition, such buildings are non-fire resistive; they have no designed resistance to collapse in a fire.

Problems with Types of Walls and Wall Components in Ordinary Construction
Hollow or Cavity Walls

Hollow and cavity walls limit penetration by rain. Although there are no recorded cases, it is possible that carbon monoxide from a fire could accumulate in the hollow space or cavity and explode with disastrous results.

In one case, gas from a leaking main outside a building found its way into the hollow of a cinder block wall in a Florida supermarket. When filtered by seeping through the ground, natural gas can lose its artificial odorant (an additive to the otherwise odorless gas). Unaware of the gas, an employee of the Florida supermarket flicked his lighter and the gas exploded.

It is an accepted practice to place sheet or foamed-in-place plastic insulation in hollow walls. These plastics have various ignition characteristics. Burning plastic produces large quantities of smoke. If the source of smoke cannot be found, it would be wise to check for plastic insulation in the walls.

Hollow walls built of hollow terra-cotta tile present a special hazard. Originally the exterior and interior wythes were connected with steel ties. The tile industry has developed special tiles that can be

used to connect the wythes. If either wythe starts to move, a tensile load is placed on a clay tile that has no tensile strength.

Composite Walls

Masonry walls were, for centuries, built only of bricks. In recent years, the composite wall—once incorporating hollow tile, now using concrete block—has been developed. Recall that both composite wall elements must react together under a load. When composite walls were first developed, bonding the wall together was accomplished by inserting brick headers and stretchers according to various design practices.

Uneven settlement between the brick and the block has caused header bricks to crack. Consequently, the **masonry wire truss** was developed. This wire truss is embedded into the mortar in specified courses. As a result, the header course is no longer necessary, and the appearance of a masonry bearing wall may be no different than that of a **veneer wall** of all stretchers. In some veneer walls, bats or half bricks were inserted to give the appearance of a bonded wall. Thus, it is often impossible to tell a bearing wall from a veneer wall by external appearance alone. Moreover, not all brick-and-block walls are composite. Brick may be veneered onto concrete block or cast concrete using the same ties that are used in brick veneer-on-wood construction.

Cast Iron

An intermediate step in the development of cast-iron-front buildings was the use of cast-iron columns and wrought- or cast-iron arches or lintels at the street floor level carrying masonry above the first floor. This architecture enables the builder to provide larger show windows or entranceways than would have been possible with masonry.

The small area of a masonry wall available for windows led to the acceptance of walls made of prefabricated cast-iron sections. Usually a unit was cast integrally in two sections. These sections were then bolted together at the crown to form the arch. There are few buildings with cast-iron fronts left, and the demolition of one today is regarded as an act of barbarism in some circles.

Over time, the cast-iron front may separate from the masonry side walls. Some walls are tacked onto the masonry with steel straps. The columns, like interior cast-iron columns, may transmit fire vertically.

Lintels

Arches and beams are the two ways to carry the wall above an opening. Such a beam is called a lintel. The lintel commonly used today is a steel "L" or channel section. When such sections were first used, the practice was to arrange the bricks above the lintel as a false flat arch. Later, the pretense of an arch was abandoned and the bricks above the window were set vertically. Today, the common practice is to carry the brickwork horizontally across the lintel. Steel lintels are tied tightly into the masonry wall. When heated, they elongate and the masonry can fail. Combustible wood lintels are also sometimes used **Figure 9-13** .

Figure 9-13 Wooden lintels used in masonry walls will have an early failure potential.
© Naomi Marois/ShutterStock, Inc.

Imitation Brick

Some imitation brick is made by spreading a coat of gray concrete on lath. A coat of red concrete is then applied. The red concrete is cut away to expose the gray concrete in horizontal and vertical lines to simulate mortar joints.

Another method is to cement thin slices of brick onto panels of gypsum board. These layers are then mounted on a steel frame and joined together to give the appearance of a brick wall. Several bank branches have been built using this method. Imitation masonry is also created by applying a thin coat of concrete over blocks of cast foamed plastic. The exteriors of many of the buildings at Universal Studios in Orlando, Florida, are constructed in this way.

Structural Stability of Exterior Masonry Walls

Traditional fire service training on the subject of falling walls provides little real guidance in actual incidents. Although some trainers speak wisely of walls falling a fraction of their height, bricks can fly in all directions and great distances. If a single brick gets you, the fate of the rest of the wall is immaterial **Figure 9-14**.

Figure 9-14 Most fire service references will direct that a collapse zone should be set up at 1.5 times the height of the structure. When a brick wall fails and collapses, the bricks can injure fire fighters at much greater distances as seen resulting from this gas explosion.

© REUTERS/Chip East/Landov

General Collapse Indicators

Is it probable that a masonry wall will collapse during a fire? Some do; many do not. No study has been conducted to determine if there is a common pattern in fire collapses. There are, however, certain general indicators of probable collapse, some of which can be observed beforehand and noted in a prefire plan. Other indicators may be observed during a fire.

Collapse may be due to a variety of causes:

- Inherent structural instability, aggravated by the fire
- Failure of a nonmasonry-supporting element upon which some portion of the masonry depends
- Increase in the live load due to firefighting operations, specifically retained water
- Collapse of a floor or roof with consequent impact load to the masonry wall
- Impact load of an explosion
- Collapse of a masonry unit due to overheating
- Collapse of another building onto the building in question

Specific Factors That Can Lead to Collapse

Some of the signs of potential collapse are derived from building elements and materials. These can be observed prior to a fire and should be noted in the prefire plan.

Bricks and Mortar

Bricks have a reputation for toughness and permanence that can be misleading. When properly made, they can be a tough and long-lasting building material. However, poorly made bricks can deteriorate readily. Poorly made bricks absorb moisture and can fail due to freezing. Poor-quality brickwork is not uncommon.

Look for signs of poor brickwork in buildings. At times, brick defects are covered with parging (sometimes pargetting). This parging may be scored to look like brick. The parging itself may be falling off—often an indicator of poor brickwork.

Mortar that reacts chemically may "grow," forcing the masonry out of alignment. Parapet walls, which project above the roof line, are particularly subject to weather deterioration. Many such walls, as well as some chimney tops, consist of bricks held in place by gravity, with nonadhering mortar simply acting as a shim that is used to wedge a component into position.

As has been noted, the use of sand-lime mortar creates a potential failure hazard because it is water soluble **Figure 9-15** . Suspect any building built in the last century or early in this century of containing this type of mortar. To get rehabilitation tax credits, sand-lime mortar must be used to repair old sand-lime mortar buildings. Consequently, even currently restored buildings should be suspect. If a wall is torn down, look at the bricks. If they are clean with little mortar clinging, sand-lime mortar was used.

A number of nonfire collapses have occurred when water leaks washed out sand-lime mortar. The old Broadway Central Hotel in New York and the Empire Apartments in Washington, D.C., collapsed because of water leaks that washed out sand-lime mortar. A probationary fire fighter noted and reported the washout of sand-lime mortar at one of these fires. The building was cleared and it collapsed shortly thereafter.

Wood Beams

Wooden beams carry an amazing load. The destruction of the wood in a fire, in turn, may cause the failure of the otherwise impressive structure. This type of deadly construction is not apparent from the exterior, but can be detected only through competent prefire evaluations focusing on collapse potential. The practice of using wooden lintels in the interior portions of masonry walls, while showing substantial masonry arches on the exterior, is centuries old.

In a number of European buildings, heavy masonry walls are carried over openings on wooden beams. This technique is cheaper and faster than building full arches. The craftsmen who knew this technique immigrated to the United States and taught it widely; thus it has been seen across the country. Look closely at masonry buildings. Check the basement and the attic where the walls are exposed. Determine whether the wall over the openings for doors and windows is carried on masonry arches or on wood. If it is on wood, expect early failure. Stone barns have had fatal collapses resulting from fires attacking wooden lintels.

Cracks

Cracks, even if patched, indicate weakness in the wall that may be due to inadequate foundations. If there is a thrust against one part of the wall, the rest of the wall may not resist the thrust **Figure 9-16** .

A horizontal crack may indicate that the wall is being pushed out by steel roof beams that are elongating in summer heat. When the same beams contract in winter, the wall may be left out of plumb and the beams may have little bearing. This happened in the Knickerbocker Theater in Washington, D.C., in 1922. Steel trusses had elongated and contracted repeatedly over the years, pushing the hollow tile walls out of plumb. On a cold winter night, the trusses, then at minimum length, were bearing on very little of the wall. A 22-inch (55.9 cm) snowfall brought down the wall. Almost 100 people died.

Figure 9-15 Fire fighters directing streams onto a fire situation can wash this mortar out of the walls, causing the wall to fail and collapse.

© MitarArt/ShutterStock, Inc.

Figure 9-16 Any crack, no matter the size or if it has been patched, will prove to be a weakness in the wall.
© zastavkin/ShutterStock, Inc.

Arches

A brick or stone may fall out of an arch. If any arch unit (vousoir) is out, there is no arch. The window or door frame then becomes an undesigned structural post and beam. Any added or shifting weight will cause the masonry to collapse.

Wall Weaknesses

All walls are inherently unstable. One stabilizing method uses intersecting or cross walls. Typically, the front and side walls are designed to brace one another. Often the walls are made of different materials, such as brick and stone or different types of brick, which expand and contract at different rates. Look at the wall joints; in many cases, the separation is readily apparent. Other ways to stabilize a wall include the use of the following:

- Buttresses or masses of masonry outside the wall
- Pilasters or columns built inside the wall
- Wall columns, built within the wall

In a concrete block wall, the wall columns may be solid concrete blocks, concrete bricks, regular clay bricks, cast-in-place reinforced concrete, or steel. Often sizable holes are cut through walls. Until a proper lintel is installed, this gap represents a serious weakness in the wall. This often happens when building owners wish to create a usable space that spans two buildings or create a passageway between two basements to enlarge a storage area. Such an expansion in Manhattan in the 1980s led to the collapse of an entire building of ordinary construction on 9th Avenue.

Any construction work to create additional openings (e.g., windows, doors, walkways) should be done under the supervision of a structural engineer. Fire fighters are also warned to be especially careful when breaching brick walls (or any load-bearing wall for that matter) during firefighting and rescue activities for the same reason.

Steel Lintels

Steel lintels have been used for years without any protection for the steel. In recent years, there has been a great increase in fire loads and the rate of heat release of contents, due principally to plastics. Steel lintels have deformed and thrown bricks off the wall.

Reinforced-concrete lintels are commonly used in masonry walls. Because the building is not required to be fire resistive, the concrete is not required to be fire resistive. The bottom concrete may spall off, exposing the reinforcing rods, which provide the tensile strength to the composite structure, causing the lintel to fail. It is a common myth that concrete is inherently fire resistive. Concrete is inherently noncombustible. It can be formulated to be fire resistive.

Bracing

Braced walls are another basic sign that a wall is in distress **Figure 9-17**. (Their presence is signaled by stars, plates, channel sections, or other spreaders, or straps tying the front wall to the side wall.) The walls may be tied together across the building or tied to floor beams. Braces made of unprotected structural steel, or worse yet, of steel cable (complete failure at 800°F [427°C]) will fail at fire temperatures.

Figure 9-17 These ornamental stars should alert fire fighters to the instability of this wall. The cable itself has been proven to completely fail at 800°F (427°C).

© tatiana sayig/ShutterStock, Inc.

Figure 9-18 Eccentric loads like this one must be properly braced or the weight of the sign can pull the entire wall down.

© Thomas Nord/ShutterStock, Inc.

Wooden walls are tied together in a similar manner. Failure of a steel tie caused the 1975 collapse of a wooden church wall that cost a Paterson, New Jersey, fire captain his life.

Braces are not always an indication of instability. Some buildings have walls that were tied to the floors to make the structure more rigid. Generally these can be identified by spreaders in a regular pattern.

In California and other earthquake-prone areas, bracing may be used either initially or as a retrofit to upgrade the earthquake resistance of unreinforced masonry walls. In this case, fire-caused failure of the steel may cause failure of the structure.

Eccentric Loads

Some walls carry an eccentric load, such as from a projecting sign Figure 9-18 . This load must be counterbalanced, usually by tying the wall to the interior structure. A large abandoned garage with masonry walls and a wood truss roof had a projecting sign; the wall was tied to the roof trusses. The trusses and the roof burned. There were no contents. When the trusses weakened, the sign pulled the wall down, injuring 17 fire fighters. There was no anticipation of collapse.

Cantilevered beams can also project from a wall to form balconies. When confronted with a cantilever, the key question is, "What is happening to the other end?" Three New Orleans fire fighters died when a balcony collapsed and pulled down the wall. Cantilevered canopies and balconies are very common. Those supported on trusses or wooden I-beams are particularly dangerous.

Suspended beams that form canopies are quite common. They are secured by tie rods connected to the interior structure. If the connection fails, the canopy becomes an undesigned cantilever. As it falls, it will bring the wall down with it. Pictures of such a collapse in Long Beach, New York, show the parapet wall also being pulled down for about 50 feet (15.2 m). The collapse zone is not merely at the canopy, but the entire front of the structure. With suspended beams, the key question is, "What are the suspension rods tied to?"

The canopy support rods of a Washington, D.C., theater were tied to the roof trusses, and a multi-alarm fire occurred in the roof void of the building. Alert fire officers noted that the top of the front wall had been pulled out just short of the collapse point and cleared the area.

An Alabama overhanging façade attached to an auto body shop collapsed during a 2002 fire, killing a fire captain who was standing directly underneath it.

The subsequent investigation determined that the load imposed by the façade—which was composed of plywood, 2 × 4's, and asphalt shingles—had caused cracks in an underlying parapet wall, necessitating the use of steel rods attached to the building's roof to stabilize it laterally. When the roof collapsed, the parapet and façade fell as well.

Unvented Voids

Carbon monoxide gas trapped in unvented voids can detonate violently and blow down walls. In one incident, fire fighters were battling a blaze in an abandoned icehouse. Carbon monoxide accumulated in a 9-foot (2.7-m) cockloft and exploded, bringing down the rear wall. Thirty-nine people watching the fire from an apartment house died when the wall fell on the apartment house. A small boy who was saved from the rubble grew up to become a fire fighter.

Masonry walls are not designed to resist lateral impact loads. Consequently, explosions have blown down many masonry walls. In one fire, an explosion of coal dust blew down the five-story masonry curtain wall of a steel-framed building. Brick or masonry walls that separate one electrical transformer from another are often specially reinforced to resist lateral impact.

Planes of Weakness

It is difficult to get floor beams level in masonry buildings due to masonry irregularities. A simple method is used to solve this problem across the country, in buildings as much as 200 years old. A wooden beam is laid in the brick wall at the line of the bottom of the floor beams. The floor beams are then set on the wooden beam. This produces a level floor without shimming. The hidden danger occurs when the wood burns out and a plane of weakness (like the scoring to cut glass or gypsum board) is created in the wall. Even a modest lateral thrust may be enough to break the wall along the plane of weakness.

In one building, a horizontal plane of weakness was noted in a two-story wall under construction. A slot was left in the wall to accommodate a concrete floor yet to be poured. When the floor was poured, if the concrete was not worked well into the gap, a hidden weakness would be left in the wall. Such a weakness may cause a collapse. This weakness will give no warning during a fire.

It is possible to put a vertical plane of weakness into a building by cutting a chase into a wall. A chase is a passageway, usually vertical, cut into a masonry wall for a pipe or conduit. Close examination of the rear wall of an old building in Toronto disclosed an unusually weakened wall. The roof downspouts of the building were often damaged by trucks moving through the alley. A chase had been cut vertically down each end of the face of the wall, from the ceiling to grade. The downspouts were set into these chases, keeping them protected from exposure to damage. When chases are built into a wall during the initial construction process, they may represent a weakness, but at least the joints can be properly laid to forestall disintegration. In contrast, a chase cut into an existing wall presents a serious weakness. Downspouts that need protection from vehicles should not be recessed into a wall, but rather should be protected by masonry or cast iron.

In 1973, the Broadway Central Hotel in New York City collapsed. The investigation disclosed that 30 years earlier an unauthorized chase had been cut into the wall for a drain. The drain leaked water that washed out the sand-lime mortar. Without being able to cite any such experience, Frank Brannigan warned of this potential in the first edition of this text—two years before the collapse. This is just one example of how fire officers can identify a hazard by analysis, not experience.

Effects of Interior Structural Elements and Building Contents on Exterior Walls

The interior structure of a building may push down a wall in several ways. Wooden floor beams fitted tightly into the wall may act as a series of levers

on the wall; when they collapse, they will pull the wall down. The usual solution is the "fire cut," in which the end of the beam is cut at an angle so the beam can fall out of the wall. This may save the wall but weakens the beam because there is less wood to burn. Floor beams are placed with an upward camber or rise, but over the years they may sag. To restore the upward camber, the fire-cut beams are sometimes turned upside down. This often results in a very small bearing surface of the beam on the wall.

In some areas, the beams are not set into the wall, but rather sit on a masonry ledge, corbelled out from the wall. In heavy timber construction, a cast-iron beam box may be inserted into the wall. The beam end is inserted into the box. If it falls out, there is no problem with the wall.

Appearances can be deceiving. The heavy masonry walls of a Montréal church were supported on an unprotected steel grillage of three I-beams bolted together. This grillage was supported by unprotected steel columns that extended down into the crawl space. Steel columns inserted into the wall rested on the grillage and supported the roof trusses. Thus, the wall was not actually a bearing wall, but rather a curtain wall. Nothing of this construction was evident. The wall appeared to be a typical solid masonry wall. When fire in the crawl space caused the steel columns to fail and drop the wall, two fire fighters died. Be careful of potentially fatal assumptions made on the fireground.

Effects of Fire Streams on Brick Walls

Some fire texts advance the concept that wall collapse is caused by cold water hitting a hot wall, causing unequal expansion. This hypothesis is suspect. Too many brick walls have been heated to high temperatures, struck by heavy cold streams, and survived. It is likely that that the factors mentioned previously—particularly the never-cited wooden components—are probably far more significant.

However, fire streams can damage brick walls. Older brick buildings have an exterior veneer like the surface of finish brick. A heavy stream can penetrate the veneer and strip it off the wall, sending deadly missiles in all directions. Heavy streams can rip loosened bricks, as from a deteriorated parapet. Very heavy streams, such as from fireboats, can smash through a brick wall.

Interior Structural Stability

In this text, interior and exterior collapses are discussed separately, although they are interrelated. An interior collapse of an overloaded floor can cause the walls to collapse. The collapse of an exterior wall into the structure or onto the adjacent structure may cause an interior collapse. The interior structure consists of floors, any necessary interior support for the floors, any suspended loads, and the roof.

Historically, consideration of interior structures assumed collapse in case of fire. Questions arose as to whether to design connections to permit easy collapse of girders (with self-releasing floors) to prevent collapse of columns or to design easy collapse of floors to prevent collapse of walls with fire-cut joists. The opposite point of view called for tying the building together tightly and disregarding the concept of releasing the floors in case of fire.

Fire-Resistive Combustible Assemblies

In recent years, combustible floor and wall assemblies of wood and gypsum board have been developed that can achieve fire resistance ratings in standard fire tests conducted in accordance with the American Society for Testing and Materials (ASTM) E-119, *Standard Test Methods for Fire Tests of Building Construction and Materials*. They are used to produce code-classified protected combustible structures.

Interior Structural Support Systems of Columns, Beams, and Girders

The list of inherent defects of interior supports is endless. Those that are most important to fire fighters,

however, are found in the structures for which you are responsible. Observe, study, and ask questions.

In older masonry buildings, the interior walls, if not of masonry, are of balloon-frame construction. Balloon-frame wall-carrying interior loads may fail very early due to the small cross-sectional area of the studs. Masonry walls are usually much thicker than wooden walls. Brick and stone interior walls can also have all the defects of exterior walls except perhaps weathering. Stone walls and interior walls also may lose their strength due to spalling during a fire.

Adjustable steel jack posts or simple steel posts may be used to support overloaded beams. They are vulnerable to early failure. Interior merchandise racks in retail stores are often of very lightweight steel construction, and subject to early collapse.

Amateur builders often use water pipe for columns and beams, porches, and canopies. This is very dangerous. Water pipe is designed for internal pressure. The metal removed to cut threads is replaced by the heavy collar around the fitting. As a structure, this connection is weak because metal has been removed where the shear load is greatest.

Interior masonry bearing walls may stand in the way of projected improvements, such as combining several rooms. If an opening is made in the wall, or if the wall is entirely removed, this arrangement will have much less stability and less inherent fire collapse resistance than the original brick wall.

Such alterations were major factors in the Washington, D.C., Empire Apartments collapse and the Vendome Hotel collapse in Boston **Figure 9-19**. A beam and cast-iron column unconnected to the beam were often used in these structures. In the remodeling of the Vendome Hotel in Boston, an air-duct opening was cut through the wall directly below a cast-iron column that supported a brick wall. Nine fire fighters were killed in the collapse in June 1972. In the Empire Apartments, the wall above an opening was supported on a beam cut into the old sand-lime masonry wall. Leaking water washed out the mortar, and the structure collapsed with several fatalities.

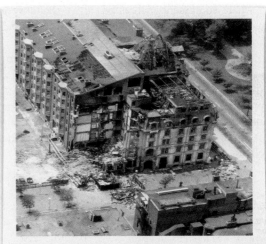

Figure 9-19 The Vendome Hotel collapse in Boston, MA.
© Boston Globe/Getty Images.

Deficiencies of Materials

Structures originally built with heavy timber columns, girders, and beams are found reinforced with added steel in a variety of ways. In such cases, the failure of the steel may precipitate collapse. Assume that any structural support that was added after the building was built was added because it was absolutely necessary.

The addition of unprotected steel to a building with a wooden interior structure is sometimes apparent, but is often fully concealed. In one case, steel bar joists had been installed between each pair of wood joists to strengthen a building. This arrangement did not become apparent until the building's demolition. Many newly built ordinary construction buildings have some unprotected steel structural members.

In structures, wood beams may have been trussed initially or at a later date. This is usually accomplished by erecting a strut downward from the center of the beam and stretching tension rods from one end of the beam to the other, over the strut. The strut or the rods may fail first. This technique has been used where heavy loads are on the floor above,

or where equipment such as belt drive shafts were suspended from the overhead, or to transfer a load formerly carried on a column.

A tea shop is located on the first floor of a commercial building in Chester, England. To clear the dining area of columns, wooden trusses were erected at the ceiling level to support the floors above. To the untrained eye, they appear to be decorations. A heavy fire in the dining area, however, might precipitate the collapse of the entire structure.

Connections in Ordinary Construction

Beam-to-Girder Connections

The only way to connect wood beams to wood girders and receive the full benefit of the time it takes the wood to burn through is to set the beams atop the girder. However, this is undesirable because it adds height to the walls. All sorts of systems were designed so that the top of the beam could be level with the top of the girder.

Old-time craftsmen are proud of their mortise and tenon joints. In such a joint, the beam was cut down from the top and up from the bottom to form a tenon or tongue. This was inserted through a mortise cut into the girder. A wooden dowel, or trunnel, was driven through a hole bored in the protruding end of the tenon. Another method of connection used notched beams on strips nailed to the side of girders.

The weakness of any method that reduces the size of the wood is readily apparent. The effective strength of the wood under fire attack is determined by the size of the thinnest portion, not the mass of the member as a whole.

Beam-to-Beam Connections

Beam-to-beam connections must be made whenever an opening is provided in a wooden floor. The joists that are cut to provide the opening, and thus do not reach to the girder or bearing wall, are connected to a header beam, which in turn is connected to trimmers. These connections are accomplished in

Figure 9-20 Joist hangers are points of weakness.
© WilliamRobinson/Alamy Images.

various ways. Mortise and tenon joints are found in older buildings. Notched beams are also common. Metal **joist hangers** or stirrups and heavy steel or wrought iron straps shaped to receive the joist and nailed to the top of the header are used. Tests made years ago showed that such stirrups are dangerous, charring the joist and softening rapidly **Figure 9-20**. More recently, very lightweight hangers have been developed for lightweight construction. The rapidity with which such hangers may fail in a fire is an important factor. The small mass of metal and the dependence upon nails do not inspire confidence.

Beam-to-Column Connections

Buildings of ordinary construction present many options for connecting girders to columns. Self-releasing floors can be seen in most old interior mercantile buildings. The girder sits on a steel shelf that protrudes from the column. A scrap of wood or a **dog iron**, like a big staple, connects the girders and imparts some lateral stability under normal conditions. Although self-releasing floors may limit the collapse area, any system that is designed to precipitate collapse must be weaker and more likely to

collapse. In any event, many of the buildings fire fighters must cope with are "vernacular" buildings, designed without benefit of professional engineers.

The principle that columns should set directly atop one another to carry loads to the ground as axially as possible has been widely violated. Where a column is offset, the girder on which it rests becomes a transfer beam and is subject to severe shear stresses unless it was designed as a transfer beam. Improperly loaded structural components are more likely to fail than similar components properly loaded.

Cast-Iron Columns

Cast-iron columns create many problems. Good practice dictated that the column be topped by a **pintle**, a solid iron pin of much smaller diameter than the column. The pintle passed through the girder, surrounded by wood, and then enlarged to meet the full width of the column above.

Pintles are often used to connect wooden columns end to end. In practice, the column passes through the girder at its full width. The ends of the girders are cut in a semicircle to make room for the column. Thus, a relatively small amount of wood rests on the cast flat shelf that extends from the top of the column. The loss of this wood or the softening of the cast-iron shelf could cause the initial collapse of the column.

Over the years, failures of cast-iron columns have been blamed for many serious collapses. It is true that cast-iron columns can fail, particularly if they are poorly cast so that the material is thin at one point. After looking at a number of buildings with cast-iron columns, it appears likely that the chief cause of their failure is the unsafe connections, rather than the questionable "cold water on red-hot cast iron" theory (although this is still a possibility). Often the cast-iron column is held in place only by gravity, so that the slightest lateral movement can cause it to kick out.

Interior Suspended Loads

Balconies, **mezzanines**, half stories, and other suspended loads, including merchandise racks, are very hazardous **Figure 9-21**. Suspended loads are becoming

Figure 9-21 Often constructed directly above the ground floor and projecting as a balcony, these overhangs can become obscured by smoke and trap heat.
© iStockphoto/Thinkstock.

more common as architects realize the space advantages of suspension over columns. In a fire, the loads in such structures may be obscured by smoke, and the suspension system may be subjected to temperatures hundreds of degrees higher than at floor level.

One grocery market was converted to a Japanese-type restaurant in which the guests sit around a grill where the chef performs as the food is prepared and cooked. Such a restaurant requires a heavy steel exhaust hood for each table. The hoods were suspended from the roof. Because of the added dead load, this roof is not the same working platform for fire fighters that it was when its only purpose was to keep the rain off the groceries.

Architects are realizing more and more the utility of hanging loads from the overhead. Interior designers are also prone to hanging heavy loads from the overhead. The overhead may be the floor above or the roof. For example, a supermarket designer wanted to have suspended shingled canopies over the checkout stands. In the resulting structure, the ends of steel rods are held by two 2×4s that extend across the bottom chords of bar joists. A fire overhead could either burn the wood away, letting the washer slip through, or cause the restrained bar joist to elongate or twist, dropping the canopy.

Another grocery supermarket was housed in an otherwise noncombustible building with a gypsum plank roof. The meat cutting area was in the rear of the structure. The health department required a washable gypsum ceiling rather than lay-in tiles. A substantial wooden structure was hung from the roof bar joists to support the gypsum board. The wood structure also supported the refrigeration units. An electrical fire occurred in the ceiling void. When a truck company was overhauling the completely extinguished structure, the company members were literally pulling the building down on themselves. Directing their attention to the hazard, Frank Brannigan asked the officer to back the fire fighters up and to have one fire fighter gingerly break away the gypsum board so he could get a picture of the poor-quality support. As the fire fighter touched the gypsum board, the entire refrigeration unit fell down.

Floors

Often, old buildings of ordinary construction housing a mercantile occupancy had open stairwells to permit unimpeded access for shoppers to move from floor to floor, including the basement. Of course, this allows fire to also travel unimpeded up through the building. In one case, heavy fire loading took the form of foam mattresses that were stacked to the basement ceiling.

When a stairway is actually relocated, the old opening is usually covered over. The closure is usually much lighter than the floor. In one observed case, 2×4s were used, whereas the floor was made of 3×10s. From above, the floor looks uniform. Such an opening has figured in at least one serious collapse.

In some store alterations, the upper floors are abandoned. The stairway is removed to provide ground-floor room. Often, a single sheet of metal is placed over the opening. A Washington, D.C., fire fighter died when he fell through a patch of sheet metal over an old opening.

A **light well** is a vertical shaft with windows that provides light and ventilation to enclosed rooms. A light well in an old **tenement** building was floored over.

The new flooring was very light and collapsed under the applied load.

In past years, floors were made of subflooring and a finished floor. In more recent construction, floors are often a single thickness of plywood and carpeting. These floors are much more quickly consumed by fire.

Floor collapse can occur early in the fire. A National Fire Academy executive, Dr. Burton Clark, was a member of a local fire company and responded with the first unit to a townhouse fire. He made a primary search of the second floor on his hands and knees, testing the floor as he went. He went through the floor into a void below. Fortunately, the active fire had moved to another area.

Roofs

Fire departments are concerned about equipment such as ropes, aerial ladders, and aerial platforms because equipment failure can cost lives. Specifications are tightly written and acceptance and in-service tests are carefully monitored. However, there is one working platform fire fighters use that has cost many lives—the roof. One of the basic purposes of a roof is to keep out the rain and enclose the structure. A secondary purpose is to stabilize the walls. A roof is not designed or constructed as a fire department working platform. Generally, fire suppression tactics assume the stability of the roof until there is warning of collapse. In fact, there are a few warning signs, but they are often too little and too late.

Fire Characteristics of Conventional Wood Roofs

The discussion that follows chiefly concerns roofs supported on solid sawn rafters and beams. Metal deck roofs are discussed in the chapter that reviews noncombustible construction. The roofs of tilt slab concrete buildings are also discussed in the aforementioned chapter. The hazards of single-ply membrane roofs are discussed in the chapter that reviews concepts of construction.

The information identified in one prefire plan simply listed architectural roof types: flat, gable, hip, gambrel, shed, mansard, dome. Such information, although useful, is incomplete. Structural characteristics as well as the appearance of the roof must be known beforehand.

Solid sawn wood contains "fat"—wood not necessary to carry the imposed load. When only this part of the wood is burning, roof strength is not greatly affected. As it continues to burn, however, the beam gradually weakens, and the roof becomes spongy. This may or may not warn fire fighters to get off the roof in time to avoid disaster.

Roof Hazards

The best roof is one in which the roof beams rest on girders. Any sort of hangars or other metal connections make the roof more vulnerable to failure. In some localities, steel U channels have been used for rafters. The effect of elongation and failure should be evaluated.

Roofs apparently supported on heavy wood beams or laminated arches can be much less reliable than they appear. Long wooden beams are scarce. Apparent long beams are often several beams spliced together with metal connectors, which can heat up and burn out of the wood.

In a 90-year-old building being renovated in Orlando, Florida, the roof joists were exposed. The joists in the center section between the wood girders (any beam supporting other beams is a girder) were short and did not rest on the girders; instead, they were nailed to joists that overhung the girders. This use of such overhanging and drop-in beams is not uncommon. The nailed connections could fail very early in the fire and collapse the center section of the roof. Even if beams resist collapse, the roof boards may be burned away.

Energy conservation and rapid completion are two important considerations in today's buildings. Some roofs are made of large panels in which foamed plastic is sandwiched between sheets of plywood. Even if the plastic is inhibited to discourage combustion, it may melt away in a fire, allowing the roof panel to fall.

In an attempt to make use of natural light, corrugated glass fiber-reinforced plastic panels are made to the same dimensions as corrugated steel. A fire fighter could easily step through the plastic when visibility is poor. When exposed to fire, the plastic will burn away. The glass fiber matting, however, may stay in place and obstruct ventilation. In 2009, a Newark, New Jersey, battalion chief fell through such a panel during a nighttime fire, falling 22 feet (6.7 m) **to the floor** Figure 9-22 . **The edge of the panel** was flush with the corrugated steel panels, making it nearly impossible to distinguish it from the steel panels.

Ventilation tactics can accelerate collapse. Be aware that the impact of an axe may be the straw that breaks the camel's back. Some think that impact is avoided by the use of saws. Not so! There is no such thing as zero impact. A fire fighter stepping onto a roof as lightly as possible still imposes a momentary load on the roof at least twice his or her at-rest weight. Of course, cutting through structural load-carrying members can initiate collapse.

Excess live loads on roofs can accelerate collapse. In one case, three fire fighters ventilating a roof died in a collapse where there was heavy ice on the roof. When they opened the first hole, the roof beams were seen to be burning. This should have been a clear warning that much of the roof's normal resistance to

Figure 9-22 A Newark, New Jersey, battalion fire chief fell through a plastic skylight panel in this roof.
Courtesy of Frank Bellina.

collapse had already been expended. Unfortunately, they went on to open another hole.

Water trapped on a flat roof because of blocked drains has caused many collapses. Snow on a roof also has caused many collapses. Sometimes wind piles snow up in one area, causing a local concentrated load and resultant collapse.

Air-conditioners and similar equipment are often mounted on roofs. Sometimes grillage is added to the roof to support the additional load. If all the grillage is affected by the fire (as is usually the case), the extra support is ineffective in a fire. In some cases, equipment is erected on steel beams that are supported on exterior walls. This is a superior construction method, but the supporting beams should be watched for any signs of overheating and be cooled with hose streams if there is direct flame exposure.

Except for fire-resistive buildings, the roofs of buildings today have little or no inherent fire resistance to early failure. This even includes the roofs of noncombustible buildings. The typical non-fire-resistive roof today is usually supported on wooden I-beams, wood trusses, steel trusses, concrete T-beams, or in some cases, cored concrete slabs, which are susceptible to failure.

Second or even third roofs may be built over an original roof. At times, the reason is simply to improve the appearance of the building. A peaked roof is considered more attractive than a flat roof. Intractable leaky roofs are often overcome by adding an entirely new roof. A "rain roof"—covering a bowstring truss roof—was a factor in the deaths of six New York City fire fighters in a Brooklyn supermarket fire in 1978.

Lightweight Wood Truss Roofs

In 1991, a neon light started a fire in the gable end of a truss roof restaurant in Orange County, Florida. The staff and patrons of the restaurant were unaware of the fire. Patrons of a nearby restaurant gave the alarm. The first unit on the scene raised a ladder to the attic scuttle hole. The acting lieutenant noted that the fire was moving rapidly through the trusses, pushed by a high wind, so he ordered his unit out.

The acting battalion chief ordered the building cleared. About 5 minutes later, a large section of the roof collapsed. Images of the restaurant show heavy equipment on the roof. There was a gypsum board ceiling on the bottom of the trusses, probably intended to protect the trusses for a few minutes from fire originating below. The ductwork was suspended below it, and a grid ceiling concealed the ductwork. This construction made the fire impervious to an interior attack.

In February 1999, three fire fighters were killed in the collapse of a lightweight wood truss church roof in Lake Worth, Texas. The collapse occurred only 10 minutes after the fire was reported. Fire fighters working on the roof, one of whom survived after falling completely through to the sanctuary below, reported that the roof deck felt firm just prior to the collapse. Heavy fire involvement of the trussloft led to extremely rapid failure of the trusses. The deadly lesson of this fire is this: working under or on a lightweight wood truss roof that is well involved in fire is extremely dangerous and must not be permitted.

Bowstring Truss Roofs

The bowstring truss gets its name from the curved shape of the top chord **Figure 9-23**. Sometimes bowstring trusses are known as **arched trusses**. This can be confusing. The thrust of a truss is downward. In contrast, the thrust of an arch is outward, and usually resisted by a mass of masonry. Some arches are **tied arches** in which a steel tension rod ties the ends of the arch together to eliminate the need for the masonry. Early Arby's restaurants featured arched roofs; some of these arches were tied arches.

Bowstring truss roofs were very popular during the mid-20th century, having found their way into occupancies that desired open floor spaces like bowling alleys, supermarkets, and auto body repair shops. They have also been a killer of fire fighters, including the truss collapse during a 1988 fire in a Hackensack, New Jersey, Ford dealership. Five fire fighters lost their lives in this collapse, which originated in an unapproved auto parts storage area in

Figure 9-23 A bowstring truss roof.
Courtesy of David G. De Vries/Library of Congress.

the trussloft. Although a major roof fire was evident to fire fighters on the roof, conditions just moments before the collapse indicated virtually no smoke in the auto repair bays below. The loss of these fire fighters led to New Jersey regulations that require placarding on buildings of truss construction.

It is important to note the observed conditions at this Hackensack fire and the fire in the abandoned Chicago laundromat described at the beginning of this chapter: there was little or no smoke visible at floor level while a fire was burning in the trusses above.

Treated Plywood Roofs

Treated plywood has been proposed as a fire barrier. Builders have long objected to the cost and leakage problems of parapetted fire walls. Many code authorities in the eastern United States accepted the use of fire-resistant treated plywood (FRTP) roofing extending 4 feet (1.2 m) on each side of the fire wall, which extended to the underside of the roof. Delamination of ordinary plywood permits fire extension over fire walls, compromising the reliability of this method.

At least some chemicals used in FRTP react and deteriorate in ordinary temperatures. This causes the roof to leak and it becomes unsafe, even for normal use. A fire fighter or a worker walking on such a roof may fall into the structure. Unfortunately, in many cases, the FRTP was used not only along the fire barrier wall, but also randomly throughout the roof.

Void Spaces

As in frame buildings, void or concealed spaces are an inherent part of buildings of ordinary construction. The voids may be totally combustible, as in a wood joisted building with wood floors and lath and plaster ceiling. Partially combustible voids are exemplified by wood stud construction with gypsum interior finish.

During construction, requirements for firestopping may not have existed or may have been ignored. Even if firestopping has been installed, the execution may have been imperfect or it may have been removed to permit the installation of ducts or wires.

As has been noted, void spaces are a major firefighting problem. The existence of combustible voids has caused many serious losses even in sprinklered buildings because of failure to comply fully with NFPA 13, *Standard for the Installation of Sprinkler Systems*. NFPA 13 requires sprinklers in many types of combustible void spaces; certain small confined areas and those protected with special ignition resistance coatings are permitted. The fire can be extinguished only if it breaks out of the void or if the fire department breaks into the void. In January 2006, a five-alarm fire broke out in the old John Jacob Astor commercial building on Prince Street in New York. The fire spread throughout the structure through numerous void spaces, despite the fact that the building was alleged to have been fully sprinklered. If circumstances are favorable, expansion of steam (created by hose streams hitting burning material) into the void may accomplish extinguishment.

The fact that a particular void was not as noncombustible as similar adjacent voids was responsible for the fire loss of a new high school. Low-density combustible fiberboard had been applied to the studs behind gypsum wallboard on interior walls to provide acoustical treatment. A fire started by a plumber's

torch found a fully combustible void space in a wall and extended quickly to the void space above the ceiling. The roof was an insulated metal deck. The noncombustible acoustical tile ceiling proved to be a hindrance to the fire department's efforts to deliver the water to the underside of the roof, the vital place to have absorbed heat.

Interior Sheathing

There is a concept, never definitively stated but acted upon nevertheless, that by providing a protective interior sheathing or finish for a structure, a substantial contribution has been made to the building's fire safety. Examples include placing sheet metal or gypsum board over combustible surfaces. This is true as long as the sheathing keeps a contents fire out of the structure. However, when the fire originates behind the protective sheathing or extends behind it, the protective sheath becomes a detriment to the firefighting effort.

Light Smoke Showing

The expression "light smoke showing" has too often been the first report of what turned out to be a disaster. It is important to recognize with fire that "what you see, ain't necessarily what you've got." Ordinary citizens and inexperienced fire fighters are generally unaware of the tremendous fire threat that can be concealed in voids. For example, three fire fighters died in a furniture store fire. Heavy smoke was reported showing from more than a mile away by responding units. Fire fighters operating inside found only light smoke. In fact, they dropped their 2.5-inch (6.4-cm) line in favor of extending a 1.5-inch (3.8-cm) line. Suddenly, fire burst out from a void and killed the three fire fighters. A number of other operational problems also contributed to this tragedy: unfamiliarity with the structure, a missing preplan, and a broken radio microphone.

In another instance, a fire broke out in an L-shaped one-story row of stores. The firefighting operations were at the short leg of the "L." A sewing machine store with fabric up to the ceiling was at the top of the "L," farthest from the visible fire. The store was absolutely clear. A brown spot appeared on the metal ceiling. Gradually, it turned red and the paint started to burn. Flaming paint fell on the fabrics and the store was soon fully involved. The entire process took fewer than five minutes.

Ceiling Spaces

Older buildings have much higher ceiling heights than those built today. False ceilings commonly are added to conserve heating and cooling costs and to modernize the building. It is not uncommon for a building to be modernized several times in the course of its lifetime with a new and lower ceiling each time.

Lowering the ceiling of old buildings is most easily accomplished by constructing a new ceiling. This provides a convenient void for the new utilities, air ducts, and so on. Firestopping is usually nonexistent in these situations. Vertical voids connect the horizontal voids so the building is honeycombed with what amounts to one big interconnected non-firestopped void. It seems as if authorities place almost magical trust in the use of fire-rated gypsum board to absolve all sins of omission and commission. In one particular case of a department store with three ceilings, code officials required the owner to extend the protection to all three "levels" as required by NFPA 13. This 100-year-old San Antonio structure became an "anchor store" in a large downtown mall.

Usual construction methods create a huge three-dimensional void across the ceilings. In the rehabilitation process, the buildings are gutted and apartments are created by removing and/or adding walls. Dropped ceiling frames are hung and the new interiors are lined with plasterboard panels. The walls are not firestopped at the ceiling line and the vertical stud spaces are unprotected above the ceiling. Old dumbwaiter and pipe shafts are retained for utilities. Such buildings burn furiously. Fire fighters have to pull down the ceilings to find the fire and occasionally are met by a carbon monoxide explosion.

Joist Spaces

Important void spaces in any multistory, wood-joisted building are the joist spaces. Containing many square feet of exposed fuel, these spaces are protected from hose streams by their construction and the ceiling below. The modern use of wood truss floors has immeasurably increased this problem. As the fire gains in volume, it spews torches of flaming gas out of every hole, opening, and utility service in the floor above.

Combustible Gases in Void Spaces

The ignition of combustible gases accumulated in voids within the building may provide the fuel for a devastating explosion, even though the building is ventilated in the accepted manner. The generation of carbon monoxide in concealed places can be as much as 50 times greater than what it would be in the open. The carbon monoxide flammability range is from 12.5 to 74%. Its ignition temperature is 1,128°F (609°C).

At a stable fire in New York in 1938, Frank Brannigan witnessed an explosion of carbon monoxide that had accumulated in the cockloft (the void between the ceiling and the roof). The violent explosion took place an hour and a half after the first alarm. It was presignaled by dense clouds of boiling black smoke. A longtime observer of fires noted, "It's going to blow." The blast caused the collapse of a side wall and the loss of one fire officer's life. There was actually a deflagration. Apparently, the gas–air mixture was just in the right proportion. The building had been vented according to standard procedures, but removing the skylights did not vent the cockloft. It was incomprehensible to the investigating committee that a backdraft explosion could occur 90 minutes after the first alarm in a vented building. It was reported that the wall simply collapsed.

Large Voids

Churches, schools, town halls, and other buildings often include vast void spaces. Often, there is a basement full of combustibles, nonfirestopped walls, and sometimes hollow columns. The voids are so extensive that handlines cannot reach the fire. Access for heavy-caliber streams is limited so there are large areas that cannot be reached by water. Some valuable fire protection features were never required by code. They resulted from provisions for some other purposes, and the fire protection benefit was unintended. An excellent example is windows at the front and rear of a building and skylights in the roof to let in light. These wall openings are absolutely necessary to the control of any advanced fire in a building of ordinary construction, but were never legally mandated.

In an effort to modernize the typical downtown store, designers often close up the front windows of the upper floors. Air-conditioning and fluorescent lights substitute for windows and skylights. The skylights and rear windows are closed up to foil burglars, and the often heavily fire-loaded upper floors become, in effect, one big inaccessible concealed space, completely impervious to fire department streams. In some cases, building and fire code authorities have recognized the fire department's problem and required fire department access panels. These are usually of limited size.

Many codes require sprinklers for inaccessible cellars or concealed spaces below mercantile buildings, so it is logical to press for sprinkler protection when the void is built above the first floor. The problem is the same. The sprinklers may prevent the loss of the building, but a substantial fire loss is almost inevitable.

Of course, the same problems of lightweight wood trusses discussed in the chapter on wood-frame construction are applicable to the truss voids of new buildings of ordinary construction: large void spaces within the collapse-prone trusses and lack of code-compliant draftstopping. As noted earlier, lightweight wood trusses are prevalent in strip malls and small office buildings.

Fire Extension

In ordinary construction, rarely is any provision made to prevent the extension of fire through

the stairways and halls. Even if the stairways are enclosed in response to legislation passed after some terrible disaster, there are often many bypasses. The interconnected voids, pipe closets, utility shafts, and elevator shafts provide fire paths.

Interior Walls

Masonry buildings with spans greater than 25 feet (7.6 m) must have interior bearing walls. In older buildings, these walls are usually of balloon-frame construction unless made of masonry. Interior masonry walls can be distinguished from wooden interior walls by their greater thickness. Large wooden buildings also must have interior bearing walls, which usually are of balloon-frame construction in older buildings.

In large buildings of ordinary construction, masonry walls may provide the interior load-carrying structure. Sometimes closures of more or less adequate fire resistance have been provided at openings. These can be useful. However, because the purpose of the wall may not be to stop fire but only to carry the interior floor loads, it would be unusual to find such a wall carried through the attic space. The combustible attic may provide a path by which fire, initially confined to one section, may pass over the top of the wall.

Voids in Mixed Construction

Many buildings are composites of older sections, possibly frame or ordinary construction, and newer sections of protected steel or concrete construction. Unless the architect is aware of the hazard, or an alert fire department can present a convincing analysis of the total fire problem to the building department, the new fire-resistive addition may be completely at the mercy of the old building, particularly when the architect wishes to make the result look like one building.

In one case, an industrial building in New England appeared to be of fire-resistive concrete construction. The view of the building from an upper story of a hotel nearby shows that it was built around an older building with a huge peaked-roof attic on which air-conditioners are mounted. It is almost certain that the attic is of wood construction. This would be quite a surprise at a fire.

Cornices and Canopies

In the 1800s and early 1900s, many buildings were not complete without a cornice. A cornice is a structure of wood, metal, metal over wood, or masonry that tops the wall and projects from it. Collapse of cornices has caused many fire fighter fatalities and injuries.

The more functional architecture of the 1920s eliminated the use of cornices. In recent years, however, there has been another change in style and tastes. Cornices, fake mansards, overhangs (sometimes called eyebrows), and other projections are being installed on many new buildings and added to old buildings to improve their appearance. They are usually made of wood and sometimes poorly attached to the structure. These decorations present a collapse hazard. New decorative cornices are made of coated plastic polystyrene and are backed with plywood **Figure 9-24**.

Figure 9-24 Decorative cornices at the roof level of two buildings.
© iStockphoto/Thinkstock.

A noncombustible motel in Pennsylvania is decorated with a huge mansard that extends around the building. The fuel load of a motel room is sufficient to pour heavy fire out the window. This fire could ignite the mansard and destroy the motel. It would be interesting to ask the architect, "Did the owner give you a criterion that a single room fire should put the entire motel out of business?" This question might interest the owner.

Closely allied to the cornice is the sidewalk canopy. In some cases, this is of cantilever construction, thus providing a serious collapse potential. Even when not cantilevered, the cornice and canopy represent substantial loads, usually supported by combustible members.

Fire Barriers

Masonry Bearing Walls as Fire Barriers

A fire wall, by definition, is erected for the sole purpose of stopping a fire, and the truly adequate fire wall can often accomplish this function unaided. Masonry bearing walls may serve to stop the fire, but conditions generally make a bearing wall less than a fire wall. When defects are recognized, tactics may then be planned to take advantage of the assistance offered by the masonry wall. "Main Street, USA" is made up of buildings of ordinary construction, usually brick and wood-joisted, built side by side. If each building has a 12- or 16-inch (30.5- or 40.6-cm) unpierced bearing wall, these two walls together form a barrier to the passage of any fire that can be generated. The fire may go over or around such a pair of walls but not through it. Unfortunately, often the walls are pierced with openings for utilities or doorways when occupancies encompass both buildings.

Fire Doors

When openings have been made in walls to connect buildings, fire doors should be provided **Figure 9-25**. The local code may not require them, however, and

Figure 9-25 Automatic closing of rolling fire doors can be activated by fusible links at the top of the door opening, or by an electronic failsafe release device connected to an alarm system or local detectors.

Courtesy of Lawrence Doors.

even if installed they may be ineffective. In some cases, the originally planned openings were properly protected by fire doors, but additional openings were made later without proper protection. In one case, two older buildings in Virginia were combined to form a restaurant. The created opening was protected with a fire door. Alongside it there was an unprotected window so the manager could see the dining room from the office.

Fire fighters should be trained in how to inspect a fire door for proper operation. Particular attention should be paid to keeping the door path clear at all times of any displays or equipment or anything that would block the door. Prefire plans should indicate the presence of fire doors. It is a truck company's function to see that they are closed and latched properly. During a fire, pay particular attention when stretching hose through an open fire door—if the fire door closes behind you, you will lose water pressure and a means of egress.

In bygone years, adjacent buildings may have been connected by doorways at one or more floors. The fire department may be totally unaware of openings that were abandoned in the past when the occupancies were separated. The closure at an opening may be an ordinary door, an inoperative fire door with merchandise stacked against it, fiberboard lath and plaster, or properly constructed masonry to restore the full integrity of the wall.

Protection From Exposures

When one building is taller than its neighbor, the owner of the taller building may install windows in the upper stories overlooking the smaller building. Fire coming through the lower roof may extend to the adjacent building via the side windows (although modern building codes require wired glass or other types of protection of openings in new structures). Burning material may fall out of the upper windows onto the adjacent roof.

Old, hidden windows may provide a surprise path for fire to travel from one building to another. A key factor in a $1 million fire in Gloucester, Massachusetts, was a window in a brick building that had been boarded up. The window was next to the wooden exterior wall of a sprinklered department store. The fire entered the ceiling of the store above the sprinkler piping. The sprinklers collapsed and the store was lost.

If two buildings are not of the same depth, the rear windows of one and the side windows of the other may expose one another. In such a case, the rear exposure may require protection, even before the fire is attacked. Otherwise, a frontal attack may drive the fire out the back windows into the exposure.

Narrow alleys between buildings present difficult defense problems against exposures. Often sheet or corrugated steel, or steel-clad wooden fire shutters, protect windows facing the alley. Unfortunately, their worth is suspect. Wired glass is also of limited value against radiant heat.

Figure 9-26 An exterior water spray system on a building to protect it from an exposure fire.

Courtesy of Tim Scroggins

At times, outside sprinklers or spray systems are installed to protect against exposure fires **Figure 9-26**. The fire department should be fully familiar with their operation. In most cases, they must be turned on manually and the valve is located inside the building. The system should be tested at least annually because the "open" nozzles can become clogged. Ensure that the water covers the entire face of the building. The system will do the work of one or more engine companies in covering the exposure and without risk to personnel.

Party Walls

In many cities, it is quite common to erect **party walls**, structural walls that are common to two buildings. In the area of the original city of Washington, D.C., fire protection regulations issued by President George Washington required that masonry party walls be built, and that the builder of the subsequent building pay the first builder for half the cost of the wall.

Usually party walls are established by mutual contract between the owners. The Washington, D.C.,

situation is probably unique. In examining the building, determine the thickness of the wall. A party wall is thinner than two separate walls. Because the joists of both buildings are supported on the same wall, the builder often found it convenient to support the joists in the same socket. Sometimes the joists were overlapped to provide greater strength. Some codes forbade this practice.

A typical party wall fire extension occurred in Ottumwa, Iowa. The fire involved two stores separated by a party wall in which joists were lapped in common sockets. The fire in the original store was under control. No examination was made of the ceiling void of the adjacent store. Suddenly, the ceiling void of the adjacent store was found to be involved. A backdraft explosion drove the fire fighters out and the building was lost. The fire had passed into the ceiling void of the adjacent store by way of the joist openings.

Fireground Safety

Fire fighter safety at the scene is a critical factor that all departments must face—not just those charged with fighting fires in certain occupancies. One method

that could significantly improve fire fighters' safety at the fireground is for the fire department to set up a Board of Building Review. Experienced fire officers who have studied building construction, assisted by consultants from the building department or private sources, could survey major buildings of ordinary construction and typical smaller buildings. Considering the available fire load, the type of construction, the nature of connections to adjacent buildings, water supplies, protection provided, and other relevant circumstances, an upper limit in minutes could be put on the time that fire fighters were permitted to stay in the building. The time rating of the building would be part of the prefire plan. The incident commander (IC) would reserve the right to order the fire fighters out before the expiration of the rated time, but strong restrictions should be placed on authority to keep fire fighters in the building after the expiration of the rated time.

The IC would be substantially relieved of an unfair responsibility (i.e., determining the boundary between stability and collapse under fire conditions). In a fire, the boundary between building stability and instability is a thin line. Ignorance of the boundary is no excuse when a failure occurs.

TACTICAL CONSIDERATIONS

1. Ordinary construction is brick masonry or noncombustible material on the exterior with interior dimensional lumber supporting the floors and roofs. It is found in a variety of structures, including apartment buildings, department stores, and light industrial structures. They are usually three or four stories, but can be as tall as seven stories. Three-story buildings usually do not have a standpipe. Four-story buildings or higher will typically have a Class I or Class III standpipe system.

2. In Type III ordinary buildings, the fire load consists of the contents and the combustible interior components of the building. With combustible material in the form of wood lath, wood furring strips, cross bridging, wood joists, and 2×4 wood studs, even a vacant Type III building will have enough of a fire load to produce a large fire. A rapid deployment of an interior attack line to the seat of the fire is necessary to prevent the structure from becoming involved. Void spaces need to be checked immediately after the fire is extinguished to stop or minimize structural involvement.

3. Ordinary construction from the 1950s and earlier typically used heavier lumber for interior structural members. Heavier lumber can withstand fire conditions longer than lightweight wooden assemblies, providing fire fighters with a wider margin of operational safety; 15 minutes is an accepted safe operational period. Remember, collapse of interior assemblies will make the exterior masonry walls less stable and more prone to collapse.

4. The major fire problem in ordinary construction is smoke and fire spreading throughout concealed spaces. The largest and most serious concealed space is the common cockloft between the highest ceiling and the roof. Rapid flame spread across this area has caused the destruction of thousands of buildings across the nation. Vertical ventilation, including the trench cut, is an effective tactic to stop the horizontal spread of the fire in long buildings.

5. With the historical preservation and renovations of older ordinary constructed buildings, concealed spaces have become larger and more numerous. The emphasis is on keeping the exterior façade as it was when it was built. There is more leeway to allow for modern and energy-efficient construction for the interior. For these reasons, the problem of concealed spaces in the brick-and-joist construction will always present a major hazard for fire fighters.

6. Other serious hazards to fire fighters with older ordinary construction are the parapet walls and cornices, some of which are very ornate, and all of which are extremely heavy. Usually located on the front side of the building, the parapet wall can collapse during structural firefighting. Over time, these walls are weakened by their constant exposure to the elements. Rain may wash away the mortar between the bricks, and freezing may cause cracks in the wall. Heavy master streams sweeping across the top of a burning roof can wash tons of loose bricks from the parapet wall down onto fire fighters below. The lateral force from heavy-caliber streams can topple the wall down a hundred feet on to the street. When the bricks hit, they can become projectiles. The collapse danger of the parapet wall is one reason why the area directly in front of the building is so dangerous, and why fire fighters should either move inside the doorway or away from the A side of the building altogether. Establish a collapse zone and look for other ways into the building.

7. More fire fighters have been killed in fires in ordinary construction than in any other type of construction. There is a variety of reasons, but one is simply age. Many of these buildings are over 100 years old. Because much of "Main Street, USA" is being lost to progress and urban sprawl, there is a push from historical societies to restore these dilapidated buildings rather than tear them down. Special loans are made available for those who wish to restore these old buildings. The primary emphasis is keeping the front façade in its original condition.

What's inside is often unknown—it can be original construction or a combination of new lightweight interior members. Remember, fire safety is not a priority in building preservation and restorations. Fire officers should become more analytical of the hazards of this type of structure, rather than relying on experience alone.

8. The firefighting forces of fire departments—not just the fire prevention division—should become involved in major building renovations and restorations. In this manner, fire fighters may make recommendations for improvements. In any event, fire fighters will be well aware of structural deficiencies after performing such inspections.

9. Many major cities, such as New York, Chicago, San Francisco, and Seattle, suffered "The Great Fire of…" around 1900. These great fires destroyed much of their downtown areas, which were populated with buildings that were constructed in the late 1800s. These buildings were rebuilt with ordinary construction that was popular at the time. Because many of these "replacement buildings" still exist, it is important for fire fighters to understand the history of the buildings in their response districts. Prefire planning and training are essential. With new CAD system programs and mobile data computers, prefire information, drawings, maps, digital photographs, and informational updates can be disseminated throughout the fire department computer network from the cab of the inspecting fire company with the touch of a button.

10. Stay alert to alterations and new construction permits. Most fire departments have a system in place to notify the local fire companies of new construction projects happening in their district. Remodels can be harder to track because many are completed illegally without a permit. Pay attention to change of use (occupancy) construction projects.

11. When it's time for fire fighters to immediately evacuate the building, the department should have a recognizable signal, such as air horn blasts in addition to radio orders. The fire communication centers also have the ability to broadcast an "emergency evacuation tone" over all the radio channels on the fireground. Drills should be held on evacuation in real time. There is very little, if any, reaction time. Frank Brannigan saw a number of images of escaping fire fighters looking back—you

should train personnel with the imperative, "Don't look back!" Look where you are going because there are many tripping hazards.

12. Many fire departments use electronic sirens on their apparatus, which usually have three settings: wail, yelp, and high-low. In cities such as Seattle, Washington, the wail and yelp are used while responding to emergencies. The "high-low" European siren is reserved exclusively for an emergency "abandon the building" tone. When the IC calls to abandon the building, the fire alarm center transmits a "high-low" tone on the fireground radio channel and the drivers of apparatus on scene turn on their "high-low" siren. The policy prohibits the high-low siren from being used during responses so it cannot be confused with any other signal.

13. Ordinary construction that houses food processing businesses is required by many health department codes to have a concrete topping over the first floor to provide a sanitary floor and to seal it from rats and mice. This additional dead weight may confuse fire fighters as to the nature of the floor. If there is a basement fire, the concrete topping may also conceal the heat below. Concrete topping on the floor was a factor in at least two building collapses that caused multiple fire fighter fatalities, including a fire in Seattle, where four fire fighters died when the heavy floor collapsed into the burning basement. The dimensional lumber that was supporting the floor burned out in approximately 20 minutes. Whenever you're responding to a commercial fire in a food processing facility, for example, that usage should raise a flag. You're probably dealing with a concrete floor, so remember to quickly check to see if anything is burning below the first floor.

14. If fire fighters must pass under a burning cornice or canopy, use a solid stream to knock off wood about to fall and drive the fire back. This may buy a short amount of time during which it is somewhat safer to pass under the cornice.

15. A possible wall collapse must be in the forefront in the constantly changing evaluation of the fire problem assessed by the fireground commander. If good prefire planning has been done, information on possible weak points in the walls will be available. In any event, a sector officer should be assigned to monitor wall stability. Remember, an exterior collapse will usually cause an interior collapse. Conversely, the

elongation of steel beams can push out an exterior wall and cause it to come down.

16. Once you have a wall collapse, the building is telling you it's giving up to the fire. This building no longer has any value and is not worth fire fighters getting injured or killed. Withdraw all crews from the building and regroup. The situation needs to be reassessed. If there are no parts of the building that have any value, you must switch to a defensive strategy. One specific tactical objective that sometimes results in futility is the trench cut. When the IC calls for a trench cut, the IC is essentially setting up a fire break—similar to a fire break established during a wildland fire. On the fire side of the trench, you're essentially writing the building off. On the opposite side of the trench, you're indicating: "Here is where we're making a stand because this part of the building is worth saving." The problem develops when crews initiate the trench cut too close to the fire. This is a labor-intensive and time-consuming operation. If you don't have enough personnel making the cut, the rapidly advancing fire will force crews to abandon the operation before the trench is completed. Another problem develops when the trench isn't long enough. The trench cut needs to extend from wall to wall; an incomplete trench will allow the fire to pass underneath the uncut part of the roof to the rest of the attic space. Finally, a trench cut needs to be backed up with a hose line. As the fire vents out the trench, there is still radiant heat attacking and preheating the unburned part of the roof and attic space. The only way to protect this area is with water.

17. The IC must be able to predict time-consuming tasks and have sufficient resources available to complete the task to have the desired effect on the fire.

18. Know your collapse indicators. Bricks have a reputation for toughness and permanence that can be misleading. Poorly made bricks are not uncommon. They absorb moisture and can fail due to freezing. Walls may come down in several ways. The worst or most dangerous collapse is when the wall breaks at ground level and falls straight out along the full height. Bricks and other debris will fly farther than the standard collapse zone (1.5 times the height of the building). In one case, bricks were measured to have traveled as far as 200 feet (60.9 m) away from the wall.

19. There should be strict control of the number of personnel in or close to the collapse zone. One way to establish a collapse zone is to use uncharged fire hose. There are hundreds of feet available at a fire scene. This provides a clear line of demarcation so there is no mistake. Though fire fighters like being up close to the action, it doesn't take four or five fire fighters to operate a portable deluge set. In fact, the equipment can be left unattended, but this "one spot" delivery is rarely effective. A heavy stream nozzle can be directed with special equipment or by means of radio signals, thus providing effective delivery without much risk to fire fighters. New apparatus are now being equipped with remote control master streams and deck guns.

20. Fire fighters directing hose streams against older brick buildings can wash out the mortar, causing the wall to fail or collapse. Watch out for leaning brick chimneys or chimneys pulling away from the structure. Be especially careful around freestanding chimneys after a fire. The safety officer should take note of this and notify the IC. A radio message should be transmitted on the fireground so all companies are aware of the safety hazard. Use flagging tape or uncharged fire hose to mark off a collapse zone and make all fire fighters aware of it.

21. Examine a wall before breaching it. If it is deteriorated, don't breach it because the entire wall may come apart. If you are breaching a wall and you find materials different from the rest of the wall, such as brick, steel, solid concrete blocks, or reinforced concrete, you are probably into a wall column and under a major load. Immediately cease operations, reposition someplace else, and start again.

22. When breaching a wall, make a triangular cut with the apex at the top. Do not cut a straight line across the top of the opening. Masonry units mortared together are not designed to hang upside down, so this action can lead to a collapse. Available shoring material (e.g., wood timbers, raker shores) should be used as soon as possible to support the opening in the wall. It's a good idea to note the location of existing construction sites. Also, have a prearranged agreement with the local lumber yard for quick access to shoring materials.

23. Whenever materials are combined in the structural support system, the fire characteristics of the weakest component should govern assessment of the collapse

potential. A building system is only as strong as the weakest connection. Never get into a position where lightweight interior structures, such as a rack storage or scaffolding, can collapse behind you.

24. Rundown theaters and auditoriums often have marquees that extend over the sidewalk and main entrance. Try and look for other entrances into these structures if they have been left abandoned and vacant for some time. If the front entrance must be used, consider shoring up the marquee overhang with one or two ground extension ladders as a temporary safety measure.

25. Steel beams, although not normally found in buildings of ordinary construction, may be used as replacement beams for sagging wooden beams. Steel elongates when heated. For example, a 100-foot (30.5 m) steel beam reaches 100 feet and 9 inches (30.7 m) at 1,000°F (538°C). Elongating I-beams and trusses have pushed down many walls. The cause of elongating steel collapse is rarely reported after a fire because all that is evident at that time is failed steel. Material that absorbs water, such as carpet, bulk cardboard, large rolls of paper, and fibrous material, should be stored at least 1 foot (30.5 cm) away from masonry walls to allow for expansion. Expanding water-soaked material can push the walls out, causing a collapse. Though this precaution is enforced during a fire prevention inspection, it can pay big tactical dividends to fire crews operating at 3:00 in the morning.

26. Even where prefire planning was not accomplished, much can be learned at the time of the fire and deduced from a good knowledge of similar buildings. Walls that are obviously deteriorated, braced, or tied certainly indicate that a floor collapse will probably bring the walls down. Watch for the ornamental stars that indicate walls are in distress.

27. The volume of the fire gives clues about how long the floors will last. Ordinary wood-joisted floors are not formally rated by any standard fire resistance test. Recent simulated (laboratory) basement fire tests indicate an *exposed* 2×10 solid wood joist lasts 18 minutes. The attachment of a ½" (12.7 mm) gypsum board ceiling raised the endurance to 44 minutes.

28. Fire fighters should not be working on any roof structure that is on fire. In such a case, the roof will shortly self-vent. When vertical ventilation is performed, the attic space is filled with hot,

"oxygen-deficient" smoke and gases. This means there is no fire...yet. Fire fighters must realize that ventilation increases the air flow to the fire. Wherever you cut the vent hole, that's where you're bringing the fire. A roof structure not involved before ventilation may quickly become involved soon after the roof is opened. Part of the vertical ventilation strategy entails redirecting where you want the fire to go.

29. The truck company officer also serves as the roof safety supervisor, with the responsibility of determining whether the roof is safe to use as a working platform during vertical ventilation operations. Departments without designated truck personnel still have to perform vertical ventilation. The most experienced member should direct the cuts and serve as the roof safety supervisor to determine when it is time to get off the roof. Always take a roof ladder to the roof; even when operating on a flat roof. The roof ladder may be needed to climb over a parapet wall or be used as a secondary means of emergency egress to an adjacent roof. Once the vent hole is cut, it is time to get off the roof.

30. A ventilation opening acts like a chimney. Chimneys exhaust smoke and heat, but they also accelerate the fire.

31. If a roof is deemed unsafe to work from as a platform, but rooftop ventilation is still critical to the offensive strategy, consider working from an aerial ladder or sectional types, which permits a bucket to be placed down on the roof.

32. Collapsing roofs often bring down masonry walls. When a roof collapse is anticipated, fire fighters should be withdrawn beyond the wall collapse zone.

33. Training officers and company officers must make every effort to present the invisible hazards of void spaces to their fire fighters with as much importance and drama as the live, visible fire.

34. Heavy volumes of boiling smoke that persist even after the visible fire has been well controlled indicate fire in a void. However, light smoke or no fire at all does not necessarily mean that there is no fire in the void spaces. In 1986, two fire fighters in Biloxi, Mississippi, were killed in a residential boarding house fire. The initial attack was over and the fire had appeared to be knocked down. The two fire fighters reentered the structure without a hose line; when fire busted out of a second-floor void space, they were killed. Always open and check the attic

and void spaces after the fire has been knocked down. The thermal imager is one of the most important lifesaving technological advances to assist fire fighters in searching for victims and checking for hidden fires.

35. In the case of row buildings, expect extension to the adjacent exposures. Get them opened up early and get charged hose lines into position inside the exposed buildings. If a shortage of personnel does not permit such lines to be staffed, leave the charged lines in place with a patrol. In too many cases, a fire department has suddenly found that the fire has extended; by the time operations can be rearranged, the exposed building might be fully involved.

36. Fire takes the path of least resistance. In addition to laying hose line at a row building, you can place positive-pressure blowers and pressurize the adjacent B- and D-side exposure buildings. If the fire building has a sufficient vent hole in the roof, by making a hole in the ceiling to the attic space of the exposure buildings, you can pressurize the attic space and reverse the horizontal spread of smoke and hot gases, channeling them back out the vent hole of the fire building. Anticipate the possible extension of fire as a result of collapse.

37. Fire fighters should be trained in how to inspect a fire door for proper operation. Particular attention should be paid to keeping the door path clear at all times of any displays, storage, equipment, or anything that would block the door from closing and latching. Prefire plans should indicate the presence of fire doors.

38. Fire units should never pass through a sliding or overhead rolling door without blocking it open temporarily. Closed sliding fire doors can be very difficult to open; overhead rolling doors may be impossible to open. A sudden burst of fire may trigger the door and the fire fighters could become trapped. If crews are operating hose lines beyond the door, they will suddenly lose water and may be trapped on the fire side of the door. If they have advanced beyond the door with an uncharged line, the door will prevent water from reaching the nozzle when water is called for. In both scenarios, fire fighters may become trapped on the fire side of the door with no way to protect themselves from fire.

39. In buildings more ornate than warehouses, fire doors may be artfully concealed. In one Rochester, New York, building, an overhead rolling fire door is concealed in the ceiling. The tripping fusible link is not conspicuous. The door lies across the attack path of the first-due engine company. Should a door like this activate after a sudden burst of heat, this door will come down like a giant hose clamp on crews' hose lines. Thus, whenever crews cross the threshold of a large rolling fire door, they must wedge it open.

40. Teach newer, inexperienced fire fighters that the purpose of a large fire door is to confine the fire to the space or building of origin. By design, if a fire is large enough to activate a large rolling fire door, the building is writing off property contents on the fire side of the door. If a crew is sent to the adjacent building or space to protect the exposures and prevent the fire from extending beyond the fire door, it may mean entering the building with a charged hose line and simply waiting—either for interior crews to extinguish the fire or for the fire to burn itself out.

41. Waiting is not easy for aggressive fire fighters. In one example, inexperienced fire fighters were manning a 2 ½" (6.4 cm) exposure line in a smoke-free, low-heat environment and felt like they had nothing to do. They realized the seat of the fire was just on the other side of the fire door. Because interior crews were having trouble finding the fire, they felt compelled to act. They concluded if they opened the D-side fire door, they would be at the seat of the fire, could quickly extinguish the fire, and would be heroes for saving the day. They had no idea of the fire behavior and thermal dynamics they were about to encounter. The crew split because it took two fire fighters to slide back the giant fire door. Superheated smoke and fire immediately started spreading horizontally across the ceiling. The radiant heat was so intense that they could not tolerate the thermal assault. They also could not move the 2 ½" (6.4 cm) hose line fast enough to push back the fire. The crew was forced to back out and the fire extended into the exposure building.

42. Fire fighters operating hose lines in a narrow alley are in serious danger. Because you cannot establish a collapse zone in an alley, study the problem in advance. Lines directed downward from the roof of the exposed building or ladder pipes directed into the alley from the street are two methods of attack. If there is a way to operate from a protected area off

the alley, that would be best. Many times there are covered entryways or covered ramps to underground parking off the alley for the C-side exposure building. You can also make entry from the next block. Take your exposure lines through the building back toward the alley. This tactic gets lines into position to support the strategy of attack and allows your crew to set up from a rear door or windows and operate from a safe position.

43. Extension through party walls is sometimes hard to detect. Examination of the adjacent premises early in the fire may show nothing. It may take time for the fire to work its way through. Smoke, which might ordinarily give cause for alarm, goes unnoticed in the general smoky condition of the fire. Without some evidence, officers are often reluctant to order the necessary opening of ceilings. When the need becomes evident, it is often too late. Thermal imagers have made it easier to detect hot spots without needlessly damaging property.

44. Tabletop drills have always been effective because they give an aerial perspective of an incident. But how effective would they be if you had to lay your head against the table? The incident command post (ICP) is usually set up in front of the building or at the nearest street intersection. Some ICs are comfortable running an incident from the inside of a command vehicle, and even insist on this arrangement because there is less distraction. On large multijurisdiction incidents, you do not have a choice. However, on a normal fire response, you need your eyes for initial and ongoing size-ups. Sometimes a floor or roof of an adjacent building can help you with a better vantage point to realize what's going on or what's going wrong. An overpass or a bridge located higher than the roof line of the involved buildings can also be helpful. If it is not practical for the IC to set up there, have a deputy IC or company officer be your aerial sector to give you a "bird's-eye view" update on the situation.

45. With mobile data computers, helmet cameras and apparatus-mounted cameras can be used to transmit digital video to the ICP in real time. Local news stations, many of which have helicopters, can also feed live video directly to the ICP for an aerial perspective of the ongoing incident. Local and state agencies may have stationary security video cameras mounted throughout a community and on highways, many of which have remote control spanning capabilities. If a fire incident is within the lens range of the camera, the video can be streamed to the ICP in real time.

Chapter Summary

- The term *ordinary construction* describes an almost infinite variety of buildings.
- In simpler days, ordinary construction was generally known as brick and wood-joisted construction. Today, this construction uses a variety of walls and interiors.
- The chief common characteristic of ordinary construction is that the exterior walls are made of masonry with combustible frame members.
- Ordinary construction is classified as Type III construction.
- Over the years, most old buildings have undergone extensive modifications. Usually, such modifications have had a detrimental effect on the structure from the fire suppression point of view, creating collapse potential or interconnected voids from which fierce fire can burst out on the unwary.
- Many buildings constructed in recent years depart from simple ordinary construction; this is not necessarily an improvement.
- In modern construction, a noncombustible void can accumulate explosive carbon monoxide gas as readily as a combustible void.
- The problems presented to the fire department by ordinary construction can be divided into the following areas:
 - The structural stability of the masonry wall
 - The stability of the interior column, girder, and beam system
 - Void spaces
 - The masonry wall and fire wall as barriers to fire extension
- Many fire texts cite indications of building failure that may be observed on the fireground:
 - Typical indicators include smoke or water flowing through walls, soft floors, a small partial collapse, walls out of plumb, and time since arrival on the scene.
 - Although these are all good indicators, sometimes they are insufficient or too late.
- The interior structure may push down a wall in several ways.
- Fire departments are concerned about equipment such as ropes, aerial ladders, and aerial platforms because equipment failure can cost lives. However, there is one working platform fire fighters use that has cost many lives—the roof.
- Fire fighter safety at the scene is a critical factor that all departments must face—not just those charged with fighting fires in certain occupancies.

Key Terms

Adobe Large, roughly molded, sun-dried clay units of varying sizes.

Arched truss A truss with an arched upper chord and a straight bottom chord, with vertical hangers between the two chords.

Ashlar masonry Stone cut in rectangular units.

Cantilever wall A freestanding wall unsecured at the top that acts like a cantilever beam with respect to lateral loads, such as wind or a hose stream.

Cast-in-place concrete Includes plain concrete, reinforced concrete, and post-tensioned concrete. This concrete is molded in the location in which it is expected to remain.

Cavity wall Hollow wall in which wythes are tied together with steel ties or masonry trusses.

Cockloft Void space between the top floor ceiling and the roof.

Composite wall Two different masonry materials, such as brick and concrete block, used in a wall and designed to react as one unit under load.

Concrete masonry unit Precast hollow or solid structural block. Sometimes referred to as cinder block.

Concrete topping Concrete placed over the first-floor wood floors for fire resistance or to provide sanitary floors.

Coping The masonry cap on top of a wall.

Cornice A projecting decorative (ledge) at the top of a masonry wall.

Course A horizontal line of masonry.

Cross wall Any wall set at a right angle to any other wall; the walls should brace one another.

Dog iron Connects the girders and imparts some lateral stability under normal conditions; resembles a big staple.

Fire limit Older code provision that would not allow a structure to be built without the use of exterior masonry walls that would limit fire extension.

Fireproof Material applied to structural elements or systems that provides increased fire resistance; usually serves no structural function.

Flying buttress Masonry pier at a distance from the wall and connected to it that resists the outward thrust of the roof.

Header or bond course Bricks laid so that the end is visible.

Hollow masonry wall Two connected wythes of masonry with an air space in between.

Joist hanger Metal angle or strap used to support an individual joist against a beam or a girder.

Light well Small court commonly placed in large buildings to admit daylight into interior areas not exposed to an open view.

Masonry column Masonry bracing incorporated into unstable masonry walls; also called piers, buttresses, pilasters, or columns.

Masonry wire truss Wire truss embedded into the mortar in specified courses, making the header course no longer necessary.

Mezzanine A low-ceilinged story located between two main stories; usually constructed directly above the ground floor, often projecting over it as a balcony.

Ordinary construction Buildings in which the exterior walls are noncombustible or limited combustible, but the interior floors and walls are made of combustible materials.

Parging (pargetting) Application of mortar to the back of the facing material or the face of the backing material.

Party wall A structural wall that is common to two buildings.

Pintle Square metal device used to transfer loads of columns on upper floors by passing the loads through intervening beams and girders to metal column caps on the column below.

Rubble masonry Rough stones of irregular shapes and sizes, used in rough, uncoursed work in the construction of walls and foundations.

Rubble masonry wall A wall composed of an inner and outer wythe of coursed masonry. The space between is filled with random masonry sometimes mixed with mortar. Such walls are unstable to a lateral thrust.

Solid masonry walls Masonry units (either solid or hollow) laid contiguously with the joints filled with mortar.

Stretcher course Bricks laid so that the long side is visible.

Strip mall Modern one-story retail occupancy building that typically has a lightweight wood truss roof and concrete block walls (Type III construction) or steel bar joists and a metal deck roof with a masonry wall (Type II construction).

Tenement Multistory working-class apartment buildings constructed in the 1800s and early

1900s, often substandard in terms of fire safety and health.

Terra-cotta tile Made of clay and fine sand and fired in a kiln.

Tied arch Arch in which a steel tension rod ties the ends of the arch together to eliminate the need for the masonry.

Type III construction Ordinary construction.

Unreinforced masonry Ordinary masonry walls are not reinforced, so they have no resistance to lateral movement.

Veneer wall A wall with a masonry facing that is not bonded but is attached to a wall so as to form an integral part of the wall.

Wythe A single continuous vertical wall of bricks, one masonry unit in thickness.

Case Study

While riding in charge of the truck company, you decide to take the crew out for a tour of the new strip mall that has gone up on the east side of town. On the way over, you talk to the crew about the hazards associated with strip malls and ordinary construction.

1. Masonry walls in ordinary construction may consist of _____.

 A. concrete block
 B. terra-cotta tile
 C. precast concrete
 D. all of the above

2. Which term best describes a freestanding wall unsecured at the top?

 A. Composite wall
 B. Cantilever wall
 C. Cavity wall
 D. Hollow masonry wall

3. True or False: Assuming this strip mall was constructed using lightweight wood trusses, you can rely on the presence of code-compliant draftstopping to prevent horizontal fire spread.

4. Cornices found on new strip malls are likely made of:

 A. Wood
 B. Metal
 C. Stone
 D. Polystyrene

Challenging Questions

1. Describe the basic characteristics of an ordinary construction building.

2. Why is the term *brick building* improper?

3. Identify the indicators of structural collapse.

4. Identify the paths of fire spread through void spaces.

5. You are a battalion chief in command of three engine and two ladder companies for a reported fire in a one-story, unsprinklered supermarket with a bowstring truss roof. Upon arrival, you note light smoke in the store but no fire. You are advised that store employees noted smoke coming from the ceiling HVAC diffusers 10 minutes earlier. A ladder company on the roof reports moderate smoke coming from scuttle openings on the roof. What are your safety concerns? What overall strategy would you employ to deal with the fire?

Noncombustible Construction

OBJECTIVES

At the conclusion of this chapter, you will be able to:

- Understand the difference between noncombustible and fire-resistive construction.
- Identify the different types of steel building components and their characteristics.
- Describe the use of masonry, including concrete, in noncombustible buildings.
- Describe different types of steel structural systems.
- Describe the hazards of a metal deck roof fire.
- Understand the hazards of high fire loads in unprotected steel structures and ways to improve the situation.

Case Study

Figure 10-1 The Memphis dollar store fire, in which the roof collapsed at the rear of the building.

In 2003, an arson fire broke out in the rear office of a "dollar store" (with items for sale ranging from clothing to automotive supplies) in Memphis, Tennessee (**Figure 10-1**). Constructed of unprotected steel bar joists supported by concrete block walls, the 8,900-square-foot (826.8 m²) building was divided by a concrete block wall into two areas. The sales area, approximately 100 feet (31 m) deep, was at the front of the building; a smaller storage and office space was located at the rear of the store. The roof deck consisted of a corrugated steel deck covered with ¾" fiberboard and asphalt. A suspended ceiling was installed in the sales area, while the bar joists were exposed in the rear storeroom and office.

Fire fighters entered the front of the store and noted a light smoke condition. Other fire fighters entered the structure through a rear door, going into the storeroom. The fire fighters in the storeroom encountered moderate smoke conditions; when they opened the door to the office, they were met with heavy fire conditions. Handlines were stretched through the front door, and fire fighters were assigned to the roof for ventilation as smoke conditions worsened.

One fire fighter was directed to open up the drop ceiling in the store. As he conducted this task, the fire intensified in the ceiling void and blew down a number of ceiling tiles and HVAC diffusers onto fire fighters below (believed to be the result of a backdraft). At about the same time, a portion of the bar-joist roof in the storeroom collapsed; two fire fighters—one in the storeroom and one in the sales area—became trapped, ultimately succumbing to the effects of the fire.

1. What are the effects of a drop ceiling in terms of locating a fire?
2. How can backdrafts occur in a void space consisting of a drop ceiling, bar joists, and a built-up metal deck roof?
3. Which type of fire load is necessary to initiate the collapse of steel bar joists?
4. How many buildings of this type exist in your community?

Introduction

Both Type II noncombustible construction (covered in this chapter) and Type I fire-resistive construction (covered in the chapter that discusses fire-resistive construction) use construction materials that will not support combustion—namely, steel and concrete. What, then, is the difference between noncombustible and fire-resistive construction? The answer is simple—the level of fire resistance (fire rating) assigned to the structural frame, walls, floors, and roof. Noncombustible construction has little or no fire resistance for its structural members, while fire-resistive construction has moderate to heavy fire resistance.

Noncombustible construction also differs from fire-resistive construction in that its allowable area and height (discussed in the chapter that reviews building and fire codes) are much smaller than the corresponding criteria for fire-resistive construction. Many noncombustible buildings are 1–3 stories in height, with the maximum height being 12 stories for certain types of occupancies. Fire-resistive construction, by comparison, is permitted to be of unlimited height.

Although the use of concrete is found in some buildings of noncombustible construction, noncombustible construction relies heavily on the use of steel for its structural system, including the roof and floor framing. In this chapter, the focus is on steel because it is the most prevalent material of choice. Nevertheless, many actual examples of fire-resistive construction can be cited that use steel for their framing systems. These buildings will have high levels of fire resistance (protection) for the steel.

The use of concrete in Type II construction is typically limited to exterior walls and shaft enclosures. For example, many strip malls use concrete block as exterior load-bearing walls. Similarly, warehouses of tilt slab construction (discussed later in this chapter) often use large exterior wall panels of concrete. Concrete block or panels are used to provide fire-rated enclosures of stair and elevator shafts. In addition, concrete is often used to construct a fire-rated floor assembly in Type II construction. A detailed discussion of concrete can be found in the chapter on fire-resistive construction.

Steel is the most important metal used in building construction. It is widely available and relatively inexpensive in the United States. Without it, construction would be limited to massive all-masonry buildings with arched floors or masonry wall-bearing buildings with wooden floors.

Steel is tremendously strong. Its **modulus of elasticity** (a measure of its ability to distort and restore) is about 29 million pounds per square inch (psi)—far more than any other material. Steel's compressive strength is equal to its tensile strength. Its shear strength is about three-quarters of its tensile strength. This great strength enables steel members of relatively small mass to carry heavy loads, particularly when used in trusses. Fire resistance, however, is a function of mass. Such strong but lightweight members have little inherent fire resistance.

Fire Characteristics of Steel

Steel has several important characteristics to consider regarding its behavior in fire. The characteristics that concern fire fighters are:

- Substantial elongation can take place in a steel member at ordinary fire temperatures (about 1,000°F). This elongation may cause the disruption of other structural components, such as masonry abutting the ends of the steel. If the steel cannot elongate because of restraint, it will buckle or overturn. This can be significant when other components rest on a steel member.

- At higher temperatures (above 1,300°F [704°C]), steel members may completely fail, bringing about a collapse of the structure.
- Cold-drawn steel, such as steel tendons used for tensioned concrete and for excavation **tiebacks** and elevator cables, will fail at about 800°F (427°C).
- Steel is a good thermal conductor, so it transmits heat readily. Heat can be transmitted by conduction through walls to combustible materials.

Unwarranted Assumptions

The fact that steel is noncombustible leads to unwarranted confidence in its "fireproofness" and suitability for all applications where fire is a problem. "I never would have believed it" is a common reaction by otherwise well-informed people when they see the ruins of an unprotected steel building.

Water on Hot Steel

The heat evolved by a fire can be **triaged**—that is, considered and treated according to priority needs. In reverse priority, this heat can be classified as follows:

- *Heat leaving the structure.* Let it go. Every British thermal unit (Btu) that leaves is one less Btu to keep the fire going. This is sometimes contradicted by the "Chicken Little" school of firefighting, which apparently believes that the sky can catch fire and the world will come to an end.
- *Heat being evolved from contents that are burning.* This is of secondary importance. This concept often comes as a surprise to those whose sole objective is to "put the wet stuff on the red stuff." Contents partially damaged are, in effect, totally destroyed. In fact, much of overhauling consists of throwing out partially or very limited damaged contents.
- *Heat being absorbed by contents or structural elements that will be ignited or caused to fail.*

This heat is the most important heat to be removed. In the case of unprotected steel, failure can occur early in a fire and the consequences can be catastrophic.

The fire department's heat removal medium is water. It seems apparent that water should be used to remove the destructive heat from the steel. Unfortunately, there is a fairly widespread myth in the fire service that water should not be thrown on heated steel because of possible dire results—this is simply not true.

If the steel is elongating, the cooling effect of water draws it back to its original dimensions. Many drawbridges, elongated by the summer sun, refuse to go back into position until the bridge is cooled with water or diminished sun. If the steel is failing, the water simply freezes it in whatever shape it is currently in. When these statements were published in the first edition of this text in 1971, they were questioned by several people. The American Iron and Steel Institute assured that this was correct.

A Texas fire department was observed fighting a fire in a steel hay shed. All attention was given to throwing water on the hay. No attention was paid to cooling the steel, and the shed was severely damaged. There is no value in burned, wet hay.

Definitions

Some necessary definitions of steel construction members are given here:

- **Angles**: These steel members have two legs at right angles to one another. They are L-shaped in cross section.
- **Bars**: Plates fewer than 6 inches in width. Bars are also made square or round.
- **Box columns**: Large hollow columns built from steel plates.
- **Box girders**: Square or rectangular steel girders that are hollow, like box columns, and used to support other beams in buildings. Large box girders are used to construct bridges Figure 10-2.

Figure 10-2 A set of rectangular box girders.
© Melinda Fawver/ShutterStock, Inc.

- **Castellated beams**: Wide flange beams that have been cut in half in a zig-zag pattern and then welded back together offset, **creating new, deeper beams** Figure 10-3.
- **Channels**: Steel structural components that have a square U-shape cross section.
- **I-beams**: Beams shaped like the letter "I". The top and bottom of the "I" are called flanges and the stem is called the web. The top flange resists compression, while the bottom flange resists tension. The web is necessary to keep the top and bottom apart. This principle has been copied in lightweight wooden I-beams. Composite beams also exist that include wooden flanges and corrugated steel web.
- **Plates**: Flat pieces of steel.
- **Purlins**: Beams, usually channels, set at right angles to trusses or roof rafters to provide support for lightweight roofing.
- **Rolled or built-up members**: Steel structural members can be either rolled or built up. Rolled members are one solid piece of metal. Built-up members are made up of different sections riveted, bolted, or welded together. A girder built from steel plate with angles riveted on each side of the top and bottom to form flanges is often called a plate girder. A column made of vertical units connected with

Figure 10-3 A castellated beam in a parking garage.
Courtesy of Glenn Corbett.

Figure 10-4 A lattice column.
Courtesy of Jack Boucher/Library of Congress.

Figure 10-5 Wide flange beams have more structural support than do standard I-beams.
Courtesy of Jason Roe, All Metals Incorporated.

diagonal pieces is called a **lattice column** Figure 10-4.

- **Spandrel girders**: Girders that tie wall columns together in a framed building. They also carry the weight of the panel wall, which can be considerable when masonry is used. In addition, spandrel girders sometimes are connected to the columns and stiffened to help resist wind shear.
- **Tees**: A standard I-beam cut lengthwise through the web forms two such beams with T-shaped cross sections. If the end of the cutoff web is thickened, it is called a **bulbtee**.
- **Tubes**: Steel structural members that are rolled in cylindrical, square, or rectangular shapes. They are most often used as columns. Some circular steel columns can be made fire resistive by filling them with concrete. There are hazards with these Lally columns.
- **Wide flange shapes**: I-beams that have flanges wider than those found in standard I-beams Figure 10-5. Some are H-shaped and, because they are square, are more suitable for columns. The letters "WF" are part of the designation of wide flange shapes.

- **Zees**: Members with a Z-shaped cross section. They are not often used in structures. Short lengths of "Z" sections, fitted over the edge of a subway platform, provide a base to shift a

subway car away from the platform. As steel technology developed, mills were built that could roll larger and larger sections. Therefore, built-up structural members of a size for which rolled members are used today are typically found in older buildings.

Steel as a Construction Material

Often the first structure that comes to mind when steel is mentioned as a construction material is the high-rise building. The development of steel framing as an engineering technique made it possible to erect tall buildings. The combined use of steel and concrete in high-rise buildings is discussed in the chapter on specific occupancy-related construction hazards.

The variety of steel shapes available has led to a specific nomenclature for specifying structural steel members. The following abbreviations are used for different-shaped steel members:

- C: channels
- CB: castellated beams
- L: angles
- S: America Standard (I-beam)
- W: wide flange beams and columns
- WT: structural tees

In addition, these abbreviations are used along with a set of numbers on architectural blueprints to designate a member of a specific size. The first number in this set refers to depth of the member (in inches), while the second number refers to the weight of the member per foot of length (in pounds), for all shapes except angles. For example, W 12 × 96 designates a wide flange beam or column that is 12 inches deep and weighs 96 pounds per foot of length. Remember the importance of the depth of a beam—the deeper it is, the larger the load it can carry. In the case of angles, three numbers are used to identify the steel member. The first two refer to the nominal depths of the two "legs," and the last number refers to the thickness of the legs. For example, an angle designated L5 × 3 × 0.375

indicates an angle with two legs, one measuring (nominally) 5 inches and the other 3 inches. The two legs are each 3/8″ (9.5 mm) thick.

The strength of steel, the consistency of its structural characteristics, and its ability to be connected to other structural elements so that loads can be adequately transferred are all important to the use of steel as a building material. Only a few specialized buildings, such as fiberglass buildings that are "invisible" to electromagnetic radiation or less affected by corrosive atmospheres, are built without steel.

The effect of fire on steel can be crucial to the stability of almost any building during a fire. Steel fire escapes are provided on many buildings. If they are in good condition, they can be useful not only in the evacuation of occupants, but also for fire department access. Steel reinforcing bars and cold-drawn steel tendons are vital to concrete construction in providing the tensile strength that concrete lacks. Steel is also used in concrete flooring systems. Corrugated steel provides "left in place" forms, which are often designed to react together with the concrete under load, thus forming a composite.

Steel is used to repair failures in concrete buildings. Even when a building is required to be fire resistive, the added steel is often improperly installed. In one case, an alert fire fighter noticed unprotected steel columns being installed and painted to match the walls in a fire-resistive-rated concrete mall. The building department was notified but did nothing for years, until a building permit was requested for another job.

Steel Buildings

Steel framing is used for many commercial and industrial buildings, usually of one-story construction. To achieve a peaked roof, the framing may consist of columns and beams with triangular trusses. The columns may support deep parallel-chord trusses spanning wide areas, with smaller trusses, often called **bar joists**, spanning the main trusses to support a flat roof Figure 10-6 . Space frames (three-dimensional trusses) provide huge clear spans in some modern buildings.

Figure 10-6 Roof bar joists.
Courtesy of Glenn Corbett.

Figure 10-7 In the rigid frame, the column is narrow at the base and tapers to its widest point at the top where it meets the roof rafter. Note the lightweight purlins on top of the rafter (acting as a girder), which will support the roof deck.

© Glen Jones/ShutterStock, Inc.

The steel in such buildings is almost universally unprotected. At best, such buildings can only be classified as noncombustible. In fact, steel is rarely protected from the effects of fire until the building is of such height or size (or occupancy in some cases) as to be required by law to be fire resistive.

Protected noncombustible sprinklered construction is found more commonly. In such buildings, major structural elements are provided with some fire resistance, but not enough to qualify the building as fire resistive. The concept is that the protected members will resist collapse until the sprinklers can control the fire.

Rigid frames are a cousin to the arch and are used to achieve wide clear spans. In the rigid frame, the column is narrow at the base and tapers to its widest point at the top where it meets the roof rafter (beam) Figure 10-7 . This rafter is also tapered so that it is narrow at the ridge and wide where it joins the column. It is the specific type of connectors—which resist movement—that gives this type of structural frame its rigidity. This wide haunch resists the outward thrust of the roof. Rigid frames are sometimes tied together under the floor to keep the legs from moving outward. If there is a basement, and the ties are exposed to a basement fire, failure of the ties may cause failure of the building.

Rigid frames can provide clear spans of about 100 feet (31 m). The area of the structure can be increased indefinitely by using Y-shaped columns, common to both frames. If such a column is distorted by fire, the area of damage may be doubled. Rigid frames can also be built of reinforced concrete or laminated wood.

Many steel-framed buildings are prefabricated. The Butler Company is a prominent manufacturer of steel "prefabs." Such buildings are often referred to as "Butler buildings," regardless of the manufacturer. Originally, the lightweight steel-framed building was exclusively for industrial use, but in recent years, schools, churches, and many other types of buildings have been built with similar designs Figure 10-8 .

Huge Spans

When huge spans (in some cases, well over 100 feet [31 m]) are achieved by using rigid frames, trusses, or space frames, collapse can be sudden and tragic. In buildings with huge spans, adjacent bents are

Figure 10-8 This Washington, D.C., Fire Department reserve apparatus building is a Butler building, incorporating an exterior insulation finishing system (EIFS) and masonry into the exterior walls.

Courtesy of Butler Manufacturing.™

tied together to resist wind load. Tying the steel units together, however, means that if one part of the building is distorted by fire, torsional or eccentric loads beyond the designed capacity may be placed on the balance of the building. This can start the progressive collapse of the building, often far beyond the area involved in fire. As a matter of fact, the better the building is tied together to resist wind load, the more likely it is to suffer progressive collapse due to fire distortion.

Designers are working wonders in developing wide-span trusses made up of very light sections. Recently, a hangar was built in Spain with a span clearing 787 feet (239.9 m), 272 feet (82.9 m) deep. The push to design large open areas with a minimum of mass is fraught with the potential for disaster. Designs may be pushed beyond the limits of steel. Hasty field changes or errors in construction can have catastrophic consequences. The weather-related collapses of the C.W. Post Auditorium in 1978, a 171-foot-span (52.1 m) steel and **aluminum** space dome located on Long Island, New York; the Hartford Civic Center in 1978, a 300-foot-span (91.4 m) steel space frame; and the Kemper Auditorium in Kansas City, Missouri, in 1979, a 324-foot (98.8 m) steel truss roof suspended from a space frame, caused no fatalities. Considering the thousands of people who had occupied these buildings, it is easy to see that there was a disaster potential that could dwarf the problems faced by rescue personnel in the Kansas City Hyatt Regency Hotel skywalk collapse in 1981, in which 113 people died and 186 were injured.

Very deep parallel-chord trusses have been used as floor beams in hospitals. Engineers borrowed a medical term and called the space produced by such trusses **interstitial space**. The hazard of such huge spaces is the temptation to use them for storage, maintenance shops, and the like. Such voids are dangerous. Automatic sprinklers should be required in them.

Heavy parallel-chord trusses have been used as **transfer beams**. A transfer beam is used to laterally relocate the vertical load of columns to clear an area, such as in a hotel ballroom. The trusses are often hidden in partition walls or ceilings.

There are also true trussed arches. The arch of a steel-arch bridge is often a truss. The arch is a compression structure. In this unique case, the top and bottom chords are both under compression.

Exterior Walls of Steel-Framed Buildings

A steel-framed building can be provided with a variety of walls. Some common wall materials and their fire characteristics are described here:

- **Cement-asbestos board**: This material is noncombustible and is often used for friable construction. Friable construction is used where an explosion is a possibility. It will break away readily, relieve the pressure, and not provide missiles during an explosion. Although such a scene appears to have great devastation, the loss is much less than would occur in a building that was not designed to resist the blast (e.g., regular construction). Concern for asbestos hazards may require monitoring before overhauling or investigating a fire where cement-asbestos board is involved. (Transite™, a trade name of such material, is often used as a generic name.) Some sandwich boards of cement-asbestos "bread" and low-density fiberboard "meat" are used in steel-framed buildings

as well; fire can burrow into the fiberboard and smolder.

- **Glass fiber-reinforced plastics**: Although glass fiber is noncombustible, the resinous binder most often used with it is flammable. This is not generally known, and the material is often thought of as noncombustible. Some reinforced plastic panels with good fire characteristics are available, but they would not be permitted in buildings of noncombustible construction.
- Aluminum: This material is noncombustible, but has a low melting point and little mass per unit of area, so it disintegrates rapidly in a fire. This characteristic is not always a disadvantage. In some instances, it may provide needed venting and access for hose streams to the interior of the building.
- **Precast prestressed concrete panels**: Panels of this type are erected in large sections, and their collapse can be particularly hazardous to fire fighters. When you are examining buildings that use precast prestressed concrete panels, study the relationship of the steel frame to the wall. If the steel expands under fire conditions, will it deflect the wall section from the vertical? If deflected, will the wall section fall freely or will it be restrained from falling? Precast concrete panels formed the side walls of McCormick Place, the large Chicago exhibition hall destroyed by fire in 1967. Distortion of the steel roof in the fire pushed the wall sections out of alignment and several of them collapsed.
- **Masonry walls**: The most common walls for unprotected steel-framed buildings are made of concrete block or a composite of concrete block and brick. Natural or artificial stone walls are also seen. Usually, the walls are only curtain walls. The exterior surface may show unbroken masonry; the interior wall may show panels of masonry between steel wall columns whose interior surface is approximately flush with the masonry. In other designs, the wall is independent of the steel

frame and is tied to it only for wind resistance. Occasionally in framed construction, one wall, or part of one wall, is a bearing wall. In either case, it is important to analyze the effect of the expansion of the steel frame on the wall. For instance, if a girder rests on a bearing plate, it may be free to slide and, therefore, will disturb only masonry above. In contrast, a beam tied to a girder that is incorporated longitudinally into a wall may, by pushing on the beam, bring down all or a substantial part of the wall.

Sheet steel-walled buildings are sometimes "improved" by covering the steel with a brick veneer. Any movement of the steel would likely bring down the brick because its stability depends entirely on the steel. The signs of masonry deterioration, which were mentioned in the chapter that discussed ordinary construction, are equally important in the steel-framed building. However, the absence of deterioration does not decrease the possibility of collapse of the masonry. A very probable cause of collapse is a steady lateral push by expanding steel.

- **Galvanized steel walls**: These walls are an industrial standby in applications where heat conservation is not important. Sometimes the metal is weather protected by an asphalt coating—so-called **asphalt asbestos protected metal (AAPM)**, often called Robertson protected metal (RPM), one proprietary name. The tar coating is combustible and would not be permitted in noncombustible buildings.
- **Metal panels**: Prefabricated metal panels are often made up in a sandwich construction to provide one unit combining thermal insulation and interior finish **Figure 10-9** . In earlier construction, low-density fiberboard will be found. Later, mineral and glass fibers were used. Today, plastics are most often employed for this purpose. Even when the outer layers are noncombustible, the insulation, vapor seal, or adhesive in the panels may be combustible. It may be necessary to open such panels far from the main

Figure 10-9 A damaged insulation panel from a cold storage building fire composed of polyurethane with metal cladding.

Courtesy of Glenn Corbett.

fire area if the interior of the panel becomes ignited. It is recommended that fire fighters overhauling such a structure work slowly and patiently, and cut with a circular saw rather than with an axe to avoid excessive destruction.

In one type of construction, polyurethane insulating panels are protected by gypsum board and stainless-steel sheathing. It is quite possible that, many years hence, alterations will be undertaken by persons not aware of the insulation. A cutting torch may be used, and a smoky, destructive fire may result.

Aluminum sandwich panels with foamed polyurethane are also available. Underwriters Laboratories, Inc. lists some of these materials as having low flame spread ratings, but their "smoke-developed" ratings may be quite high. Metal wall panels can be used on any framed building and, in fact, are used on many concrete buildings.

The design of panel walls, the method of installation, and the degradation of insulation or expansion of metal under fire conditions are some of the reasons the closure of the wall panel to the floor slab can fail and permit the extension of fire from floor to floor. This can

be particularly significant in high-rise buildings beyond the range of effective ground attack.

Prefabricated metal panels forming the front wall of some motel rooms have neoprene gaskets between the panel and the wall separating the rooms. A fire in one room would quickly destroy the gasket and extend toxic gases and fire to the adjacent rooms.

- **Light gauge steel-framed wall**: The use of lightweight galvanized steel studs for exterior walls has surged Figure 10-10. Shaped like the letter "C," these studs are held in place with U-shaped tracks on top and bottom. A variety of exterior finishes are applied to these steel-framed walls, including brick veneer, EIFS, and stucco. In some cases, these walls are load bearing.

High-Rise Framing

For many years, steel framing stood unchallenged as a method for high-rise buildings. (This statement applies to both "short" high-rises of Type II noncombustible construction and "tall" high-rises of Type I fire-resistive construction.) In recent years, concrete

Figure 10-10 A light gauge steel-framed exterior wall under construction.

Courtesy of Glenn Corbett.

is finding more use for framing high-rise buildings, and the two materials compete. In Chicago, the reinforced concrete South Wacker Tower is 946 feet high. The principles of construction are the same, however, and much of what is discussed here applies equally to concrete-framed high-rise buildings.

When steel-framed buildings were first built, builders were hesitant about publicizing the fact that it was the frame that supported the building. Traditionally, most people associated strength and permanence with solid masonry. Consequently, tons of brick, stone, and terra-cotta, all structurally unnecessary, were loaded onto framed buildings. The heights of buildings were also suspect. Rather than arrange the exterior so that vertical lines led the eye upward, as is common today, buildings were belted with masonry corbels or trim. Different materials were used on successive floors, and all possible architectural devices were used to reduce the apparent or perceived height of the building.

The masonry served another purpose in early construction. The **spandrel space** (the distance between the top of one window and the bottom of the one above) was made of masonry. It was sufficient to prevent the extension of fire from floor to floor on the outside.

Today, the pendulum has swung the other way. Glass and metal exterior panels are common, and the firestopping capabilities of such members are dubious. High-rise buildings should be evaluated for the dependability of perimeter firestopping. Tactical plans should be developed to deal with the problem.

Tilt Slab Hazards

There is special hazard in tilt slab construction (also known as tilt-wall construction or tilt-up construction)—that is, buildings in which the roof holds the tilted-up concrete wall panels in place Figure 10-11. This method of construction is found in industrial buildings that are very popular around the country. When these structures are being constructed, the builder carefully braces the walls with tormentors or braces. This is required for stability until the roof is in place and tied in, thus stabilizing the building.

Figure 10-11 Precast wall panels, delivered to the construction site on a flatbed trailer, being lifted prior to being "tilted" into place.
Courtesy of Glenn Corbett.

Although a roof of steel joists is the most common, the roof may be of any type of construction. If the roof is being lost in the fire, however, fire fighters should be wary of wall collapse, as has happened in warehouses from California to New Jersey. Wood roofs may simply burn away, leaving the concrete panels freestanding as a vertical cantilever. Concrete may spall from the bottom of T-beams, exposing the tendons to their failure temperature. Steel may elongate significantly before failing. This elongation may push the walls down.

In tilt slab construction, examine openings in the wall carefully. Ordinary-size doors are usually placed in an opening cast into a panel. Loading doors may take the full width of a panel. The concrete panel above may be suspended from the roof and subject to collapse.

If the building is sprinklered, make careful observations of the density and volume of smoke. If heavy smoke is being generated, then the sprinklers are probably not controlling the fire, and the roof is vulnerable. Massive and, in some cases, total-loss fires in a number of sprinklered warehouses have been especially disturbing to those familiar with the exemplary record of successful sprinkler operation.

Steel-Framed Buildings Under Construction

Except for the problem of breaching shear walls around elevator shafts, the method of wind bracing in a building is of no concern to fire fighters after the building is completed. While the building is under construction, however, wind forces must be resisted because the building is not fully connected. Often, cables are strung diagonally across the steelwork, and at times these braces are not properly installed. If there are no high winds, nothing happens.

Plastic Design in Steel Construction

In an effort to seek economical steel design and to compete with concrete, plastic design has been developed. The typical steel-framed building is pinned. In a pinned building, loads are delivered to the nearest columns. In plastic design, the connections are built to transfer loads beyond the column. This is analogous to monolithic construction in concrete. Because of the rigid connections, a load that might be locally excessive is redistributed over the structure. As a result, the weight of steel in the building can be decreased. Beams are lighter and columns are smaller than they would be otherwise. This is of more than academic interest. The lighter the steel, the less fire resistance. Thus, the provision and maintenance of adequate fireproofing for the steel is even more important in a plastic designed building than in one of conventional design.

More on the Fire Characteristics of Steel

Steel Conducts Heat

The fact that steel transmits heat is well known. A suit of armor is noncombustible—but no one would ever attempt to fight a fire in it. Despite this, many building codes previously required so-called **tin ceilings**, which are actually embossed steel, in certain occupancies **Figure 10-12**. These can transmit fire in either direction.

Figure 10-12 Although tin ceilings are typically installed in Type III ordinary construction, they illustrate the heat transfer characteristics of metal.

Courtesy of the estate of Francis L. Brannigan.

In earlier years, sheet steel doors were commonly used on vaults, some of which still guard valuable governmental and private records. The chief disastrous result of using steel doors could be the loss of records. Across the country, businesses and homeowners alike seem to rest secure in the knowledge that their vital records are "protected" in uninsulated steel files. Tin boxes are often sold in stores as fire-resistant chests.

The conductivity of steel can be a factor in spreading fires. For example, heat was conducted through **steel expansion joints** in a concrete floor at a fire in a multistory post office. Mail bags resting on the joints in the concrete floor were subsequently ignited by the heat.

In many cases, steel sheets directly attached to combustible surfaces to provide fireproofing have acted to transmit or retain heat. In one case, a steel plate was attached to the wood-joisted ceiling above the end of a rotary kiln in a building of ordinary construction. When it was ordered to be removed, plant personnel were surprised to find that all the wood behind it had carbonized to charcoal.

It is generally understood that metal smoke pipes that lead from a furnace or stove to a chimney or metal stock must have clearance from wooden members. However, this concept is often ignored in

the case of other metal ducts that may have fire in them, despite requirements to the contrary. Grease ducts are a good example. A grease duct fire often extends to adjacent combustible construction by conduction.

Ships are not buildings, but the practice of using ships as buildings, like the Queen Mary in Long Beach, California, is growing. A number of ferry boats have been converted into restaurants. Ships have steel walls known as **bulkheads**. Many people associated with ships confuse noncombustibility with nonconductivity. Typically, welding operations will be performed on one side of a bulkhead without any concern for heat transmission through the steel to ignite combustible material on the other side. This can cause a problem when the other side is covered with wood paneling, which could be ignited from within the concealed space behind the paneling.

Self-storage facilities built of steel have many of the characteristics of ships Figure 10-13 . The Tualatin Valley, Oregon, Fire and Rescue Department fought a stubborn fire in a self-storage building that involved 97 units. Some of the units contained flammables and explosives. The fire spread from unit to unit by conduction and radiation. Scalding

Figure 10-13 Self-storage facilities have similar characteristics to those of ships, including the extensive use of heat-conducting steel, allowing fire to spread from unit to unit through conductive heat transfer.

© John Spink/AP Images.

steam was generated when water hit the hot steel sheets.

Steel Elongates

Steel will expand 0.06 to 0.07% in length for each 100°F (37.8°C) rise in temperature. The expansion rate increases as the temperature rises. Heated to 1,000°F (538°C), a steel member will expand 9 1/2 inches (24 cm) over 100 feet (30.5 m) of length. As you will see in the next section, at temperatures above 1,000°F (538°C), steel starts to soften and fail, depending upon the load.

Elongating steel exerts a lateral force against the structure that restrains it. If the restraining structure is capable of resisting the lateral thrust, the expanding steel structure may be even stronger, at least for a period of time. Some floor assemblies have a different fire resistance rating depending on whether they are restrained or unrestrained.

If steel beams are restrained, as by a masonry structure, and the temperature of the fire is sustained at around 1,000°F (538°C), the expansion of the steel may cause the displacement of the masonry, resulting in a partial or total collapse. A one-story commercial building in Dallas was constructed of concrete block walls and a steel-joist roof; the steel joists were restrained by the walls. A metal deck roof fire sent heat rolling through the roof void. Elongating bar joists pushed down the walls, dropping six fire fighters into the structure. Fortunately, all six were rescued.

In hot, fast fires, failure temperatures are reached rapidly and the lateral thrust against the wall is minimized. In a one-story and basement noncombustible building, a heavy girder was supporting a one-wythe-thick decorative brick wall in the reception lobby. There was a hot fire in the basement. The steel beam attempted to elongate but could not because of the presence of the restraint. It then took the only other course available to it to absorb the increased length—it twisted, dropping the wall completely across the lobby. Fortunately, no one was in its path. This overturning can be anticipated when unprotected steel girders are used to support wooden floors in combustible buildings.

Steel Fails

When steel is raised to temperatures above 1,000°F (538°C), it starts to lose strength rapidly Figure 10-14 . Temperatures above 1,000°F (538°C) are quickly reached in fires. Recognizing this, the standard fire test used to test building materials and assemblies (described in the American Society for Testing and Materials [ASTM] E-119) reaches 1,000°F (538°C) in 5 minutes. Recent tests conducted by the National Institute of Standards and Technology (NIST) of a typical basement room fire reached 1,500°F (816°C) in 5 minutes, reflecting the presence of today's heavy plastic fire loads rather than any special significance of basements.

In standard fire tests of the fireproofing of steel columns, the column fails when a temperature of 1,200°F (649°C) is exceeded at one point or 1,000°F (538°C) is exceeded on the average in the column. How fast the fire temperature will be achieved in the steel varies. The principal variable is the weight or mass of the steel unit. For instance, a bar joist or lightweight steel truss will absorb heat very rapidly. In a fire equivalent to the standard test fire, the bar joists generally will fail in about 7 minutes. In one observed fire in a one-story and basement bar-joist building, the bar joist–supported floor was collapsing before the people could get out of the first floor. Very heavy steel sections might have survived longer. Fireproofing is

Figure 10-14 Steel will begin to lose its strength at temperatures of 1,000°F (538°C).

© Harald Koch/AP Images.

sometimes found in low-rise multistory buildings and certain occupancies of Type II construction.

The amount of ventilation may also be a significant factor. It appears that if a fire is well ventilated, the temperature in the steel may be less than if the fire is confined. This may help to explain observed differences in the effect on steel in different fires. The heavier the steel is loaded, the faster it will fail.

Overcoming the Effects of a Fire on Steel

A number of options are open to the designer to deal with these characteristics of steel as they relate to fire risk:

- Ignore the problem.
- Rely on inadequate code enforcement.
- Take a calculated risk.
- Protect (insulate) the steel.
- Protect the steel with sprinklers.
- Protect the steel with an internal water cooling system.
- Locate the steel out of range of the fire.

Ignoring the Problem

The potential for fire damage to steel buildings is not clearly understood. Often, unwarranted confidence is placed on the fact that the steel is noncombustible. A major insurance carrier found it necessary to conduct demonstrations of actual fires for plant managers in which the only fuel was the hydraulic oil in presses to show the seriousness of the fire problem to plant managers without any fire experience.

In one case, a metal building intended as a site for thawing frozen drums of uranium residue was under construction at an atomic energy facility. The shed and all the roller conveyors were destroyed by a fire. The only fuel was the 1-inch fiberboard lining of the building. The builder was astonished—and chastened: he had been so convinced that there was no fire problem that he had not insured the job.

Steel Highway Structures and Bridges

Bridges and overpasses on highways provide a good example of unprotected steel that is vulnerable to an occasional gasoline or oil truck fire. Such fires can result in enormous damage, possibly accompanied by major traffic disruptions. A bridge on the Trans-Canada Highway in Nova Scotia was damaged when struck by a fuel truck that exploded. The bridge was out of service for 69 days and the repairs cost $350,000.

The Chicago Fire Department fought a truckload of magnesium on fire under a concrete overpass. The structure was protected by hose streams as the fire burned out. The bridge was saved.

Seventy years ago, when prefire planning was generally unknown, the New York City Fire Department developed very specific preplans to get water as soon as possible onto fires on the East River bridges, recognizing the vulnerability of the unprotected steel. If assigned units are unavailable when incidents involving these bridges occur, the dispatcher must cover them and notify the replacement units of their specific function to be performed.

The Hazards of Concentrated Fire Loads

Even when average fire loads are low, an unprotected steel building may be endangered by a highly concentrated fire load in one area. Occupants then break up the big open space to meet operation needs. Although unprotected steel industrial buildings are usually only "one-high story," it is not uncommon to find a mezzanine built into the building to provide office or storage space.

The term *one-high story* is used to designate buildings of greater than usual height from floor to ceiling. A description such as "one equals five" describes a one-story building of a height equal to five ordinary stories. Often a mezzanine within such a building is built entirely of wood (not permitted in a noncombustible building but often done without a building permit). Where problems such as noise, temperature, pilfering, or privacy exist, structures may be built within the building to provide office space, lunch rooms, a controlled environment, privacy for secret processes, or security for valuable merchandise. Unfortunately, such structures are often combustible. The structure then represents a hazard not only to the building, but also to the contents that it is designed to protect.

Experiment halls at research facilities are often built of lightweight steel construction to provide the greatest free floor area at the least cost. The steel is usually unprotected; the fire load is expected to be minimal. In recent years, however, the practice of purchasing ordinary house trailers to use as offices or to house telemetry equipment inside such halls has grown. From the fire protection point of view, they provide a serious potential source of heat. When grouped closely together, the trailers represent exposures for one another, increasing the possibility of a serious loss in the event of a fire.

Prefabricated metal buildings are sometimes used in place of the trailers. Often these are of sandwich construction with an insulator placed between the inner and outer walls. The nature of the insulation should be determined. Many kinds of insulation present severe fire problems.

Excavation Bracing

Because of the requirement for underground parking, building excavations are now being made much deeper than in previous years. When a hole is opened in the ground, the earth under the adjacent structures may tend to shear sideways into the hole, causing a structure to collapse. Often it is necessary to shore up the excavation. Such shoring literally holds up the neighborhood.

Rows of vertical steel beams called soldier beams are driven into the ground. They are tied together by a horizontal beam called a **waler**. Diagonal columns called **rakers** brace the entire structure. These columns may be loaded to twice the load that would be permitted if they were permanent structures. They are, therefore, subject to early failure in a fire—and the consequences of such a failure could be catastrophic.

An excavation is often loaded with combustibles, such as plywood, plastic, and fuel. This array of rakers is a nuisance in the erection of the building. If ground conditions permit, tiebacks are used. With this technique, cold-drawn steel cables are inserted in holes driven into the rock and anchored with epoxy. They are then tensioned, similar to posttensioned concrete. The end of the cable sticks out, so it can be easily heated in the event of a fire. If the tensioned portion of the cable reaches 800°F (427°C), it will fail.

The unprotected steel tower crane is similarly vulnerable. This hazard does not appear to be covered in any code, but it should be explained to the building department and to the builder. "Nothing like that has ever happened. You guys are always thinking up busy work," is a likely response to this warning. However, the fire department will be in a much better position if the disaster occurs.

The Redmond, Washington, Fire Department, with the help of mutual aid units, fought a fire in a fire-resistive concrete building under construction. The fire heated the tower crane and put it as much as 4 inches out of plumb. If uncontrolled, this shift in balance could have caused a collapse. Heavy hose streams reduced the out-of-plumb gap to 2 inches, averting such a disaster.

Buildings Under Construction

During construction, steel that will later be fireproofed may be unprotected for extended periods.

When the first World Trade Center was under construction, original plans to use fire-retardant treated plywood for formwork were scrapped in favor of ordinary plywood and $1 million in savings. When fire protection of the project was discussed at a National Fire Protection Association (NFPA) meeting, the Chief of the New York City Fire Department, John O'Hagan, warned the Port Authority of New York and New Jersey about the hazard of such a "7-block-long lumber yard" to the structural steel of the World Trade Center. Fortunately, no fire occurred.

Relying on Inadequate Codes

Building codes generally classify steel buildings as "unprotected noncombustible," or "protected noncombustible." In this latter case, major columns and beams have some protection. The word *protection* is often confused with *sprinkler protection*, but actually refers to physical protection of the steel with gypsum board, spray-on fireproofing, or the like.

In studying a potential fire in an unprotected steel building, the first consideration should be the basic type of construction. Column, girder, and beam construction is used for both single-story and multistory construction. If a masonry bearing wall is substituted for some of the exterior columns, the building is wall bearing. Column and girder buildings are characterized by relatively short spans, so the failure of one column may affect only one portion of the building. As with all such generalizations, however, watch for the exceptions.

A Maryland church is built of steel columns that support steel roof trusses. The roofing is wood. The 40-foot-high curtain walls, which do not carry the load of the roof, are composite brick and block. Should there be a severe roof fire, the heat absorbed by the roof steel would cause it to elongate, pushing the columns out of alignment. The moving columns might bring down the walls.

Another unfortunate assumption about fires is that they will only burn upward. Fuel in a church balcony might include hymn books and cushions of foamed plastic. A fire in these fuels would readily burn down into the void. A fire involving the wooden balcony or the metal deck roof could well cause the steel framing to move and thus cause one of the brick-veneered walls to fall. A well-advanced fire in such a structure should be treated with caution.

Steel High Above the Floor

In some building codes, steel used to support roofs at a certain distance above the floor (usually about 20–30 feet [6–9 m]) does not require protection. Atria roofs over 55 feet (16.8 m) are given this exemption as well. It is generally believed that flame

impingement is not likely at this height. The concept is probably valid if there is no significant fire load in the area below the steel. Such a situation is unlikely, however. For example, the situation that existed at Chicago's McCormick Place is far more common.

The McCormick Place Fire

McCormick Place was a huge Chicago exhibition hall built in 1960. It burned on January 16, 1967, with a loss of $154 million **Figure 10-15**.

The main exhibit area of McCormick Place provided a clear area of 320,000 square feet (29,729 m²). The roof was supported on 18 steel trusses, each 60 feet (18.3 m) on center. Each truss had a 210-foot (64 m) central span and a 67-foot (20.4 m) cantilever on each side from the column to the exterior wall. The trusses were 16 feet (4.9 m) deep at the column, tapering to 10 feet (3 m) at the center of the building. They were made of heavy steel members.

The columns were trusses themselves. In accordance with the previously mentioned provisions common to many building codes, the columns were fire protected up to a height of 20 feet (6.1 m) above the floor. The roof trusses were unprotected.

The exterior walls consisted of freestanding steel columns supporting precast concrete sandwich wall panels. The sandwich consisted of concrete interior surfacing, foamed plastic insulation, and exterior sculptured concrete.

Figure 10-15 The 1967 Chicago McCormick Place fire.
© Paul Cannon/AP Images.

At the time of the fire, preparations were being completed for the exhibit of the National Housewares Manufacturers Association. Exhibit booths crowded the floor, with some exhibits being two stories high. The fire loading at the same time has been estimated at 80,000 to 120,000 Btu per square foot. In addition, the combustible material, both by nature and arrangement, was fast burning, and provided a high rate of heat release. Fortunately, the fire started in the early morning hours. The loss of life might have been staggering had the fire occurred during the day.

The fire was discovered early, but soon spread beyond the capabilities of the maintenance employees. There was about a 6-minute delay in the alarm. A second alarm was transmitted on arrival. The first-due engine company had penetrated the hall a few feet when they heard a great roar as the flames approached. At this time, their water supply failed. It became known later that the hydrants in the area that were out of service were the responsibility of the exhibit hall management. In retrospect, at least one close observer thought this was a blessing. Had the water been available, it would have made little or no difference to the loss, but it is possible that a number of fire fighters would have been in serious danger while advancing lines into a collapsing building.

Within a half hour after the start of the fire, portions of the massive roof trusses began to fail. The failing trusses pushed out a number of the wall panels, making close approach hazardous. Several of the panels collapsed.

Based on the size and occupancy classification of the building, the construction of McCormick Place apparently met the fire protection requirements of the Chicago Building Code and most other widely recognized building codes. Most contemporary codes would classify the building as an assembly occupancy. However, at the time of the fire, it more closely resembled a mercantile occupancy such as a department store, discount house, or shopping center. In this type of occupancy, different fire protection measures would have been required by most modern building and fire codes.

If McCormick Place was viewed as a mercantile occupancy, modern building codes would have required a building of this size to be equipped with an automatic sprinkler system throughout. Additional or supplementary fireproofing of the structural members would not have been required.

Automatic sprinklers installed throughout the exhibit areas probably could have limited the spread of this fire and enabled the fire department, with a properly operating water supply, to bring it under control with a minimum of damage.

A building that can accommodate 50,000 people and is the equivalent of a tinder-dry forest cannot be protected by mere compliance with codes that did not contemplate such a building. Neglecting installed equipment and assuming that employees would take proper emergency action also critically affected fire protection at McCormick Place. At least this point became apparent after the fire. But how apparent is it to your local armory, hotel ballroom, or exhibit hall management? Is your fire department satisfied that the extent of their responsibility is to see to it that all "rubbish" is removed and all fire code "exhibit" requirements are being met?

The New McCormick Place

In the new McCormick Place, structural steel is protected with directly applied fireproofing delivering 1-hour fire resistance. The entire building, except electrical enclosures and enclosed stair towers, is sprinklered. Sprinkler systems protecting the high fire load exhibit area are hydraulically designed to deliver 0.30 gallons per minute (gpm) (1.1 Lpm) per square foot over areas as large as 6,000 square feet (557.4 m²). Ninety-eight percent of the floor area can be reached with 100 feet (30.5 m) of hose from standpipes. Elaborate provisions have been made for smoke venting. Supervisory control has been provided for water supply. Training programs for employees and prefire planning with the Chicago Fire Department are provided.

Important Test Experience

Testing undertaken at Underwriters Laboratories, Inc., in the aftermath of the McCormick Place fire examined whether sprinklers would have controlled the fire. The data developed in these tests is worth examining, especially as it relates to the temperatures developed at quite ordinary fires.

The fire tests were conducted in a building 30 feet (9.1 m) high, the height above a floor at which it is presumed that steel in the roof needs no protection. The fuels used were lightweight, readily combustible materials typical of those used for exhibition hall displays. Plywood, tempered hardboard, cardboard cartons, and packing materials made up the fire load average of 20 pounds per square foot. The materials had a high surface-to-mass ratio and thus would produce a fast, hot fire. The display material was set up over a 20 by 30 square-foot (6.1 m by 9.1 m) area. Thus, the total weight of combustibles was about 12,000 pounds (5443 kg), which would yield almost 100 million Btu if completely consumed. This was a rugged test, but not unrealistic.

The first of the test fires is of great interest. The plan was to keep the sprinklers turned off for 6 minutes after ignition, corresponding to the reported gap between ignition and time of discovery of the McCormick Place fire. The report gives no indication that even the experienced engineers running the test had any qualms about the plan of holding back the water for 6 minutes. After 5 minutes and 45 seconds, temperatures of 1,500°F (816°C) were recorded at the ceiling. The sprinklers were turned on at this time to avoid damage to the test facility.

Steel beam temperatures were monitored by thermocouples **peened** into the surface. A bar joist at the ceiling reached 1,540°F (838°C) and an I-beam reached 1,355°F (735°C) at a little over 5 minutes. Temperatures at the ceiling level exceeded temperatures in the booth area by a slight margin.

The fire load used in the test cited is not excessive in either quantity or rate of heat release. There are many unprotected steel buildings with such a great fire load that collapse will have started prior to the arrival of the fire department, even with prompt alarm and normal response.

Despite the McCormick Place fire and the tests cited above, buildings are still being designed in accordance with the concept that steel needs to be fireproofed only below a certain height. Plans should include the necessity of using hose streams to cool affected steel.

Taking Calculated Risks

The lack of fire protection for unprotected steel may reflect a calculated risk taken by the building's developers and managers. For our purposes, there are three classes of calculated risks:

- Financial or economic
- Engineering
- Forget it

The basis of the calculated risks must always be closely examined.

Financial Calculation

One calculation for a government-owned building went like this: This noncombustible building is of low value and is slated for demolition in the 5-year plan. In the meantime, it is used for the reserve storage of foamed plastic shipping containers. The value is low and the loss of the containers would not hamper production. The cost of any special protection could not be justified. "In case of fire—let it go."

Engineering Calculation

Here's an example of an engineering calculation: The fire load in this noncombustible building is minimal. Unpackaged noncombustible heavy metal sections are stored on pallets. The combustible load averages about 2 pounds per square foot and is dispersed. There is no apparent source of ignition. No special protection is recommended. This calculation is adequate.

The steel industry is in constant competition with the concrete industry. When steel is required to be fire proofed, the cost shows up as a line item in the budget. The cost of fire-resistive concrete, in contrast, is buried in the cost of the concrete. This gives concrete a competitive advantage.

Some steel-framed parking garages are large enough to require fireproofing under some codes. The steel industry conducted fire tests to demonstrate that steel-framed parking garages do not require fire protection of the steel. As a result, some codes permit unprotected steel to be used if a substantial portion of the garage is open to the air, which permits the escape of heat. This engineering calculation could not be applied to the parking garage in a Texas city in which the floors are supported on bar joists, rather than on heavy steel.

In another example, a steel garage was built over a railroad track that carried flammable liquid tank cars. It was decided rightly that the steel columns of the garage should be protected. This was done literally—the columns and steel directly at the track were protected. If a fire occurs, however, it will roll out beyond the protected elements and destroy the building. The same sort of questionable engineering was seen on a Canadian car–passenger ferry. Originally, a deck was fitted out with passenger cabins. The overhead steel in the cabin area was spray-applied with fire protection. The beams extended out to form the overhead of the outside deck, but this steel was not protected. When the cabins were removed to make more space, the entire area was enclosed. The newly enclosed steel was not protected. Thus, only half the length of each beam is protected; the other half is bare.

Forgetting It

In taking "calculated risks," keep in mind the well-known computer programmers' acronym GIGO—garbage in, garbage out. No calculation is any better than the information used in it.

Fire protection interests are often hampered by management obsessed with the short-term results. Industrial managers may defer fire protection expenditures to show a better quarterly profit picture. Politicians may vote to defer the expense of sprinkler installation to erect a new structure that will carry a bronze plaque.

Some managers are fond of using a calculated risk as an excuse to do nothing, particularly when the cost of providing protection interferes with other objectives. They get upset when it is pointed out, however, that they do the calculating while others take the risk.

Insulated Metal Deck Roof Fire Problems

In August 1953, the General Motors transmission plant at Livonia, Michigan, burned **Figure 10-16**. Its loss of $32 million was the largest industrial fire loss to that date. The metal deck roof was the principal contributing factor to the destruction of the plant.

Metal deck roofs consist of metal sheets laid over steel bar joists. The sheets are crimped together, making a joint that is not gas-tight. Usually insulation is added—often a low-density fiberboard, although plastics currently are making inroads into this application.

Insulation that has absorbed moisture is useless, so it must be protected from moisture driven through the roof by capillary attraction. An adhesive is also necessary to secure the insulation to the roof deck and to prevent loss of the roof covering in a windstorm. A bituminous (asphalt) coating serves as the adhesive and sometimes as a moisture-stopping vapor barrier. On top of the insulation are successive layers of bituminous (bitumen) material and roofing felt or fiberglass mats. From the presence of these successive layers comes the term *built-up roof*. While the construction of "hot applied" built-up roofs using tar kettles and hot asphalt still occurs, roofs of this type are now typically "cold applied" using fiberglass sheets with special adhesives. Some roofs are built with a layer of concrete laid on the corrugated metal and the roof built up above the concrete. Such roofs are not of concern here.

In recent years, the "rubber roof" (composed of ethylene propylene diene monomer [EPDM]) has increased in popularity in Type II construction (although this roof may also be found in other types of construction). It is composed of a single ply of this rubber-like material (to minimize water leakage in the roof) laid over fiberboard and plastic insulation, held in place with metal fasteners, seam tapes, and flammable adhesives.

A five-alarm wind-aided fire in Brooklyn in the 1990s highlighted a number of problems with this construction technique, including the fire's ability to spread like a brush fire; burning sections of the roof were picked up by the wind and deposited downwind, starting new fires. The entire 750-foot (228.6 m) length of the roof was consumed in about half an hour. The lessons of this fire include the application of heavy-caliber streams ahead of the fire; checking for extension inside the building in the form of a flaming, dripping liquid coming through the metal deck seams and falling on to combustibles below (the same problem as a metal deck roof fire in a traditional built-up roof); the need to apply cooling streams to the underside of the roof deck to prevent this problem; and the significant safety concern of ensuring that no fire fighters are placed in the path of the fire.

Another relatively new type of roof is the modified bitumen ("mod bit") roof. In this construction

Figure 10-16 The Livonia, Michigan, General Motors fire in 1953.

© AE/AP Images.

technique, a traditional built-up roof is modified by the addition of synthetic polymers to reinforce the roof. The sheets of the roof may be laid down with adhesives liquefied by a propane torch; more recently, a cold application technique has been developed to lay down these sheets.

Approved roof is the accepted term for roofing that meets a testing lab (e.g., Underwriters Laboratories, Inc.) standards for roofing resistant to the propagation of fire from building to building by flying brands. A roof can be listed but still be a combustible metal deck roof—the two problems are entirely different. The listing points out that fire under the roof is not a consideration.

When a fire occurs below a combustible metal deck roof, the metal deck heats up. Tests show that exposure to a temperature of 800°F (427°C) for 5 minutes is enough to start the process. The heat is conducted through the deck to the bituminous adhesive; this adhesive liquefies and then vaporizes. The gas cannot escape through the roofing material, so it forces itself down through the joints in the deck. When the gas mixes with the air below, it ignites from the fire below.

Such a gas fire rolls along under the roof, heating additional roof areas, which generate more gaseous fuel. Almost unbelievable quantities of thick black smoke are generated. This type of fire becomes self-sustaining, independent of the original fire, and spreads rapidly in all directions.

The roofing insulation and some of the roofing felt may burn, showing fire on top of the roof. This is of little consequence; the problem is at the underside of the roof.

Two solutions are available to the builder to prevent such occurrences:

- Use Factory Mutual Class I roofing or a UL Classified Roof. These have been tested in accordance with NFPA 256 and will not provide self-sustained combustion—the problem with the ordinary combustible metal deck roof.
- Provide adequate automatic sprinkler protection for the roof, even though the contents may be noncombustible. If a ceiling is installed, the sprinklers must be located above and below the ceiling. Several buildings have sprinklers above and below the ceiling.

Fires of Interest

In 1985, U.S. taxpayers suffered a $195 million loss in a fire at a production building at Tinker Air Force Base in Oklahoma City. The fire was started by roofers. The sprinklers were below a wire lath and plaster ceiling. The water did not hit the underside of the roof deck, and the fire burned unimpeded. An attempt was made to cut the fire off by removing the built-up roofing, but the fire passed the cuts.

Even Class I roof construction, if combustible, may not always be acceptable. For example, a fire in a large concentrator building at the Wabush Mines complex in Labrador, Canada, badly damaged a metal deck roof 90 feet (27.4 m) above the floor. Investigators theorized that even if the roof had been of Class I construction, damage still would have been severe. Failure of unprotected steel supports undoubtedly would have let the roof sag, exposing to the fire asphaltic materials in the built-up covering over the roof insulation.

The term *Class I* may be taken to mean "completely satisfactory under all circumstances" by persons not fully familiar with this subject. If a Class I roof has any combustibles at all, it may not be satisfactory for use over a high-value installation, such as an area filled with computers. Total noncombustibility should be sought in such a case.

The Atomic Energy Commission (AEC) studied the problem of fire protection for thousands of acres of metal deck roofs covering some of the most vital production plants in the United States. Many proposals were examined. The final decision was to add sprinklers to the plants. Over $20 million was spent on sprinklers and water supply. A fire occurred in an AEC facility in Paducah in a sprinklered building; it was confined to the area of origin. The fire was a very hot chemical fire, and the sprinkler operation caused a steam cloud that rolled through the plant. Over 2,000 sprinklers opened, causing a water flow of millions of gallons. Thereafter, all sprinklers in

these plants were replaced with higher temperature sprinklers in order to drastically lower the number of sprinklers that would open.

An Unrecognized Problem

Careful reading of fire reports indicates that the metal deck roof is a factor in many losses where its significance is not understood by the fire suppression group. In May 1977, the Beverly Hills Supper Club in Kentucky burned with a loss of 164 lives **Figure 10-17**. An extensive study of the circumstances of that fire indicated that a significant factor in the huge loss of life in the Cabaret Room was the tremendous amount of black, choking smoke developed by the fire in the metal deck roof. The initial investigators discounted the possibility of fire moving through the voids because they were cut up by walls. It is true that the roof was pierced by parapetted walls. The building had been added to many times, but undoubtedly there were penetrations through the walls. A metal deck roof fire does not need a big opening. A tongue of flame passing through a small hole, impinging on the metal deck

Figure 10-17 The Beverly Hills Supper Club in Kentucky, May 1977.
© RF/AP Images.

of the next section would be sufficient to extend the fire. (Factory Mutual tests on metal deck roof assemblies have shown that it takes only 800°F (427°C) for 5 minutes for heat impinging on the surface of steel decking to start a self-sustaining roof fire. This fire is independent of the original fire.)

It is not at all clear that if the Supper Club building had been sprinklered according to the usual practice, a substantial fire could have developed. Under typical practice, the void would be regarded as noncombustible and left unsprinklered. Because it was above the ceiling insulation, a separate dry-pipe system would have been required to prevent freezing. Of course, had code requirements for exits and fire resistance been met (a fire-resistive structure would not have had a metal deck roof), there would probably have been few lives lost.

It appears that in most jurisdictions, an ordinary combustible metal deck roof would be permitted on a noncombustible building. Sprinklers below the ceiling in such a building will not protect against a metal deck roof fire moving through the void. However, they might prevent a contents fire from reaching the metal deck roof. They would also operate on contents fires started by dripping, flaming tar.

Four elderly persons lost their lives in a one-story nursing home in Dardenelle, Arkansas, in 1990. The building had been described as a protected noncombustible structure by the Arkansas Department of Health. The building had a combustible metal deck roof. Extremely thick black smoke, typical of a metal deck roof fire, drove the staff from the west wing. The NFPA Fire Investigation Report declared that "flammable vapors from the heated roof assembly contributed to the fire extension in the concealed space, and the smoke and toxic gases generated in the concealed space filled the occupied patient rooms."

Fighting a Metal Deck Roof Fire

Frank Brannigan first became aware of the metal deck roof problem in 1946. A fire occurred in a 250-foot (76.2 m) section of a 1,500-foot (457.2 m) by 100-foot (30.5 m) warehouse at the Marine Corps

Supply Depot in Norfolk, Virginia. The warehouse was loaded with field supplies. The fire originated in the center of the section, involving stock on both sides of the center aisle. Brannigan was supervising handlines inside the building, hitting fire in the burning stock. Blue flames extended down six feet or more from the underside of the roof, accompanied by dripping, flaming tar. When the heat and tar became hazardous, the streams were directed upward to put out the fire. The fire fighters made a "good stop," cutting the fire off right in the piles of stock 20 feet (6.1 m) high.

The pictures of the building, however, told a different story. The roof was gone from fire wall to fire wall, and it appeared as if the fire department had played no effective part in stopping the fire, but that rather the fire walls had done the job. After much discussion and study, the conclusion was that the roof had burned overhead, independently of the main body of fire in the contents.

It appeared that a fire involving a metal deck roof could be controlled only by the continuous application of water to the underside of the roof. By cooling the steel, the tar would be prevented from generating gas. Literally, the fire would be extinguished by removing the fuel.

The chance to try the plan came in about 15 months. In a similar warehouse section, the Marines had built a plywood office, a sick bay, and theater, while the balance of the area was loaded with combustible stock. Like the first fire, this fire was well advanced upon fire fighters' arrival, and an immediate second alarm was sounded.

Looking up along the underside of the roof, from the door at the end opposite the fire, Frank Brannigan observed the characteristic blue waves of fire several feet deep advancing rapidly. A hose wagon with deck pipe was ordered into the building, taking a position in the center aisle opposite the door. The orders were to direct a master stream at the underside of the roof, keep it moving, and ignore the burning contents. The operation was successful. The warehouse looked as if a knife had cut through the roof and walls; the building was lost only in the

area where the huge body of initial fire had done its work before arrival.

Metal Decks on Nonmetal Buildings

Metal deck roofs are not found exclusively on steel-framed buildings. Many are built on masonry buildings and more recently wood-frame buildings, but the problem remains the same. In December 1970, in Kensington, Maryland, fire fighters responded to a serious fire in the contents of a one-story building housing a janitor supply business. There was heavy smoke, and the ceiling was not visible through the smoke. The chief felt hot tar dripping on his neck. Later he said: "I immediately thought metal deck roof, and though we couldn't see the overhead, we directed a big line up to the ceiling area to cool everything." When the smoke cleared away, the fire fighters were able to see and photograph icicles of chilled tar hanging from the joints in the steel.

Types of Protection for Steel Structures

Prefire planning for steel structures must begin with a determination of the type of protection provided for the steel. Steel structures can be divided into four types:

- Unprotected
- Dynamically protected
- Passively protected
- Passive/dynamic combination protection

Unprotected Steel

Unprotected steel structures can be extremely hazardous because of the potential for early collapse. Trusses are covered elsewhere in this text, but the story of a fatal collapse should serve as a grim warning of the dangers of buildings in which clear spans are achieved by unprotected steel trusses.

Four fire fighters lost their lives in the collapse of a Wichita, Kansas, automobile showroom that

occurred about 10 minutes after they arrived. The showroom had repair bays that required large open areas and was built of lightweight steel-truss construction. Fire was visible to units turning out, and a second alarm was transmitted immediately. This fact alone should have signaled caution. A large amount of heat was being produced and the steel would absorb the heat. Early collapse should have been anticipated. Unfortunately, the fire fighters followed the standard aggressive offensive procedure of advancing a handline to the seat of the fire.

Through constant training, personnel can become accustomed to the need to cool all heated steel. The quantity of water needed at any point is not excessive. Water supplies for spray systems designed to protect tanks and steel supports from flammable liquid spill fires are calculated on a requirement of 0.25 gallon per minute per square foot. It is most important to cool all the steel that is within reach of hose streams and to give special attention to columns. The stream should be kept moving; overcooling in one spot obviously means undercooling, perhaps disastrously, someplace else. It is important that the stream reach the steel. When a long reach is required, a solid stream tip might be better than a fog tip (even on the fog tip's most solid setting). A fog stream may absorb a greater amount of heat in penetrating the flame zone. Converting all the water to steam as it passes through the fire, leaving none to cool the steel, is not the best use of water, particularly when the fire is uncontrollable.

A Washington, D.C., junkyard consisted of several identical unprotected steel buildings joined together without walls. A fire in wastepaper exposed two identical steel girders above the fire. One was successfully cooled with a master stream. The other, shielded by the first one, heated, overturned, and dropped the roof. Fire fighters conditioned to hit the fire sometimes neglect to protect the steel—clearly a mistake in this case.

Safe locations for cooling the steel should be preselected. Personnel should not be endangered by being placed in the collapse zone, which may cover a much greater area than the fire zone.

Water Damage

The use of water in the manner urged here might bring the criticism of unnecessary damage. However, the damage potential was created by the person who placed the combustible loading in a thermoplastic building. If the building can be saved, it will probably be possible to save some of the contents. If the building collapses, the contents are lost, too. In this day of close attention to municipal revenue sources, note that it is the building—rarely the contents—on which the tax base rests.

To the argument that "we never throw water except when we see fire," the reply can be offered that "If the owners had sprinklered the building, as they should have, water would be discharged on the steel, whether or not the sprinklers could see fire."

Where cooling-stream tactics are deemed necessary to prevent collapse, it would be beneficial to explain them to management in advance. Then, if the tactics are successful, the water damage to contents may be severe. The necessity of using the tactics might not be apparent after the fire, and the fire department may find itself ridiculed or even sued.

The possibility of severe water damage from hose streams might cause management to take a second and closer look at providing automatic sprinklers. The secondary objective of the preplanning for any major hazard should be to reduce the severity of the hazard.

Curiously, this also works in a reverse way. Highly protected risk (HPR) insurance companies usually require sprinklers for combustible metal deck roofs, even over noncombustible contents. This mandate often surprises management, who in turn may decide that it would be better to install a more costly roof that would not sustain combustion, so as to avoid the initial cost and maintenance of sprinklers.

Dynamic Fire Protection

Dynamic protection is generally accomplished with various types of automatic sprinkler systems. In some cases, a building is simply conventionally sprinklered, though this system may not be adequate to protect the steel.

It is important for the fire department to know whether the sprinkler system is adequate for the fire load. Rolled paper has been the fuel for a number of fires that destroyed sprinklered steel buildings. In one case, the building collapsed on the piles of paper, but the sprinkler pipes didn't break and the sprinklers operated continuously.

Hydraulically calculated sprinkler design can produce water flows adequate for the fire load, but this is no guarantee of success. Sometimes the procedure is used to design the system down to the bare minimum to keep costs low. Increases in fire loads overwhelm the system.

Heavy structural steel is sometimes protected by special lines of sprinklers. Flammable liquid hazards have long been covered by deluge sprinkler systems. More recently fog/foam systems have been installed to provide a foam blanket over the flaming liquid on the floor.

Passive Fire Protection

Passive fire protection is provided by protecting steel in one of the ways discussed earlier in this chapter. If a building is rated as fire resistive, two considerations are vital:

- Is the legally required level of fire resistance adequate for the fire load as it exists in the building?
- Has the protection of steel been completely provided, and is it maintained?

If the answers to these two questions are negative, three other critical questions should be asked:

- Is there any legal relief?
- Is it up to the fire department simply to do the best it can in the event of a fire?
- If the fire department estimate of the situation indicates potential or inevitable disaster, who, if anyone, is notified?

All personnel who perform inspections should be made aware of the requirements for fire resistance as applied to specific buildings. The manner in which fire protection of steel can be degraded by the actions of persons completely unfamiliar with its vital purpose is also crucial.

As noted, sprayed-on fireproofing and membrane protection are the two methods most likely to be degraded. Sprayed-on protection is often removed by mechanics accomplishing other work or disposing of an asbestos hazard. The fire department should insist on its being replaced with an adequate safe material. Sprayed-on protection is also vulnerable to being washed away by hose streams. Caution should be exercised when returning to areas from which the fire department was driven away by the fire—hot steel as a result of missing fireproofing can still collapse even after the fire is out. Spray-on fireproofing is discussed in further detail in the chapter that reviews fire-resistive construction.

There are numerous ways a membrane ceiling can be tampered with, so a prefire plan for any building in which one is installed should draw special attention to the possibility that collapse might occur. This is particularly a possibility where steel columns are unprotected in the plenum space.

One scenario for catastrophic damage to a column or columns could go like this: An interior stockroom has a characteristically high fire load. A ceiling tile has been removed for replacement, is being used elsewhere, or the ceiling fails. The column is heated to failure along a limited length and the column drops a foot or two. The effect would be staggering.

The fire department should be aware of the extent to which membrane fireproofing is permitted under the local code. Discussions should be held with the building and fire department and local installers to reinforce the need for strict adherence to installation details as specified by the manufacturer's listing. In one case, poke-throughs for wiring permitted a fire to spread to several floors. There was extensive collapse of bar joists.

All must be aware of the difference between fire resistance or flame spread listings of tile and the fire-rated tile as part of the fire resistance system of the floor and ceiling assembly. Similar-appearing tile is often used in these two quite different categories.

Not all suspended ceilings are a part of the fire resistance rated system. The ceiling may be installed for any reason, such as to reduce the room volume for better air-conditioning, to hide ducts and pipes, to conceal an old ceiling, for acoustical treatment, and so on. At one fire, it was found that the fire-resistive membrane was provided by suitable lath and plaster, that the visible tile ceiling was only acoustical, but that to install the tile ceiling, many holes had been punched in the fire-resistive ceiling to accommodate the support wires.

If the ceiling, including light fixtures and air diffusers, is an integral part of the fire resistance rating of the building, the owner and occupants should be put on notice that the fire department or building department must be consulted before any changes are made. Inspecting and preplanning personnel must be aware of the hazard of tampering with the ceiling and should check them during inspections, with an appropriate item provided on checklists.

Passive/Dynamic Protection

Protected combustible construction combines partial static protection of the steel with automatic sprinklers. One major insurer approves a spray-on coating that does not meet the standard for rated fire resistance, but would insulate the steel for a period of time that might be adequate to permit the sprinklers to control the fire.

Code Problems

In any discussion of prefire planning, it is difficult to separate firefighting procedures from fire protection requirements for steel structures or proper maintenance of the passive or dynamic protection.

The fire department must develop competence in the many facets of the overall problem. It is impossible for all officers to be equally competent and informed. One person or a small group should be designated to develop expertise in each relevant area. Becoming familiar with current and past local building and fire codes, and most importantly, the manner and degree of this enforcement, are prime areas of concern to effective fire departments.

Some building officials are less than enthusiastic about fire department interference in "their business." Incredibly, in one metropolitan-type county, fire inspectors were not permitted to inspect a building until the certificate of occupancy was issued. This practice not only ignores the hazard of fires in buildings under construction, but also allows many defects to be hidden from view in a completed building.

Some fire prevention managers are equally unenthusiastic about any interest by fire suppression managers in "their business." The ultimate fiasco is to have information that is vital to fire suppression locked up in fire prevention office files at 3 a.m. when a fire occurs.

Code Variances

Building and fire prevention officials sometimes assume that they have the authority to grant exceptions to the code, based solely on their own undocumented action. Actually, codes do not permit waivers to specific requirements but do permit the use of alternative, equivalent methods. Political influence or worse may be involved when unapproved variances are undertaken. All modifications to requirements of the code should be fully documented, meet the intent of the code (including the level of protection), and the fire suppression forces should be involved in the process. It is possible that the beneficiary of the unrecorded variance may at some future date drag the municipality into a lawsuit. Without proper documentation, the municipality is very likely to be found liable.

Preplanning Your "McCormick Place"

With the advantage of hindsight and a study of the McCormick Place fire, consider a possible preplan of a fire in an unprotected steel building being used

as an exhibit hall. For discussion purposes, assume the building is a typical armory (i.e., brick-bearing walls with unprotected steel trusses supporting the roof). Also assume there is a balcony running down both sides that is used for sports events but not for exhibits. Recommendations for automatic sprinklers have not been carried out. No money is available. The state will not permit the city to dictate to it, and most of the time the fire load is low.

The first move might be to set up a "bruising battle" for either sprinkler protection or denial of permission for the facility to house high fire-load exhibits. This battle is set up not to be won, but rather with the full expectation that it will be lost. In losing, however, the fire department may gain some concessions. Without being too specific, get in writing an agreement that the fire department can take whatever steps necessary to protect life and property during exhibits. Having beaten back the greater financial threat, the managers of the armory can only be magnanimous and yield a point to the fire department. When they learn the proposed precautions, they may be quite upset, but the fire department has the written agreement. In such a public battle, it is not the fire department that is willing to risk the lives of our citizens.

Consider a worst-case scenario in this armory. Suppose an exhibition such as a housewares, home, or furniture show is scheduled. Calculations of the combustibility of typical exhibits together with the floor area indicate that a fire with a high rate of heat release would develop. The surface-to-mass ratio of the material and its arrangement will provide a fast fire with early high-heat release at a rapidly increasing rate. The same fuel load in a more restricted area might give off heat more slowly due to restricted airflow, but an unlimited supply of oxygen can be assumed in the exhibit hall.

The fact that a hot, fast fire is anticipated indicates a more severe test of the structure than a slower fire. The radiant energy of a fire is proportional to the fourth power of the absolute temperature (actual thermometer reading in degrees Kelvin plus 459); thus a 2,000°F (1093°C) absolute fire of 20 minutes'

duration could have a radiant energy potential 16 times that of a 1,000°F (538°C) absolute fire of the same duration. Because of the large void above the bottom chord of the trusses, a huge smoke and heat storage area is provided, and for quite some time there might be little smoke to protect the bottom chords of the trusses from radiant heat. If a bottom chord fails, the truss fails.

On first glance, such a plan appears ridiculous. "Fire fighters and equipment cannot be tied up to fight a fire before it happens. The precautions would upset the patrons of the show. We never did anything like that before. Who will pay for the time? We have a fire, we call you, you come. That's the way it's always been."

However, this concept may not be as farfetched as it seems. Instead of overhead piping, perhaps the concept of an alternative automatic system is feasible.

The Alamodome in San Antonio has installed just such a system. Flame detectors at roof level "view" the floor below. When the safeguards built into the system are satisfied that there is a serious fire, water is started from the two closest of six fixed nozzles located around the perimeter of the area.

The suggestions made above thus are no longer in the realm of fantasy. If a fixed system can be installed, it is certainly reasonable to improvise a similar system. For example, a department store in San Jose, California, has a fabric roof protected by sprinklers suspended from the roof cables.

If the building is sprinklered, study the potential water supply requirements, particularly the possibility of the opening of more than 20 sprinklers. It might well be advisable to make a standby hookup to the sprinkler system Siamese connection so that the supply could be augmented in terms of both quantity and pressure immediately upon the start of a fire.

Be Proactive

Frank Brannigan recalled lecturing to Navy personnel, "There is no law saying you must have a fire before you open a can of foam. Foam can be used to

suppress a flammable liquid hazard before the fire starts." Although it was a revolutionary idea then, the use of foam to control a flammable liquid spill is now commonplace.

There is nothing the human intellect resists more than a new idea. Step by step, logic leads inexorably to the practical solution—prepare for fire before it happens. Learn from those fires that have occurred.

Fire fighters should be eternally curious. Just about everything that happens has some impact on the fire protection situation. Evaluate it and take action to contain or perhaps eliminate any problem.

TACTICAL CONSIDERATIONS

1. Fire-resistive and noncombustible construction, both of which use steel and concrete, may lead new recruits to believe that they shouldn't expect a significant fire. Not so! There are more hazards to fire fighters in Type II construction than just the fire. Make that clear to new recruits. They need to know the difference between Type I and Type II construction. Type II construction—especially nonsprinklered Type II buildings—are actually more dangerous because little or no fire resistance protection for their structural steel members is provided. Fire resistance is a function of mass. Strong but lightweight steel members have little inherent fire resistance.

2. If you are a student of building construction and fire, certain temperatures are worth memorizing. Probably 80% of the fire service knows the temperature recommended for baking cookies (350°F [177°C]), but how many fire fighters and company officers know the temperature at which substantial elongation can take place in a steel beam? It is about 1,000°F (538°C). This temperature is easily attained at any structure fire. The elongation can push out masonry walls abutting the ends of the steel, causing a roof or wall collapse. If the steel cannot elongate because of restraint (such as an exposure building), it will buckle or twist, causing a roof collapse.

3. Everyone knows that water boils at 212°F (100°C), but what is the temperature at which steel members may fail under their own weight, bringing about collapse? Temperatures above 1,300°F (704°C). You must know this to operate safely on the fireground when dealing with a building using steel as structural components.

4. Steel is a good thermal conductor of heat. Heat can be easily transmitted by conduction through the cold-drawn steel cables often used in tensioning concrete, excavation tiebacks, and elevator cables. These cables can fail at about 800°F (427°C). Fire fighters riding in elevators should have this temperature memorized or take the stairs.

5. In a noncombustible building, if the fire cannot be quickly located and extinguished (remember—putting the fire out eliminates all kinds of problems), the next important tactic is applying water to cool the exposed steel. Cooling the steel stops the detrimental effect of the heat and helps maintain the structural integrity of the building. Some incident commanders (ICs) are concerned with water damage, but if the building is sprinklered and the system is operating, water damage is taking place whether the fire department is on scene or not. Keep your priorities straight: life safety, followed by incident stabilization, and then property conservation. The effects from high temperatures on steel happen quickly. If you lose the building to structural collapse due to failure of steel members, does it matter how much water is being dumped into the building? At this point, it is probably a surround-and-drown defensive operation.

6. Plans for cooling steel should provide for a safe location that personnel can operate from. If a safe position is not practical, consider unmanned hose streams or unmanned monitors. Aerial ladder pipes are ineffective while the roof is intact because the roof will operate as designed—it will shed and keep water out of the building. If this is the case, aerial master streams should be shut down because they rob water pressure from ground-level appliances that can still be effective in controlling the fire.

7. If it becomes too dangerous to cool the steel, pull fire fighters back, establish a collapse zone, and change to a defensive strategy. Notify the building owners of the probable outcome of such a fire. Don't risk your fire fighters on buildings that no longer have any value.

8. In the case of unprotected steel, if temperatures in excess of 1,000°F (538°C) are reached (easily attainable in major structure fires), failure can occur early in a fire and the consequences can be catastrophic. It is still important to try and remove the heat, however. Vertical rooftop ventilation operations can be time-consuming depending on the roofing material, unforeseen obstacles, crew resources, available equipment, and the speed and experience

of the assigned crew. A metal deck roof with a significant fire below is a very dangerous place. Roof ventilation in such cases should be avoided. The fire department's fastest and safest heat removal medium is water. Water should be used to remove the destructive heat from the steel. If the steel is elongating, the cooling effect of the water will draw it back to its original dimensions. If the steel is failing, the water will simply freeze the steel in whatever shape it is currently in.

9. It's often difficult for fire fighters and officers to believe how fast unprotected steel can be destroyed. The distortion of steel due to fire can weaken the integrity of the building so that it may no longer support superimposed or its designed loads. Unexpected stresses, particularly torsional, may be created by buckling, which lead to structural failure if the condition is not reversed. This condition can be reversed only with the application of water. Structural steel that has been sufficiently cooled by water can retain the same structural strength of steel that has not been exposed to fire.

10. During new construction of Type II noncombustible buildings and Type I fire-resistive buildings, the practice known as "field bolting" may be used, in which temporary bolts are placed instead of permanent rivets. Whether at 2:00 p.m. or 2:00 a.m., most fire fighters would be unlikely to be able to spot the difference or identify this condition. Fire fighters should carefully examine temporary bracing before going up to extinguish a fire in a steel building under construction, particularly when the incident occurs in conditions of high winds. Consider giving a wide berth to field-bolted structures during high winds.

11. If you have exposed steel, then the building is in the early phases of construction. Once the civilian life hazard has been mitigated, there isn't much property value at stake yet. It's not worth risking fire fighters' lives over steel framing. Because many noncombustible buildings are likely to be 1 to 3 stories in height, with the maximum being 12 stories in height (fire-resistive buildings are even higher), consider using elevated master streams operated from the perimeter. This type of cooling sustains the structural integrity of steel.

12. Elongating steel bar joists may potentially push a block wall out of alignment. If fire fighters can determine which way the trusses run, this is critical information that needs to be passed on to the IC. This can often be determined by using thermal imagers (TIs). By scanning the roof with a TI, whether it is on fire or not, the trusses will be visible on the screen. For example, if the trusses run from side B to side D, the elongation will take place from B to D, so the walls in danger of being pushed out are the B and D walls. The IC can establish collapse zones or cordon off access to these sides of the building.

13. When cooled steel shrinks back to its original position, the trusses may potentially pull away from the walls, thereby causing a roof collapse due to truss failure. The point to remember is to use the reach of the hose stream to cool the steel from a safe position. A 2 ½" (64 mm) hose line is best. The more gpm delivered along with reach, the faster the cooling effect. A 2 ½" (64 mm) hose line can easily throw a horizontal stream for more than 80 feet (24.4 m)—a lot of reach.

14. Stay alert. Be aware of the hazards to fire fighters who are fighting the fire in an unprotected steel building. Steel fire escapes still exist on the exterior of older buildings all around the country. If they have been well maintained and certified (every 5 years per code), they can be useful not only for the evacuation of occupants, but also for fire department access. Some exterior standpipes run up adjacent to the fire escape and can be accessed only from the fire escape landings. These standpipes are exposed to the elements all year round, however, so they age faster than interior standpipes. The tie rods supporting the standpipes are vulnerable to temperatures above 800°F (427°C). Therefore, it is essential that these fire escapes be inspected on a regular basis and certified by a structural engineer every 5 years. You may have to trust their structural integrity with your life.

15. A prefire plan for a metal deck–roofed building should be based on taking a safe position where all areas of the underside of the roof can be reached with heavy stream appliances. Do not hesitate to throw water onto smoke, even though the target will probably be obscured—it's still there and needs to be cooled.

16. You cannot cut off a metal deck roof fire by implementing vertical ventilation. The fire can move very fast. Tests show that only huge vent holes—far beyond the capability of fireground tactics—would be effective in such a case.

17. Working on a roof over a metal deck roof fire would be extremely dangerous. An article discussing collapse of wood trusses makes the point that wood trusses collapse, whereas steel trusses sag. This is not necessarily a better alternative because the sag may cause fire fighters on the roof to slide downward on a slick, liquid tar surface. For this reason, you should place a secondary ladder for escape. In fact, feel free to throw additional ground ladders on every side of the building if extra personnel are waiting for an assignment. Ground ladders at an incident aren't doing any good stored on the apparatus.

18. A ladder pipe or ladder tower master stream directed downward onto a traditional built-up roof is useless. Such traditional built-up roofs are doing what they are designed to do—shed rain and keep it out of the building. The fire is always many feet beyond the point at which the roof is opened up. A fire that can be seen on the top of the roof is in the fiberboard, the plastic insulation, or the upper layers on a built-up roof and is unimportant. That part of the roof no longer has any value. If you stop the progressive heating of the "frying pan" from underneath with hose streams, the roof fire will die out or decrease to easier manageable proportions. "Rubber roofs" in which the roof surface itself is on fire, in contrast, call for the use of aerial master streams because they are the only way of stopping the fire.

19. When construction operations present unacceptable hazards to unprotected steel, fire departments are well advised to put the building owners on notice—even if there is no applicable code.

20. Preplans should provide for the rapid cooling of all exposed steel in an excavation fire. Apparatus placement and operating positions should take into consideration the fact that adjacent buildings might collapse into the excavation. The occupants of such buildings should be evacuated.

21. Fire departments should survey all highway structures for vulnerability from building fires. Prefire plans for buildings that expose highway structures should take into account protection for these structures; the highway structure may be far more valuable than the building on fire. In such a case, cooling the steel of the highway structure might be more important than trying to extinguish the building fire. Unorthodox tactics, such as driving the fire down into the building, may be necessary.

22. Accidents under highway structures and underpasses may result in massive fires. Fire departments should preplan to protect the steel (or concrete) and anticipate a flammable liquid fire in addition to a vehicle fire. An adequate water supply must be planned for. Generally, the water must be delivered by units located off the highway. Water used with foam to suppress the fire may not be as effective as using the water to cool the steel and concrete of a bridge structure. The best tactic depends on the value of the structure and the ability to maintain an uninterrupted water supply.

23. When faced with a massive fire involving highway structures, consider making mutual aid requests of nearby airports. Often, airport fire departments can send a crash truck to help extinguish a large vehicle on a major highway. This apparatus may carry as much as 3,000 gallons (11,356 L) of water, 300+ gallons (1,136+ L) of foam, and 500 pounds (227 kg) of Purple K powder (PKP) dry chemical ready to disburse from its turrets.

24. If water is used to cool steel and mixes with burning gasoline, it may produce a flowing stream of fire. Will the fire flow harmlessly down the road or will it enter a storm drain and potentially cause environmental damage to nearby streams and watersheds—or start a fire further away? Could the burning stream enter the sewer system or flow into nearby buildings? In the latter case, crews may have to build a dike to contain the runoff and extinguish it with foam. If no exposures are threatened and there are no environmental hazards, it may be best to let this stream burn. Once the fuel is consumed, the fire will go out.

Chapter Summary

- Steel is the most important metal used in building construction. Without it, construction would be limited to massive all-masonry buildings with arched floors or masonry wall-bearing buildings with wooden floors.
- Steel has several important characteristics to consider regarding its behavior in fire:
 - Substantial elongation can take place in a steel member at ordinary fire temperatures (about 1,000°F [538°C]). This elongation may cause the disruption of other structural components.
 - At higher temperatures (above 1,300°F [704°C]), steel members may fail, bringing about a collapse of the structure.
 - Cold-drawn steel such as steel tendons used for tensioned concrete and for excavation tiebacks and elevator cables will fail at about 800°F (427°C).
 - Steel is a good thermal conductor, so it transmits heat readily. Heat can be transmitted by conduction through steel to combustible materials.
- The strength of steel, the consistency of its structural characteristics, and its ability to be connected to other structural elements so that loads can be adequately transferred are all important to the use of steel as a building material.
- To achieve a peaked roof in a steel-framed building, the framing may consist of columns and beams with triangular trusses.
- A number of options are open to the designer to deal with the characteristics of steel as they relate to fire risk: ignore the problem; rely on inadequately enforced codes; take a calculated risk; fireproof (insulate) the steel; protect the steel with sprinklers; fireproof the steel with an water cooling system; or locate the steel out of range of the fire.
- The metal deck roof is a factor in many losses where its significance is not understood by the fire suppression group.
- Steel structures can be divided into the following types:
 - Unprotected
 - Dynamically protected
 - Passively protected
 - Passive/dynamic combination protection

Key Terms

Aluminum A lightweight metal that is both malleable and nonmagnetic. This material has very good conductivity. This noncombustible material has a low melting point and little mass per unit of area, so it disintegrates rapidly in fire.

Angles Steel members that have two legs at right angles to one another.

Asphalt asbestos protected metal (AAPM) Asphalt coating that is combustible and used as a weather-protective coating on galvanized steel walls.

Bar joist A joist that generally runs in the same direction as a beam and forms a lightweight, long-span system, used as floor supports and built-up roofing supports.

Bars Plates fewer than 6 inches in width; may be square or round.

Box column A large hollow column built from steel plates.

Box girder A large girder, which is hollow, like a box column, and often used for highway bridges.

Bulbtee A tee where the end of the cutoff is thickened.

Bulkhead An upright partition that divides a ship into compartments and is meant to prevent the spread of leakage or fire.

Castellated beam A wide flange beam that has been cut in half in a zig-zag pattern and then welded back together in an offset manner, creating a new, deeper beam.

Cement-asbestos board Noncombustible material often used for friable construction.

Channel Steel structural component that has a square U-shaped cross section.

Galvanized steel walls Walls made of weatherized steel. Can conduct heat easily.

Glass fiber–reinforced plastic A composite material made of plastic reinforced with glass fibers.

I-beam Beam shaped like the letter "I".

Interstitial space Void space made by utilizing deep parallel-chord trusses.

Lattice column Column made of vertical units connected with diagonal pieces.

Light gauge steel-framed wall An exterior wall constructed with lightweight galvanized steel studs.

Masonry walls The most common walls for unprotected steel-framed buildings; made of concrete block or a composite of concrete block and brick.

Metal panel Prefabricated metal structure that is often made up in a sandwich construction to provide one unit combining thermal insulation and interior finish in a steel-framed structure.

Modulus of elasticity A measurement of the ability of steel to distort and restore.

Peened (peening) Embedded into the surface.

Plates Flat pieces of steel.

Precast prestressed concrete panels Concrete panels that are precast and brought to the construction site.

Purlins Beams set at right angles to trusses or roof rafters to provide support for lightweight roofing.

Rakers Diagonal columns that brace an entire structure.

Rolled or built-up members Steel structural members; rolled members are one solid piece of metal; built-up members are made up of different sections riveted, bolted, or welded together.

Spandrel girders Girders that tie wall columns together in a framed building.

Spandrel space Distance between the top of one window and the bottom of the one above.

Steel expansion joints A metal connection that allows for movement of floors.

Tee A standard I-beam cut lengthwise through the web forms two such beams with T-shaped cross sections.

Tiebacks Cold-drawn steel cables inserted into holes driven into the rock and anchored with epoxy. Also refers to braced sheeting used in soil walls to protect against collapse.

Tin ceiling Embossed steel; will transfer heat in either direction.

Transfer beam Used to laterally relocate the vertical load of columns to clear an opening area.

Triage To evaluate and categorize.

Tube A steel structural member that is rolled in cylindrical, square, or rectangular shapes.

Waler A horizontal beam that ties rows of soldier beams together.

Wide flange shapes I-beams that have flanges wider than standard I-beams.

Zees Members with a Z-shaped cross section.

Case Study

During some down time at the firehouse, you decide to talk with your fire fighters about steel frames and some catastrophic incidents that have happened in the last 60 years. You explain that we

must learn from past experiences and that if we can understand how buildings are put together, we can understand how they will come apart.

1. What is a W 12 × 22 steel member?

 A. A wide flange beam with a weight of 22 pounds
 B. A wide flange beam with a depth of 22 inches
 C. A structural tee with a weight of 12 pounds
 D. A structural tee with a depth of 22 inches

2. McCormick Place used unprotected steel roof trusses. We know from research that steel will expand _____ % in length for each 100°F rise in temperature.

 A. 0.06 to 0.07
 B. 0.07 to 0.08
 C. 0.08 to 0.09
 D. 0.09 to 0.10

3. The General Motors transmission plant fire in Livonia, Michigan, is a good example of a:

 A. deadly collapse of a walkway.
 B. tilt wall collapse.
 C. mechanic knocking off spray-on fire-proofing.
 D. metal deck roof fire.

4. An abbreviation for a "rubber roof" is:

 A. EFIS
 B. EIFS
 C. EISF
 D. EPDM

Challenging Questions

1. Detail the differences between noncombustible and fire-resistive construction.

2. Describe how a building of tilt slab construction is constructed and the associated hazards when it is exposed to a fire.

3. Identify and describe the general ways of protecting steel from the heat of a fire.

4. Describe the indicators and hazards of a metal deck roof fire in an unprotected steel joist warehouse as well as the techniques to deal with the problem.

5. You have been assigned to the bureau of fire prevention. You attend a meeting of architects, engineers, and building code officials concerning the design and construction of a new convention center in your city. As a follow-up to the meeting, you are to prepare a list of concerns that you have about the new structure. What specific fire protection and firefighting concerns would you have?

Fire-Resistive Construction

OBJECTIVES

At the conclusion of this chapter, you will be able to:

- Recall the difference between noncombustible and fire-resistive construction.
- Describe different types of concrete structural systems.
- Describe the two types of prestressing.
- Describe the hazards of formwork.
- Describe methods of fireproofing steel and of ensuring a level of fire resistance in concrete.
- Describe how concrete and concrete structural elements react to fire.

Case Study

Figure 11-1 Note the significant sagging of the steel beams in the One Meridian Plaza fire.

Courtesy of USFA/FEMA.

While high-rise fires may be rare compared with other types of fires, they can prove to be the most challenging of all. They can also prove to be deadly, which was unfortunately the case in the nighttime One Meridian Plaza fire in Philadelphia in 1991 (**Figure 11-1**). Fire broke out in a 22nd-floor office under construction, most likely when a pile of oil-soaked rags spontaneously combusted. A delayed initial alarm slowed response to the nighttime fire. Fire fighters encountered a series of problems, most notably the woefully low pressure settings on the stand-pipe's pressure-reducing hose valves, which was totally inadequate to handle the department's 1¾-inch hose and automatic nozzles. The fire spread quickly upward, through penetrations within the building and through autoexposure on the exterior; it compromised exterior granite panels on a stair enclosure, sending parts of them to the ground. Three fire fighters on the 28th floor became disoriented and ran out of air, succumbing to the effects of smoke inhalation. As the fire consumed the contents of floor after floor, significant concerns were raised about structural collapse. Finally, the fire was stopped by 10 sprinklers on the 30th floor. A subsequent investigation revealed that some floor beams and girders had sagged 3 feet (91 cm) as a result of the heat of the fire.

1. What are the problems with delayed alarms in a fire in a nonsprinklered or partially sprinklered building?
2. Once control has been lost of a high-rise fire, how much concern is there for structural collapse? Which kind of fuel load could initiate a collapse?
3. Do you have unsprinklered high-rises in your district?
4. Which kind of standpipe hose valve is installed? When and how were these valves inspected?

Introduction

From a fire protection perspective, fire-resistive construction is considered to be the best. It is the type of construction most resistant to collapse and does not contribute fuel to a fire. With these attributes, fire-resistive construction is given the largest permissible area and heights, including unlimited area and height. Fire-resistive construction often uses concrete or substantially fire-protected steel (more than noncombustible) as a structural system. We have already covered steel, so we will now analyze concrete in detail.

Concrete

Concrete is a cementitious material produced by a chemical reaction of Portland cement (primarily limestone) and water to which inert materials called aggregates are added. Shortly after it is mixed, concrete sets into a solid mass, but it continues to cure indefinitely **Figure 11-2**. Construction specifications set a date by which concrete must reach its required compressive strength. For instance, concrete required to reach design strength in 28 days is sometimes referred to as 28-day concrete.

While curing, concrete generates heat of hydration. During its initial curing, concrete must be protected from freezing. Low temperatures retard the curing of concrete, and freezing is harmful to the material. Good concrete results from proper handling of carefully controlled materials. Because there are so many ways to create a poor concrete product, high factors of safety are used in concrete design.

Concrete is weak in tensile strength and has poor shear resistance. Its compressive strength is good, particularly when compared to the cost of steel to resist the same load. Fire service personnel should avoid the common error of using the term *cement*, when *concrete* is the proper term. Cement is a component of concrete.

Figure 11-2 Concrete being poured. For most types of concrete, it will take 28 days to obtain its full strength.
© Christina Richards/ShutterStock, Inc.

Figure 11-3 The Guggenheim museum in New York, NY.
© Adriano Castelli/ShutterStock, Inc.

Concrete Structures

Until the post–World War II era, structural concrete was regarded as suitable only for massive, low-rise utilitarian structures in which aesthetics played little part. Fire-resistive high-rise buildings almost universally built of steel frames were often fireproofed with concrete and used cast-in-place concrete floors. Concrete warehouses and factories built before World War II are easily recognizable. Their panel walls, usually of brick or brick and block composite, were built on each floor between the exterior columns.

Precast concrete principally was in the form of cinder blocks. Cinder blocks use cinders as the aggregate to produce a light, relatively inexpensive block. Concrete blocks use other materials for aggregate. Heavier concrete blocks usually have smoother surfaces than cinder blocks.

<u>Underwriters blocks</u> are concrete blocks produced under Underwriters Laboratories, Inc.'s classification. The manufacturer is authorized to issue a certificate giving the type and number of units delivered to a specific job. These blocks are used in assemblies that must meet fire resistance standards.

Today, there is a bewildering variety of mix-and-match building construction elements. Steel-framed buildings now often have cast-in-place concrete floors made with forms that are removable or that can be left in place. The floors may or may not contribute to the stability of the building. Precast concrete and prefabricated metal wall panels are common.

In some cases, an architect may wish to emphasize concrete construction. Instead of considering form marks unattractive, architects have accentuated them in some building designs Figure 11-3 . In other cases, decorative brick veneer is used to conceal the concrete construction.

Over time, substantial brick veneer has been lost from some reinforced concrete buildings. High early strength concrete (special concrete that cures quickly to high strengths) tends to shrink as it matures, but the brick veneer overlying it does not. If proper provision is not made for the contraction, the brick veneer may crack and fall off.

Steel Versus Concrete Framing

Many designers prefer to design in steel, whereas others prefer concrete for framed buildings. Some buildings present a unified exterior appearance, yet have one section concrete framed and another section steel framed. Building owners are sometimes less interested in the material of construction than in

speed and cost of construction. When a building is put out for bids, the price often depends on which contractor has readily available personnel and other resources.

Fire Department Problems with Concrete Construction

The problems that fire departments face regarding concrete construction can be divided into three distinct areas:

- Collapse during construction with no fire.
- Fire during construction.
- Fire in completed, occupied buildings.

Types of Concrete Construction

There are two basic types of in-concrete construction:

- *Cast-in-place concrete* includes plain concrete, reinforced concrete, and post-tensioned concrete.
- *Precast concrete* includes plain concrete, reinforced concrete, and pretensioned concrete.

Concrete Definitions

- **Aggregate**: This is material mixed with cement to make concrete. Common aggregates are both fine and coarse. Fine aggregate is usually sand. Coarse aggregate may be one of a great variety of materials, depending upon availability and the characteristics desired in the finished product. Concrete often is described by the name of the aggregate used. Crushed stone, gravel, cinders, breeze (a coke by-product no longer used), expanded slag, shale, slate, clay, vermiculite, and perlite (natural lightweight materials), lightweight stones such as lava and pumice, and fly ash reclaimed from boiler plants have all been used as aggregates.

- **Cast-in-place** concrete: This concrete is molded in the location in which it is expected to remain **Figure 11-4**.
- **Casting**: This is a process of placing the fluid concrete into molds, generally called forms, in which the concrete is permitted to harden to a certain shape.
- **Chairs**: These are small devices designed to keep the rods up off the surface of the form, so concrete will flow underneath the rods.
- **Composite and combination columns**: These columns use steel and concrete combined into one unit. As in all composite construction, the different elements must react together, and any failure of the bond between them

Figure 11-4 Concrete will be poured into this form, creating a cast-in-place wall.

© Dick Stada/ShutterStock, Inc.

creates a distressed column. Composite columns were formerly used on the lower floors of a heavy-duty building to avoid surrendering floor space to large-diameter concrete columns made solely of reinforced concrete.

- **Composite construction**: This is a different term than a composite structural element. Reinforced concrete is one of several composite structural elements. The term *composite construction* is often used to describe buildings in which different load-bearing materials are used in different areas, such as exterior columns being made of concrete and interior columns being constructed of steel. Sometimes the top floor of a concrete building is constructed of lightweight steel. This lighter structure may be called a penthouse, meaning any structure above the roof line. The term *penthouse apartment* properly refers only to an apartment built on the roof. Apartments on lower floors that have terraces are properly called terrace apartments.

- **Continuous casting**: This is a process for casting or pouring concrete without interruption from start to finish. This avoids the problem of joining new concrete to concrete already poured. If such a joint is not properly made, cold joints, which are planes of weakness, result.

- **Continuous slipforming**: This entails pouring concrete continuously as forms move upward so that continuous casting may be accomplished.

- **Drop panels**: These are thicker sections of floor near columns to assist in resisting the natural tendency of the floor to shear off at the column **Figure 11-5**.

- **Flat plate structural system (or continuous beam)**: A cast-in-place floor in which there are no beams supported by columns; the floor plate itself rests directly on the columns.

- **Footings**: Typically thick concrete pads, blocks, or strips of concrete below the surface of the surrounding soil, sometimes heavily reinforced, that transfer the loads of walls, piers, or columns to the ground. Footings

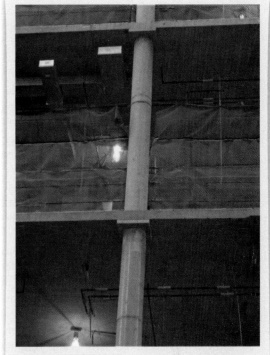

Figure 11-5 Drop panels atop two concrete columns.
Courtesy of Glenn Corbett.

attempt to minimize building settlement by spreading the load over the large soil/concrete interface.

- **Lally columns**: These are steel pipes filled with concrete to increase their load-carrying capacity. In the 1800s, cast-iron Lally columns were usually drilled to provide a vent for the steam generated in the concrete. Today, steel pipe concrete columns are only sometimes drilled. When exposed to heat, they may explode. In one example in a fire in a concrete plant, expanding steam from concrete blew out one end of the Lally columns, creating rockets.

- **Lift slab**: This is a type of construction in which the columns are erected to their full height, then the first-floor slab is cast on the

ground. Each subsequent floor slab and then the roof slab is cast, one directly on top of the other. A bond breaker is used between each slab so there is no adhesion. Starting with the roof, the slabs are successively lifted into position. Steel deck roofs have been assembled on the ground and lifted into position in a similar manner. L'Ambience Plaza, a lift-slab building under construction in Bridgeport, Connecticut, collapsed in 1987 with a loss of 28 lives; a single component in the lifting system failed and caused the collapse Figure 11-6 .

- **Monolithic construction**: A method where all the concrete in a building is properly bonded together. The resultant structure can be likened to one piece of stone.

- **Mushroom caps**: These are tapered extensions at the tops of columns that assist in transfer of loads from floor to column. Columns in heavy-duty concrete construction may show both mushroom caps and drop panels.

- **One-way structural system**: This system incorporates floors that have beams running in one direction only Figure 11-7 .

- **Plain concrete**: This term refers to concrete that has no reinforcement, except possibly light reinforcement to resist temperature changes. A sidewalk slab and a concrete base for a child's swing set are examples of plain

Figure 11-7 A one-way structural system.

© Brad Wynnyk/Alamy Images

concrete. Some early concrete bridges, such as the Taft Bridge in Washington, D.C., were built of plain concrete. Necessarily, the entire bridge is in compression, as were the early stone arch bridges developed by the Romans.

- **Precast concrete**: This concrete has been cast at a location other than the place where it is to remain. The precasting may be done at a plant miles from the construction site, at the construction site but remote from the ultimate installation location, or immediately adjacent to the point of use. The common cinder block or concrete block is precast plain concrete. Other precast concrete units are reinforced or tensioned.

- **Pretensioning and post-tensioning**: These are processes by which steel **tendons** or rods are placed under tension, drawing the anchors together. The tensioned steel places the concrete in compression.

- **Reinforced concrete**: This is a composite material made of steel and concrete. Steel provides the tensile strength that concrete lacks. Steel may also be used to provide compressive strength. The term *reinforced concrete* is actually a misnomer because it suggests that

Figure 11-6 The L'Ambience Plaza collapse of 1987.

Courtesy National Institute of Standards and Technology

concrete has a certain strength and that steel reinforces that strength. This assumption sometimes leads to a lack of recognition of the hazard of a failure of the bond between the concrete and the steel. Reinforced concrete is a composite, and both elements are equally important. To accomplish the designed function of reinforced concrete, the concrete and steel must remain in contact. Most importantly, exposed steel reinforcement lacks fire resistance, as does any unprotected steel.

- **Reinforcing bars or rods**: These are steel rods or bars that are usually intentionally deformed. A raised pattern in the surface of the steel aids in the transmission of stress from the concrete to the steel. It would be more accurate to call them tension rods. There is also interest in glass fiber reinforcing rods. The effect of fire on fiberglass-reinforced concrete is uncertain.
- **Slipforming**: This is a technique by which forms are moved, usually upward, as the concrete is poured.
- **Spalling**: This term describes the loss of surface material when concrete (or stone) is subjected to heat **Figure 11-8**. It is due to the expansion of moisture in the concrete. Some concrete and certain aggregates are more subject to spalling

Figure 11-8 Spalling of concrete.
© Marek Pawluczuk/ShutterStock, Inc.

than others. Concrete made with ground-up fire brick is resistant to spalling and is often used at fire testing facilities. Explosive spalling occurs violently, throwing out bits of concrete projectiles. Note the correct pronunciation is *spalling*, not *spallding*.

- **Temperature rods**: These are thin rods installed near the surface of concrete, usually at right angles to the main reinforcing rods to help the concrete resist cracking due to temperature changes. However, failure of the bond between the temperature rods and the concrete does not materially affect the strength of the structure.
- **Two-way structural system**: This system incorporates floors that have beams running in two directions.

Concrete Structural Elements

Columns

A chief virtue of concrete is its high compressive strength and low cost, which is why concrete is used for columns, even in structures otherwise built of steel, such as highway bridges. A short column with a large cross section in proportion to its length is called a pier. It may be of plain concrete if it is loaded only compressively; otherwise, columns are of reinforced concrete to provide the tensile strength necessary for loads created by eccentric or torsional loading.

The steel reinforcing rods ("rebar") not only provide necessary tensile strength, but also carry some of the compressive load. The compressive strength of steel is many times that of concrete. In early concrete construction, it was the practice to design larger diameter columns successively on lower floors to cope with the increasing loads. At the lowest level, the size of the columns might be so great as to interfere with the use of the building. In some cases, standard steel columns were used within the concrete to take advantage of the far greater compressive strength of steel and reduce the width of the column.

The steel rods are provided with ridges to strengthen the bond between the rod and the

concrete . Over time, however, the steel inevitably begins to corrode within the concrete. Epoxy and stainless steel coatings are applied to minimize this problem. In addition, rods entirely of stainless steel are available. Recently, glass fiber polymer reinforcing rods, with twice the tensile strength of steel, have been put in use to minimize corrosion; they are not capable of being welded, however.

In modern buildings, increasing column sizes is an unsatisfactory solution to the problem of dealing with ever-greater loads because the useable area would vary from floor to floor. This is overcome by increasing the size of the reinforcing steel as the loads increase so the outside dimensions of columns are identical on all floors. In some cases, the columns have so much steel in them that a special additive is mixed into the concrete to permit it to flow into the relatively small spaces between the steel.

Reinforcing rods are long with a relatively thin diameter. Sleeves, reinforced overlaps, or welding is used to connect the ends of rods and transfer the load. The rods in a column are joined by lateral reinforcements called ties or hoops. (In round columns, the ties connecting the outer rods are often formed in a helix. Such reinforcement is wrongly termed *spiral reinforcement*.)

Figure 11-9 Steel reinforcing rods. Note the ridges, which enhance the adhesion of the concrete to the steel rod.

© hxdbzxy/ShutterStock, Inc.

The ties or hoops serve to cut the long slender column into a number of relatively short columns, one on top of the other. Together with the concrete to which they are bonded, the rods form a composite structural element. If the unity of the composite fails, such as by loss of concrete either under normal circumstances or as a result of fire, the steel rods may buckle under the load. Buckling will cause the rods to protrude from the column.

Beams and Girders

Plain concrete is strong in compression but weak in shear, and it does not have any tensile strength. As noted, when a beam is loaded, it deflects. This deflection brings about compression in the top of the beam and tension in the bottom of the beam. Reinforcing bars are placed in the bottom of the concrete beam to provide the necessary tensile strength. Precast beams often have the word *top* cast into the top of the beam to ensure that the beam is installed right side up.

In a cantilever beam, however, the tension is in the top of the beam, and thus the reinforcing (tension) rods are in the top of the beam. The far end of the beam bends downward like a diving board.

In a continuous beam supported at more than two points, there is tension in the top of the beam in the area over the tops of the columns, and tension in the bottom of the beams between columns.

In all cases, reinforcing must be properly designed. Vertical reinforcing bars in concrete beams designed to prevent cracking under shear stresses are called stirrups. Some tensile reinforcing bars may be bent up to accomplish the same purpose.

T-Beams

Concrete below the neutral plane (where there is neither tension nor compression in the beam) serves little purpose. Sufficient concrete is needed to bond with the reinforcing rods for the relatively slight compressive stresses and to protect the steel from fire.

This leads to the characteristic T-beam design in which the neutral plane coincides with the bottom of the wide, thin floor slab **Figure 11-10**. Double

Figure 11-10 Double T-beams being shipped to a construction site.

Courtesy of the estate of Francis L. Brannigan.

T's are floor slab and beam combinations with two beams. Floor beam combinations with four beams are also manufactured.

As in columns, where space requirements are paramount, steel may be substituted for some of the concrete to handle compressive loads. For any unit area, steel has 15 times the compressive strength of concrete. Steel is, however, more costly.

Autoclaved Aerated Concrete Units

Although autoclaved aerated concrete (AAC) has been in existence for nearly 100 years, it has been in use in the United States only since World War II Figure 11-11. This type of prefabricated concrete uses all of the traditional components of concrete, but adds aluminum powder as well. When the concrete is mixed together, the aluminum powder reacts with the lime in the mix, creating tiny cells in the final concrete product. It is these cells that make the AAC relatively lightweight when compared with traditional concrete (one-fourth the weight by volume). Once the AAC has set, it is placed in a autoclave (kiln) under high-pressure steam where it gains its ultimate strength.

AAC may be formed into panels, blocks, and floor slabs. It can be cut easily, thereby obtaining the desired dimensions. In terms of fire protection, the material has held up well when tested in an

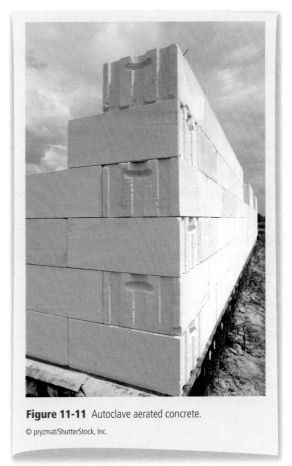

Figure 11-11 Autoclave aerated concrete.

© pryzmat/ShutterStock, Inc.

American Society for Testing and Materials (ASTM) E-119 furnace, achieving a 4-hour rating with an 8-inch-thick (20.3 cm) block wall.

Concrete Floors

Many varieties of concrete floors are used in concrete and/or steel-framed buildings and in ordinary construction. Concrete floors may be cast-in-place or precast.

Concrete was first used for leveling brick and tile arch floors. This led to segmental and flat concrete arches sprung between unprotected wrought iron or steel beams. Other early floors were built of individual beams supporting a floor slab all poured at

the same time. The appearance of the underside is reminiscent of wood joists.

Hollow tiles were used in early efforts to lighten concrete floors by eliminating concrete below the neutral plane. The tiles are visible from the ceiling side of the floor. This technique often puzzles modern engineers.

Many concrete floors are designed as continuous beams. The entire floor is one big beam, just as a wooden floor might be made of one huge sheet of thick plywood. In heavy construction, mushroom caps and drop panels are used at columns to cope with high shear stresses.

Concrete floor structural systems may also utilize individual beams. In a one-way structural system, beams run in one direction. Logically, two-way systems have beams running in two directions.

In so-called **waffle concrete**, which comprises a two-way structural system, closely spaced beams are set at right angles to one another, and unnecessary concrete is formed out. The lower side resembles a waffle **Figure 11-12**.

In lighter construction, the floor may be just a flat plate, with no projections below the floor line. This gives a smooth surface that is easily finished; thus, it is a popular choice for offices and apartment buildings.

When concrete floors are cast onto corrugated steel, the steel is a **left-in-place form**. The steel provides necessary tensile strength. If the bond fails, the floor section may fail. In one parking garage, the steel was eaten away by deicing salt, and a collapse resulted.

When precast T-beam units are used, additional concrete is often cast-in-place on top of the units, which are left rough on top to ensure a bond. The entire unit becomes an integral beam-and-floor element. Precast cored concrete floor units have cylindrical openings cast lengthwise through the units to remove unnecessary weight and to provide channels through which building utility services can be run **Figure 11-13**. Concrete floors may be simply load bearing or may be an integral part of the structural stability of the building.

Older building codes in some cities required that the street floor of buildings of ordinary construction be of fireproof construction. Consequently, a concrete floor or, more likely, a concrete topping over a wood floor, may be found in a building of ordinary construction. In one case, a concrete topping over wood beams was required by health department regulations for a washable floor in a restaurant. This concealed the destruction of the beams by fire. Four fire fighters died when the floor collapsed without warning.

Figure 11-12 A waffle concrete floor.
Courtesy of Glenn Corbett.

Figure 11-13 This type of floor takes away unnecessary concrete, which decreases the overall weight while allowing openings in which utilities can be run.
Courtesy of the estate of Francis L. Brannigan.

Some 1-hour-rated designs of wood floors include a lightweight concrete topping as much as 1–1.5 inches (25–38 mm) thick. This retards the passage of heat through the floor and delays the temperature rise on the unexposed surface, a condition of the ASTM E-119 fire resistance standard.

Building a cast-in-place concrete floor into ordinary wall-bearing masonry buildings sometimes creates a hazard during construction, and perhaps for the life of the building. As part of the construction, the wall is carried up a story or more beyond the point where the floor is to be located. A slot is left in the wall at the point where the floor is to be cast. The concept is that concrete will fill the slot, thus uniting the floor and wall.

If a windstorm occurs during the time that the slot is open (usually a few unmortared bricks set on end are relied on for temporary support), a collapse may result. A ladder or hose line over the top of the wall may also cause collapse. For all practical purposes, the wall is undercut, as is done in felling a tree.

Unless the concrete is carefully pushed up into the gap when the floor is poured, the wall will always be weak at this point. The finished floor may give no evidence that the continuity of the wall is interrupted by a void that creates a permanent plane of weakness.

Concrete floors in steel buildings may be precast or cast-in-place. They may be only load bearing or they may provide structural stability by being designed to resist lateral loads such as wind. In the latter case, studs are welded to the steel beams and the reinforcing bars are attached to the studs.

Concrete floors in cast-in-place concrete-framed buildings are cast integrally with columns, providing a monolithic rigid-framed building. Precast units may be pinned or can be connected as a monolithic unit. In this case, reinforcing bars are designed to protrude from the finished precast slabs or other units. When the slabs are laid down, a space is left between them. The protruding bars of one slab extend past the ends of the protruding bars of the other slab. They overlap without touching. An 18-inch space, for instance, may be left between slabs with bars that are usually hooked, protruding 12 inches. A form is made, and concrete is poured to level. The sections, thus joined by a wet joint, become monolithic.

Concrete floors in precast pinned concrete buildings may be only load bearing and may not contribute to the building's structural stability. Precast columns are often built with haunches or shelves cast into their sides on which beams are set. Steel plates embedded in the concrete may be welded together. This, of course, is pinned, not monolithic construction. In some cases, the beams are held in place only by gravity.

Prestressed Concrete

Conventional reinforced concrete has been used since the invention of the reinforcing bar in 1885. In recent years, prestressed concrete has been developed. **Prestressing** places engineered stresses in architectural and structural concrete to offset the stresses that occur in the concrete when it is placed under load. The concrete is "preloaded" with stresses to help strengthen it for tensile loads.

Consider a row of books side by side. Used as a beam across two chairs, such a row will fail of its own weight, without any superimposed load. There is no shear resistance between the books. However, drill a hole through the row of books laterally, pass a wire through the books, and tighten the wire against the end books. The row of books would be compressed by tensile stress in the wire and compression in the books. The "beam" could then be placed across two chairs and loaded. The "beam" has been prestressed sufficiently to counteract the stresses placed on it by the load.

Special high-strength, cold-drawn steel **cables**, similar to those used for suspension bridges, or alloy steel bars are commonly used. In building construction they are technically known as tendons, but are called **strands** or cables by those working with them.

High-tensile-strength wire, ordinarily used for prestressing, is more sensitive to high temperatures than structural steel. There is virtually complete loss of prestress at 800°F (427°C)—less than the

temperature of a self-cleaning oven. Protection of the tendons from fire is of paramount importance.

There are two methods of prestressing: pretensioning and post-tensioning. The prefixes *pre* and *post* refer to whether the concrete is poured before or after the tension is applied.

Pretensioning is done in a concrete plant in the process of creating some precast concrete building components. High-tensile-strength steel strands are stretched tightly between the ends of a form. Concrete is then poured in the form. As the concrete sets, it bonds to the tensioned steel. When the concrete reaches a specified strength, the tensioned strands are released from the ends of the form. The stretched strands attempt to draw back, thus compressing the concrete. This creates a built-in resistance to loads that produce tensile stresses. The completed units, such as the common T-beams, are shipped to the job site.

Post-tensioning is done on the job site **Figure 11-14**. High-tensile-steel strands (tendons, wires), encased in plastic or paper tubing, or wrapped to prevent any adhesion between steel and concrete, are positioned in the forms. The concrete is poured. After the concrete is set and reaches a specified strength, the steel tendons are stretched and anchored at the ends of the unit. Technically, this is called **stressing the tendons**. On the job site, it may be called "jacking the cables." A major advantage of post-tensioning is that floors can be thinner. Thinner floors will allow several more rentable floors in a tall building. Sometimes post-tensioning is performed in increments, as the load of the building increases with additional stories.

Some bridge girders are tensioned enough to make shipment possible, then post-tensioned after being placed. After tensioning is completed, some designs call for pressure grouting in which a cement paste is forced into the space between the tendon and the concrete to provide a bond. Some tensioned floors consist of small precast units through which tension cables are run, then tensioned.

Reinforced Masonry

There is an economic limit to the height of an ordinary brick bearing wall building. This exists because of the requirement that the walls increase in thickness as the building's height increases. The limit is generally about 6 stories, but there are brick bearing wall buildings as tall as 16 stories. In recent years, it has become possible to build brick bearing wall buildings 20 or more stories in height, with no wall thicker than 12 inches (30.5 cm). Concrete block can be used in a similar fashion.

In one method of construction, two wythes of brick are built. The width of one brick is left between them. Reinforcing rods are placed vertically in the cavity. After the brick masonry sets, concrete is poured into the void. Floors in one such building are of precast tensioned concrete double slabs, but other systems can be used.

Reinforced masonry construction is widely used to resist earthquakes, particularly on the West Coast. Reinforced masonry is unsuitable for multistory buildings in which large clear spans are required, but can be used for buildings where multiple interior walls are acceptable. The interior and exterior walls are equally load bearing. Apartment houses and motels with their characteristic repetitive box construction are well adapted to this method.

In low-rise buildings up to about 70 feet, some recent designs have eliminated reinforced concrete in the wall. High-compressive-strength bricks and

Figure 11-14 This concrete floor is post-tensioned.
Courtesy of the estate of Francis L. Brannigan.

special mortar are used along with masonry reinforcement trusses and placed between courses and to tie cross walls to the exterior walls. Together with precast concrete floors integrated into the building by the use of wet joints, these materials can produce a masonry wall bearing building several stories high with no wall thicker than 8 inches.

Similar construction, usually of concrete block, has become popular for some resorts. Some recently built seaside motels consist of individual concrete boxes. When these are serviced by outside open-air stairways and balconies configured in accordance with the Life Safety Code (NFPA [National Fire Protection Association] 101), or the egress-related provisions of the building code in effect, the epitome of life safety in a multiple-unit occupancy is achieved.

Collapse During Construction

From time to time, concrete structures under construction collapse. The fire department is almost invariably called to rescue construction workers. The following suggestions are offered as aids in planning a department's role in rescue operations when a building collapses and there is no fire. Although the authority of the incident commander is often clearly spelled out, the authority to act or to direct others at emergencies other than fires is rarely defined. Fire officers should be well informed on the legal position of the fire department, particularly once all victims have been rescued.

Many building codes assign the power to order the removal of a dangerous structure to the building commissioner or similar official, thus making it quite clear that the fire department has no right to demolish a structure just because it represents a hazard in the view of the senior fire official. Politics, grandstanding, glory grabbing, turf protection, and potential movie rights have all played a part in disaster operations that are improvised. Every community should have clear plans setting forth areas of responsibilities at emergencies other than fire.

Lawsuits are common today. Often after the dust settles in a building collapse, there may be a series of legal actions as the owner, architect, general contractor, subcontractors, and victims attempt to determine financial responsibility. Collapses do not just happen, they are caused—and some of the people at the scene may well know, or at least fear, that the blame may be put on them. They may actively advise or assist the fire department. Be careful. Somebody may be trying to cover up. It may not be your duty to prevent this, but it certainly is not part of your duty to assist. Be wary of such suggestions as "Let's knock down the rest of this before it falls." A study of the surviving portion of a structure may show why the collapse occurred.

An Industry Warning

Some experts have warned of the collapse hazard of concrete structures. They argue that design engineers should use construction loads as governing loads in structures. At times, the loads on buildings under construction exceed the design load and can approach failure load.

Problems of Falsework

Falsework is the temporary structure erected to support concrete work in the course of construction. It is composed of shores, **formwork**, and lateral bracing **Figure 11-15**. Formwork is the mold that shapes the concrete. Shores are members that support the

Figure 11-15 These shores and formwork will serve as a temporary structure that will support the load of the concrete during the course of construction.

© photoslb/ShutterStock, Inc.

formwork. Lateral bracing is usually made up of diagonal members that resist lateral loads on the formwork. Formwork can represent 60% of the cost of a concrete structure.

Concrete formwork is designed without the extra strength calculated into a building to compensate for deterioration. It is usually built at the lowest possible cost, out of low-quality material, by people rarely versed in basic engineering and mechanical principles. With formwork, the builder hopes to contain a fluid load that can provide a head pressure of up to 150 pounds per square foot for each foot of height. Large fluid loads are placed at the tops of slender columns, and the loads are then vibrated. The structure must often absorb the impact load of the sudden stopping of motor-powered buggies carrying heavy loads of concrete. It is not surprising that formwork failures occur; what is surprising is that they are relatively rare.

Falsework for walls or columns must have adequate strength to resist the tremendous pressure of the heavy fluid concrete. As concrete sets, the pressure is reduced due to internal friction. If a wall is designed for several successive pours and an over-pour occurs, collapse may result. The setting of concrete is temperature dependent; cold weather slows it down. Concrete poured at 50°F (10°C) will develop one-third more pressure than at 70°F (21°C).

Reshoring

Concrete requires time to cure; the length of time depends upon the type of cement used and the temperature during the curing period. After a proportion of the design strength is reached, formwork is removed for use elsewhere. Some shores are set back in place to help carry the load of the still-curing concrete in a process called **reshoring**. The presence of reshoring is an indication that the concrete is not yet up to full strength. It may be unable to handle large superimposed loads, such as debris removed from a collapse, excess numbers of people, or additional apparatus or construction equipment. Avoid adding equipment and construction material to existing loads.

Collapse of Floors

Many collapses involve formwork supporting newly cast floors. Sometimes the collapse is of framework supporting a high bay floor, where two sets of normal floor-high shores are used one on top of the other. Such structures have all the instability characteristics of long, narrow columns. Proper cross-bracing is vital, as is proper footing.

Often, formwork rests on the ground, sometimes on unconsolidated backfill. The planks on which the shores rest are called **mudsills**. If water liquefies the soil and turns it to mud, the bearing may be inadequate. Repeated vibrating of the concrete to provide a dense smooth surface may be the straw that breaks the camel's back, causing a collapse.

A Widely Believed Fallacy

In the first edition of this text, it was stated, "Reinforced concrete which has set hard to the touch usually has developed enough strength to be self-supporting, though it may not be capable of handling superimposed loads." This assertion was based on information from experienced concrete designers, and it is still widely believed.

Unfortunately, this assumption was disproved in the Skyline Towers collapse in Arlington, Virginia, in 1973. In this collapse, shoring was removed from the topmost floor. The floor collapsed, and the collapse was progressive, ending at the ground level **Figure 11-16**.

The removal of the shoring by laborers is no different than its removal by fire; the result would be identical. In short, any concrete formwork failure presents the likelihood for catastrophic collapse, though the potential is even more pronounced in the case of post-tensioned concrete. Few concrete buildings are designed to withstand the collapse of one floor onto another. When this occurs, progressive collapse is almost a certainty.

Hazards of Post-Tensioning

The tendons for post-tensioned concrete are laid out on the formwork before the concrete is poured.

Figure 11-16 The Skyline Towers collapse in Arlington, Virgina.

Courtesy of the Fairfax County Fire and Rescue Department.

When the concrete is being poured, samples are taken in standard sample cylinders and are sent to a testing laboratory for subsequent quality verification and strength testing. When the concrete reaches a designated fraction of its ultimate strength, hydraulic jacks are used to tension the tendons or jack the cables. Until this is accomplished, the entire weight of the concrete is on the formwork. There is no bond between the tendons and the concrete—quite the contrary, the tendon is contained in a sheath to *prevent* bonding. The weight of the concrete is transferred to the columns only when the tensioning is completed, several days after the concrete has been poured.

The 22-story Skyline Towers was made of conventional reinforced concrete. The garage, located behind the Towers, was of post-tensioned construction. When a crane hit the corner of the garage, both floor slabs sheared completely off the columns. Shear resistance of post-tensioned slabs is poor.

Collapse of Reinforced Masonry

Because large numbers of masonry units are used in construction, it is easy for workers to overload a portion of a floor with brick or block. This happened in Pittsburgh, Pennsylvania. The excess load caused the partial collapse of several stories of precast floors.

Collapse of Precast Concrete

Precast concrete buildings under construction are unstable until all connections are completed. Temporary bracing, such as tormentors (adjustable poles set diagonally) or cables, holds units in place. In some cases, wooden temporary shoring is used. The completed structure may be of pinned construction, with bolt-and-nut or welded connections, or it may be made monolithic by the pouring of wet joints to join precast units together. In Montgomery County, Maryland, a three-story parking garage nearing completion totally collapsed at a single connection because an oversized washer was used on a bolt.

Lift-Slab Collapses

In lift-slab construction, the ground floor slab is cast first; later, the other floor slabs and the roof are cast on top of the ground-floor slab. Bond breakers are used between the slabs so they will not adhere to one another. When the slabs reach the designed strength, they are raised up the columns by hydraulic jacks placed on top of the columns. All the jacks are controlled from a central console. As each floor reaches its final position, it is temporarily connected to the columns.

The fatal collapse of the L'Ambience Plaza concrete building under construction in Bridgeport, Connecticut, which resulted in the loss of 28 lives,

focused greater attention on this method of construction. This collapse was also due to the failure of a single connection.

Many of the accidents that have occurred while buildings were being constructed using this method have happened while the slabs were being lifted. In some cases, however, failure occurred when no lifting was being done. In San Mateo, California, a roof slab was being lifted 16 feet (4.9 m). The columns were not braced—only bolted to the foundations. As the slab reached the top, the columns were 3 inches (76 mm) out of plumb. As an attempt was made to pull the slab back into place, it collapsed.

Fire Problems of Type I Buildings Under Construction

Concrete buildings under construction can present serious fire problems. Fire in formwork can easily result in major collapse with potential for massive life loss. There is little reserve strength in formwork. There seems to be little understanding by designers, constructors, or even fire officers of the potential catastrophe that could be caused by a fire in formwork. A few knowledgeable people are astonished that fire fighters would place themselves below burning formwork that is supporting the weight of a concrete floor.

Fire Potentials

Numerous fire causes are present at a construction site, including welding, cutting, and plumbers' torches; temporary electrical lines; and arson. Fuels, such as wooden formwork, flammable liquids, and plastics for insulation, are readily available. Glass fiber formwork is also combustible, which is often a surprise to those handling it.

The most dangerous hazard may be from heating. Sometimes workers burn scrap wood in steel barrels or use kerosene heaters to keep warm. By far the most dangerous, and perhaps most common, heating method is the use of liquefied petroleum gas (LPG). To retain heat, the building is sheathed in plastic. Inside the heated area, unsecured gas bottles are present with the nearby open-flame salamanders (heaters) in which the gas is burned. When this is the case, the stage is set for a tremendous disaster. It is important that construction sites using portable heating equipment be inspected by qualified fire inspectors.

Most codes for the use of LPG require storage of the gas away from any open flames, for good reason. In 1963, 74 people died in an LPG explosion at the Indianapolis Coliseum. A gas-fired cooker was being used under the concrete stands. A leaking cylinder exploded when the gas reached the flame of the heater. According to a rescuer, one victim was found in the center of the arena with a huge piece of concrete on him.

In one West Coast city, the practice is to store the gas bottles at ground level and pipe the gas with plastic tubing to the salamanders. When this system is used, an **excess flow valve** should be installed. Such a valve senses a sudden increased flow of gas, as from a broken line, and shuts off the gas, preventing a tragedy. Such installations also must be inspected for fire code compliance.

Hazards of Post-Tensioned Concrete

Post-tensioned, prestressed concrete construction presents catastrophic fire collapse potential. It is vital for a fire department to be aware of any post-tensioned structures being built in its area and recognize the clues that indicate post-tensioning. Possible post-tensioned structures include bridges and parking garages.

Post-tensioned concrete presents a greater catastrophic collapse hazard during a fire than does conventional reinforced concrete. As noted, the entire weight of a post-tensioned concrete floor or beam rests on the falsework until the tensioning transfers the load onto the columns. A falsework fire, therefore, could cause the sudden collapse of an entire concrete floor slab, or the entire structure, onto fire fighters.

After tensioning, the ends of tendons are left hanging and exposed for varying periods of time. In the case of floor slabs, several feet of excess tendon may protrude from the socket until they are cut off and the socket is dry packed. In the case of beams, the anchors may be left exposed for further tensioning or due to normal construction delays. In either case, this represents a serious threat to the stability of the building during a fire. Hanging tendons can act as heat collectors and will fail at about 800°F (427°C).

When a floor is tensioned in increments, the excess tendons are rolled up and attached to a wooden rack. In a fire, the rolled-up tendons would act as a heat collector, transmitting heat to the previously tensioned portion of the tendons. The substantial fuel loads present at construction sites or from exposed buildings may provide the heat to cause tendons to fail. Failure of tendons will cause the collapse of that part of the structure.

Protection of Tendons

There is one opportunity to control a disaster before it occurs. Insist on protecting tendon anchors immediately after tensioning is completed and on the provision of temporary protection for incrementally tensioned tendons. Instruct all personnel in the hazards of to-be-tensioned concrete.

A Total Collapse

A post-tensioned building under construction in Cleveland, Ohio, suffered a falsework fire. The fire department operated from the adjacent completed section, extinguished the visible fire, and notified the contractor that he had a problem. A second fire occurred. As the fire department was setting up, the entire 18-story building collapsed. No fire fighters were injured because they had been kept out of the collapse zone.

Precast Buildings

When precast concrete units are being erected, temporary bracing or support is provided until the full space-frame relationships of the building are completed. Columns are usually braced by telescoping tubular steel braces (tormentors), but wood braces are also used. Wooden falsework is often used to support horizontal elements until connections are made. The loss or upset of any such temporary support may precipitate a general collapse. Cold-drawn steel cables (which fail at 800°F [427°C]) often provide diagonal bracing in precast buildings.

Cantilevered Platforms

Cranes are often used to deliver construction materials to buildings under construction. This necessitates a temporary cantilevered platform extending out from the floor. A common method of securing such a platform is to brace it into place by using wooden shores to deliver a compressive load to the floor above. A fire involving such a platform would destroy the shores and drop the platform to the street. Keep all personnel and apparatus clear of the collapse zone.

Tower Cranes

The high tower crane, a common sight in modern construction, seems to climb up the building. As it climbs, it is supported on the building's structural frame. The weight of the crane may be distributed over several floors by falsework. A fire involving this falsework can bring down the crane.

Falsework on an otherwise apparently completed floor should be investigated. It may be supporting a patch over a hole that was left in the building to handle or transport materials or it may be supporting a heavy load such as the crane.

In one example, formwork for concrete placement burned on the 23rd to 25th floors of a highrise under construction. Flaming debris fell to the street, and the tower crane operator was trapped in his cab above the 28th floor. Fire fighters protected him with a heavy-caliber stream from a nearby roof until an engine company fought its way through the fire to save him.

Fire Problems in Finished Buildings

Fire Resistance

When concrete construction was first developed, it was regarded by some as the answer to all fire protection problems—it was believed to be truly fireproof. After a series of disastrous fires, however, it became evident that concrete, like any other non-combustible material, can be destroyed by fire given sufficient fire load and fire duration.

Concrete is inherently noncombustible. It may have been fabricated to meet a fire resistance standard. Unfortunately, many people, including some who should know better, confuse noncombustibility with fire resistance. Neither is synonymous with fire safety. Concrete construction can be made to deliver different levels of fire resistance. There are a variety of ways of protecting steel, including concrete.

Concrete construction carries no guarantee of life safety. The reinforced concrete Joelma building in Sao Paulo, Brazil, burned in 1974 with a loss of 179 lives. The structure escaped with relatively minor damage. Factors other than the structure itself, such as interior finish and the inability of the occupants to quickly reach a safe nontoxic environment, may be far more important to the safety of the occupants.

Methods of Providing Fire Resistance for Steel, Including Concrete

Fireproofing (Insulating) Steel

Steel is fireproofed by one of several systems. Steel qualifies for a fire resistance rating if the protection system previously passed a standard fire test (ASTM E-119, discussed in other chapters). Although there is no such thing as a truly fireproof building, the term has survived as the designation of the system by which steel is insulated (fireproofing the steel). Steel that has been fireproofed is known as protected steel (which is also the term for weather-protected steel).

Types of Fireproofing

Fireproofing of steel is classified as individual or membrane. Individual fireproofing provides protection for each piece of steel. In one method, steel is encased in a structure, such as concrete, terra-cotta, metal lath and plaster, brick, or gypsum board. This method is called encasement. In another method, fireproofing is directly applied to the steel, usually by spraying. Materials used in this method include an intumescent coating containing noncombustible fibers that swell and char when exposed to flame and cementitious coatings. Asbestos fiber was once used in building construction but is now no longer used.

Membrane fireproofing does not protect individual members. In one method, wire lath and cement plaster are used. In another method, the fireproofing of the floor is accomplished by a rated floor–ceiling assembly. In rating such a fire protection system, Underwriters Laboratories, Inc. tests and lists a floor and ceiling assembly or a roof and ceiling assembly.

In such floor or roof assemblies, the ceiling, hangers, electrical fixtures, bar joists, left-in-place formwork for the concrete floor (usually corrugated steel), air ducts and diffusers, and the concrete floor are tested as a whole unit as it would be built. Therefore, it is incorrect to speak of a fire-rated ceiling because the ceiling is one part of the total assembly.

The assemblies are tested under the ASTM E-119 standard fire test. The original designers of the ASTM E-119 test did not envision an assembly that provides a void trussloft (like a cockloft) in every floor. Fire can and has extended through this void. The designers of the test also did not anticipate the problem of otherwise unprotected columns passing through the trussloft.

Hazards of Floor–Ceiling Assemblies

Buildings with floor–ceiling assemblies, particularly where the columns are unprotected passing through the plenum space above the ceiling, can present a serious menace to the safety of fire fighters who are, generally, unaware of the dangers. The observations and considerations that lead to this opinion are many.

The building trades sometimes disregard detailed instructions, particularly when they do not understand them or they find them inconvenient. A fire inspector found aluminum wire being used to support the ceiling channels. "What's the matter with aluminum?" asked the mechanic.

The value of listing building materials is contingent upon the field assembly being accomplished exactly as performed in the laboratory. It is impossible for inspectors to have the time to follow and examine every item and inch of an assembly.

The ceiling system is at the mercy of all owners, operators, tenants, tradespeople, and mechanics who have reason to remove ceiling tiles. Access to utilities and additional storage space for tall items such as rugs or ladders are only two common reasons to remove tiles. Where the plenum space is part of the air-conditioning supply, employees soon learn that displacement of a tile improves airflow at their location.

All fire codes require that fire-rated assemblies be maintained. Enforcement may be difficult. Replacement acoustical tile, which appears to be identical to the original tile, may be combustible, rated for fire hazard only, rated as part of a fire resistance system but not the system installed in the building in question, or may be the proper tile, which meets the specification of the listing. All electrical fixtures, air-duct openings, and other penetrations of the ceiling must be rated as part of the ceiling system. Proper fire protection and building and fire codes require that fire ratings be maintained. Unfortunately, there are many lapses between code directives and actual practice.

The term *fire rated* is used quite often in the fire protection and building construction fields. By itself it is nonspecific and, therefore, meaningless. With respect to ceiling tiles, it may refer to tiles that have merely met the requirements for flame spread or to tiles that are part of a listed fire resistance system. As has been noted elsewhere, no part of a listed fire resistance system stands by itself.

Owners, managers, and occupants—even of buildings where fire protection equipment is well maintained—are usually totally unaware of the significance of absolute integrity of the ceiling system. A common practice is to remove tiles from the storeroom to replace tiles in public locations. (Who needs a fancy ceiling in the storeroom?) Tiles are also damaged by water leaks. Tiles are removed by tradespeople. Holes are cut through tiles. Displays are hung from the metal grid. Until recently, floor–ceiling systems were not tested with superimposed loads.

The list of ways in which the integrity of a floor–ceiling assembly can be compromised is endless. There are many pictures to prove it. Laboratory fire tests are conducted under a slight negative pressure to remove smoke and fumes. Fires generate positive pressure, and lay-in ceiling tiles may be easily displaced by fire pressures. When a tile is displaced by a worker, it is often dropped back down into place without any restraint to upward motion.

The addition of insulation that is not part of the specifications of the listed ceiling assembly, particularly in a roof and ceiling assembly, can significantly affect the fire performance of the assembly. The insulation will cause heat to be retained in the channels supporting the tiles, causing them to fail earlier than expected. A membrane protection system must be perfect—the failure of even a single tile is the equivalent of a pinhole in a water-filled plastic bag.

It has been noted that, in effect, there is a cockloft between the ceiling and the floor. In one incident, fire starting in one room traveled across a hallway above the ceiling and came down through joints in the tile ceiling of another room to ignite books on a top bookshelf. Only alert fire fighters prevented the full involvement of the floor.

Some code provisions provide for firestopping, particularly along demising walls or tenant boundaries and around columns, but the use of the plenum space for various services makes it probable that the firestopping will only continue to conform to the definition of legal firestopping.

In some buildings, the use of deep long-span trusses to provide clear floor areas creates plenum spaces several feet in height. Wasted space is often

intolerable. What is more logical than to provide access to this space and use it for storage? Engineers closely associated with hospital construction call this void interstitial space. Whether the membrane fireproofing is provided by suspended tile or plaster on metal lath is immaterial, the fire load has been placed adjacent to vulnerable unprotected steel, where firefighting will be almost impossible.

Not all floor–ceiling arrangements are intended to be fire resistive. A steel bar-joist floor with concrete topping with flame-spread-rated tiles below may appear to be fire resistive. In fact, it may only be noncombustible construction.

The fact that tiles are obviously missing does not necessarily mean that a fire resistance system has been violated. The building may not be required or intended to be fire resistive; it may be simply of noncombustible construction. In such a case, ceiling tiles are at the option of the owner and need merely to meet flame spread requirements, if any.

The building may be of concrete construction. There are some listed concrete assemblies in which suspended tiles are incorporated, but, in general, the suspended ceiling is installed to provide a hidden void for utilities as well as an acoustical and aesthetic treatment.

Fireproofing and Building Codes

Fireproofing is applied to meet the standards required by the local building code. If, for instance, the code requires 3-hour fire resistance on columns, the designer selects a system that has been approved by the local building department. As a supplement to the building code, the building department will indicate which systems tested at which laboratories are acceptable. After a manufacturer develops a fire resistance system, it is tested at a laboratory to determine whether it meets the standard. Testing laboratories (like Underwriters Laboratories, Inc.) publish lists of tested assemblies, not just for fireproofing applications but for all fire-rated assemblies.

The efficiency of fireproofing depends first on the competence of the subcontractor and the willingness of the builder to demand quality work. It further depends on the building department staff who inspect the original installation, and on the fire department inspectors who determine that fireproofing and fire protection are not compromised. This is particularly important in the case of spray-on fireproofing and membrane floor and ceiling assemblies. There are advantages and disadvantages of each method.

Encasement Method

Terra-cotta tile was an early method used for encasement. The cast-iron columns of the Parker Building in New York City (destroyed by a fire even though it was regarded as the best type of fireproof building) were protected with 3-inch (76 mm) terra-cotta tiles. Electrical conduit was laid against the columns inside the tile. When the heat elongated the conduit, it tore the fireproofing off the column. A major column failed catastrophically. There are buildings still standing that may have this same defective design.

Another error in the encasement was to leave the bottom web of beams unprotected. After a number of failures, this error was corrected by the development of tiles that were shaped to fit around the steel. The tiles (called **skewbacks**) below beams are easily removed to access the beams during alterations, such as to hang equipment from the overhead. As a result, the fireproofing is often seriously defective.

Fireproofing that is easily removed is at the mercy of the person making alterations, who often has no knowledge of why this material is present. In one case, while altering an operating department store for the construction of a subway entrance, a contractor removed the protection from a major column. The sprinklers were shut off in this area. About a hundred cylinders of propane gas were stored adjacent to the column. A fire in the propane could have collapsed the building. Fortunately there was no fire.

At one time, concrete was quite popular as a protective covering for steel, particularly where concrete floors were being used. The wood falsework

required for forming concrete provides a high fuel load and has been involved in a number of serious construction fires. Concrete has the advantage of being the most permanent fireproofing because it is so difficult to remove. However, concrete can be knocked off beams and columns by vehicle impacts.

The fireproofing of steel and concrete beams is integral; that means it is accomplished by a specified mix of concrete in a specified thickness. Some concrete, particularly that below the reinforcing rods of a concrete beam, is not necessary for structural strength but is necessary for fireproofing. Because there is no budgetary distinction regarding the material, this may give concrete an economic advantage.

The disadvantage of concrete is its weight. In the effort to reduce the dead weight of a building, and thus its cost, fireproofing is often a tempting target. Encasement systems of gypsum board or wire lath and plaster were developed to save weight. Wire lath covered with cement plaster has been used for fireproofing for many years.

Sprayed-On Fireproofing

The search for lighter fireproofing led to the development of various sprayed-on coatings. Unfortunately, sprayed concrete spalls badly when exposed to fire, so many combinations of asbestos fibers and other materials were developed. Although some of these materials can pass laboratory tests, some serious questions have been raised about their reliability in the field. Ensuring that the material is applied at the specified thickness and density is another problem with sprayed-on coatings. Adhesion of the material to the steel is also of concern; the base steel must be "clean"—free of oils and rust scales.

As a result of the disaster at the World Trade Center in 2001, attention was focused on the reliability of fireproofing. Changes were made to the model building codes to require "stronger" fireproofing (in terms of adhesion to the steel as well as its internal strength to stay together, known as cohesion. Sprayed-on material is often easily knocked off by members of other trades doing their own work **Figure 11-17**. Many years ago, Frank Brannigan was invited by an insurance

Figure 11-17 It is important for inspectors to conduct routine inspections and to catch problems such as when the fireproofing has been removed from the steel.

Photo courtesy of Blaze-tech Fire Protection

executive to the 30th floor of the nearly complete Bank of America Building in San Francisco. The fireproofing had been stripped from all the columns by the plasterers.

A state office building in California was being sprayed with fireproofing material by a contractor who had no appreciation of the importance of the work. Overspray was swept up from the floor, laid on the lower web of the beams, and covered with a fresh coating. The building was not insured, but an insurance engineer took photos of the work. When a fire occurred in the building, it cost almost as much to repair the damage to the steel as it had to build the building.

On the other hand, the First Interstate Bank Building of Los Angeles, which suffered a multistory fire in 1988, was found to have excellent spray-on cementitious fireproofing. Had the fireproofing been as poorly applied as observed elsewhere, a catastrophic collapse might have occurred.

Asbestos fiber fireproofing is under attack on another front—a serious health hazard arises with its use. It is difficult, if not impossible, to sell a building with asbestos fireproofing. Health regulations have eliminated the use of asbestos fireproofing, and asbestos is being removed from existing buildings.

Published accounts are invariably silent as to what is being done to replace the asbestos.

Signs of Trouble

There are clear signs of possible trouble with concrete elements that can be identified in a prefire plan. Deteriorated concrete, spalling that exposes reinforcing rods, and cracks in concrete that can admit corrosive moisture to the reinforcing rods are indications that the building is in distress.

Many chemical processes give off fumes that can damage or destroy concrete. Chlorine corrosion caused the collapse of a concrete roof of a swimming pool in West Roxbury, Massachusetts. Aluminum also reacts with concrete—aluminum electrical conduit buried in concrete did severe damage to a sports stadium.

Parking garages in areas where salt is used to melt snow and ice are particularly vulnerable to corrosion. The extent of damage is often difficult to determine. In some nonfreezing areas, calcium chloride added to the concrete has caused problems. Preventive measures include sealing the concrete, providing adequate drainage, and flushing surfaces with fresh water in the spring. Rehabilitation of concrete has included removal and replacement, installation of cathodic protection, and using additional steel beams to shorten the span of concrete slabs. Some concrete garages have been demolished when their repair was found to be untenable.

Unprotected Steel

When a concrete structure is in trouble, it is often repaired with steel. In the case of fire-resistive buildings, often no consideration is given to providing equivalent protection for the steel. If steel cables are used, failure will occur below 800°F (427°C).

Fire fighters should be ever watchful about what is being done to buildings. Almost none of what is done to a building after it is completed, except for installation of automatic sprinklers, benefits the fire suppression effort. One fire fighter student saw steel columns being brought into a 3-hour-rated fire-resistive shopping mall. The columns were pre-painted to match the wall color in the mall. The roof was failing and the columns were provided to shore it up. The owner did not want the building department to know of the problem.

If steel is initially designed into the structure, the proper degree of fireproofing is specified. If installation of steel is an afterthought, or part of a repair job, it will usually be unprotected. In particular, steel connectors for precast units are often unprotected.

⊕ Branniganism

We must go beyond experience to competent risk analysis. We must learn that "fire resistance" is at best just a hope, not a mathematical given, as is often assumed. We must learn that the failure of a single vulnerable connection can precipitate disaster.

Ceiling Finish and Voids

One of the fire protection advantages of concrete construction is that it lacks inherent voids. While voids may be created in the finishing of the building, they are not a necessary part of concrete buildings, although they are useful to the occupants. In one case, a fire in the void between the hanging plaster ceiling and the concrete floor of a former department store proved difficult to find. The fuel for the fire consisted of an 80-year accumulation of dust.

Waffle slab concrete was originally considered ugly and was improved aesthetically by installation of a suspended ceiling. Later architects appreciated the coffered effect and emphasized it. In one mall, imitation plastic waffle concrete is suspended below the structural slab.

Unfortunately, when ceilings were installed, combustible tile with a high fire hazard rating was often used. The combustibility is not intended; it

just happens that combustible tile is cheaper than tile with a low fire hazard rating. The interconnected voids above the tile make it possible for the tile to burn on both sides simultaneously.

When suspended ceilings are installed as part of initial construction, it is more likely that they will have satisfactory fire hazard characteristics. But it is necessary for fire officers to be aware of the local code requirements. The tile usually will be only as safe as the law requires.

Combustible tile need not be suspended to create a serious hazard. Tile applied with a flammable adhesive to a noncombustible surface was a key factor in a fatal hospital fire. Installing new ceilings that meet flame spread requirements below old combustible tile ceilings presents a serious hazard. Of particular note is the 1989 fire in the John Sevier Retirement Center in Johnson City, Tennessee, where 16 occupants died in a fire-resistive concrete building. A key factor in this incident was the concealed combustible acoustical tile.

Combustible voids can be created in a variety of ways. In one case, a heavy wooden decorative suspended ceiling was installed in a restaurant of otherwise concrete construction. Sprinklers are below the ceiling. Fire could burn unchecked in the void. The fire resistance of the ceiling hangers, which are attached to anchors (possibly lead), is uncertain. A fire might cause the collapse of the heavy false ceiling. Decorative or functional wall paneling and screens may also create combustible voids.

Although voids are not inherent in concrete construction, they can be useful. In one case, a church was built of concrete without voids. When the church was to be air conditioned, the only solution was to run the ducts on the roof. This created costly maintenance problems. In a modern office building, with its huge communications and other utility requirements, as much as one-third of the height from floor to ceiling may consist of in-ceiling or under-floor voids.

Even when the void is noncombustible, it may still present problems. Combustible thermal or electrical insulation and combustible plastic service piping may be concealed in the ceiling void. Because the fire resistance of a building is achieved by the concrete construction, the hung ceiling, though it may be identical to the ceiling used for membrane fireproofing of steel, is generally not required for the structural integrity of the building. The owner is free in this case to remove or modify the building within code requirements for flame spread. However, there are some rated fire-resistive assemblies that combine precast concrete elements with suspended ceilings. In these cases, the maintenance of the ceiling is necessary for the fire resistance of the floor–ceiling assembly.

The Integrity of Floors

If a building is fire resistive, it is presumed that the floor will act as a barrier to the extension of fire. More and more, however, codes are requiring sprinkler protection in addition to the passive protection provided by compartmentation. Many older buildings were constructed under the less costly but inadequate assumption that compartmentation was all that was required to limit fires. The concept of isolating the fire to a limited area has great appeal, and at first glance appears to have great merit. However, true compartmentation is rarely achieved. Even in a limited area, a high fire load may yield a difficult, costly fire.

Building use often requires that the floor be penetrated for utilities, plumbing, heating, air conditioning, electrical service, and communications wiring. There are proper ways to do this—using listed "through penetration protections systems" (commonly referred to as firestopping) or through the use of fire-rated shafts—so that the integrity of the floor is not compromised. Unfortunately, they are not always used. Further, some of the methods used are open to question.

For whatever reason, enclosures around ducts may be inadequate and have failed in fires. This can permit transmission of fire and/or smoke to other floors. In addition, failure of the enclosure may expose fire fighters to the danger of falling down the shaft.

Poke-throughs are holes provided to draw utility services up to a floor from the void below.

The penetration of the floor for poke-throughs may negate the fire resistance rating of the floor.

In addition, any holes that violate the integrity of concrete floors may provide passage for deadly fire gases. It is not safe to assume that there cannot be an extension of fire from floor to floor in a building because the floors are of concrete. Be aware of the local practices in providing utility services, such as penetrating a floor for cable TV and computer cables. Even if a floor was properly constructed when it was built, it can be compromised by contractors retrofitting utility equipment years after the floor was built.

As penetrations of floors for services increase, the floor may be unable to resist the passage of fire adequately, and the building department may require a suspended ceiling, which, together with the floor, should ideally develop the necessary fire resistance. In this case, the owner is not free to modify the ceiling. Many tested and rated assemblies are floor–ceiling assemblies and, therefore, cannot be modified. Building designers must submit a new design with supporting evidence (e.g., tests, reports) for approval.

Concrete floors require expansion joints. In one case, steel expansion joints transmitted fire from floor to floor in a huge postal building. Aluminum expansion joints dropped molten aluminum into the lower level of McCormick Place in Chicago and extended that 1967 fire. Combustible fiber expansion joints can provide a difficult fire problem.

Concrete shrinks, and this shrinkage may create large cracks. Such cracks may permit the passage of fire from floor to floor, even though the concrete is structurally sound. Shrinkage, distortion, or settlement may put door frames out of alignment, thus preventing the closing of fire barrier doors.

Imitation Materials

Imitation concrete panels are commonly used, particularly on the exterior of buildings. In one product, they are made of two sheets of asbestos board, with a combustible foamed plastic interior. Epoxy glue is painted on the surface, which is then coated with small stones.

The top panels of some buildings are genuine precast concrete. The wall panels are usually imitation concrete. At the base, however, concrete block panel walls are painted with epoxy adhesive and stones are added. The entire building appears to be of identical finish. The fasteners that hold the panels on the building are held in the plastic. If the plastic burns or melts, the panels will drop off the building. Most imitation concrete panels are made of a steel framework with asbestos millboard. Real concrete is applied in a thin coat or stone is epoxied on.

Energy conservation has brought about the use of exterior insulation finishing systems (EIFS; discussed in other chapters) that give a building the appearance of sand-finished concrete. One type starts with exterior gypsum board screwed to steel studs. Glass fiber insulation is placed between the studs. A foam panel is then glued or mechanically attached to the gypsum board and a woven glass fiber mat is laid over the plastic panels. A layer of adhesive secures the glass fiber. Usually, a coat of organic polymers is added to the finish coat.

Buildings can be finished in this manner when constructed, or EIFS can be added to the building for energy conservation or to change its appearance. A Fort Worth, Texas, courthouse was finished in this manner to make its appearance more consistent with the 19th-century courthouse next door.

Concrete's Behavior in Fire

The concrete in fire-resistive construction serves two purposes—it resists compressive stresses and it protects the tensile strength of steel from fire. The latter function is accomplished in proportion to the thickness of the concrete cover. By sacrificing itself, the concrete provides time to extinguish the fire; however, it may also lose its strength.

Impact loads to damaged concrete, such as from sledges or jackhammers, should be carefully evaluated. All areas where spoiling has reached the reinforcing steel should be shored before any work is

done in the area. Concrete floors that are seriously damaged on one side may give no clue to the distress on the other side.

Cutting Tensioned Concrete

Fire tactics sometimes include cutting through a concrete floor to operate distributors on an otherwise inaccessible fire. One department's high-rise unit carries a hole cutter. It is possible to cut a hole in conventional reinforced concrete and, in the process, to cut through reinforcing rods without doing serious structural damage.

This is not the case in tensioned concrete structures. With this approach, the tensile strength is provided by steel cables under tremendous tension, and these cables are not in contact with the concrete. Cutting such a cable would be like cutting any steel cable under tension. The ends might whip about with tremendous energy. The cutting of a cable also might cause a massive structural collapse.

Precast Concrete

A cast-in-place, monolithic concrete building is resistant to collapse. Because this type of building has a rigid frame, the loss of a column does not necessarily cause collapse. The load will be redistributed and other columns, if they are not overloaded, will be sufficiently strong to carry the load.

Precast concrete buildings are another story. The individual columns, floors, girders, and wall panels (load-bearing or non-load-bearing) are usually pinned together by connectors. The fire resistance of the completed structure is dependent upon the protection afforded to the connectors. Connectors are set into the precast element, and mating connectors are provided on the structures to which it is to be attached. Bolts and nuts may be used or the connections may be welded. The recess provided for the joints is then dry packed with a stiff mortar. Often no protective covering is provided for the connectors; it may not even be required.

If an ordinary or noncombustible building is permitted by the code, a builder is free to raise a structure that may appear to be fire resistive, but is not. Typically, in such a case, the connectors are unprotected. This problem should not be overestimated. Connectors must resist wind and other forces, and the fire load usually must be quite severe to cause their failure.

The possible effect of an explosion in a precast, pinned building should be evaluated where such possibilities exist. Such buildings have none of the redundancy that characterizes the rigid-framed monolithic concrete building.

Probably the most spectacular disaster involving precast concrete construction took place in England in 1968 with the Ronan Point collapse **Figure 11-18**. A 24-story apartment building was constructed of precast floor and bearing wall panels—literally, a giant concrete house of cards. A gas explosion high up in one corner blew out one wall-bearing panel. This caused the collapse of the floor above, which fell to the floor below. The impact load was too much for the next

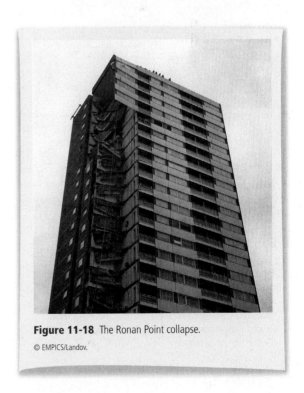

Figure 11-18 The Ronan Point collapse.
© EMPICS/Landov.

lower exterior bearing wall panel and it buckled. Thus, the corner of the building progressively collapsed down to ground level. Five people died in the collapse.

Concrete Trusses

Concrete trusses are not common, though huge concrete trusses are seen in the Tampa, Florida, and Dallas/Fort Worth, Texas, airports and some high-rises **Figure 11-19**. An American Airlines hanger at Dallas/Fort Worth has a 1,280 × 280-foot (390 × 85 mm) clear span, with doorways 74 feet (22.5 m) high. Concrete towers 80 feet (24.4 m) high support concrete trusses 76 feet (23.1 m) high. Each pair of concrete trusses supports steel box trusses 20 feet (6.1 m) wide and 280 feet (85.3 m) long.

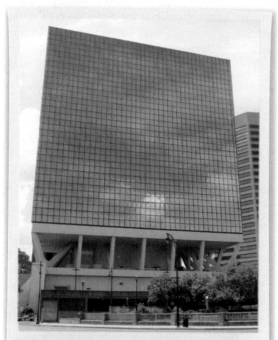

Figure 11-19 Although not constructed with traditional concrete trusses, this unusual high-rise uses laterally braced angled steel columns encased in concrete to support the floors above. The columns are sandwiched between two structural diaphragms, the top diaphragm in compression and the bottom one in tension.

Courtesy of Glenn Corbett.

Fires in Type I Buildings

In 1986, the Los Angeles Central Library suffered the largest loss of any library fire in the United States. An estimated 200,000 books and numerous periodical collections were destroyed, along with the largest collection of patents in the western United States. Built in 1926, the building was of massive concrete construction. Its bearing walls are as much as 3 feet (91 cm) thick, with massive columns and thick roof and floor slabs. The book stacks were on seven tiers of unseparated, unprotected steel loaded to an estimated 93 pounds (42 kg) of fuel per square foot. This would be the equivalent of about 11 hours of fire exposure to the standard fire curve. The fire required a tremendous commitment of apparatus, more than 350 fire fighters, and 7.5 hours to extinguish.

A Christmas tree fire in a 16-story reinforced concrete high-rise apartment building in Dallas, Texas, resulted in $340,000 in damage and the potential for a large loss of life were it not for competent preplanning by the fire department. Utility and vent pipes were punched through the ceilings. The gypsum and steel stud enclosures extended only to the suspended ceiling, not to the underside of the slab. Lighting fixtures and air-conditioning grills pierced the ceilings. Thus, smoke had an unobstructed path from the eighth floor of origin to the floors above.

The Military Records Center near St. Louis, Missouri, was severely damaged in a fire in 1973 **Figure 11-20**. The building was of concrete construction, six stories in height. The top floor, where the fire occurred, had over 200,000 square feet (18,581 m²) of undivided floor space. The incredible fire load included over 21 million military personnel files in cardboard boxes on metal shelves. The fire destroyed the top floor, which was subsequently removed from the building. The fire caused the roof to elongate several feet almost all around the perimeter. The elongation caused top floor columns to shear off

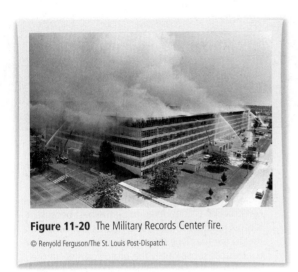

Figure 11-20 The Military Records Center fire.
© Renyold Ferguson/The St. Louis Post-Dispatch.

column or floor are not so obvious during a fire. It is very difficult to observe that a concrete column or floor has been so severely spalled in a fire that its rebar is exposed. Therefore, instead of relying solely on visual and audible signs (you may still experience some of them in a fire-resistive building), it is equally important to anticipate collapse based upon conditions such as:

- Knowledge of problematic existing building conditions (poor fireproofing, vulnerable connections, etc.).
- Dangerous loads (concentrated, heavy, etc.).
- Cutting tensioned cables.
- Heavy fire conditions over an extended period of time; such fire conditions will ultimately cause any building (or portion of a building) to collapse, regardless of construction type.

at either the top or bottom. It should be noted that the portion of the roof that collapsed held together for about 22 hours after the start of the fire. The building did all that could be expected of it and more, considering the huge fire load.

Signs of Potential Collapse in Type I Construction

In this chapter, a variety of problems that can result in a disastrous collapse have been identified. In terms of visual indicators of collapse of fire-resistive buildings, it is important to realize that instead of the "recognizable" indicators found in other types of construction like smoke pushing through brick walls (ordinary construction) or noises from cracking wood (wood frame and ordinary construction), an impending collapse it is not always so clear in fire-resistive construction.

While a sagging fireproofing-encased steel beam–supported floor is a clear signal that the floor is in distress, problems with a reinforced concrete

Know Your Buildings

For several years, many fire departments complained, "They are building so many buildings so fast here, we can't keep up with them." The 1990 to 1991 recession showed that in many areas of the country there was a glut of buildings of various types, causing a slowdown or even a cessation of construction. Slowdowns present an opportunity to get current on the hazards of specific buildings and to educate fire officers in the hazards of construction before the building boom picks up again.

A fire fighter's nightmare is a fire department charging headlong into a formwork fire in a post-tensioned building under construction, that collapses completely with scores of fire fighter casualties. Fire fighters must learn not to wait for experience. Ben Franklin told us, "Experience keeps a dear school, but fools will learn in no other." In the fire service, the price of experience is blood and grief. The post-tensioned collapse hazard is entirely credible. Make sure your department is not the one to provide the convincing.

TACTICAL CONSIDERATIONS

1. Reinforced concrete is a composite material. Any failure of the bond between the two materials means that the composite has failed and it no longer has the designed intended strength, only dangerous dead weight.

2. Concrete may be hailed as fireproof, but nothing is fireproof. Concrete adds nothing to the fire load. Though concrete is merely noncombustible, fire experience has shown that concrete is subject to serious damage. At a fire, it is impossible for the company officer to know if he is dealing with normal-strength concrete (NSC), or high-strength concrete (HSC). NSC loses about 25% of its compressive strength at about 575°F (302°C) and 75% of its compressive strength at about 11,007°F (6,097°C). (These are temperatures within the concrete, not ambient temperatures.) HSC loses much more strength than NSC in the same temperature ranges. The thing to remember is that concrete spalls (pieces break off from the surface), and when it does, it loses compression strength.

3. Many landscaped recreational fire pits are constructed with curved concrete blocks to form a circle. Additional concrete blocks are stacked forming a three-layered fire pit ring. Moisture expands within the concrete blocks when exposed to sustained heat and begins to lose its surface material. After a season or two of bonfires, if a concrete block is struck with a rock or heavy log, it crumbles apart. The compression strength concrete block is lost after being subjected to the sustained temperatures generated by the bonfire. The concept applies to concrete in structures. At first, they can withstand the radiant heat of the fire, but after the actual concrete is heated to its critical temperatures, it will spall and eventually fail.

4. Remember that when reinforced concrete spalls, it can expose steel tie rods and tendons of prestressed concrete, which can lose total tensile strength at temperatures in excess of 800°F (427°C) (that's cooler than a self-cleaning oven). This can cause a catastrophic collapse in a "good, solid, fire-resistive concrete building" during firefighting operations.

5. From a fire protection standpoint, Type I fire-resistive construction is considered to be the best because it is most resistant to collapse and does not contribute fuel to the fire. Type I construction often uses concrete or substantially fire-protected steel (more than Type II Noncombustible) as a structural system. This gives the incident commander and fire crews the widest safety margin for the time that can be spent on interior fire attacks and support operations. Because it can take so long to establish adequate fire flows in high-rise buildings, interior fire operations have proceeded for nearly 2 hours without the structural compromise.

6. Not every alarm is going to be a fire in an occupied building. We are often called to construction sites for collapse during construction or fires during construction. Rescue operations may be required during the concrete pouring phase of a project. When concrete is being poured, usually there are workers under the structure, bracing and wedging any shores that appear to be distressed. Determine from the supervisor whether all of the workers are accounted for. In one case, the fire department was assured that all the workers were accounted for. While the fire department was preparing to leave the scene, a worker wriggled out from under the plywood.

7. Concrete-on-wood floors require special attention in a fire. The insulating concrete may deceive fire fighters above the fire as to the fire below their feet. The added dead weight may also be a factor in the collapse of the floor–ceiling assembly. This is exactly what happened at the 1995 Mary Pang Fire in Seattle, Washington. Four fire fighters died when the floor collapsed into the burning basement below them. One of the survivors distinctly remembered the lack of heat for the amount of smoke that was on the first floor. Visibility was practically zero, but it wasn't very hot. Then, suddenly, the floor collapsed.

Be advised, these concrete-on-wood floors are typically found in food preparation or food packaging plants. This requirement comes from the health department for sanitary reasons and to prevent entry from rats. This condition is especially dangerous if the fire originates in the basement or the floor below a concrete-on-wood floor.

8. When responding to construction site emergencies, your first contact will usually be with a foreman or supervisor to determine the nature of the emergency. It's critical to the life safety of your fire fighters to understand industry nomenclature inserted into critical fireground information such as, "They haven't jacked the cables yet."

9. It is important to perform a risk–benefit analysis, especially with buildings under construction. Our mission of protecting property doesn't quite extend to construction projects. The property value hasn't been established yet. It is easy to rebuild at that phase of the project. There is no point in risking the lives of fire personnel in a concrete building under construction. All concrete involved in the fire will be torn out and replaced. It is almost impossible to predict the effect of the loss of even a small portion of the falsework in a fire. Fire officers should err on the side of caution. Heavy-caliber streams from safe locations deserve serious tactical consideration. The falsework structure cannot survive the loss of material caused by a fire.

10. When a wall formwork bursts, it usually goes out at the bottom. Workers may be trapped, particularly between formwork and the sides of the excavation. There will probably be no reinforcing steel to interfere with the rescue. Rescue involves uncovering faces, supplying oxygen, digging the victims out, and avoiding further injury. If the patient is able to use his arms, throw a small shovel down to him. It may be possible to lower a roof ladder down to the patient. Crews can then pull the ladder back up to safety with the patient holding on. A ladder may be all he needs to self-extricate and climb out on his own. Bear in mind that the concrete can transmit pressure. Avoid standing on concrete that might transmit the weight of your body to the victim.

11. In a collapse, most of the victims will be on top of the falsework. Use bolt cutters to cut reinforcing rods to manageable lengths. Thicker rods may be required to be cut with a Sawzall, a circular rescue saw, or an oxy-acetylene torch. It would be wise to drill ahead of time in situations where cutting tools are most effective. Rods are tied together in a matting. Further harm may be done by attempting to remove sections that are too large to safely manage.

12. In all buildings under construction, move carefully and deliberately. Floor openings may be inadequately guarded or covered over with a flimsy piece of plywood. Side rails may be nonexistent or of poor quality. Pay special attention to the possibility of elevator shafts. With requirements from the Americans with Disabilities Act (ADA), almost every new construction except for private R3 residences requires an elevator, even for two-story structures. Elevator shafts are serious fall hazards that can kill fire fighters and are present at the start of the project when the foundation is being laid. It is a reasonable risk–benefit conclusion for an incident commander to refuse to place fire fighters on or under new construction structures until all the floor-column connections are permanent. Once workers are accounted for, the remaining life hazard at construction sites are the fire fighters.

13. When a concrete building is finished, there are no clues to help the fire officer determine whether the floors are of conventional reinforced to tensioned construction. If the stranded cables that signal tensioned concrete are found, they must not be cut. The cutting of even a single cable can be disastrous. This is why the boring of holes between concrete floors in high-rise buildings has been abandoned. Besides being time-consuming and labor-intensive work for fire fighters, it's just too risky. Just make it a rule; don't cut cables in concrete Type I fire-resistive construction.

14. The core cutter or concrete blade must never be used to open a hole in a known or suspected post-tensioned floor. Beyond the danger of cutting the cable, tendons are cold-drawn steel and fail at 800°F. Even if the tendon isn't cut, opening a hole over the fire would probably subject one or more tendons to temperatures that would cause failure. In short, all cutting into post-tensioned concrete floors in a fire is unsafe. Abandon all consideration of this tactic. It is too risky; don't do it.

15. In a fire, it is vital that any exposed steel connected to stressed tendons be cooled immediately with hose streams for the duration of the fire, if this can be done from a safe position.

16. Do not fail to examine the ceiling void early on when seeking an unknown source of smoke. Two tools must

be carried by units performing this reconnaissance: a thermal imager (TI) and a 6-foot pike pole. A TI can reduce the time it takes to locate a heat source. Other forcible entry tools like an ax or a Halligan lack the reach to lift up ceiling tiles to check the ceiling space. You need a pike pole. If there is fire overhead, you don't want to be underneath it and you don't want to keep advancing further into the building with fire in the ceiling space. Eventually, the fire will bank down into the hallway behind you and you'll be trapped. This scenario has happened numerous times with tragic results. Don't repeat errors that have already resulted in fatal consequences.

17. Concrete structures represent a massive heat sink compared to other structures. This is beneficial because every British thermal unit (Btu) that heats the concrete is one less that is available to support the fire. However, heat—fed back from the concrete—can seriously affect fire suppression forces. This problem should be recognized and anticipated. Call for additional alarms so you can rotate and rehab crews. It was a serious problem when *The Castle* of the Smithsonian Museum burned in the 1865. Despite the upgraded "fireproof" construction, the fire caused extensive damage to the upper floors of the building and the roof collapsed. The heat-retention characteristics of concrete were also a problem at the 1986 Los Angeles Central Library fire.

18. The Los Angeles Central Library was constructed in 1926 and used the most modern technology available at that time, making it a truly a fire-resistive building. The exterior walls are 16-inch-thick (40.6 m) steel-reinforced concrete. The interior is supported by steel-reinforced concrete columns and cross beams. The roof and floors are 6-inch (15.2 cm) steel-reinforced concrete. The rotunda is 3-foot-thick (91 cm) steel-reinforced concrete. Ventilation was extremely difficult and at times impossible. Visibility was zero. The heavy fire load and the concrete shell made the building act as a heat sink, absorbing a tremendous amount of heat, and allowed interior temperatures to develop in excess of 2,000°F (1,093°C).

 It took 60 companies and 350 fire fighters (45% of the LAFD's [Los Angeles Fire Department's] on-duty strength) 7 hours and 38 minutes to knock down the fire. It was impossible—for hours—to enter the fire area after extinguishment due to the extreme

retention of heat. Walls had to be cooled by hose streams for hours after the fire was out and still, full firefighting ensemble and self-contained breathing apparatus (SCBAs) had to be worn to endure the residual heat. There were 50 fire fighter injuries; there were no deaths or critical injuries. Total property loss $22 million: $2 million for the structure, $20 million for the contents. Of the 1.2 million books, only 350,000 received fire or water damage.

19. In any fire-resistive building, the existing fire loads should be compared with the fire resistance rating of the structure. Where high fire loads exist, and where fire suppression efforts will be inadequate, severe structural damage is a distinct possibility. Furthermore, because of their great structural strength, concrete buildings can be heavily loaded with combustibles and, thus, the fire load may be excessive. You'll be forced to implement a defensive strategy.

20. Even if the interior compartmentation were to be perfect, the hazard of exterior extension still exists. Those who saw huge volumes of fire pouring out the windows of the 1988 First Interstate Bank Fire in Los Angeles, California, will have no difficulty understanding the potential of exterior extension in an unsprinklered building. The building is a 62-story steel-frame tower. The exterior perimeter is entirely covered with glass. The fire started on the 12th floor and extended to the 16th. It took 383 LAFD fire fighters, 10,000 feet (3,048 m) of fire hose, one-half million gallons of water, and 3 hours and 39 minutes to knock down the fire.

21. Impact loads to damaged concrete, such as from sledges or jackhammers, should be carefully evaluated. All areas where spoiling has reached the reinforcing steel should be shored before any work is done in those areas. Concrete floors that are seriously damaged on one side may give no clue to the distress on the other side.

22. Protruding rods are a serious sign of possible collapse. A column failure may have catastrophic results. Fortunately, column failures are rare. Any protruding structural members are signs that the building is deteriorating. We are fire fighters, not structural engineers. Recognize what the building is telling you and immediately back out your crews and operate from a safer position.

23. Remember, of all the five construction types, Type I fire-resistive buildings are the most stable to fight a

fire in. They have the best collapse resistance record from the perspective of fire fighter injuries and deaths. There are two basic types of fire-resistive construction: reinforced concrete buildings and structural steel buildings. Both are designed to resist a fire that burns out an entire floor without spreading flames to other floors or collapsing the structure. With limited resources, your best strategy may simply be to protect the exposure floors above the fire and let the fire consume the fuel on the fire floor; in other words, let the fire burn itself out.

24. With serious fires in fire-resistive buildings, the collapse danger lies in the concrete. In reinforced concrete buildings, heated concrete ceilings collapse on fire fighters working underneath. In steel skeleton buildings, heated concrete floors explode upward. Both of these structural failures are caused by spalling.

 Spalling is when small amounts of moisture, normally trapped inside the concrete, expand when heated by fire and increase the internal pressure within the concrete. This pressure can cause heavy sections of concrete to crack away from a ceiling and collapse down into the fire or on top of a fire fighter. This type of collapse occurs in a building without a suspended ceiling where the concrete ceiling is directly exposed to flames below.

25. In steel skeleton construction, the underside of each floor is not concrete—it's light-gauge corrugated steel sheets. These steel sheets support several inches of concrete on the floor above. When heat from a fire reaches the underside corrugated steel, it radiates to the concrete above. As the temperature of the concrete increases, so does the internal pressure from the moisture trapped inside the concrete. When spalling occurs, the steel sheets prevent the concrete from collapsing downward. The energy takes the path of least resistance and, therefore, it explodes upwards (an upward collapse). A sudden failure of pressurized concrete erupts and can send chunks of flying concrete 6–12 inches (15-30.5 cm) in the air. If this situation occurs, fire fighters working on the floor above the fire must quickly evacuate to the safety of the stairwell. It means interior crews on the fire floor are not making progress and the fire is getting worse or the fire is being left to burn itself out. Protect the exposure floor above from the safety of the landing. Use the reach of the fire stream to wet down the walls, the floor, and other surfaces.

Chapter Summary

- Concrete is a cementitious material produced by a chemical reaction of Portland cement and water to which inert materials called aggregates are added.
- There are two basic types of in-concrete construction: cast-in-place concrete and precast concrete.
- A chief virtue of concrete is its high compressive strength and low cost, which is why concrete is used for columns, even in structures otherwise built of steel such as highway bridges.
- Prestressing places engineered stresses in architectural and structural concrete to offset the stresses that occur in the concrete when it is placed under load.
- Concrete buildings under construction can present serious fire problems. Fire in formwork can easily result in major collapse with potential for massive life loss.
- When concrete construction was first developed, it was regarded by some as the answer to all fire protection problems—it was believed to be truly fireproof.
- After a series of disastrous fires, it became evident that concrete, like any other noncombustible material, can be destroyed by fire given sufficient fire load and fire duration.
- Clear signs of possible trouble can be identified in a prefire plan. Deteriorated concrete, spalling that exposes reinforcing rods, and cracks in concrete that can admit corrosive moisture to the reinforcing rods are indications that a building is in distress.
- One of the fire protection advantages of Type I construction is that it lacks inherent voids.
- The concrete in fire-resistive construction serves two purposes—it resists compressive stresses and protects the tensile strength of steel from fire. The latter function is accomplished in proportion to the thickness of the concrete cover.
- By sacrificing itself, the concrete provides time to extinguish the fire; however, it may also lose its strength.

Key Terms

Aggregates Any of a variety of materials, such as sand and gravel, added to a cement mixture to make concrete.

Cables Special high-strength, cold-drawn steel cables. Also referred to as strands or tendons.

Cast-in-place concrete Includes plain concrete, reinforced concrete, and post-tensioned concrete. This concrete is molded in the location in which it is expected to remain.

Casting A process of placing fluid concrete into molds, generally called forms, in which the concrete is permitted to harden to a certain shape.

Chairs Small devices designed to keep the rods up off the surface of the form so concrete will flow underneath.

Composite and combination columns Columns that use steel and concrete combined into one unit.

Composite construction Buildings in which different load-bearing materials are used in different areas of the building.

Continuous casting Process for casting or pouring concrete without interruption from start to finish.

Continuous slipforming Pouring concrete continuously as forms move upward so that continuous casting may be accomplished.

Drop panel Thicker section of floor on top of columns to assist in resisting the natural

tendency of the floor to shear off at the column.

Excess flow valve Valve that senses a sudden increased flow, as from a broken line, and shuts off flammable gas.

Falsework Temporary shoring, formwork, beams, or lateral bracing to support the concrete work in the process of construction.

Flat plate structural system (continuous beam) Cast-in-place floor in which there are no beams supported by columns; the floor plate itself rests directly on the columns.

Footing Typically thick concrete pads, blocks, or strips of concrete below the surface of the surrounding soil, sometimes heavily reinforced, that transfer the loads of walls, piers, or columns to the ground. Footings attempt to minimize building settlement by spreading the load over the large soil/concrete interface.

Formwork Mold that shapes the concrete.

Hollow tile Tile unit composed of vertical hollow cells, utilized in early efforts to lighten concrete floors.

Lally columns Steel pipes filled with concrete to increase their load-carrying capacity.

Left-in-place form Concrete floors that are cast onto corrugated steel.

Lift slab Type of building construction where concrete slabs are cast on the ground and lifted into place.

Monolithic construction Method in which all the concrete in a building is properly bonded together and acts as one.

Mudsills Planks on which formwork shores rest.

Mushroom cap Tapered extension at the tops of columns that assists in the transfer of loads from floor to column.

One-way structural system System that incorporates floors that have beams running in one direction only.

Plain concrete Concrete that has no reinforcement, except possibly light reinforcement to resist temperature changes.

Precast concrete Concrete that has been cast at a location other than the place where it is to remain.

Prestressing Engineered stresses placed in architectural and structural concrete to offset the stresses that occur in the concrete when it is placed under load.

Pretensioning and post-tensioning Processes by which steel tendons or rods are placed under tension, drawing the anchors together. Tensioned steel places the concrete in compression.

Reinforced concrete In concrete masonry structure, steel reinforcement that is embedded in such a manner that the two materials act together in resisting forces.

Reinforcing bars or rods Steel rods or bars used to reinforce concrete.

Reshoring Shores that are put back into concrete to help carry the load of the still-curing concrete.

Skewbacks Tiles shaped to fit around steel.

Slipforming Technique by which forms are moved upward as the concrete is poured.

Spalling Loss of surface material when concrete is subjected to heat.

Strands Special high-strength, cold-drawn steel cables. Also referred to as cables or tendons.

Stressing the tendons After concrete is poured and reaches its specified strength, steel tendons are stretched and anchored at the ends of the unit. Also referred to as "jacking the cables."

Temperature rods Thin rods installed near the surface of concrete, usually at right angles to the main reinforcing rods, to help the concrete resist cracking due to temperature changes.

Tendons Special high-strength, cold-drawn steel cables. Also referred to as strands or cables.

Two-way structural system System that incorporates floors that have beams running in two directions.

Underwriters blocks Concrete blocks poured per Underwriters Laboratories, Inc.'s classifications.

Waffle concrete Two-way structural system with closely spaced beams set at tight angles to one another in which unnecessary concrete is formed out; lower side resembles a waffle.

Case Study

You are the lieutenant of a downtown fire company and are on the site of a new Type I high-rise under construction. As you preplan the building, your fire fighters have several questions for you to answer. You consider all the possible answers.

1. By definition, falsework is composed of:

 A. Formwork
 B. Shores
 C. Lateral bracing
 D. All of the above

2. The word for loss of surface concrete over time is:

 A. Spalding
 B. Fissuring
 C. Spalling
 D. Smalling

3. Which ASTM standard addresses the standard fire resistance rating tests that qualify steel?

 A. ASTM E-119
 B. NFPA 252
 C. NFPA 351
 D. ASTM E-84

4. Which type of fireproofing does not protect individual structural members?

 A. Asbestos fiber
 B. Intumescent coating
 C. Cementitious
 D. Membrane

Challenging Questions

1. Identify the ingredients necessary to produce concrete.

2. Detail the differences between flat slab, one-way, and two-way floor systems.

3. List the problems associated with membrane fire protection for steel.

4. Describe the purpose of prestressing concrete and the two methods used to accomplish this goal.

Flames: © Dx/Dreamstime.com; Steel texture: © Sharpshot/Dreamstime.com; Chapter opener photo: © Pavel Vakhrushev/ShutterStock, Inc.

Firefighting Concerns of Green Construction

OBJECTIVES

At the conclusion of this chapter, you will be able to:

- Define green (sustainable) construction.
- Explain the role of green certification rating programs and green construction regulations.
- Identify and describe the four general categories of green construction.
- Identify firefighting hazards associated with each type of green construction material or green building system.

Figure 12-1 Note the solar panel array on the roof of the fire building.

Courtesy of Bill Tompkins.

Fire fighters expect the unexpected. Such was the case in 2010 when fire broke out in New York's Chinatown in the rear of a first-floor store in a six-story tenement of ordinary construction. Fire spread vertically rapidly through unprotected voids all the way to the roof. Fire fighters on the roof, however, found something that would not typically be found on a 100-year-old structure—an array of solar panels (**Figure 12-1**). The array extended over a large portion of the roof, covering the building's light shaft and exposing the array's frame to fire, endangering fire fighters working below. With fire spreading throughout the building, a defensive strategy was taken. The seven-alarm fire claimed the life of an elderly resident.

1. What is the impact of green technology in buildings?
2. How would you keep track of green "upgrades" to buildings in your community?

What Is Green Construction?

While the term *green construction* is relatively new, the use of environmentally friendly "technology" has been around for thousands of years. The sod-roofed houses of Ireland, the windmills of Holland, and the old grist mill water wheel are all examples of using nature to generate and conserve energy. It is, however, the more recent concerns over the environmental damages inflicted by fossil fuels that have driven the "green movement" for the last several years.

Green construction is also known as **sustainable construction**, essentially meaning construction that is deemed to be efficient and environmentally friendly in terms of both the materials used and their overall impact. This includes construction materials and methods that save energy or systems that generate energy. From a firefighting perspective, however, *do not assume that all materials used in green construction are "natural."* Numerous synthetic materials—particularly plastic foam insulation—are included in the "green" category because they are used to conserve energy. We have discussed some of these synthetic materials in previous chapters.

The Trend Toward Sustainable Construction

The demand for green construction has led to the creation of several related advocacy groups and certification programs. Most notable is the U.S. Green Building Council, which established a rating program, the Leadership in Energy & Environmental Design (LEED) certification, for actual buildings that meet certain minimum green criteria. The stated objectives of the LEED program are as follows:

- Lower operating costs and increase asset value
- Reduce waste sent to landfills
- Conserve energy and water
- Be healthier and safer for occupants
- Reduce harmful greenhouse gas emissions
- Qualify for tax rebates, zoning allowances, and other incentives in hundreds of cities

Another popular rating system is the Building Research Establishment Environmental Assessment Method (BREEAM), a worldwide program intended to assess the "greenness" of buildings of all types.

As these examples suggest, there are literally dozens of green rating systems, covering all types of buildings and occupancies. For the most part, these

rating systems do not deal with fire safety issues. Instead, it is assumed that fire safety is dealt with by local building and fire codes.

This lack of focus on fire safety led to concern within the code enforcement community. In response, the *International Green Construction Code* was issued in 2012 in an attempt to regulate green construction. This code is coordinated with the *International Building Code* and the *International Fire Code*, which were discussed in the chapter that reviews building and fire codes.

It's important to remember that with green construction, the traditional building hazards still exist. In fact, fireground hazards are sometimes *increased* by green construction. For example, a photovoltaic (PV) cell array on the roof of a wood-frame building may delay or eliminate vertical ventilation as a firefighting option. *Do not equate green construction with increased safety.*

For the purposes of studying green construction from a firefighting perspective, we can group these green systems and materials into four general categories: eco-friendly materials, energy conservation, water conservation, and energy generation.

Eco-Friendly Materials

The following "natural" materials are considered green, given their availability, relatively minor impact on the environment, and ability to be recycled:

- *Bamboo*—a quick-growing plant (compared with hardwood trees) that is now used not only to create furniture but also in structural laminated beams, curtain wall panels, and even trusses. In terms of fire resistance, there do not appear to be any American Society for Testing and Materials (ASTM) E-119 type tests conducted of structural bamboo materials. Expect early failure of unprotected bamboo beams and trusses.
- *Adobe*—a composite material made of clay, a fibrous substance (e.g., straw, manure), sand, and earth, baled into bricks.
- *Straw bale*—bales of straw covered with cement plaster or "earth" plaster. While examples of both cement- and earth-coated bales have been tested in accordance with ASTM

E-119 (for 2 and 1 hours, respectively), there is little information on the performance of these coatings over time, particularly when abused. A loss of coating will expose the straw, while still densely packed, to a fire.

- **Cob**—a composite material composed of clay, straw, sand, and earth; similar to adobe; typically laid in courses. Plaster is often added to improve its appearance.
- **Biopolymers**—in terms of construction materials, naturally produced "green plastic" made from renewable, living sources (e.g., starches, wood), instead of nonrenewable sources such as oil. Biopolymer plastics are being used to make insulation materials, plastic "building blocks" in which concrete and rebar are placed (similar to traditional insulated concrete forms [ICF] that use petroleum, except in this case, the foam is a natural biopolymer), "lumber," roofing shingles, and even tilt-up walls. Fire experience and fire testing of these materials are lacking, but we should monitor combustibility and flame spread experiences.
- *Ammonia*—an "old" refrigerant in use since the 1870s. When Freons and other "greenhouse gases" were banned in the 1980s, ammonia reemerged as a primary refrigerant replacement for large-scale refrigeration systems **Figure 12-2**.

Figure 12-2 A modern ammonia refrigeration system.
Courtesy of John Piho/Piho Engineering.

While considered eco-friendly, such systems pose not only a deadly toxicity hazard, but also an explosion hazard. Ammonia will burn in concentrations as low as 16%.

Energy Conservation

The energy crisis of the 1970s led to the creation of many synthetic—that is, plastic—insulating materials. While these materials have beneficial energy-saving characteristics (making them green), they have been criticized because of their health (smoke toxicity) and recycling problems.

More recently, there has been a move to incorporate vegetation on top of and on the walls of structures to save energy. While such vegetation can help save heating and cooling costs, the materials are still combustible if they dry out due to lack of maintenance **Figure 12-3**.

Other energy conservation measures that may be found in green construction include the following:

- *Polystyrene insulation*—a spray-applied foam or rigid plastic insulator. Polystyrene was spray-applied to the surface of older cold storage buildings, while rigid foam panels are used to insulate an enormous number of buildings each year.

Figure 12-3 A green roof structure with vegetation.
Courtesy of Raul A. Angulo.

- *Polyurethane insulation*—a spray-applied foam or rigid plastic insulator. It is very often used in the stud channels of walls.
- *Radiant barriers*—a flexible reflective membrane added to the surfaces of attic spaces to reflect heat, cooling the space and reducing cooling energy costs.
- *Glass*—its use as an insulating material makes it a green material. As discussed earlier in the text, multipane insulated windows also play a role in accelerating fire development, delaying fire-induced breakage in many cases. *Daylighting* advocates for the increased use of glazing to light spaces naturally, reducing energy costs.
- *Green roof*—an insulating roof cover composed of vegetation and materials used to support plant growth. A descendant of the sod roofs used for centuries, this modern version is used to insulate a building's roof. It comprises a complex set of layers below the grasses and shrubs, including soil, filtering fabrics to prevent the soil from being washed away, plastic or metal soil restraints to keep the soil from washing down drains, a drainage layer made of plastic panels or mats to allow water drainage, foam plastic insulating boards, and a waterproofing membrane to keep water out of the building. Collectively, these layers add up to a substantial amount of weight, potentially leading to a collapse. Be wary of installations in existing buildings.

With green roofs, the obvious problem of roof ventilation must be considered when preplanning such structures. In addition, the "mini-wildland fire" is not a far-fetched concept: a recent fire on the roof of a Manhattan high-rise created numerous problems for responding fire fighters, who found it difficult to access the roof under heavy fire conditions. The 2012 edition of the *International Fire Code* now incorporates provisions to limit the size of such gardens, provide nearby standpipe outlets, require more fire-resistant roof surfaces, and require ongoing maintenance of these gardens to avoid the above wildland scenario experienced in New York City.

Green Walls

Walls play a significant role in making buildings green. From a firefighting perspective, these walls can influence fire behavior. It is important to study these green walls (and any other types you may encounter) to anticipate their role in a fire. There is little actual fire experience with them because they are relatively new; therefore, we often need to rely on our intuition to deal with them.

Heavily insulated walls, such as **structural insulated panels (SIP),** are designed for energy efficiency; SIP are "stud-less" wall assemblies using a core of polyurethane. The more recent use of super insulated panels with higher amounts of polystyrene, polyisocyanurate, and polyurethane plastics add even further to the problem because of their combustibility. **Structural insulated sheathing (SIS)** is installed as an exterior sheathing in wood-frame dwellings, providing structural stabilization (similar to plywood and particleboard sheathing over wood studs), insulation, and a barrier to moisture. Its use of polyisocyanurate and polystyrene is an issue of combustibility. While some of these materials may have a low fire spread rating according to ASTM E-84 (discussed earlier in the chapter that reviewed the features of fire protection), they still provide fuel for a fire.

Fires involving sprayed polyurethane foam (SPF) insulation (the foam is applied with a spray gun in the channels between studs and beams) have occurred while the foam was being applied. It is believed that an exothermic chemical reaction occurred after the individual foam chemical components were mixed.

Green concrete includes ICF wall panel systems. Although not combustible, ICF wall panel systems offer the potential of trapping the heat of a fire inside the structure. However, some newer green walls are designed to trap and move heated air up and around a building. The **trombe wall**—essentially a set of glass exterior panels and concrete block wall with a void space in between (used in smaller buildings)—is used to trap sun-heated air and deliver the hot air into the adjacent living spaces through a series of one-way vents. A similar arrangement is known as a double-skin façade, used for ventilation in some modern high-rises. While this void space itself is surrounded by the noncombustible materials, it could provide a vertical path for a fire coming from within the building.

We are all familiar with ivy- and vegetation-covered buildings, often referred to humorously as Chia® buildings because of their similarity to the small clay Chia Pets® knickknacks that sprout grass on their exterior. Today, green advocates have brought the vine-like plants to the interior of the building, using them as a wall covering. Of course, plant vegetation is combustible, particularly when dry. Expect the plants to aid in fire spread.

To complicate this issue even further, some of these living walls are part of a larger biofilter, in which the plants become part of the air-handling system. Air is drawn through a wall of plants attached to a wall growing medium, improving air quality in the building. The plant combustibility problem now becomes even greater; burning plant material can spread fire and smoke through the HVAC system until it is shut down (by a duct smoke detector or manually). Although the filters used in a traditional HVAC system are fire tested, these plant "filters" are not.

Plastic Lumber

Recycled plastic waste have found use in outdoor applications such as decks, boardwalks, furniture, trash cans, and several other products. They exhibit the same combustibility characteristics as the plastics from which they are made. Refer to the chapter on methods and materials of construction, renovation, and demolition for the discussion on thermoplastics and thermosets.

Firefighting Accessibility and Green Construction

Although not a construction issue per se, the use of vegetation for energy conservation has led to the increased use of trees and shrubs to shade buildings

from the heat of the sun. In addition, green hardscape soils are used to absorb rainwater runoff. In both of these cases, the accessibility to a burning building may be reduced. Trees can block the use of ladders while large swaths of soil (used to absorb the runoff) can place the building further away from access roadways, increasing hose stretches and putting buildings out of the reach of aerial devices.

Be wary of the use of fire access roadway paving other than asphalt, including concrete pavers. It is best to visit and test these alternative roadway materials somewhere else before accepting them for use. Verify that the weight of your apparatus will be properly supported, all outriggers can be used properly, and that the roadway will not become a muddy mess during a fire when large quantities of firefighting water are used.

The trend toward community walkable streets, in which motorized vehicles are limited or eliminated altogether, will make firefighting more difficult. The only solution is to ensure firefighting access is provided prior to construction. Don't wait until the complex is built.

Water Conservation

Many green buildings use reclamation systems to capture and store rainwater. Of concern from a firefighting perspective are the associated tanks used to store the water. Be wary of the dead load presented by the tank for any structures it may be mounted upon.

Given the water shortages seen in different areas across the United States, a trend toward using nonpotable (nondrinkable) water for firefighting operations is on the rise. From a firefighting perspective, the use of **gray water** for firefighting is becoming more commonplace. Gray water is wastewater from industrial processes, laundries, and other sources. It should not be confused with sewage, which is termed *black water*.

Large industrial plants often use gray water for a variety of purposes, including for supplying private fire hydrants (colored differently than normal hydrants). An aluminum-rolling mill in San Antonio, Texas has such gray water–supplied hydrants. While such water can be considered for firefighting purposes, all apparatus must be thoroughly flushed with potable, fresh water. Gray water is not drinkable; it has not been treated to potable water standards.

The retention of water for conservation can pose additional problems for fire fighters. Tanks may be located on roofs or in buildings, which adds a substantial dead load. Water collection tanks may also be buried under roadways; ensure that fire apparatus access roads do not have water tanks below the driving surface.

Energy Generation

Generating electricity independent of fossil fuels has been a national priority for many years. Making the United States oil independent and protecting the environment are at the heart of this effort. Recent financial incentives, including tax breaks, are making many Americans think about installing their own energy-generating systems, including within existing buildings. A variety of technologies may be used for this purpose.

PV Roof Panels

PV roof panels incorporate individual semiconductor cells that react to energy from the sun and generate direct current (DC) electricity. Each individual PV roof panel—also known as a module—contains many semiconductor cells Figure 12-4. A group of modules is known as an array. The arrays typically face south (sometimes west) to get the maximum exposure to the sun.

PV systems are proliferating across the United States. These systems may be ground mounted or roof mounted. Some systems are actually incorporated into the roof and wall surfaces of a building. There are also freestanding PV structures that are used for covered parking Figure 12-5. Preplanning is critical to locate these systems before an incident occurs.

Obviously, there are a number of potential hazards and firefighting problems associated with these systems. These threats include electric shock, the

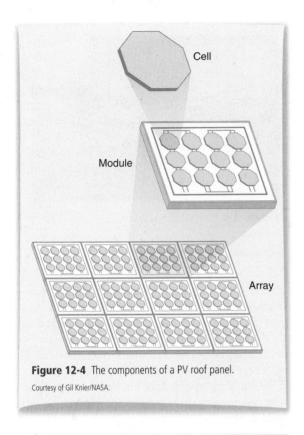

Figure 12-4 The components of a PV roof panel.

Courtesy of Gil Knier/NASA.

Figure 12-5 A ground-mounted PV array used as a parking space cover at the Bergen County, New Jersey, fire academy.

Courtesy of Glenn Corbett.

Figure 12-6 A PV inverter.

© Mark Boulton/Alamy Images

When fire fighters determine that a PV system is present in a building on fire, they should take specific precautions to mitigate the hazards associated with this type of energy-generating system. *Do not break or cut through the panels or associated equipment.* The system should be deenergized, including all circuits leading from the PV system, any switches to the building's primary electrical panel, any switchgear on the roof, and, on both sides of the **inverter**, the system device used to convert the DC into the alternating current (AC) used by the building Figure 12-6. Fire fighters should also verify that there are no other electrical supplies (e.g., generators) for the building. Of course, the batteries used to store the electricity will still be energized even if disconnected.

Recent fire incidents have exposed significant problems due to the lack of disconnect switches in these systems. In a 2010 big box retail store fire in Bakersfield, California, fire fighters were unable to deenergize the system due to lack of a specific disconnect switch. A 2012 Trenton, New Jersey, manufacturing plant caught fire when workers installing the system activated it prematurely, sending electricity to the rest of the incomplete system. In 2013, a fire starting at the roof level of a 300,000 square-foot food warehouse in Delanco, N.J.— covered with thousands PV panels and a polyurethane insulated EPDM roof—destroyed the building, in what may be one of the largest PV panel-related fires in U.S. history Figure 12-7.

dead load of the system on the roof as a fire burns below, the release of a variety of hazardous materials contained in the individual modules, and the batteries used to store the generated electricity.

Figure 12-7 This New Jersey food warehouse was totally destroyed in 2013 when a fire apparently began at the roof level near the center of the structure. Thousands of PV panels covered the roof, making it impossible to access the fire. It ignited the rubber roof surface and plastic insulation panels underneath, creating an unstoppable roof fire burning above the sprinkler heads throughout the 300,000 square foot structure. It also started other fires at floor level which destroyed the contents, racks, and steel frame structure.
Courtesy of Glenn Corbett.

Figure 12-8 A PV roof shingle system.
Courtesy of NIST.

Of particular concern is the conduit (piping) containing the conductors (wiring) coming from the PV panels. During daylight, these conductors will remain energized, creating a significant shock hazard, even if all of the system switches have been disconnected. It's important to locate these conduits and avoid them. They may be labeled, but don't bet on it. A standardized labeling system for all important components of a PV system is only just coming into use across the United States; there are many existing systems without any such labeling.

PV systems are sometimes incorporated into the building itself in the form of either PV shingles Figure 12-8 or PV wall panel systems, often referred to as **building-integrated photovoltaics (BIPV).** As with roof-mounted arrays, these systems present problems for fire fighters, particularly when attempting to ventilate a building.

The PV shingles can prove to be extremely slippery and will make ventilation difficult or impossible

because it is not safe to break the panels or associated equipment. In some cases, the shingles have been added without consideration of the weight they add to the roof. Some new buildings incorporate PV panels into the entire façade of a multistory building. These wall panels will make portions of the face of the building inaccessible for fire fighters. These BIPV systems also share the same problem discussed above in that these systems remain energized while the sun is shining on them.

Flexible laminate solar panels are finding their way on to buildings and other structures, such as boats and tents. These thin panels can simply be unrolled and stuck (self-sticking in many cases) to a flat surface and connected, generating electricity. Many of these panels are small in size; however, when used and installed by individuals with little technical knowledge, these panels can be dangerous to fire fighters. The same rules that apply to PV panels apply here, as well.

Off-grid systems are PV systems that are not connected to a traditional public electrical grid and are often located in rural areas. Although they share the same problems identified above for PV systems, they also pose the additional problem of large-scale battery storage, because they are not connected to the public grid. Because they often have numerous banks of batteries that are interconnected, they pose an electrical shock hazard. The batteries are sometimes

Figure 12-9 A roof-mounted wind turbine system.

Courtesy of Glenn Corbett.

Figure 12-10 An active solar water heating system.

Courtesy of Glenn Corbett.

stored in a separate structure from where the PV panels are located. Often, the rooms in which they are located are not designed for that purpose; significant quantities of hydrogen gas can be generated by the batteries.

Roof-Mounted Wind Turbines

A wind turbine is a device used to capture the kinetic energy of the wind, convert it into mechanical energy so as to drive a generator, and subsequently convert the energy into direct electric current. Small roof-mounted wind turbines have found their way onto the roofs of residential and commercial buildings across the United States Figure 12-9 . While these systems present similar hazards to fire fighters to those discussed for PV systems, they have the added danger of the spinning blades of the turbine.

Solar Water Heating Systems

Solar water heating systems use the sun's energy to heat a fluid (typically antifreeze in cold climates and water in warm climates) within a closed loop of tubes, which in turn heats a storage tank of water. Two general types of solar water heating systems are distinguished "active" and "passive."

Active systems use a roof-mounted collector (containing a network of closely spaced tubes; they look similar to a PV panel, but they are not the same) and a pump to move the sun-heated fluid to the water tank Figure 12-10 . A heat exchanger then heats the water in the storage tank.

Passive systems heat the water directly through the use of a batch collector, essentially a large well-insulated tank designed to heat the water within. These tanks are often roof mounted, creating substantial dead load and potential for collapse. Consider this risk when the fire is directly underneath the tank.

TACTICAL CONSIDERATIONS

1. Fire departments that perform preplans (transferred to computer files, accessible by a mobile data computer in the fire apparatus) using a notation of "GC electrical roof hazards" can help alert responding companies that there may be an array of PV roof panels that could create an unfamiliar electrical shock hazard. Since most of these panels face south or southwest to capture the maximum exposure to the sun, it is best to ladder the building from the north side of the building to access the roof. Laddering from the opposite side of where the arrays are aimed allows the fire fighters to look underneath the panels to see how they are attached to the roof. It may also reveal the location of disconnect switches.

2. When array PV disconnect switches are encountered, they should be flipped to the "off" position. However, similar to conventional electrical installations, flipping the main circuit breaker on an electrical panel doesn't guarantee the building is deenergized. There may be other panels or subpanels that are still energized. Remember, as long as the sun is shining, the PV panels themselves will be energized. The power utility company has more experience with green energy-generating systems than the fire department does. Dispatchers should contact the power utility crews as soon as the fire department resources are responding. Even after the power company claims the building is deenergized, keep away from power units and electrical wires as much as possible.

3. Many green constructed buildings incorporate living vegetation on the roofs to save on heating, cooling, and insulation costs. Some of these roofs look like manicured lawns, while others look like wild fields. One problem that occurs is when maintenance is neglected or when there hasn't been any precipitation (rain) for months. The shallow root system causes the grass to dry out easily, making ideal conditions for a dry grass fire on the roof. Holidays that are celebrated with fireworks, such as the Fourth of July, are particularly dangerous. A stray bottle rocket can be the source of ignition. Fire brands from a structure fire in close proximity can also land on the dry vegetation and start a fire.

4. Green roofs that use grass and natural vegetation are planted in a layer of soil over a metal roof. There are also other filtering fabrics to keep drains open and to prevent the runoff of the soil. These green systems create additional weight on the roof. They also create impractical, if not impossible, conditions to vertically ventilate the roof using powers saws. The good news is, these green roofs often have skylights or light shafts that can quickly be broken out for vertical ventilation. Break open the skylights or light shafts and get your crew off the roof.

5. Green construction is fairly new to the fire service; at this time, we do not have a lot of experience in dealing with the various roof structures related to energy sources, water conservation, and energy conservation. The roof is the wild card—there is nothing conventional about them. Unless we are very familiar with the green buildings in our district, what we find on the roof may be a surprise. The best way to remain effective and efficient in our firefighting efforts as it relates to the ventilation of green buildings is to rely on horizontal ventilation. There are few, if any, hazards related to breaking out windows. The fastest and safest way to ventilate a green building is to select the exit point for the direction we want to exhaust the smoke, then pressurize the building with PPV fans or an MVU. Unless we need to vertically ventilate a green construction roof, positive pressure horizontal ventilation allows us to bypass the roof altogether. This is the safest option for fire fighters when we are dealing with the "unknowns" of green roof construction and related eco-energy roof fixtures.

Chapter Summary

- Green construction is proliferating in response to environmental and energy concerns.
- The materials used in green construction are not necessarily "natural" materials.
- Do not equate green construction with safe construction.
- From a firefighting perspective, green construction can be categorized into four categories: eco-friendly materials, energy conservation, water conservation, and energy generation.
- Each green construction system and material presents different potential firefighting hazards; preplanning is key to locate, identify, and understand these dangers.

Key Terms

Biopolymers In terms of construction materials, a naturally produced plastic from renewable, living sources (e.g., starches, wood), instead of nonrenewable sources such as oil.

Building-integrated photovoltaics (BIPV) PV systems that are incorporated into the building itself in the form of either PV shingles or PV wall panel systems.

Cob A composite material composed of clay, straw, sand, and earth; similar to adobe; typically laid in courses.

Gray water Wastewater from industrial processes, laundries, and other sources.

Inverter A device used in a PV roof panel to convert DC into AC.

Off-grid systems PV systems that are not connected to a traditional public electrical grid; they are often located in rural areas.

Structural insulated panels (SIP) "Stud-less" wall assemblies using a core of polyurethane.

Structural insulated sheathing (SIS) Installed as an exterior sheathing in wood-frame dwellings, providing structural stabilization (similar to plywood and particleboard sheathing over wood studs), insulation, and a barrier to moisture.

Sustainable construction Construction that is deemed to be energy efficient and environmentally friendly; also known as green construction.

Trombe wall A set of glass exterior panels and concrete block wall with a void space in between (used in smaller buildings).

Case Study

As a fire lieutenant, you are taking your company out to preplan a new residential subdivision containing hundreds of new homes. Since each home contains PV roof panels, you see the opportunity to assess your company's knowledge of such systems.

1. A group of PV modules is known as a(n):

 A. Relay
 B. Inverter
 C. Array
 D. None of the above

2. True or False: PV semiconductor cells convert the sun's energy into AC.

3. A device used to convert DC into AC is known as a(n):

 A. Converter
 B. Inverter
 C. Current director
 D. Battery

4. True or False: Conductors within conduit from the PV array will be energized when the array is exposed to sunlight.

Challenging Questions

1. Identify the different green construction systems that might present a dead load hazard to fire fighters operating underneath them.

2. Develop a list of tactical objectives for dealing with a PV energy-generating system. Include safety concerns.

Specific Occupancy-Related Construction Hazards

OBJECTIVES

At the conclusion of this chapter, you will be able to:

- Determine the unique design and construction details found in buildings based on occupancy type.
- Establish how occupancy-specific building code requirements dictate particular safety features.
- Identify the unique details and hazards associated with specific occupancies.
- Understand how occupancy specifics affect firefighting operations.

Case Study

Figure 13-1 The storage area on the 12th floor of the Cook County Administration Building where the fire started.

Courtesy of NIST.

Fire broke out in the Type I fire-resistive high-rise Cook County Administration Building in Chicago in 2001 (**Figure 13-1**). Starting around 5:00 p.m. in a small storage area on the 12th floor, the fire spread throughout the office suite in the unsprinklered building. Due to the lack of partitions that extended above a suspended drop ceiling, smoke was able to spread horizontally and then vertically to upper floors via HVAC ducts and penetrations. Fire fighters quickly ascended to the fire floor, encountering building occupants. The occupants were advised to go up the stairwell and exit on to another floor. Unfortunately, the occupants could not leave the stairwell—magnetic door locks prevented them from exiting the stairwell and gaining reentry to another floor. Fire fighters, believing the stairwell to be clear, opened the door to the fire floor and began to attack the fire. Six occupants trapped in the stairwell succumbed to the effects of smoke inhalation. The fire itself consumed about 14% of the 12th floor and never spread to any other floor, in contrast to the smoke migration.

1. How many unsprinklered high-rises are in your district?
2. How common are magnetic door locks on stairwell doors in high-rise buildings?
3. Which standard operating procedures would apply to the use of a stairwell for fire attack, specifically with respect to the possibility of civilians in the stairwell above the fire floor?

Introduction

Although we have classified buildings into five general categories based on construction, the type of use—the **occupancy**—also plays a role in how a building is constructed. Specific hazards affect the design and construction of a building and are often regulated by building and fire codes.

The following discussion of particular occupancy types and building hazards provides you, the fire fighter, with useful details when dealing with structures of this type. Note that the details related to codes are city specific. Although the "model" building and fire codes have attempted to standardize requirements, local jurisdictions make local amendments (changes to customize the codes for the city). In addition, a variety of editions (codes are rewritten and issued every 3 years) have been adopted. This means that you need to familiarize yourself with the actual codes in effect in your locale.

Apartment Buildings
Garden Apartments

For 10 years, Frank Brannigan investigated low-rise, combustible multiple-dwelling fires for the Center for Fire Research of the National Bureau of Standards (NBS). The NBS is now known as the National Institute of Standards and Technology (NIST). The purpose was not to uncover the fire cause, but rather to determine why the fire extended beyond the area of origin in a protected combustible construction building. Much of the material in this section is based on these fire investigations as well as the study of buildings under construction in many distinct jurisdictions.

Combustible multiple dwellings include garden apartments as well as modern row houses and townhouses and similar structures **Figure 13-2**. For convenience, the term *garden apartments* will be used to describe these structures. Many motels and small office buildings share the same construction characteristics.

Figure 13-2 Difficulties in accessing garden apartment fires delay fire department response and hinder rescue efforts.

© Jones & Bartlett Learning. Photographed by Christine McKeen.

The term *condominium*, although commonly used for many structures, is a financial/legal term. It is not a usefully descriptive term for fire fighters. It may be important legally in determining ownership or where the building and fire code holds individually owned units to a higher standard than rental units. The fallacy of considering condominiums as a separate type of building is shown by "lollipop" condos. Additional floors are built above an existing nonresidental building. In some jurisdictions, by law the condominium must be connected to the ground. The elevator shaft is "commonly owned" and serves as the tie to the ground.

Characteristics of Garden Apartments

Exterior structural walls of garden apartments are made of various materials:

- Solid masonry
- Brick veneer over **platform** wood frame
- Partially solid masonry, partially brick veneer on wood
- Wood

The usual height limit for these structures is three stories, although four stories is now becoming commonplace. As a consequence, many fire departments may not be able to reach victims at the rear windows of top floor apartments. Ladders may not be long enough, response strength may be inadequate to raise long ground ladders, and terrain conditions may make it impossible to move an aerial apparatus into position.

Usually individual living units are confined to one floor; however, townhouses and row houses often are multifloor units. Some apartment houses have duplex or triplex units. In some cases, one structure may have both one floor and multifloor units.

Balconies are customary in many apartments, either extended or within the exterior bounds of the building. Balconies may be made of combustible or noncombustible construction. The balcony may also provide normal access to the apartment. If the balcony is cantilevered, an interior structural fire may destroy one end of the cantilever ("seesaw") and collapse the balcony. Combustible and even partially combustible balconies have been responsible for many vertical extensions of fire.

In many of these structures, gable roof attics extend over the entire structure, with areas separated to the maximum permitted by the local code. Attic fire barriers—draftstopping—are frequently not effective because they have been compromised.

Fire department operations on peaked roofs are dangerous! Although some buildings appear to have flat roofs, they must have a pitch to drain rainwater. This pitch creates a void between the tops of horizontal ceiling beams and the sloping roof. Fire can spread laterally through this space **Figure 13-3**.

Regardless of the exterior construction of garden apartments, the interior construction is almost totally of wood, using construction techniques similar to those used in single-family homes. When these dwelling units are stacked atop one another, this structure multiplies the fire extension potential through the voids inherent in combustible construction. When steel is used in these structures, it is often not protected by any listed material or approved technique, so it must be considered unprotected.

In garden apartments, plumbing fixtures are vertically aligned, one above the other, for economy

Figure 13-3 Working from the bucket provides a safe environment for fire fighters performing roof ventilation.

© Nathan DeMarse/ShutterStock, Inc.

and speed of construction. The necessary piping is run through vertical voids. Plumbers often weaken structural members during installation by cutting into them. The result is that a fire that starts in or penetrates this void could extend rapidly, both vertically and horizontally, as well as to the attic above. The weight of fixtures, structure cuts, and the air supply due to interconnected voids make early collapse a real hazard in these structures.

A person who must escape from a single-floor ranch home is infinitely better off than the person who must escape from the top floor of a combustible multiple dwelling. In too many cases, stairways and enclosures as well as attics overhead are combustible.

Stairways are not a place of refuge for occupants or a safe operating platform for fire fighters. For example, roof air conditioners mounted above a stairway collapsed on the first-arriving units at one apartment fire.

Educating the Management and Tenants

Most fires start in tenant spaces. All tenants in apartment complexes are at the mercy of an individual tenant who, for whatever reason, disregards basic fire prevention measures or who even starts a fire. In addition,

tenants are at risk from fires starting in the public or common use spaces, whether accidental or incendiary.

Tenants should be advised to help themselves in four basic ways:

- Be fully insured for the value of all personal property.
- Keep property of unique value in a bank vault.
- Call the fire department immediately if a fire or gas leak is suspected.
- In a fire, evacuate immediately, even if the fire seems inconsequential.

Parking

Parking often presents a problem at this type of structure. Space is generally limited when it comes to parking. The preincident plan review should address this problem, including calling for a minimum of 20 feet (6.1 m) of clear width, proper turning radii, red striping of curbs, and signage. If fire officers on alarm runs or inspections observe illegal parking that would interfere with firefighting, the police should be summoned immediately. The effort to eliminate this hazard should be vigorous and followed up on until occupants understand and comply.

Building Location

Building location can be confusing and may present problems not necessarily related to street address. Even though maps are provided by the management, a map drill makes a good indoor planning exercise. In this activity, the fire officer draws an outline of the area on the board. Special difficulties, such as gullies and fences that make it difficult to get from point A to point B, should be discussed and solutions worked out beforehand. Building owners should be encouraged to provide lettering and numbering on buildings to ensure quick identification of locations in an emergency.

Natural Gas Service

Natural gas service can provide special hazards in apartment structures. Where gas service is provided,

the system should be studied by the fire department. Often the service layout is made for the convenience of plumbers and meter readers, with little or no thought given to the safety of fire fighters. Usually all the gas meters in an apartment complex are located in one basement location Figure 13-4 . Individual lines are run, usually on the basement ceiling, from these meters to each apartment. Grouped together, these pipes represent a substantial weight and the hangers are often flimsy.

In one case, gas lines were located below the ceiling of the passageway between the chicken wire lockers provided to each tenant. A fire engulfed some of the lockers and one of the hangers gave way. Twenty gas pipes overhead fell down to the floor. The pipes failed, gas was released, and another fire ignited. Fortunately, the fire units were on the near side rather than the far side of the collapse. Had the units been on the other side of the collapse, a catastrophe would have occurred. In this case, the mass of pipes formed a barrier protecting the units.

Where there is no utility gas service, the apartment developer may install a large liquefied petroleum gas (LPG) storage tank. In Anne Arundel County, Maryland, an apartment complex was served from a single large gas tank. Its regulator failed, and many fires arose simultaneously throughout the complex.

Water Supply

The hydrants in many garden apartment developments are often on private mains. All hydrants should be checked periodically Figure 13-5 . Sometimes this practice raises objections from tenants because it may cause rusty water to flow. Often the fire department is forced to abandon testing.

Older apartment complexes often have undersized mains—in some cases, as small as 4 inches. The best way to establish the amount of water available is to have the owner conduct a flow test using two hydrants, a Pitot tube, and a cap gauge. With the test results graphed, you can calculate the total amount of water on hand and plan accordingly.

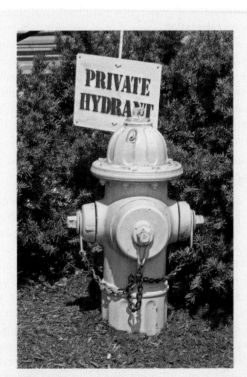

Figure 13-5 Private fire hydrants are sources of major concern, typically due to their limited maintenance.

© blindsquirrelphoto/iStockphoto.

Figure 13-4 Knowing where utilities are and which apartments they serve before an incident can save fire fighter lives—preplan, preplan, preplan.

© Jones & Bartlett Learning. Photographed by Kimberly Potvin.

Protected Combustible Construction

If you were to ask the designer or building management how a fire in one apartment would be confined and not extend to the neighboring apartments, the answer probably would be: by the fire-rated gypsum board sheathing or shell of the structure. Gypsum has excellent fire protection characteristics. It is the only construction material that does not yield heat when burned in pure oxygen. However, when binders and paper are added to make the familiar gypsum board, it yields some heat.

Effect of Fire on Gypsum Board

When gypsum board is heated by fire, it starts to deteriorate as it gives up its moisture. This process (called **calcination**), once started, appears to be irreversible. Sometimes insurance adjusters will order the burned gypsum board painted over. In a short time, paint will flake off the deteriorating walls. Removing all burned gypsum board in overhauling makes the most sense. If you do so, however, avoid charges of needless destruction by explaining the benefits of this practice.

Fire Rating of Gypsum Board

Wood-stud and floor beams combined with gypsum board can be used to construct floor–ceiling assemblies or walls that pass the American Society for Testing and Materials (ASTM) E-119 fire resistance test and earn a 1-hour or greater rating. Unfortunately, the use of fire-rated gypsum board in fire-resistive enclosures has created a myth about an almost magical ability of gypsum board to limit the extension of fire. The rating of the gypsum board cannot be separated from the test structure of which it was a part. By itself, the fact that gypsum board is rated means only that it is better than nonrated board. Gypsum board is a good material, but its applications do not always produce the results many have come to believe that it will, nor can it overcome a building's sins of omission or commission.

Underwriters Laboratories, Inc. (UL) warns that its rating is assigned to the entire tested and listed assembly, not to individual components. There are no rated ceilings, only rated floor–ceiling assemblies. The rating also assumes the structural integrity of the supporting structure.

Gypsum Board Installation

There are four principal deficiencies in the installation of gypsum board. Gypsum board commonly is nailed up over voids with a large or even infinite air supply behind the board. As heat penetrates the board and ignites the wood, there is ample air to continue combustion.

Nail heads often are not properly cemented over. Unfortunately, this heat barrier is often regarded as merely cosmetic. All nail heads in tested assemblies are covered with cement.

In some assemblies, joints are not properly taped. This also is often regarded as cosmetic and not required for unpainted surfaces. However, taping is part of the fire resistance system.

Deviations from the listing are permitted by building departments. Yet every item in a listing can be crucial. This becomes clear when we understand the procedure. If a 1-hour rating is desired and the sample lasts 1 hour and 15 minutes, the assembly is redesigned to just earn the 1-hour rating. This eliminates the added economic penalty of making the assembly "too good." Every item of a listing was put in by the manufacturer to reach a specific rating. There is nothing added for extra margins of safety.

Protective Sheathing

Though never stated explicitly in any code, the concept behind the use of gypsum board is that it provides a protective sheathing or membrane to protect the combustible structure from a fire in the contents. Some membranes can tolerate a small leak, but the gypsum sheath of a combustible structure must be as sound as the plastic bag that carries the goldfish home from the pet shop—a single pinhole can cause disaster. The penetrations of the gypsum sheath range from pinholes to sizable openings. Any penetration allows the fire to spread

to the structure, thus converting a contents fire to a structural fire.

The penetrations commonly discovered in structures vary. One commonly cited penetration in the sheathing is caused by failure to close the gypsum sheath around utilities such as electrical or gas service. There are listed devices or methods for passing utilities through fire-resistive barriers, but these are rarely used in residential construction.

Failure to install the gypsum sheath behind the bathtub presents another penetration problem. Heat is conducted through the steel. Often bathtubs in adjacent apartments are located back to back. This provides a common void and breaches the supposed compartmentation. In some cases, where plastic bathtubs are used, the code often requires that the sheath be complete, but numerous untaped joints and uncovered nails have been observed.

Open doorways are sheathed in gypsum. When a door is installed, the thin wood door casing is the only sheath. This can fail rapidly. Interior vents, such as from bathrooms and kitchens, not only penetrate the sheath, but also deliver fire to the vertical voids of the building.

Attics are ventilated to prevent condensation. The lower vents are located in the soffit below the projecting eaves. There is a constant natural air flow upward through the lower vent and out the upper vent, generally at the highest point of the gable ends. Fire reaching out the window will ride this air flow "escalator" into the attic, bypassing the protective sheathing or shell.

Where a fire-rated partition wall meets a fire-rated floor, the "compartmentation" satisfies the building department; however, the floor is easily penetrated downward by a fire in today's fuels. The fire may enter the floor void and come up on the other side of the partition. Further, soffits, such as the box dropped down to provide built-in kitchen cabinets, connect the wall voids to the floor voids.

The fire-rated gypsum board myth allows for huge interconnected voids. Often the biggest void is the attic. There is no such thing as a fire-rated top floor/attic ceiling. The constant flow of air makes a fire-rated structure impossible here.

The creation of dangerous, hidden voids is prevalent in the rehabilitation of older buildings, especially when the ceiling heights were greater than today's acceptable 8 feet—a characteristic that easily allows for renovations by dropping a ceiling. Carbon monoxide gas can accumulate in these voids. When the ceiling falls, the gas can ignite as either a backdraft or a detonation.

"Protected Combustible" Is Not "Fire Resistive"

From time to time, buildings incorporate floor–ceiling assemblies and interior walls that are fire rated according to ASTM E-119. Test methods are sometimes referred to as being "fire resistive." Such characterization should be emphatically rejected. The limited application of a complete gypsum sheath or shell, and the many defects cited here, should make it clear that the term *fire resistive* should never be applied to such a structure. In fact, the term *protected combustible* appears to be overly generous in describing this situation.

Fire Walls/Barriers and Draft Stops

Fire walls are often used to separate units in multifamily residential structures **Figure 13-6**. Ideally, a masonry fire wall should be able to stop the extension of the fire without assistance from the fire department. Unfortunately, there are many possible defects

Figure 13-6 Bringing the wall through the roof line limits fire advancement.

Courtesy of the estate of Francis L. Brannigan.

in such walls. These require prompt action by the incident commander (IC) to reinforce the fire wall and cover its weaknesses.

One of the primary defects involves not bringing a masonry fire wall through the roof with a masonry parapet. Builders often resist this requirement because it is costly, and unless the wall is flashed properly, there can be leaks. In North Carolina, a builder finished off a block fire wall with brick veneer that merely rested on the wood roof. It collapsed suddenly into top floor apartments.

It is almost impossible to fit masonry tight to the roof. There will be at least small gaps. In many fires, expanding gases can force fire through the smallest opening. Furthermore, plywood delaminates and can lift up and enable the fire to pass.

Permitting overhangs or mansards to project beyond the fire wall is another potential defect in fire walls. This provides a gap for fire to pass around the end of the wall. Apparently, some codes read "the fire wall must extend to the exterior wall," and, unfortunately, this is interpreted as a limitation.

Permitting the fire wall to end at the interior of a combustible exterior wall simply enables the fire to pass around it. Often there is no exterior evidence of the fire wall's location. The fire wall should be extended straight out or in a "tee" or "ell" shape to cut off the fire.

Utilities often are passed through the fire wall. In turn, openings around these pipes may permit fire to penetrate the fire wall. The best practice is to run utility mains parallel to the building with branches into each unit. Unprotected openings frequently are cut into fire walls, particularly in attics, for maintenance purposes; they compromise the fire walls.

Openings at the basement level provide access to storage and laundry areas. These are usually designed and built with proper self-closing doors. All too frequently, however, these doors are blocked open. Ironically, in some cases, the door closer has been removed and used as a doorstop.

Permitting the fire wall to act as a party wall, or structurally as part of both buildings, creates problems. Party walls often have beams or girders from both sides

in the same opening. These common openings provide a path for fire extension. Look for this in townhouses, particularly where units have adjacent doorways. Stairway headers also may be in the same opening.

Older row-frame buildings often had brick laid in the party wall stud voids as a fire wall. It doesn't work because the barrier is incomplete. The **brick nogging** (brick and mortar filling between studs) does not block the floor or attic voids **Figure 13-7**.

Fire barriers (theoretically fire rated) and draft stops (nonrated) are intended to limit the combustible void area in the attic to which the fire has access. In some cases, plywood or chipboard is used for the draft stop. By far, most construction makes use of gypsum board on wood studs.

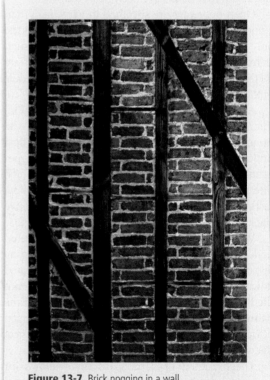

Figure 13-7 Brick nogging in a wall.
© Washington Imaging/Alamy Images.

Some barriers are now being made of 2-inch (5 cm) gypsum plank in 2-feet-wide (61 cm) vertical sections. The sections are held in a steel channel. They are cut in the field to meet the slope of the roof. However, the cuts are often only approximate, producing openings for fire access.

The fire barrier may range in effectiveness from temporarily reliable to totally useless. In no case, however, can a fire barrier be regarded as even approximately equal to a parapetted masonry fire wall.

There are several types of defects in these fire barriers and draft stops. Consider a fire barrier that extends completely from one side of the attic to the other, including the overhang. Nails are set, and joints are backed by two by fours and taped, with no penetrations at all. Yet a fire delaminates and raises up the plywood so that the fire passes over the barrier. A similar problem exists when the fire wall is ended at the underside of the roof.

Another defect exists when the barrier does not extend out to the eaves but stops at the wall line, thus allowing the fire to roar around the end. Gypsum board improperly backed by wood at joints and intermediate points presents a compromised fire barrier.

Omitted nail covering and joint taping, utilities or structural elements passing through, doorways cut into fire barriers, and the use of scraps of gypsum board all affect the integrity of fire barriers. Draft stops, although they are not rated and don't have to meet the fire-rated assembly requirements, nevertheless are required to be tight and complete without openings.

When fire barriers have been placed above the midpoint of a room, both sides of the barrier are exposed to fire coming out the windows. This comes about because of code provisions limiting the size of the attic void. The builder wants the maximum area in each section, and the fire barrier is so placed even if it doesn't continue a fire separation below. There are many other defects in fire barriers produced by a wide variety of builders, maintenance personnel, and tenants. It is wise to examine buildings in use and under construction for other fire barrier inadequacies.

A Word About Sprinklers

Automatic sprinklers that protect occupied spaces will most likely only extinguish contents fires. They will rarely control any fire that originates in, or extends to, the voids because there are no sprinklers in these void areas. However, as far as tenant life safety is concerned, a system complying with the National Fire Protection Association (NFPA) 13R, *Standard for the Installation of Sprinkler Systems in Low-Rise Residential Occupancies*, is a tremendous asset.

A note of concern about NFPA 13R systems and garden apartments complexes: they often share the same water main with the hydrants that are in the complex. If this is the case and you were to hook up to a hydrant, you would be taking water from the sprinkler system. Most NFPA 13R–protected apartment buildings are required to have a 1.5" (38 mm) fire department connection (FDC)—make sure that you use it and monitor water usage.

Serving the Citizens

There is probably no class of building in which good fire planning and training can pay off more than in garden apartments. Fire departments that study why multiple-dwelling fires spread and develop tactics to cut off these fires serve the citizenry well.

Individuals who own their own homes usually have personal property or homeowners' insurance, often purchased to satisfy the mortgage holder's requirements. Renters' insurance, however, is often too inexpensive to be actively sold. The result is that a high proportion of fire victims often suffer crushing financial blows due to a wide range of fire causes, including inadequate building codes, the builders' negligence, the owners' incompetence, other tenants, or inadequate fire suppression efforts.

The knowledge the fire department gains from the day an apartment development is planned should be disseminated and become the organized property of the fire department. A system should be in place to make the information available to the IC at the fire scene. This information and the ability to make effective plans provide increased safety for all concerned.

Older Row-Frame Buildings and Townhouses

In the older parts of many cities, frame buildings were often erected in rows. These structures are contiguous and often have a common attic or cockloft; they may even have party walls that provide support to both buildings. In older construction, there may be no fire barrier between buildings. Crude attempts at making a fire barrier by using brick nogging have been made in row houses. The brick nogging does not cut the floor voids or the cockloft. Brick or stone nogging is also found in old individual houses. It served as a heat sink for winter warmth—but it also presents an additional hazard in collapse.

More recently, row houses have been given a dignified new name, *townhouses*. No matter what the name used, unless there is an adequate masonry fire wall between the separate buildings all the way through the roof, the entire structure is all one building and should be so described in stating the size for planning or fire report purposes Figure 13-8 . It doesn't matter to the fire that these houses are separately described and owned on the land records. As long as fire can move from unit to unit, totally or partially unimpeded, the structure is one building.

Tenements

Tenements got their start in the 1800s. Over the years, this type of multistory multiple dwelling of ordinary construction has come to mean substandard housing. This type of building poses many problems for fire fighters, including vertical fire spread through voids such as bathroom pipe chases, horizontal fire spread through cocklofts, compromised open stairwells linking multiple floors, and fire spread via shafts Figure 13-9 .

During the 1800s and early 1900s, a series of devastating tenement fires led to fire safety "improvements." New York City—the home to thousands of tenement buildings—changed its regulations covering these structures. In turn, fire fighters in New York City began to describe tenements as "new law tenements" and "old law tenements" based on their age and type

Figure 13-8 A row of townhouses with a masonry fire wall between each home.
© Lissandra Melo/ShutterStock, Inc.

Figure 13-9 Tenement buildings pose many hazards to fire fighters.
© Daniel M. Silva/ShutterStock, Inc.

of compliance. (The terms *new law* and *old law* are unique to New York City.)

Some of these fire safety changes included better enclosures of stairwells and more substantial apartment doors. A notable change came in the form of exterior fire escapes, which served as second means of egress for occupants and a useful platform for firefighting. By 1968, the use of fire escapes on new buildings was no longer acceptable in New York City.

The Three-Decker

This three-decker type of apartment building is typically found in New England, particularly in Massachusetts and Rhode Island Figure 13-10. These three-story flat-roofed structures have three apartment units, one on each floor, which are accessed by interior stairs. They typically do not have fire escapes. They are characterized by porches on each level, on the front and rear of the building.

It is these porches that play a critical role in fire spread. The porches offer a large surface area on which the fire can burn, providing for a large volume of fire. Rather than just having to deal with fire extension from a window on one level to the levels above, fire fighters must be prepared for a large fire that can ignite other surrounding structures via radiant heat.

Atria

Although **atria** have been around for centuries, it was the Hyatt Regency in Atlanta that brought them into the modern era. The architect of the hotel, John Portman, designed the 22-story structure in 1967. Atria are built for aesthetic reasons, and aesthetic reasons only.

From a fire protection standpoint, of course, atria are large voids that pass through multiple floors, allowing smoke and heat to move vertically through the building Figure 13-11. A few years ago, building codes considered atria to be openings that connected a minimum of 3 floors and were a minimum

Figure 13-10 A New England three-decker.

© Jones & Bartlett Learning. Photographed by Kimberly Potvin.

Figure 13-11 Atria provide avenues for fire and smoke spread.

© Jones & Bartlett Learning. Photographed by Christine McKeen.

of 20 feet (6.1 m) on each side. Today, building codes only require that atria connect a minimum of 2 floors without a minimum size requirement.

Building codes specify numerous requirements for atria: full sprinkler protection throughout the building (with the exception of the top of the atrium itself if it is more than 55 feet (16.8 m) above the floor because it is felt that sprinklers will not activate so high above the floor), a smoke control system to minimize the movement of smoke onto the adjacent floors to the atrium and to move the smoke out of the building, and standby power for the building. In addition, the fire code requires that the floor of the atrium be limited to low fire hazards. Violations of this provision are seen often; moderate and high fire hazards such as upholstered furniture, cut Christmas trees, propane grills in food kiosks, and large plastic displays are encountered. This provision is difficult to enforce—and hence the desire for sprinklers at the top of tall atria.

Building codes also typically allow up to three floors to be open to the atrium—in essence, three levels with no physical barrier separation. Separations in the case of atria mean nonaesthetic materials like gypsum board and concrete block or, more commonly, wired glass or (most common) tempered glass with closely spaced (every 6 feet [1.8 m]) sprinklers to "wash" the glass during a fire. (An exception: another area allowed to be open to an atrium is the walkway found in many hotels surrounding an atrium.)

The (maximum) three levels open to the atrium are required to be included in the calculations of the smoke control system design. In essence, the designer must ensure that the volume of the open space is included in the sizing of the exhaust system, and the system must exhaust smoke from these areas. Often, the designers forget to provide a means for exhausting these areas; they forget to provide a means for moving the smoke off the floor and into the atrium itself. Exhaust fans at the roof of the atrium are incapable of drawing smoke off of the open floors, so that any smoke stagnates on the floor. It is critical that the smoke is exhausted off the floor itself or that make-up air is introduced from the remote areas of the floor to allow for movement from the floor into the atrium.

Activation of the smoke control system is usually triggered by water flow and smoke detectors. **Projected beam detectors**, which can cover large expanses of area with a single light beam, are often used for this purpose. They are often mounted on the side of the atrium, near the top.

The bottom line on atria smoke control: fire fighters should know what the system can and can't do. Establish the location of the system control panel and how it works. Ask lots of questions. Witness a test of the system if possible.

Sprinkler protection is usually fairly straightforward in buildings with an atrium. Often, the atrium and floors open to the atrium are zoned separately from the sprinklers in the rest of the building (with a separate riser for each).

Nationally, we have limited experience with actual fires in atria. In 1991, a fire occurred in the Polo Club high-rise in Denver. Although the building had an atrium, it had no sprinklers and no smoke control system (only skylights at the roof). An apartment fire filled the atrium with smoke, nearly to the bottom floor level. The mostly elderly occupants could not evacuate because of the smoke, and fire fighters had great difficulty extinguishing the fire. No one was killed despite the danger presented.

Three civilians were not so lucky in a 2007 Houston, Texas, atrium fire. An arson fire set throughout the fifth floor spread vertically to the sixth floor. Unable to access the stairwells, several surviving occupants were forced to break exterior windows and climb down aerial ladders. A fire captain almost lost his life. Unsprinklered atrium buildings, particularly office buildings, are incredibly dangerous and non-code compliant.

A recent fire with a more successful outcome occurred at the Grand Californian Hotel at the Disneyland resort in Anaheim, California, in 2005. Lights on an artificial Christmas tree ignited, burning off the top third of the tree. Sprinklers contained the fire as 2,000 hotel guests were evacuated during the 3 a.m. fire.

Houses of Worship

Houses of worship—churches, synagogues, and temples—are found in every community. They are also often some of the oldest buildings in a city. They span the five basic types of construction and can present a myriad of challenges for fire fighters.

In most cases, these buildings are open-area structures. They can have large occupant loads, often beyond code-permitted limits (especially during religious holidays). Fortunately, the occupants are typically familiar with their surroundings.

Holidays often bring candles and combustible decorations. In one case, a San Antonio, Texas, church erected a "living Christmas tree." In essence, the tree was a multitiered wood frame with evergreen branches stapled to it and lights strung around it. Platforms at each of the three levels allowed the choir to stand in the tree (hence the living tree). Upon a closer look, fire inspectors found a snake pit of wires under the tree. Unbelievably, a fan was installed to cool off the junction box and wires that were hot to the touch!

Older churches sometimes have multiple levels of seating. Galleries surround the main sanctuary, including the choir loft. These levels often have a single narrow stair, impeding egress. Another feature of older places of worship is stained glass windows. Although valuable for ventilation, such windows also are invaluable in terms of cost and heritage. Some are faced with Plexiglas™. However, the stained glass rose window (the circular window over the main entrance) does provide a location through which large-caliber streams can be trained on the roof area.

The collapse of steeples and bell towers on churches has cost many fire fighters their lives. In 2004, a stone bell tower collapsed during the overhaul stage of a Pittsburgh church fire, crashing through the vestibule and killing two fire fighters Figure 13-12.

Perhaps the most dangerous time for a place of worship, in terms of being vulnerable to a fire being initiated, is during renovations. Cutting and welding operations, burning off old paint, and other construction activities have ignited numerous churches and synagogues. One of the most spectacular fires

Figure 13-12 The stone bell tower collapse during the overhaul stage of the 2004 Pittsburgh church fire.
© John Heller/AP Images.

was the 1998 New York City Central Synagogue blaze that resulted from workers installing an air-conditioning system.

Covered Mall Buildings

A **covered mall** is essentially (under the building code) a single building enclosing a number of tenants, including retail stores, drinking and dining establishments, entertainment facilities, offices, and other similar uses where the tenants have an opening onto one or more malls. **Anchor stores**, large stores (often department stores) attached to the mall, have all of their required exits independent of the mall. Recent building codes have allowed covered mall buildings to be of unlimited area (meaning they could be measured in units of square miles, not square feet!) if they are of Type I, II, III, or IV construction and have 60 feet (18.3 m) of open space around them (for exposure protection).

The covered mall presents a problem similar to that posed by the atrium—vertical spread of heat and smoke in a fire. Malls have the added problem of horizontal spread of heat/smoke because none of the tenant spaces have a fire-rated separation from the mall. In addition, malls have large occupant loads.

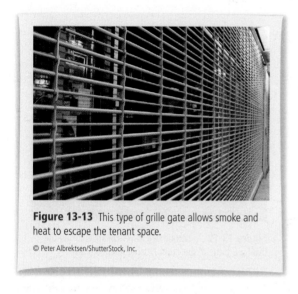

Figure 13-13 This type of grille gate allows smoke and heat to escape the tenant space.

© Peter Albrektsen/ShutterStock, Inc.

Figure 13-14 This 2.5" hose outlet is located at the entrance to a mall.

Courtesy of Glenn Corbett.

The physical separation between tenants is required to be fire rated, but typically need not go to the floor/roof deck above. Because no separation is required between the mall and the tenant space, a roll-down grille-type gate will allow smoke to move into the mall proper **Figure 13-13**.

Today, fire protection in malls includes complete sprinkler protection, a smoke control system, a standpipe system, an emergency voice communications system, and standby power. Leasing plans, showing tenant space locations, are also required for the fire department. *Note that older malls may have some or none of these systems—so preplanning of malls is crucial.*

The various fire protection systems in covered malls require closer analysis during your preplanning. For example, the sprinkler system will often be separately zoned for the mall proper and the tenant spaces (each will have its own risers). Most often, the feed main supplying the tenant spaces will run along the front of the store. Note the location of all control valves and what they supply.

The standpipe system, a Class I system, will have hose outlets in the mall at the entrance to each corridor and **exit passageway** (a more substantially fire-rated corridor with limitations on openings into it) at each floor level in stairwells and at exterior public entrances

to the mall **Figure 3-14**. Some of these systems are only filled with water; thus fire fighters will be required to pressurize the system through the FDC.

The smoke control system in a covered mall is also similar to that found in an atrium. It will attempt to minimize horizontal movement of the smoke and to exhaust the smoke through the roof over the mall. You must be familiar with the operation of this system and the panel used to control the system. The emergency voice communications system will allow you to make announcements, particularly regarding which exits to use during an emergency.

Some cities have required a standardized lettering and numbering system to identify tenant locations. With letters designating blocks/rows of stores and numbers indicating particular tenants, fire fighters can more easily and quickly find the location of a store in a large covered mall building.

Factories

Although the manufacturing base is shrinking in the United States, there are still a sizable number of factories across the country. They come in all shapes and sizes, from the earliest multistory mill from the 1800s to the modern one-story automated production facility. The

Figure 13-15 A factory fire. Which potential hazards are present in this building?

© Keith Muratori/ShutterStock, Inc.

key to safe firefighting operations in a factory is first to understand what is made in the structure and how it is made, and also how the structure has been designed/ altered to meet these manufacturing needs.

One common trait of most factories is the production equipment, which can pose a safety risk to fire fighters. These hazards include large moving parts, confined spaces, pressurized vessels, and heavy-duty energized electrical equipment. In a smoke-filled environment, fire fighters can easily get injured. The weight of the machinery in a building on fire could cause a collapse. Hazardous materials may be present. Preplanning, as always, is critical **Figure 13-15**.

How about the building itself? It may be difficult to navigate circular stairwells and dangerous **ship's ladders** to access equipment. In addition, there are open loading docks, limited access (including lack of windows), and adjacent storage/warehouse areas.

Hazardous Materials Production and Storage

Hazardous materials present obvious hazards to the fire fighter. "Hazmats" include the more familiar flammable and combustible liquids and more exotic substances like silane, a **pyrophoric gas** (gas that

ignites in air without the introduction of an ignition source). The buildings that house these materials have come under great scrutiny in the model fire codes, especially within the last 15 years. The regulations are extensive, and it is important to understand the rules that apply to hazmat structures in your response district.

Most fire codes require the creation of a **hazardous materials management plan (HMMP)** and a **hazardous materials inventory statement (HMIS)**. In essence, the HMMP explains how the materials are to be safely used and stored, and the HMIS lists the materials present at the site, including their hazards and quantities.

The fire code sets limits on the quantities of specific types of hazmat that may be stored or used in a building. Some "exempt" quantities are permitted without requiring extensive safety features for their use and storage—the actual exempt quantity is based on the type of hazmat. Groups of these low-quantity hazardous materials are permitted in **control areas**, which are essentially areas surrounded by 1-hour-rated fire assemblies. The fire code in turn allows for multiple control areas on a given floor of the building containing hazmats, depending upon the story on which it is located. The number of control areas per floor is reduced if the location is on higher floors or on lower levels below grade.

If the amount of a hazardous material present exceeds the exempt quantity, then numerous construction requirements apply. Depending upon the type of material and its form (solid/liquid/gas), special systems or building features may be required. Mandated systems can include gas detection equipment and alarms, ventilation systems, specialized suppression systems, and the like. Construction features can include drainage and containment, classified electrical equipment for hazardous locations, fire-rated assemblies, and explosion venting.

It is best to consult the HMMP, the HMIS, and your fire inspector to understand how hazardous materials are being stored and used in a facility. Have the inspector point out the special construction details.

When it comes to explosives, building and fire codes typically permit larger size structures to store these materials when built with substantial fire-resistive construction, almost guaranteeing the restraint that increases explosive destruction. Explosion venting required under today's codes for specific explosive hazards are provided in an attempt to minimize this damage. In the past, "industry" often took an opposite tack when dealing with hazardous materials. Buildings housing hazardous processes were isolated and built of **friable** (easily disintegrated) construction elements. Years ago, one such method employed a steel frame covered with an easy-to-replace material such as cement-asbestos board. If an explosion occurred, the structure vented, and the board became dust-like particles, not missiles. After such an explosion, however, the asbestos particles represent a health hazard and all necessary precautions should be taken. Today, given the requirements for fire-resistive construction, much of the equipment is likely to survive an explosion, though pictures of the scene may appear to show total devastation.

Special-purpose buildings may be designed to channel the force of an internal explosion in a desired direction (explosion venting). Transformer stations are sometimes built with heavy walls to protect one transformer from an explosion or fire in an adjacent transformer and to direct any blast upward or outward Figure 13-16.

Figure 13-16 A hazardous materials storage warehouse.
© Marcin Robert Balcerzak/iStockphoto.

High-Rises

There are many definitions of high-rise buildings. At the International Conference on Fire Safety in High-Rise Buildings convened in 1971 by the U.S. General Services Administration, the generally accepted definition of a high-rise was a building beyond the reach of aerial ladder equipment. Frank Brannigan protested that this definition was inadequate. The aerial ladder is a limited piece of equipment. In addition, the principal life safety question in any building is not its height but the time it takes occupants to reach a safe environment. Only a few feet from safety, 164 people died in the one-story 1977 Beverly Hills Supper Club fire.

When applied to fire department tactics, the preceding definition of high-rise buildings is acceptable and valid. Buildings of any height can present many of the problems associated with high-rises. For example, airport terminals and large shopping malls present many of the problems found in high-rise buildings. There are lessons to be learned from the study of high-rise buildings. Do not skip this section just because there are no buildings of more than a few stories in your area.

It is also a serious error to consider all high-rise buildings as a single problem. There are fire-significant construction differences among high-rises. The particular buildings in your area must be studied in detail to determine the particular potential modes of building failure.

With a few specialized exceptions, all high-rise buildings are designed to resist the effects of fire on the structural frame of the building and the floors. Whether the design concepts used are adequate to cope with all these possible effects is quite another matter.

General Classifications of High-Rise Buildings

Fire-resistive high-rise buildings have evolved over the past 150 years. The following classifications are broad and intended only for general guidance. The dates given are approximate.

Early Fire-Resistive Buildings, 1870–1930

There were many fire safety defects in buildings of this era. Some of the major defects of high-rises from this era of construction were the following:

- There were no standards for the protection of steel.
- Cast-iron columns were often unprotected.
- Terra-cotta fireproofing was compromised by concealed lightweight conduit, which expanded and tore the tile off columns.
- Segmental (curved) brick or tile arch floors were tied with exposed steel ties.
- Wooden floor beams were placed on piers creating a void under the floor, connected to all other voids by the hollow columns. In some cases, floors were leveled with cinders, often overloading the structure.

Structures in this era were built without any appreciation of the impact of vertical openings. Ornate open stairways and light wells were a standard feature, and in many cases, still remain. Open elevator shafts—a common feature in these buildings—have been replaced in many cases because new elevators must have shaft doors.

The Southern Building in Washington, D.C., is a typical early era fireproof high-rise building. It took about 70 years for an inherent defect to be made evident by a fire. A florist shop occupied the first floor, and an access stairway had been cut through the shop to the basement. A basement fire completely polluted the building with smoke via the unenclosed stairway. Hundreds of occupants had to escape down the outside fire escapes. It was feared that the fire escapes would fall off the building. Fire department ladders had to be used to brace the platforms.

Many other hazards are present in older high-rise constructions. In some cases, segmental brick and tile arches were supplanted by flat terra-cotta tile arches. In early construction, no protection was provided for the underside of the steel beams. Many such buildings still stand. Later, tile soffit blocks, or "skewbacks," were developed to provide protection

to beams. These are often removed by workers, leaving the steel unprotected.

Tile arch floors were delivered on site, already manufactured to specifications. Often the specifications for the floor units were wrong and the floor had to be laid in an improvised manner, so that the finished structure lacked adequate protection.

Older buildings were built with high ceilings to provide summer ventilation. Subsequent to original construction, the ceiling may have been lowered one or more times—thereby creating a combustible void. The tiles used may have a high flame spread.

In many cases, the most recent alterations to older buildings may have been done with relatively noncombustible construction; however, this may often conceal old combustible materials. In 1963, in Jacksonville, Florida, 22 people died in the Roosevelt Hotel. A new fire-safe ceiling had been installed in the ballroom—but the old combustible ceiling had been left in place during the renovation.

Although most codes require newly installed ceiling tiles to meet flame spread requirements, it is doubtful that any code requires the removal of the old tile, despite the demonstrated hazard. One of the most common alterations of older buildings is to install new ceilings, often with void spaces to hide new utility lines, heating, ventilating, and air conditioning (HVAC), and other building systems. All too often, the original, hidden ceiling is of combustible tile. Fire inspectors and fire officers should look above dropped ceilings.

Standpipes in older buildings may also be inadequate in size. It is possible that the standpipe has not been tested and could fail in a fire. It is also possible that the standpipe outlets do not match the fire department's hose. The standpipe is a fire department tool as much as the pumper that supplies it.

Almost all older buildings have been altered at some time—walls opened up and doorways sealed up. The closure during such renovations is often not as fire resistive as the original masonry walls. Stairways may be cut through and others cut off. High-bay rooms may be made into two stories. Sometimes a wooden

mezzanine is built in. In other cases, a steel bar joist floor is installed to make an intermediate story.

Live loads in older buildings may have increased over the years to a point where the building is distressed. Safes to protect computer tapes from fire, for instance, are much heavier than other safes. In some cases, unprotected steel supports may have been inserted, as might be the case in a high-rise alteration to accommodate a mechanized filing system that necessitated strengthening of the floor.

Fire loads also may have grown to high levels. This may be particularly true in the case of public buildings. Records and files accumulate. Many public buildings are domed, and the space under the dome makes a good void in which to store tons of paper. Basement rooms and attics are often similarly loaded with documents. Fire protection features are lacking.

Later High-Rise Buildings, 1930–1940

The high-rise buildings constructed after these early developments and before World War II are excellent buildings. High-rises built after World War II, in general, have poorer fire protection features when compared with many of these older high-rises. Pre–World War II buildings were universally of steel-framed construction. Floor construction and fireproofing of steel were often of concrete or tile— both good insulators that are slow to transmit heat to the floor above. The construction was heavy but no feasible alternative existed.

Relatively small floor areas were dictated by the need for natural light and air. Each floor was a well-segregated fire area in these buildings. Wall construction was frequently of wet masonry, joined to the floor so that there was an inherent firestop at the floor line. Masonry in the spandrel area (the space between the top of one window and the bottom of another) was adequate to restrict outside extension.

In these buildings, vertical shafts were enclosed in solid masonry with openings protected with proper closures. Fire department standpipes of adequate capacity were usually provided. These were wet and immediately pressurized by gravity from a tank in the building.

Exterior fire tower stairways with an atmospheric break between the building and stairway (the finest escape device available) were provided in many of the buildings of this era. Such a stairway can be compared to an enclosed tower located away from the building that is reached by a bridge open to the weather, so that smoke cannot pollute the tower **Figure 13-17**. In the Empire State Building, the fire tower was built inside the building with an adjacent smoke shaft. The inadequacy of this construction was demonstrated during a fire in 1990, when the interior smoke shaft became a chimney. The fire was confined to one office suite, but five alarms were required for fire suppression and evacuation.

In these high-rise buildings, the typical office was quite spartan, though executive suites and eating clubs often were paneled with huge quantities of wood. Nevertheless, most fire loads were low. Windows could be opened in buildings of this era, which provided local

Figure 13-17 A high-rise fire tower, as seen with the arched openings in the exterior wall of the high-rise in the center of the photograph.

Courtesy of Glenn Corbett.

ventilation and relief from smoke migrating from the fire. The windows leaked, often like sieves; in turn, there was no substantial **stack effect**.

Modern High-Rise Buildings

After World War II, a number of significant developments occurred. Fire departments, apparently bemused by the lack of problems in the previous generation of structures, generally failed to realize the dimensions of the new problem. City administrations, eager to get the spectacular new buildings onto the tax rolls, didn't question the hazard of cramming thousands of people into totally enclosed megastructures that created a potential for catastrophe.

The development of fluorescent lights and air conditioning helped to remove limits to the floor area. Thus, building populations could be enormously increased. As a result, many floors have substantial areas beyond the reach of hand hose streams.

Significant Construction Characteristics

There has been a definite push to lighten and thus reduce the costs of buildings. The Empire State Building, for example, weighs about 23 pounds (10.4 kg) per cubic foot. A typical modern high-rise, by comparison, weighs only 8 pounds (3.6 kg) per cubic foot.

Because of the development of better reinforcing steel and new techniques, reinforced concrete became a serious competitor to steel as a construction material in the post–World War II era. No longer could the building industry be indifferent to the weight penalty caused by concrete fireproofing of steel. In addition, the necessity for fireproofing is an apparent cost disadvantage relative to steel. Fireproofing of a steel building is a separate item in the cost estimates, whereas making concrete fire resistive is included in the cost of concrete. In addition, to serve modern high-rises, the spray-on materials and the floor–ceiling assemblies were developed to provide lighter construction.

One of the results of reducing the amount of concrete in high-rise buildings is the loss of a valuable heat sink. Every British thermal unit (Btu) that is absorbed into the concrete is one less Btu available to keep the fire from extending. This retained heat can be very distressing to fire fighters, and personnel may require frequent rotation when fighting fires in such structures. On the positive side, though, the fire's growth is slowed by this loss of heat.

There has been an extensive increase in the requirements for electrical service and communications systems in modern high-rises. Insulations are combustible, and the products of decomposition and combustion are toxic. Communications systems can breach fire and **smoke barriers** from floor to floor and from vertical shafts to horizontal voids.

Steel-truss floor–ceiling assemblies provide useful voids to carry utilities and communications or as a plenum for conditioned air, and unintentionally for fire and smoke. A most serious deficiency is the omission of fireproofing on columns within a fire-rated floor–ceiling assembly.

Slab floor concrete buildings do not have an inherent void. By dropping ceilings, useful voids are created and then connected via utility shafts. In a modern office building, possibly 25% or more of the floor volume is located in the ceiling void. Raceways for wiring are inserted in some concrete floors. Unless properly fireproofed below, heat can be transmitted to the floor above when this type of construction approach is used.

Gypsum rather than masonry is often used to enclose elevator and other shafts. In the One New York Plaza fire, gypsum enclosures were displaced, leaving the shafts unprotected and therefore hazardous to fire fighters.

In earlier buildings, there was little concern about using the exterior of the building for fire towers. Modern construction, however, discovered the value of exterior offices, particularly corner offices. With this arrangement, utilities are relegated to an interior core structure. Stairways are often located in the core, thereby eliminating the principle of remote exits. Stairways are frequently "scissor"

stairways—two stairways in the same shaft. The stairways are thus quite close together. Often there is no fireproofing on the underside of the stairways, so heat or fire in one stairway can make the other untenable. If there is a standpipe, the outlets will be in alternate stairways, floor by floor. Elevators are grouped together in the core; thus, it is possible that all will be affected by the same fire.

Modern design often emphasizes an uncluttered exterior. Prefabricated panels, glass, aluminum, and other curtain wall developments make it difficult to provide a real barrier to the vertical extension of fire between the edge of the floor and the skin of the building. The enclosure of balconies by tenants or condominium owners may increase the potential for exterior vertical extension.

Windows and Glass Exteriors

Many high-rises have glass exteriors. As a consequence, it is often impossible to determine from the exterior which floors are involved in a fire. Sudden massive glass failure may cause a serious change in conditions on the fire floor. It also presents other hazards. A fire fighter was severely injured by falling glass in the One Meridian Plaza fire in Philadelphia. It appears reasonable that the building code should require an entry to the building and water supply connections that are safe from falling glass and furniture and not leave this vital requirement to chance.

Air conditioning requires fixed windows to control loss of treated air. Many buildings have no operable windows. Breaking glass is extremely hazardous because glass pieces can travel for several blocks. If windows are covered with a sun-screening plastic, the glass tends to hold together when broken and can be pulled into the building. Newer high-rises provide some windows that can be opened, and keys are located in strategic but inconspicuous locations. In other buildings, a smoke management system or tempered windows that can be broken are marked with a designated fire department reflective sticker (discussed in

more detail under "Smoke Management Systems" later in this chapter).

Energy-efficient windows consist of double or sometimes triple panes of glass. They resist failure in a fire and eliminate the ventilation provided by the heat-caused failure of single-pane sashes. They are more difficult to vent from the exterior and can cause severe heat conditions to be retained in the fire area. They do resist exposure from fire below.

The Fire Command Center

Since the major overhaul of high-rise provisions in our building and fire codes in the 1970s, fire command centers have been required in high-rises. They act as the nerve center where all fire safety features are monitored and controlled.

The command center is a designated enclosed room, typically accessible from the lobby. A door is often provided directly to the exterior of the building.

A variety of equipment is provided in the fire command center: fire alarms, emergency communications to alert and direct occupants, elevator status monitors, emergency generator status and transfer switches, and fire pump status monitors. Equipment is provided to manually release magnetic door locks in stairwells, to recall elevators, and to manually control smoke management and HVAC systems.

The fire department communications system is located here as well; in the past, these systems consisted of telephone handsets that were carried up into the building and plugged into jacks in elevator cabs, stairwells, and elevator lobbies to communicate with the command center and its fire fighter communications panel. These older systems are not user friendly, however, and made it impossible for the lobby commander to immediately notify fire fighters to evacuate the building (unless a fire fighter is continually assigned to stay on the plugged-in phone with the command center—an unlikely

scenario given today's personnel availability situation). As a result of the World Trade Center disaster in September 2001, the fire department is now given the authority through the model codes to require any type of communications system (typically radio signal amplifiers).

General Problems and Hazards with High-Rises

Many of the hazards presented by high-rise buildings are not solely associated with buildings of one particular era. Due to their size and form, these buildings share some common problems that fire fighters must address.

Exits

All exits should provide a clear path to the outside. In one high-rise building, the exits opened to a rear yard. The yard was then planted and enclosed with an unbroken 10-foot-high (3 m) masonry wall.

For years, the model building codes have permitted 50% of exit stairwells to terminate in the building's lobby, rather than outdoors. Although other fire protection requirements have to be met, this arrangement is confusing to occupants. During the 1993 bombing of the World Trade Center in New York, for example, the lobby filled with smoke, making the exterior door impossible to find for trapped occupants on a mezzanine onto which the stairwells discharged.

Occupancy

The occupancy of a particular building changes its problem. Offices, hotels, apartments, homes for the elderly, factories, and showrooms all pose different scenarios. Some buildings have mixed occupancies. One rigid code interpretation provided smoke detectors in occupied apartments but none in an architect's office that encompassed two apartments and had the high fire load of exposed paper contents.

Accommodation or Access Stairs

Many tenants occupy more than one floor in a building. They often find it convenient to have an accommodation or access stairway installed to connect their various office floors. Such a stairway is usually installed as an alteration and is rarely enclosed. The result is that two or more floors become one fire area, completely negating the concept of floor integrity. In the 1991 One Meridian Plaza fire in Philadelphia, fire was seen falling down through such an accommodation stairway from the 22nd to the 21st floor.

Duplex and triplex apartments (two- or three-story houses) can be found in high-rise apartments. Often, there are no exits from the upper levels. If there is an elegant restaurant located at the top of a structure, it may be a multilevel venue with an open stairway connecting the levels. All of these encroachments on good containment increase the fire area and can multiply the fire problem. In some codes, and among some designers, there is little concern about open stairways, provided the required exits are properly enclosed.

Forcible Entry

Security against crime is a serious consideration in all occupancies. Heavy doors with multiple locks may delay entry. In most cases, the forcible entry may be achieved by gypsum wallboard on studs, or gypsum block or terra-cotta tile, all of which are easily penetrated. Conversely, if the building is of reinforced masonry construction, the corridor walls are likely to be of reinforced masonry, which would be difficult to breach.

Some stairways are locked against reentry on all floors. Some codes require doors to be unlocked for reentry on certain floors, typically no more than four intervening floors between reentry floors. Other codes require automatic release of locks upon a fire alarm system activation, manual release from the lobby command center, and telephones in the stairwells every five floors. This was not the case at the Cook County Administration Building in Chicago in 2003, where six people were trapped and killed above the fire floor in the stairwell.

Elevators

The development of safe elevators that would automatically stop if the cables failed helped to make high-rise buildings practical. A complete discussion of the problems that elevators present is beyond the scope of this text. Extrication of persons trapped in stalled elevators, for example, requires a detailed knowledge of elevator construction and operating systems. Nevertheless, several facets of the "elevator problem" should be touched upon here.

In the past, the use of elevators by occupants for fire evacuation was considered dangerous and was universally warned against by signs placed in or near elevators. However, since the World Trade Center disaster in 2001, this issue is being reconsidered and a new generation of "hardened and robust" elevators and shafts are being developed for evacuation purposes. In addition, legislation making discrimination against the handicapped in employment subject to severe penalties has created a serious evacuation problem for handicapped persons who cannot use stairways.

Some Additional Points

All experience points to the possibility that, at some fires, all elevators may be totally unavailable to the fire department. This happened in the One Meridian Plaza fire. All power for lights and elevators failed. In total darkness, all equipment had to be hauled up 20 or more floors. Just imagine the problem if the loss of power occurred in a fire on the 60th floor. Preincident plans should embrace this contingency.

The fire department should be thoroughly familiar with the elevators in all high-rise buildings. They were probably installed under different codes and may have markedly different operating systems. It might be useful to appoint an officer to serve as elevator information coordinator in such a scenario.

Very tall high-rise buildings have blind shafts in which elevators serving upper floors pass many floors without door openings. In some occupancies, commercial and residential elevators open directly onto the floor, without any vestibule.

Shaft doors are often left for last-minute installation due to possible damage during construction. Inadequate makeshift barriers may be used to close these openings. The opening of one high-rise Florida hotel was rushed to accommodate a big banquet. The decorator wallpapered over the open shaft doors!

Elevator door restrictors, usually taking the form of metal angle irons, have been installed on many elevator car doors to prevent them from being opened from the inside. This approach is intended to prevent people who are in the car from falling down the shaft if the car is outside the __landing zone__ (18 inches [46 cm] above or below the floor landing). It is difficult to open these doors; many people likely died at the World Trade Center on September 11, 2001, in the North Tower because they could not get out, even though they were at the lobby level.

Smoke Movement in High-Rise Buildings

A number of mechanisms affect the movement of smoke in a high-rise building.

Thermal Energy

The principal smoke-moving mechanism is the thermal energy of fire. This thermal energy can be massive. The burning rate of the fuel in the MGM Grand Hotel fire in Las Vegas, for example, was estimated at 3 tons per minute.

Atmospheric Conditions

When the atmospheric temperature is constantly decreasing as height increases, the condition is called __lapse__. Under lapse conditions, smoke will move up and away from the fire. If there is a layer of air warmer than the air below it, the condition is called __pause__, and the layer is called the __inversion layer__. It acts as a roof to rising smoke. A high-rise building may be tall enough to penetrate the inversion layer. This would cause substantial differences in the smoke conditions above and below the inversion layer.

Wind

Wind is very powerful and influences smoke movement. Note the tremendous bracing required in a high-rise to overcome the force of the wind. The wind exerts a pressure on the windward side of the building and suction on the leeward. If the windows are out and the fire is on the leeward side of the building, the suppression may be "a piece of cake." Given the same fire, the windows out, and the fire on the windward side of the building, it may be impossible to move into the fire floor.

Closely spaced high-rise buildings can create a canyon effect, where the wind increases velocity as it squeezes through a narrow opening. The wind can shift direction many times during the fire. Ground-level observations are not valid. At upper levels, the wind can blow harder and from a different direction.

Wind blowing against a building seems to split about two-thirds of the way up the building. The upper portion flows up and over the roof. The lower portion flows downward, forming a vortex next to the building. It increases in velocity as it flows downward. The effect of this wind on fire pouring out windows appears to be unevaluated in the design of high-rise structures. Refer to the chapter on fire behavior for information on wind-driven fires.

The effect of wind on the structure is becoming better understood. As previously mentioned, modern high-rise buildings weigh about 8 or 9 pounds (3.6 or 4.1 kg) per cubic foot. The result is that the top stories of some of these buildings sway noticeably in the wind. To counter this, some buildings have **tuned-mass dampers**—heavy weights installed high up in the building that are adjusted by computer to counter the wind-induced oscillations.

Stack Effect

Stack effect is the term used to describe the movement of air inside a tightly sealed building due to the difference in temperature between the air inside the building and outside the building Figure 13-18. Stack effect is most significant in cold climates in the wintertime because of the great difference between the inside and outside temperatures. Stack effect

Figure 13-18 Note how the top of a high-rise may be above the inversion layer.

also occurs, with opposite airflow direction, when the outside temperature is greater than the inside temperature. This condition is, however, much less significant because the amount of stack effect is proportional to the difference between the two temperatures. There is a much greater difference between the inside temperature and outside temperature during the winter than during the summer. The winter differential lasts for longer periods of time. In a cold climate, a differential of about 70°F (21°C) might exist for weeks on end, whereas in the heat of summer the difference might be only about 15°F (9.4°C), mostly in the afternoon on days in which the building is occupied. Stack effect is not caused by a fire. The products of combustion ride the stack effect currents.

Winter Stack Effect

Under winter conditions, stack effect causes a movement of air from the floors into the vertical shafts, stairways, elevators, and so on in the lower portion of the building Figure 13-19. The greatest flow will be at the first floor, with the flow gradually decreasing as the height of the

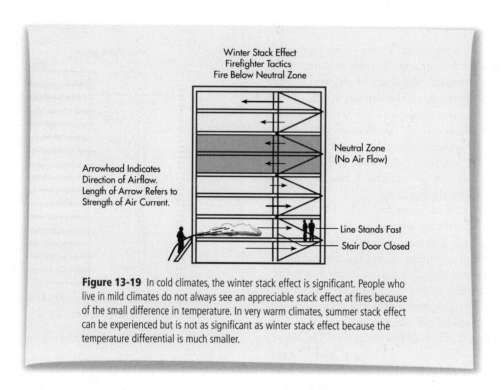

Figure 13-19 In cold climates, the winter stack effect is significant. People who live in mild climates do not always see an appreciable stack effect at fires because of the small difference in temperature. In very warm climates, summer stack effect can be experienced but is not as significant as winter stack effect because the temperature differential is much smaller.

floor above ground increases. At a point about one-third to one-half the building's height, the flow is reduced to zero; this is called the neutral zone. Above the neutral zone, the flow reverses and travels out to the floors from the shafts. The pressure increases floor by floor and is greatest at the top of the building.

If floors were completely cut off from one another, the stack effect would exist separately on each floor. Air would flow in at openings at the bottom of each floor and out the top. There would be no overall stack effect because there would be no connection between the floors. The buildings in which fires are fought do have interconnected vertical openings, stairways, elevator shafts, and pipe openings, however, so the stack effect can be significant throughout the building.

Smoke that has lost much of its thermal energy can be delivered to upper floors by the stack effect. A fire that vividly demonstrated winter stack effect occurred in a federal office building under construction in Cleveland, Ohio **Figure 13-20**. The fire was ignited in a pile of rubbish on the ninth floor. Fumes

from degraded and burning plastics traveled to the top floor by the stack effect. The workers started down the stairways, but were so affected by the fumes that they went into the 25th floor for refuge and started to break windows despite the hazard to those below.

Summer Stack Effect

In a sealed air-conditioned building in the summertime, the stack effect is reversed (and is often called a "reverse stack effect"). In such a case, the flow of air is downward. The situation is not as extreme as the winter situation because the differential is usually much smaller, but it still can be serious, particularly in very warm climates. In a Puerto Rican hospital in the summer, smoke pollution occurred several floors below the fire floor due to the summer stack effect.

Fire fighters experienced in fighting refrigerator plant fires have long been acquainted with **cold smoke**, which falls downward. In 1946, naval base fire fighters, under Frank Brannigan's command,

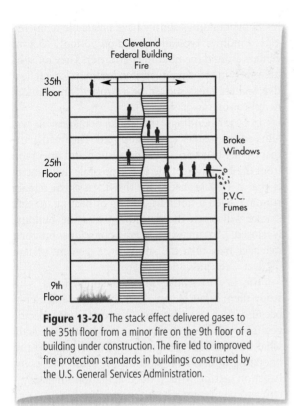

Figure 13-20 The stack effect delivered gases to the 35th floor from a minor fire on the 9th floor of a building under construction. The fire led to improved fire protection standards in buildings constructed by the U.S. General Services Administration.

assisted the Norfolk, Virginia, fire department in a refrigerator plant fire. The fire was in fungus-treated construction lumber in a cold box. The boxes surrounding the box on fire were at 0°F (-17.7°C) temperature; the box where the fire was located was at 30°F (1°C). By rigging salvage covers and setting up fans as preplanned for a similar building on the naval base, the smoke was pumped into the elevator shaft, where it fell down the shaft to the street. At that time, the term *stack effect* was unknown to fire fighters, but this was a perfect example of extreme reverse stack effect conditions.

Air Conditioning

Whether air conditioning is a significant factor in smoke movement depends on the type of system.

Individual Room Units

Individual room units that recirculate room air or draw air only from the outside, as in many recently built motels, will have no effect on the general situation. In contrast, if the individual units draw air from the corridor, then polluted air from the corridor will be drawn into the rooms.

Single-Floor Systems

Single-floor systems are seen in many smaller office buildings and in retrofitted buildings built before air conditioning became commonplace. In a typical system, the air is treated on each floor in a fan unit. Air flows from the corridor to the fan unit and is distributed through ducts that spill into the ceiling void. The air comes into the offices through openings in the ceiling tiles, then returns to the fan unit via the corridors. As a result of this arrangement, smoke from any fire on the floor will be distributed throughout the entire floor.

There may also be vents from the ducts feeding the floor above. If the building code has permitted the dangerous situation of assuming that the floor–ceiling assembly will protect the columns in the plenum space, then smoke can move along the reentrant space of the column to the voids on other floors.

Telephone service is often provided by cabinets located one above the other, with holes in the floors connecting the cabinets. Openings from the cabinet to the ceiling void provide another path for smoke distribution.

In short, the fact that the building has single-floor separate systems will not prevent smoke migration throughout the building.

Multifloor Systems

When the entire building is air conditioned by one or more building systems, the problems become severely complicated. When total building systems are used, the treated air is regarded as a precious commodity. Only certain exhausts are passed to the outside; all other air is treated and returned to the occupants. This means that any smoke from a fire would be recycled

through the building. As technology progressed and energy costs rose, people realized that there are many days when the outside air is quite satisfactory and could be pumped into the building. More modern systems often have a true full-exhaust capability; they also recirculate air to save on energy costs.

Smoke Management Systems

Complete knowledge of smoke management systems is particularly important for fire fighters. Will the fire department operate this system? Will the building engineer operate it? In one high-rise, the operating station for the engineer who supposedly will operate the system in an emergency is within the main plenum of the system. Operator instructions, however, specify that the engineer should leave immediately in the event of a fire. The control room for the air-conditioning smoke management system must be invulnerable to the effects of the fire with which the system must cope.

There are those who have insisted that high-rise fires can be controlled simply by manipulating the air supply. If such a system is proposed as the primary defense, proponents should be required to provide proof of the system efficiency in a realistic fire—perhaps with the windows out and a 35-mile-per-hour (56 kph) wind blowing into the building.

One such air supply advocate argues, "Let the fire burn, but let it burn clean"—presumably so that there will be no dangerous products of combustion. This is a fallacy. Heat is a dangerous product of combustion. There is no such thing as a clean-burning, hostile fire. Even if the fire itself is burning freely, above, below, and on all sides of the fire, materials will be in various stages of degradation, generating toxic and explosive gases.

The design of smoke management systems is an extremely complicated task. A well-designed and properly maintained system can certainly supplement the primary fire defense—automatic sprinklers—but it is certainly no substitute for adequate fire protection.

Some buildings simply have breakable tempered glass windows, typically spaced on 50-foot (15.2 m) centers around the building perimeter. They must be identified in some manner; it is common for a reflective Maltese cross sticker, or a "frosted" inscription, to appear on the glass.

In larger buildings, the smoke management system can be complex and customized. The fire department must be aware of its installation and operation from the beginning. One of the possible deficiencies is that it may be necessary that the system remain running, even if not refrigerating or heating, so that smoke will reach the detectors when the building is closed. An unaware energy-conscious building manager may turn the smoke management system off over weekends and holidays.

The key to smoke management systems is to study them *before the fire*. Each system is different—because each is designed to meet some specific goal(s). It is critical that you understand exactly what the system is designed to achieve, your role in operating it, and how to shut it down if necessary. Insist on getting written documentation for the system and be present when it is tested.

Compartmentation

Some people assume that fire-resistive buildings automatically provide compartmentation that limits the transmission of fire. This is often the case in older buildings with masonry panel walls built on concrete floors, without air conditioning and without the huge number of penetrations necessary for current communications requirements. Modern buildings often have poor perimeter firestopping as well as multiple penetrations for wiring. Air-conditioning ducts also sometimes serve as smoke paths.

Pressurized Stairways

One or more of the stairways in a building may be equipped to be pressurized when fire occurs. In such a case, fans are installed to pump outside air

into the stairway so that the pressure differential will keep the stairway free of smoke. This approach requires a dependable source of power and intakes located so that they will not pick up smoke from the fire. Occupants must be trained to use the proper stairway, and should be drilled on this procedure, because the noise of the fans may frighten people who are already upset and trying to escape. Because the temperature should be close to the outside temperature, there should be no stack effect. The overpressure may make opening doors difficult, and mechanical assistance may be required. The system does nothing to clear corridors.

Air Standpipes

In the wake of the World Trade Center disaster in 2001, some high-rises across the United States have been retrofitted with internal breathing air systems ("air standpipes"), essentially pressurized air piping systems for SCBA bottles. The stainless steel piping runs vertically up through the building and provides SCBA filling stations on selected floors (usually every few floors). The systems are capable of getting air from a fire department's air cascade truck. These systems allow fire fighters to avoid carrying hundreds of SCBA cylinders up to the staging area of a high-rise fire **Figure 13-21**.

Installation of Special Equipment

Any equipment designed to function in case of fire is, in effect, fire department equipment. Technically qualified members of the fire department should approve its installation, be familiar with its operation, and supervise its testing and maintenance by the owner or owner's contractor. The basis for this arrangement, which is opposed by some designers and building officials, is that the fire chief has unquestioned authority over firefighting operations. Of course, there is no guarantee that the concept will be upheld, but it is the position with which the fire department should start.

Figure 13-21 An air standpipe filling station in a high-rise building.

Courtesy of Rescue Air Systems, Inc.

Fire Load and Flame Spread

Consideration of the fire problem of a high-rise building must go beyond the basic structural considerations to the question of interior trim and contents. A developer who was a vociferous opponent of improved high-rise fire protection in New York City argued that his buildings could not be burned with a blowtorch when finished. He argued that the problems arose from what the tenants did, adding interior trim and contents.

In general, codes specify adequate controls on interior finish with respect to flame spread and smoke development. The difficulty is that the interior finish is often installed by the occupant without a building permit. Even when flame spread requirements are adequate to protect life safety during evacuation, fires can gain great headway in combustible trim in

the interval before an effective attack can be mounted. Multiple layers of wall coverings were a major factor in an Atlanta office building fire in which 10 people died.

Contents

The wide-open floor populated with scores of computers and associated equipment that is required by many of today's businesses has introduced a new flame spread problem. This issue was graphically illustrated by the spectacular First Interstate Bank fire. The fire spread rapidly while the guards mistakenly reset the alarms. The Los Angeles Fire Department received the first alarm from an observer in a building several blocks away. By that time, heavy fire was coming out of many of the building's windows.

Heavy fire loads may be found in special locations in high-rises such as club rooms and restaurants. Wood paneling and imitation wood beams and heavy loads of plastic are common. The wood paneling is installed on furring strips, leaving a void behind where fire can burn untouched by hose streams.

Storerooms for office supplies and telephone rooms are other high fire load areas **Figure 13-22** . Not only is the wiring insulation combustible, but a typical telephone room is usually the supply room and contains boxes of equipment packed in foamed plastic.

Maintenance Operations

Maintenance operations can result in unexpectedly serious fire loads. If arranged for by a tenant, the building management may be completely unaware of this hazard. Such was the case at the Union Bank Building fire in Los Angeles on July 18, 1988. A contractor employed by a tenant was refinishing book cases with flammable liquids. When a worker brushed a solvent over an electrical outlet, fire erupted. The fire was quickly contained by the fire department, but the loss was estimated at $500,000.

Rubbish

When rubbish is gathered in high-rise buildings, it is often concentrated in one location, creating

Figure 13-22 Stacks of paper, unprotected file cabinets, books, and binders will allow fire to spread rapidly.

© Jones & Bartlett Learning. Photographed by Christine McKeen.

another fire load problem. The condition of the material provides a high rate of heat release. The concentration of such rubbish takes place in halls, elevators, basements, and lobbies, all particularly bad locations in terms of fire and smoke spread. Collecting rubbish on elevators tempts the potential vandal and provides the circumstances where an otherwise inconsequential fire can pollute an entire building. In one case, a rubbish fire in an elevator parked at the ninth floor contaminated the 26 floors above.

Many apartment buildings have trash chutes that deliver rubbish to huge compactor units located in the basement. Seven people died when a fire roared 35 stories up a blocked trash chute. The rubbish handling system is important. *It should be properly planned with the anticipation that there will be fires.*

Alterations to Occupied Buildings

A serious hazard exists when a building is altered or rehabilitated while it is still occupied. In a university hospital being altered, an untreated plywood wall was erected to separate the construction area from the maternity ward and nursery. The construction area had the typical fire hazards of any construction job. Fortunately, no fires occurred.

Hotels are renovated periodically. This usually involves replacing the furniture, particularly the beds. The new furniture is often stored wherever space is available, including the basement. Sometimes a floor is placed out of service and used for furniture storage. Mattresses removed from rooms may be stored in halls, awaiting disposal. Two serious fires occurred in Maryland suburban motels from such mattresses, which were inviting targets to an arsonist. The disastrous Dupont Plaza Hotel fire in San Juan, Puerto Rico, started in a large pile of new furniture that was temporarily stored in a ballroom.

Partial Occupancy of Buildings Under Construction

It is a dangerous situation to have a building partially occupied and partially under construction. Fire protection systems may not be fully operational in such structures; doors may not yet be installed on stairways and elevators. Temporary heating using LPG or combustible liquids may be used in some areas, and all the hazards of a building under construction exist. The strictest special precautions should be demanded if the building department permits partial occupancy.

Automatic Sprinklers

An examination of the disaster potential inherent in the typical enclosed structure with its occupants, fire load and fuels, and construction materials and configuration leads to one conclusion: there must be absolute assurance that the toxic gases released in a fire will be severely limited and not dependent on control by a smoke management system. In theory, this could be accomplished by limiting the quantity of fuel present. However, the combustible materials are necessary to the functions performed in the building. The only practical solution is to provide a system that will limit the amount of gas and heat produced.

The only such system available is automatic fire sprinklers. Fire department reaction time (from the time the alarm is received until water is put on the fire) is rarely the fabled 1 ½ minutes, but more typically 20 minutes or longer. Burn time between ignition and alarm, however, can be any length. The relatively short reaction time of the sprinklers, plus the fact that they can automatically transmit the alarm, makes them infinitely superior to any other fire protection system. If the sprinkler system water supply is properly tied to the domestic water system, such a system is almost foolproof. This is not to say that other systems are not necessary to supplement the sprinkler protection, but rather that sprinklers are the core of fire safety for the occupants of high-rise buildings.

What gives the developer of a high-rise the right to erect or maintain a structure that places an enormous burden on the public purse and strips the city of normal fire protection by omitting the automatic sprinkler system? At one point during the February 1991 Meridian Plaza high-rise fire in Philadelphia, there were only 16 companies available to protect the rest of the city.

The argument against sprinklers is usually an economic one, buttressed at times with arguments from some construction material suppliers and devotees of managing fires by controlling smoke. The basic argument for sprinkler protection is one of

equity: the builder is creating the problem for profit, so it is up to the builder to provide the solution. This parallels what is done in the case of parking facilities, sewage facilities, and other amenities.

When the improvement of existing buildings is being discussed, many attorneys immediately dismiss any effort to improve the fire safety of an existing building as being unconstitutional. The U.S. Fire Administration asked Frank Brannigan to research the subject. He found no U.S. appellate court case that accepted a constitutional barrier to the enforcement of a new fire code to an existing building.

Checklist

There are innumerable items of information that would be useful for the IC to have at the scene (Figure 13-23). The following is merely suggestive of useful building information. Even though a sophisticated information retrieval system may not be available, this valuable building-related information should be gathered. The day is past when fire departments can depend on information known only to "Captain Joe, the old-timer," who won't always be around. The information must be institutionalized—available to all. The building may be there for a century and may outlast several generations of fire department personnel.

Structural Frame

- *Cast iron:* What is the estimated value of "fireproofing"? Are there connections from column voids to floor voids?
- *Steel:* What type of fireproofing? Are columns unprotected in plenum space?
- *Concrete:* Are there visible rods and cracks? Has there been any fire experience in this building that might have caused structural damage? Is corridor fire resistance achieved by gypsum on studs, making breaching easy?
- *Reinforced masonry:* Are corridor walls reinforced? If so, breaching would be difficult.

Floor Assemblies

- *Bar joist floor–ceiling assemblies:* What is the integrity of the floor? Early collapse is possible.
- *Concrete:* Are the floors of solid reinforced construction (good insulator and heat sink) or composite decks (Q-floors or cored slabs)?
- *Post-tensioned:* Don't drill or cut.
- *Tile arches:* Dangerous to cut because the entire panel may collapse.
- *Overloaded floors:* Potential for multistory collapse.

Floor Containment

- *Floor–ceiling assemblies:* Ceiling tile failure may permit at least partial collapse and open fire and smoke passage.
- *Unprotected columns in plenum:* Smoke and fire may pass to voids above via reentrant space.
- *Firestopping:* Is there effective firestopping in void spaces?
- *Masonry floors:* Utility openings may allow fire to pass.
- *Accommodation or access stairways:* These create huge fire areas.

Perimeter Integrity

- *Joints:* Are masonry wall to masonry floor joints solid (good inherent firestops)?
- *Panel walls:* How are exterior panel walls joined to the floor? What is their value as firestops?
- *Glass walls:* Is vertical extension of fire probable?
- *Exterior architectural panels:* Voids between panels and walls may negate firestops.
- *Contents:* High rate of heat release furnishings near windows or combustible tile ceilings increase the potential for exterior extension.
- *Lower levels:* What is the potential for exterior hose streams to block the extension of fire?

Figure 13-23 A cutaway of a high-rise fire. Although very unlikely in a fully-sprinklered high-rise, fire extension via autoexposure and in the gap between the curtain wall and floor above are possibilities. Note the avenue of potential vertical smoke spread via the typically unpressurized elevator shaft. In the stairwell, pay particular attention to the location of floor control valves for the sprinkler system on each floor, the type of standpipe hose valves and their settings, and any door locks preventing reentry on the office floors. The first floor fire command center—described on page 380—provides a number of fire protection systems, communications, and elevator, generator, and fire pump status panels and controls, as well work space for reviewing building plans stored on site.

Compartmentation

- *Integrity of compartmentation:* Are there utility openings or underfloor openings such as for computer cables? Do barriers extend to the underside of the floor above? Do fire doors function properly?
- *Gypsum board compartmentation:* Consider penetration of relatively lightweight gypsum partitions as a substitute for forcible door entry.
- *Stairways:* Are there deficiencies of stair enclosures? Are there proper operating door closures on all floors? Are stairways locked to deny entry to floors? Do all units carry forcible entry tools?
- *Elevators and other vertical shafts:* Gypsum board enclosures may be destroyed by firefighting operations.

Fire Protection Systems

- *Sprinklers:* Is the building fully sprinklered?
- *Standpipe:* Which type of standpipe system is provided and where are the hose outlets?
- *Smoke detection:* Where are smoke detectors located?
- *Smoke management:* Does the building have a smoke management system? How does it work? What is it intended to achieve? How do you operate it?
- *Fire fighter communications:* Which kind of fire fighter communications system does the building have?
- *Occupant communications:* Which type of communications are possible with building occupants? Can this system be controlled by floor or floors?
- *Equipment status:* What are the indicators for the fire pump, generator, and elevators? How are these pieces of equipment controlled from the command center?

Hospitals and Nursing Homes

Hospitals and nursing homes have a common characteristic—they are home to numerous **nonambulatory people**—that is, individuals who are not capable of self-preservation. It is up to the building, the staff, and you to protect them. Many older facilities also lack sprinkler protection, the most basic of equipment in these types of facilities. Many have relied on passive protection, such as fire doors and rated corridors. Many modern nursing homes and hospitals have sprinkler protection and an alarm system. How well protected are the facilities in your community?

In the late 1980s, a working fire broke out in an unsprinklered hospital in San Antonio. The fire erupted in the medical gas piping system, which included oxygen, inside a corridor wall. The fire, accelerated by the oxygen, incinerated the corridor wall. Nurses had followed proper protocols, evacuating the patients in the immediate area and closing all fire doors. The fire doors held the fire in check until fire fighters extinguished the fire. Thankfully, no one was killed.

The key to patient safety is to move patients horizontally, rather than vertically. Hospital and nursing home buildings are equipped with smoke barriers, 1-hour fire-rated walls that subdivide each floor into 2 or more separate areas. If a fire occurs on one side of the smoke barrier, patients may be moved through the barrier to the other side. To complement the smoke barriers, corridors are a minimum of 8 feet (2.4 m) wide to accommodate the size and width of rolling hospital beds.

Hospitals and nursing homes usually have highly trained staff. (This is one of the few occupancies for which codes recognize such training as part of the "protection package.") Hospital and nursing home staffs practice **RACE**:

- *R:* Remove all people in immediate danger to safety.
- *A:* Activate the manual pull station and have someone call 911.
- *C:* Close doors to confine the spread of smoke and fire.
- *E:* Extinguish the fire, if possible.

Hotels and Motels

On a national level, U.S. fire fighters have experienced many hotel and motel fires over the last 75 years. Some of the more notable were the fires

involving the Atlanta Winecoff Hotel (1946; 119 dead); the Chicago Hotel LaSalle (1946; 61 dead); the Clark County, Las Vegas, MGM Grand (1980; 84 dead); and the Puerto Rico Dupont Plaza (1986; 87 dead). All of these incidents took place in unsprinklered hotels **Figure 13-24**.

The 1990 Hotel and Motel Fire Safety Act encouraged improvements in fire safety for such facilities nationwide. This encouragement was accomplished by withholding travel expenses from federal employees who did not stay in approved facilities that met minimum fire safety standards. Although sprinklers were considered critical for larger hotels, small motels were not required to meet this requirement and could get away with just smoke detection, alarms, and other passive equipment.

Fire fatalities continue to occur in hotels and motels. Although there hasn't been a "spectacular" multifatality hotel fire in recent years, numerous fires in motels have claimed the lives of small numbers of people.

Older motels typically have open walkways to access the units along the face of the building. Even without an interior corridor system, fire spread occurs through voids between the floors and particularly the space underneath the top floor and roof.

Newer hotels and motels have interior corridors that have proved to be a conduit for smoke travel. The Westchase Hilton fire in 1982 killed 12 people. Besides the front desk clerk who kept resetting the alarm (for the real fire on the fourth floor), a contributing factor was a door that did not close properly and let heat and smoke escape into the connecting hallway and up the vertical shafts.

Jails and Prisons

Like hospitals, jails are considered institutional occupancies. The inmates of a jail are, of course, restrained and are incapable of getting out of the building to save their lives. They rely on prison staff and the building for their safety.

Jails and prisons run the gamut from old to new, big to small. Some use the old technology of individual key-operated doors, whereas others have automated control systems. Some have full sprinkler protection, whereas others have none.

Currently, there are five occupancy classifications of detention facilities, depending upon their level of restraint:

- *Use Condition 1*: Free movement is permitted within the building from smoke compartment to smoke compartment as well as to the exterior.
- *Use Condition 2*: Free movement is permitted within the building from smoke compartment to smoke compartment.
- *Use Condition 3*: Free movement is permitted within the building within a smoke compartment; movement to other areas is by remote control release.

Figure 13-24 The Dupont Plaza hotel fire.
© Jose Fernandez/AP Images.

- *Use Condition 4*: Free movement is permitted within an occupied space within a smoke compartment; movement to other areas of the smoke compartment or other smoke compartments is by remote control release.
- *Use Condition 5*: Free movement is restricted from the occupied space; manual release is necessary to allow movement from the occupied space to other areas of the smoke compartment or other smoke compartments.

Numerous jail fires have occurred over the years, killing many prisoners. A fire in 2002 at the Mitchell County, North Carolina, jail killed eight inmates. This was a 1950s-era facility that required the manual opening of doors.

Some jail fires start accidentally, but others are intentionally set by the inmates. Several fires over the years have involved the use of polyurethane foam in a padded cell.

In any case, you must understand how the restraint system in the jail building that you respond to works and the "use condition." Also, does the building have a sprinkler system? (Is the sprinkler system automatic? Some old jails have dry pipe systems that must be opened manually to let water into the system.) Is it a modern jail with a smoke management system? Does it have smoke barriers like those described earlier for hospitals? Are the inmates evacuated from the building, and if so, where are they placed?

Museums and Libraries

Valuable collections of historic artifacts, books, and pieces of art have been lost to fire. Recent examples include the 1986 Los Angeles Library fire in which $22 million worth of collections were lost and the structure was damaged; an arson fire in a Holocaust museum in Terre Haute, Indiana, in 2003; and a massive fire that destroyed the Biblical Arts Center Museum in Dallas in 2005.

Unfortunately, museums and libraries are not always protected with a fire suppression system. For years, many institutions mistakenly rejected sprinklers as causing too much water damage. (They never considered the damage from a well-involved fire and subsequent water from a hose line.) In some cases, these institutions did install other types of suppression such as halon systems (which now are being phased out).

Life safety is, of course, the primary concern of fire fighters. Many museums and libraries have extensive security systems, including magnetic door locks. These systems are illegal; model building codes do not permit these systems in places of public assembly like museums and libraries. If you find them, notify your fire prevention office and/or the department of buildings.

A preincident plan for these facilities will assist greatly responders when a fire occurs. Besides the typical details of construction and fire protection systems, ensure that this plan includes salvage operation details for saving the collections. Most museums will have a detailed "disaster plan" (often required by their insurance company or grant providers), which will provide many of these details.

Library Stacks

Libraries are the original high-stack storage buildings. We are not referring to the familiar branch library here, but rather to the large main libraries that have multilevel stack areas where the books are stored Figure 13-25 . There are libraries in which the stacks are open to the public, but usually, patrons do not go into the stacks; instead, stack attendants retrieve the requested books.

Most stack areas are built to a common pattern. Stack levels are about 6 feet (1.8 m) high so attendants can reach the **shelves** without a ladder. The floors are open to air circulation to prevent mildew. A better system to guarantee the spread of fire and destruction of the books in case of fire can hardly be imagined. In addition, the stacks are often structurally part of the building, so that fire damage to unprotected steel may have structural consequences.

Figure 13-26 Contributing factors to the fire at The Station nightclub included interior finishes, overcrowding, and the arrangement of exits—just to name a few.

© Robert E. Klein, File/AP Images.

Figure 13-25 Old multi-level library stacks with glass floors. Fire can spread easily in this situation.

Courtesy of Glenn Corbett.

Nightclubs

Nationally, nightclubs have been the scene of many large disasters. The most famous and largest was the 1942 Cocoanut Grove fire in Boston, which killed 492 people. The Happyland Social Club arson fire in the Bronx, New York, killed 87 people in 1990. A 2003 fire at The Station nightclub in West Warwick, Rhode Island, killed 100; it was initiated by an indoor pyrotechnics display during a concert **Figure 13-26**. A similar nightclub fire in 2013 involving indoor pyrotechnics and acoustic foam insulation in Santa Maria, Brazil, killed 242 people.

Nightclubs are places where people go to have fun. The use of alcohol is a key element of that fun, resulting in patrons who often are not fully aware of their surroundings. In addition, these clubs are often overcrowded beyond code-imposed limits. Locked egress doors (often at the rear) complete the potential for a disaster.

Many fire departments have nightclubs in their response district. It is important for fire inspectors to stop by on a Saturday night while in uniform and on duty to see what is happening. Look for locked and blocked egress, dangerous decorations or performances (including indoor pyrotechnics), and, of course, overcrowding.

Many clubs operate out of old, worn-out structures. Other than in Rhode Island, many of these existing clubs are not required to retrofit sprinklers. During the course of preincident planning, note the location of exits and potential points of ventilation. Ensure that during an emergency, all exits, including the rear, are opened fully—you may possibly find people trapped there.

Figure 13-27 Large open rooms have no barriers to hinder fire advancement.

© Jude Lazaro/ShutterStock, Inc.

Office Buildings

Every community has an office building. Office buildings are built using all five types of construction, large and small, high-rise and low-rise. These buildings have changed over the years, from highly compartmentalized structures with offices for every worker that were once popular to the **open office plan** in which low-height partitions create cubicles for personal space but the office is essentially one large, open room Figure 13-27 . Fire spread through a compartmentalized space is much different than a fire in an open office plan.

Fire fighters have been killed in fires in large office buildings, such as the One Meridian Plaza high-rise fire in Philadelphia where three fire fighters lost their lives. In this incident, the fire fighters became trapped and disoriented, giving the wrong floor as their location. Fire fighters have also been trapped and killed in smaller low-rise office buildings, such as a Denver fire fighter killed in 1992 when he ran out of air inside the building.

Open-Area Structures

Many large open-area structures, such as churches and arenas, are built of wood or have an exposed wood plank roof that makes up the ceiling. Such a valuable building should be fully sprinklered. The objection to sprinkler piping in an open decorative wood structure is understandable—the exposed piping detracts from the overall look of the decor. However, in some cases, the sprinkler piping is run close to beams, and sidewall sprinklers are used. A competent painter charged with making the pipes less apparent can work wonders.

If a building is not fully sprinklered, try to keep out the kindling—the minor light combustible structures or elements that can ignite the whole building. If a structure, such as a robing room, sacristy, ticket office, hot dog stand, or the like, is to be built within such a building, verify current building code requirements that likely call for the use of noncombustible construction materials and possibly sprinklers.

Frank Brannigan recommended sprinkler protection, even if only a couple of sprinklers taken off the domestic system are installed to protect these types of small enclosures in heavy-timbered buildings. Don't call it partial sprinkler protection; call it a super-extinguisher. If there is no kindling, the open-area timber structure should stand for generations.

Parking Garages

A parking garage may be partially or totally above grade and open to the atmosphere. All garage areas under buildings should be sprinklered. Note that many garages have dry standpipes without any permanent water supply. What this means is that the water you provide through the FDC will be its only supply. This also means that air will be trapped in the system and it will take some time to get water to the nozzles of your hose lines. Beware of open hose valves, opened by mischievous individuals.

Restaurants

One of the most common places for fires in a restaurant is in the kitchen. Interestingly (and oddly), the model building codes do not require

a fire-rated separation between the kitchen and the dining area. Some older codes even allowed exits to pass through the kitchen and out the back door!

Although codes require cooking hood extinguishing systems, our experience with them has not been all positive. Several of the systems have failed (often due to poor maintenance), allowing a grease fire to spread into the kitchen and to the structure around the ductwork going to the roof. Broken ductwork adds to this problem. It is important to track the spread of fire into void spaces around the ducts, especially if the ducts have developed holes or breaks in the seams and allowed grease to coat these locations **Figure 13-28**.

When preincident planning, pay particular attention to the location of the utilities, especially gas. Also make note of the use of propane, which is (unfortunately) permitted in some circumstances for tabletop cooking. The use of propane introduces the possibility of an explosion and fire. Such an event happened along the New Jersey shore in a restaurant, where patrons (including an off-duty fire fighter) were forced to break windows to escape the fire.

Figure 13-28 Not all cooking hood systems are as well maintained as this one. Pay particular attention to them when conducting inspections.

Courtesy of Glenn Corbett.

Schools

Building code regulations concerning schools have been shaped by one particular fire—the 1958 Our Lady of Angels fire in Chicago, which killed 92 students and 3 nuns. This fire led to numerous egress improvements in schools, such as better and more frequent fire drills, lower and more accessible windows for escape, abatement of open stairwells, alarm systems, and fire-rated corridors and doors.

Schools often have some unique features. Their corridor widths are much larger than normal, 6 feet in most cases. In addition, the egress systems may be unusual. Although no longer accepted in new buildings, many old structures, particularly in warmer climates, still have these features.

Pay particular attention to potential hose stretches. In many older schools without standpipe systems, corridor lengths are particularly long. The preincident plan should also make note of special hazards such as woodworking and machine shops, chemical storage areas, and flammable/combustible liquid storage for art classes.

With concerns over violence in schools, many school districts have instituted lockdown procedures. Find out if this is the case and how it will affect your firefighting operations. As a result of the 2012 Newtown, Connecticut, school shootings, it is likely that new emphasis will be placed on increasing the physical protection features of existing schools. These enhancements will likely include bullet-resistant glass, secured lobbies, increased use of magnetic door locks, locks on exits and room doors, thicker walls, and secure classrooms.

Self-Storage Buildings

In recent years, the number of self-storage facilities has increased dramatically. These structures may be found in a variety of construction types, most notably Types I–IV.

Generally, facilities of the one-story drive-up variety are either Type II or Type III construction. Apparatus accessibility and water supply tend to be

Figure 13-29 Many multistory self-storage facilities have limited points of horizontal ventilation—windows are few and far between.
Courtesy of Glenn Corbett.

the issues of concern in these buildings, in addition to the unknown contents of the units.

It is the multistory Type I and Type IV facilities that present more of a challenge Figure 13-29 . These are often repurposed buildings that are essentially windowless (the old windows have been blocked up; sprinkler protection is mandated). Corridors are often confusing and maze-like. Stored materials are often placed as high as possible in the units. Hazardous materials are found in many units. Ascending a folding ladder may make it possible to see over the units in some buildings. Search ropes and thermal imagers are crucial pieces of equipment.

A smoky fire in one of these buildings will make it very difficult to establish which units are involved. Try to use sprinkler discharge as an indicator of the fire's location. Ventilation will be very difficult—be sure to identify your options during preplanning.

Single-Family Homes
The California Bungalow

The California bungalow has become popular all across the country. This one-story structure has a peaked roof with rafters as small as 2 by 4 or 2 by

6 inches. Often there is no ridge beam in these homes. The attic often contains a high fire load of stored materials and is reached by a small scuttle hole. The house can have one or more dormer windows.

The Cape Cod

This is traditionally a 1½-story home with a steep-pitched roof Figure 13-30 . These are platform-framed structures. The stairway to the second floor is near the front door. Bedrooms may be located on both the first and second floors.

The Ranch House

One-story ranch houses are very popular Figure 13-31 . These houses have open interiors, large attics, and extended overhangs. In some areas, these homes are spaced so close together that they create a serious exposure hazard. Fast-spreading fires in these types of houses, heavily involved on arrival of the fire department, are not uncommon.

Properly installed and maintained smoke alarms can provide a high degree of occupant life safety in these buildings because escape from them is relatively easy. Many bedrooms in the

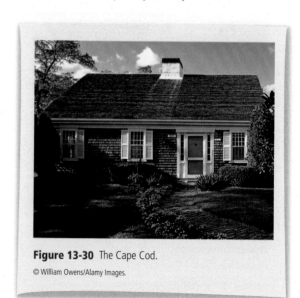

Figure 13-30 The Cape Cod.
© William Owens/Alamy Images.

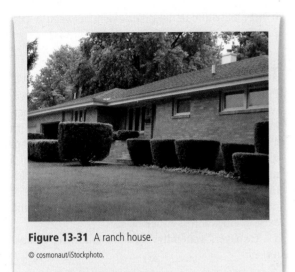

Figure 13-31 A ranch house.

© cosmonaut/iStockphoto.

Figure 13-32 A split-level home.

© Djordje Zoric/ShutterStock, Inc.

ranch house have direct routes to the outside because their windows are small and high above the floor.

One common deficiency in these dwellings is failure to place a heat detector in the attic. Heating and air-conditioning units are one potential cause of attic fires. There are known cases of attic fires in which smoke alarms in the living spaces did not activate until the attic was well involved and collapse was imminent.

The Split Level

The split level is just that—three floors joined by short-run stairways Figure 13-32. The top level usually contains the bedrooms; the middle level, the dining room, living room, and kitchen; and the lower levels, the recreation room and laundry room. This structure is platform framed.

The Victorian

This group is a catch-all of many styles of homes of the mid- to late-1800s—the Queen Anne, Italianate, and so on Figure 13-33. Victorians are characterized by significant amounts of ornamentation, steep-pitched roofs, and balloon-frame

Figure 13-33 A Victorian house.

© photos.com.

construction. Many of these homes have been converted to multiple dwellings because of their large size.

Stadia

Most major American cities have a stadium, which can vary in seating capacity from a few thousand to more than 80,000. Many are open-air stadia; some are enclosed with fixed roofs or domes. Stadia are almost always of Type I construction. The type and amount of fire protection features varies, particularly with the age of the facilities. It was not until the 1980s that more rigorous code criteria were developed for these structures, particularly enclosed stadia.

One notable issue with stadia is egress travel distance—most enclosed stadia cannot meet the traditional limits on travel distance from the most remote seat to an enclosed fire-rated stairwell. New code requirements were created to deal with this problem by requiring a smoke management system in addition to sprinklers and standpipes.

Some enclosed stadia used for purposes other than sporting events—tractor pulls, trade shows, and concerts, among other things—often necessitate additional fire suppression equipment because of the inability of traditional sprinkler systems to respond to and control a fire more than 150 feet (45.7 m) below the sprinkler. In some cases, special flame detector-activated water cannons have been installed to provide fire protection.

Taxpayers and Strip Malls

As development sprawl overtook the American suburban landscape a few decades ago, a new type of neighborhood corner store appeared—the strip mall. Like its urban ancestor, the taxpayer, the strip mall provided the convenience of having a variety of shops under one roof. Although the taxpayer and the strip mall share common construction characteristics, they also have distinct differences that must be considered. In some cases,

it may be difficult to distinguish between a taxpayer and a strip mall.

Taxpayers

Fire service legend tells us that the term *taxpayer* was developed to describe the practice of constructing "temporary" revenue-generating buildings on a tract of land along a city street. These buildings were designed to pay the taxes on the land until a more desirable multistory apartment or office building could be built. Whether this legend is true or not, the fact remains that thousands of taxpayers still exist in cities across America.

Taxpayers generally have the following characteristics:

- They most often are of ordinary (Type III) construction, usually with brick bearing walls and wood-joist roof members.
- They commonly are one story in height, although the two-story variety can be found in many jurisdictions with apartments on the second and third floor.
- Depending on the region, many taxpayers can have full or partial basements. Access to the basement is through metal sidewalk doors and stairs or interior "trap doors" in the stores themselves.
- They usually are limited to approximately 6–10 small stores (or other commercial establishments).
- They most often have common cockloft or attic spaces.

Many fires have been fought in taxpayers. Fire spread is typically through the cockloft above all of the stores. There are specific hazards to be wary of. The parapet above the stores is usually unsupported or poorly supported. Movement of the structure below, such as when fire causes expansion of the steel beam that usually runs across the front of the stores above the windows and doors, can cause the parapet to fall.

Hazards of inner-city taxpayers include the steel plates on the roof of check-cashing establishments (to deter entry of burglars). Rotted wood floors

under the washing machines in Laundromats are also dangerous—a fire in the basement can cause the first floor to collapse.

Strip Malls

Strip malls have the following characteristics :

- They may be of noncombustible, ordinary, or even wood-frame construction (Types II, III, and V construction, respectively).
- When of noncombustible construction, strip malls have exterior walls of concrete block or concrete tilt walls (possibly with brick veneer) with steel bar joist roof members and a metal deck supporting a built-up roof.
- Contemporary ordinary construction strip malls use concrete block for exterior walls and solid wood joists or lightweight trusses to support a wood roof deck.
- Older ordinary construction strip malls may have large bowstring trusses.
- Some smaller strip malls may be built entirely of wood-frame construction.
- Nearly all strip malls are one story, although a two-story strip mall occasionally may be encountered.
- Based on region, strip malls may or may not have basements.

Figure 13-34 A strip mall.

© Jones & Bartlett Learning. Photographed by Kimberly Potvin.

- They usually are larger than the taxpayers and may have as many as 15–20 small stores and a large anchor store or two.
- Strip malls usually have greater store depth than taxpayers. Strip mall anchor stores, such as supermarkets, may be more than 150 feet (45.7 m) deep.
- They most often have common cockloft or attic spaces.

Strip Mall Surroundings

Another distinguishing feature of strip malls is the property surrounding them, which most often is a parking lot out front and along the ends, and a delivery truck driveway at the rear. The strip mall setback can be both an advantage and a disadvantage.

The parking lot can be helpful, especially from an apparatus-placement standpoint. In many cases, apparatus can be placed on four sides of the structure, and additional apparatus can be placed in close proximity in the parking lot. In some cases, a staging area can even be set up in a remote area of the parking lot, relieving the public streets of tie-ups.

Among the disadvantages of the strip mall setback are the following:

- On-site private hydrants must be used because the strip mall is too far from the public hydrants surrounding the property. In some cases, the main may be 6 or 8 inches (15 or 20 cm) in diameter and of the dead-end type. These private mains are often poorly maintained. In addition, if the building has an automatic fire sprinkler system, it often will use this same water main. It is critical that fire departments establish the supply layout (including FDC, tie-in location, and underground gate valves) and flow capabilities during preincident planning.
- The delivery truck driveway at the rear of the strip mall is often dotted with crater-size potholes that create hazards for fire apparatus. These delivery driveways also may have obstructions that can present additional hazards.

For example, these areas often are used as an easement for utilities. Fire fighters must be aware that overhead power lines may be present. Retail establishments often use these driveways for storing pallets, bins, bales, cardboard, and dumpsters, which may block the access for fire apparatus. It is best to designate these areas as fire lanes and to inspect them frequently.

Strip Mall Occupancy Types

Strip malls may have a variety of tenants. In addition to the typical delicatessen, bakery, and hardware store, occupancies can include childcare centers, nightclubs, gyms, spas, auto-body shops, and even storefront churches. From a building and fire code standpoint, it is very possible to have assembly (A), mercantile (M), business (B), and, unfortunately, hazardous (H) occupancies all in the same strip mall.

What this means to you as a fire fighter is that you will be confronted with a variety of hazards as well as a variety of occupancy types in strip malls. Hazards can include flammable liquid storage/use/sales in a paint store, a dry-cleaning store using combustible Class II cleaning liquids, flammable spray operations in a body shop, and oxidizer and corrosive material storage/sales in a swimming pool supply store. Occupant count can range from dozens of young children in a childcare center to hundreds of shoppers in a supermarket to a large group of patrons in a nightclub.

Most building codes require a fire-rated separation (also known as an occupancy separation) between different occupancy types (separating a business occupancy from an assembly occupancy, for example). These walls, which must completely separate the different occupancies, can be 1-, 2-, 3-, or 4-hour fire rated, depending on the types of occupancies being separated. A higher rating is required as the hazards and threat to life safety of the specific types of occupancies increase. Depending on the particular building code being applied, the installation of automatic sprinklers in the building can reduce the hourly rating (this practice is prohibited in some building codes).

Often, no rated separation is required between a group of strip mall tenants if all are the same occupancy type. A group of small shops selling fish, greeting cards, and books requires no separation between each shop, for example. The "demising" walls between the shops may be made of rice paper, be full of holes, and still meet the building code.

Depending on the size of the strip mall, another type of fire-rated separation—a fire wall (also known as an "area separation wall")—may be required. Building codes contain provisions that limit the size (area and height) of all buildings based on the type of construction and type of occupancy—the better the type of construction (fire resistive being the best, wood frame the worst) and the less hazardous the occupancy, the larger the building can be. When the building area becomes too large, fire walls often are required to break the structure into smaller fire areas.

These fire walls may have a 2-, 3-, or 4-hour rating and also must be constructed to allow partial collapse on either side without pulling the entire wall down. Often, these walls must be continuous from the foundation to above the roof line—from "dirt to sky." Unfortunately, building codes grant exceptions to this continuity rule by allowing some fire walls to terminate on the underside of the roof deck.

Only on-site preplanning and review of your local building codes will untangle this mess of code requirements concerning fire-rated separations. Look above the ceilings and ask questions of your building and fire inspectors. Know how the building will work for you and against you.

Strip malls of Type III and V construction that use a combustible structural roof (or floor members, in a multistory building) require the use of draft stops to compartmentalize the attic or floor void spaces. The draft stops are not required to be fire rated; they may be made of plywood or gypsum board. The intent is to have the draft stop slow the fire down and keep it confined to a given area for an indeterminate period of time. This will buy fire fighters time to get ahead of the fire if it is spreading through the attic (or floor) void space.

The requirements for the draft stops have changed over the years. Currently, draft stops are

required to limit the size of attic compartments to 3,000 square feet (278.7 m²) and floor void compartments to 1,000 square feet (92.9 m²). Again, depending on the building code, the installation of automatic sprinklers can eliminate the requirement for draft stops. Conduct your own research to determine what rules apply to draft stops in your jurisdiction. When properly installed and maintained, draft stops help fireground activities. Often, however, they are compromised by holes punched in them for electrical wiring and the like. Perform frequent inspections and order repairs made where holes are noted.

One last note about preventing fire spread: some building codes require firestopping at a maximum of 20-foot (6.1 m) intervals in combustible architectural trim (projections such as cornices, mansards, overhangs, and so on). Firestopping will help prevent the fire from doing an end run on the outside of the building around your fire attack. Know where the firestopping is located and how you can use it to your advantage.

Structural Fire Resistance

Building codes limit the size of any building based on the type of construction and occupancy group type. One way to increase the allowable area is to increase the fire resistance of the structural members—a Type II building, for example, may use unprotected (bare) steel bar joists for the roof's structure. "Fireproofing" the steel by applying a fire-resistive coating can increase the allowable area. Doing this may be necessary to permit certain occupancy types into the building. Probably the most common example is a place of assembly, especially a nightclub. Nightclubs of substantial size usually require a higher level of structural fire resistance to remain within the building code's allowable area limitations. It also may be the only occupancy within the building to have fireproofing; surrounding tenant spaces may have no structural protection.

During preincident planning visits, look at the structural members in each occupancy. Note which members are protected and which are not. Record these facts in your preincident plan to establish where this structural protection will help your attack.

Sprinkler Systems

Building and fire codes do not require automatic sprinklers specifically for strip malls. However, they do require sprinklers for mercantile and other types of occupancies under certain conditions. Basements also have to be sprinklered in some cases. Under some building codes, sprinklers can also dramatically increase the allowable size (including area) of a building.

Looking more closely, some building and fire codes require automatic sprinklers for retail sales rooms larger than 12,000 square feet. Most codes, therefore, require that sprinklers be installed in large anchor stores such as supermarkets. In addition, most basements in mercantile occupancies must be sprinklered. Today, new nightclubs with occupant loads over 100 (a new lower threshold because of The Station nightclub fire in Rhode Island) or over 5,000 square feet are sprinklered. Keep in mind that only certain areas of the strip mall may be sprinklered, and this only applies to new strip malls and tenants.

While conducting a preincident planning walk-through, determine the following:

- The areas of the strip mall that are sprinklered.
- Whether the system provides complete protection (including in combustible attic spaces) or protects only certain areas.
- The type of system—wet or dry (many strip malls with unheated attics often leave unleased tenant spaces unheated, necessitating a dry system).
- The location(s) of the main riser control valve(s) (which could be hidden in the closet of a tenant space).
- The location of the FDC and the areas of the building it supplies.
- Closest hydrant to the FDC.

Utilities

Fortunately, most modern strip malls have multiple utility meters/cutoffs, one for each tenant space or group of tenant spaces. The meters/cutoffs should be identified by "suite" number (the number assigned to individual tenants). Using this system, utility service

to particular tenants or areas of the mall can be shut off. Note the location of the utility meter bank in your preincident plan. In many cases, the meters are on the exterior of the building, although gas meters may be in the basement in some buildings.

Forcible Entry

Another important consideration when developing firefighting strategies for strip mall fires is how you will gain access through the front and rear doors. Each can pose problems. Front doors usually are aluminum doors with glass panels. Using a through-the-lock entry technique usually is called for here and can be readily accomplished. In some cases, it is what is behind the front door—and even the large display window—that can impede access.

Electronic and computer stores and similar occupancies now are installing roll-down metal shutters, located on the inside of the store, behind the front door and display window. These shutters are secured with a padlock. This creates a double forcible entry situation. Consider the implications of ventilating the front display window, particularly how to coordinate the opening of this window with your interior attack.

The rear door of a tenant space poses an even more formidable entry problem. These doors are heavily secured to prevent break-ins and may even have drop-in metal bars. In some cases, the doors may actually be welded shut. Consider your operations for forcing these doors, including the use of metal-cutting saws. Also keep in mind that it is important to open these doors to provide a secondary means of egress for crews operating inside who are cut off from the front door. Another issue involving rear doors is identification. It is difficult to determine which door corresponds with which tenant space without the presence of a name or suite number. During your preincident planning visits, make sure the doors are identified, at least by suite number.

Firefighting Considerations

Fighting fires in strip malls raises many concerns. The following points deserve special discussion.

Lateral fire spread throughout the strip mall is probably the greatest concern. Without the presence of fire-rated separations or draft stops that are continuous through the plenum space to the roof deck above, fire can spread readily from tenant space to tenant space. The ease with which this fire will spread depends on the construction of the roof members themselves.

A roof of solid wood joists with a dropped ceiling below will allow fire to spread in this void space. A roof structure of wood trusses or I-joists creates a large void space, a virtual lumberyard. Fire extension through the truss will be rapid.

A strip mall with steel bar joists and a built-up roof can be subject to a metal deck roof fire. This fire (in what is classified as a noncombustible building) can spread laterally, independently of the main body of fire below. The asphalt in the built-up roof becomes heated and then ignites, spreading fire above the roof deck and dripping hot tar into the store below through the deck seams.

The key to stopping a fire from spreading through the roof area is to get ahead of it. You must pull ceilings and apply heavy streams through the openings created. Consider the time needed to gain entry into adjacent tenant spaces—where will the fire be when you gain access?

Fires in wood truss or I-joist voids will spread rapidly due to the inherent openness of the truss. Consider the quick collapse potential. In a metal deck roof fire, cooling streams must be applied to the roof deck to cool the asphalt and stop the fire's progress.

Besides assisting the interior hose teams, ventilation can help slow horizontal fire spread through the roof area. Ventilation procedures still must be conducted safely in accordance with the type of roof. Fire fighters operating on metal decks are in danger of falling through if the cantilevered end of the deck gives way as the vent hole is cut. The wood bowstring truss poses the danger of large area collapse. In many cases, it is too dangerous to operate on top of or below these lightweight structural members when fire has taken possession of them. Consider a trench cut only when the "distance for time"

is in your favor. Also, do not forget the numerous air-conditioning units and other concentrated dead loads on the roof.

Theaters

To understand the fire protection details in a theater, it is important to first understand the difference between a **stage** and a platform. By definition, a stage has a **proscenium arch and wall** (the large ornamental opening and wall that separates the audience from the stage); hanging curtains, drops, and scenery; lighting; and support rooms (dressing rooms, etc.) Figure 13-35. A platform, in contrast, is a raised area in a building for presentations, wrestling matches, and the like, where there are no hanging curtains, drops, or scenery other than lighting and sound effects Figure 13-36.

Stages have much more extensive fire protection requirements than platforms. Specifically, they must have a fire-resistant proscenium curtain, flame-resistant scenery, heat vents over the stage, 2-hour-rated separations between the stage and appurtenant rooms (dressing rooms, property rooms, etc.), sprinkler protection over the stage and accessory rooms, special stage exits, and a Class III standpipe with 1.5-inch hose and nozzle.

Some terrible fire tragedies have occurred in theaters. The worst was the 1903 fire in the Iroquois

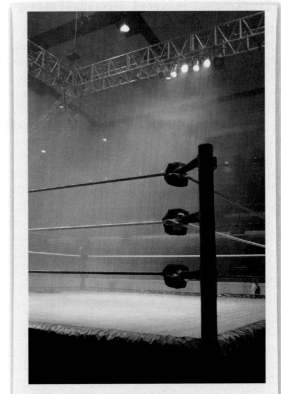

Figure 13-36 A platform used for boxing.
© Photodisc.

Theater in Chicago, where 602 people were killed in minutes. Despite the fact that the theater was only weeks old, many of the fire safety features did not work when a lamp ignited part of the velvet proscenium arch drape. The fire curtain did not drop to the stage floor, and the heat vents above the stage were nailed shut.

Figure 13-35 A proscenium arch in a theater.
© Frances Roberts/Alamy Images.

⊕ Branniganism

We must recognize that the key element in life safety of building occupants is not how high the building is but how long it takes an occupant to reach an area safe from all the products of combustion.

Bolt: © iadams/Shutterstock, Inc.; Texture: © Eky Studio/Shutterstock, Inc.; Steel: © Sharpshot/Dreamstime.com

Underground Buildings

Many buildings have basements, often used for mechanical equipment, storage, and the like. Some high-rises and other types of buildings have multiple "subbasements." Underground buildings are a relatively new concept and are beginning to proliferate across the United States.

Building codes define underground buildings as having an occupied level at least 30 feet (9.1 m) below the level of "exit discharge" (an exit discharge is typically a grade-level exit to the exterior). The code calls for the underground building to be provided with automatic sprinklers, standpipe, emergency power, a fire alarm system, a public address (PA) system, a smoke management system, and smokeproof enclosures (pressurized stairwells). When the occupied level is more than 60 feet (18.3 m) below the level of exit discharge, a smoke barrier that splits the floor level(s) roughly in half and that runs vertically up through all underground levels, must be provided.

Among the most common underground buildings are rail transit stations. Some recent examples of these are an incredible 16 stories below grade. The 9/11 museum at Ground Zero in New York is 70 feet (21.3 m) below grade. Because such buildings may house thousands of occupants at any given moment, moving these people quickly up stairwells to grade is virtually impossible. Tremendous reliance is placed on the built-in fire protection systems, attempting to protect the people where they are. It is imperative that fire departments charged with protecting underground buildings be fully informed on all aspects of their design and be able to voice their concerns.

While we have extensive experience with basement fires, we have limited experience with these super-deep buildings. The 1987 King's Cross Underground (subway) station fire in London provides some insight. In this incident, a fire developed in an escalator with wooden components, eventually flashing over and spreading fire upward through the multiline transit station **Figure 13-37**. A total of 31 people were killed, including a fire officer. While the King's Cross station lacked

Figure 13-37 The King's Cross fire.
© Keystone/Stringer/Hulton Archive/Getty/Alamy Images.

sprinklers and other fire safety features, this fire is illustrative of what may happen when these systems are impaired.

Warehouses

Once there was a simpler time when small fire departments had small fires, and big fire departments had big fires. Those days are gone forever! The advent of warehouse rack storage has brought potential major fire problems to Anyplace, USA. Materials may be stored in **solid piles** (either box on box or pallet load on pallet load), on shelves, or on **racks**.

Warehouse Fire Problems

Many serious fires have occurred in the long history of warehouse occupancies. The principal contributing factors have remained the same:

- Huge concentrations of fuel
- Tremendous dollar values
- Few employees per unit of area
- Failure to segregate extra-hazardous materials such as flammable liquids
- Failure to raise the bottom layer of stock above the floor, thereby preventing water damage and possible collapse
- Vulnerability to arson

- Failure of management to give serious attention to the potential fire problem
- Inadequate fire protection, either in initial design or in maintenance

High rack storage has made its way into big box discount warehouses. This introduces a significantly increased life hazard with potentially over 1,000 people in the store at any given time. The potential major fire problem of large storage is not limited to huge warehouse complexes. Storage areas of manufacturing facilities and smaller retail establishments, shipping and receiving areas, general rental storage warehouses, and outside storage adjacent to buildings present similar problems.

Such was the case in 2007, when a fire erupted on the loading dock between the Sofa Super Store retail showroom and its associated warehouse in Charleston, South Carolina. Fire spread to the unsprinklered showroom, which was constructed of concrete block with an unprotected steel bar joist. A variety of factors—high fuel load, hidden build-up of heated gases and smoke between a dropped ceiling and the roof, and breaking of the front showroom windows—led to rapid fire spread through the store and the deaths of nine firefighters.

Pallets

The problem of fire in warehouses was already significant when material was being stacked only by hand. About 50 years ago the lift truck became available. With this device, stock could be stacked on pallets, and pallets could be stacked one atop the other. This development provided a massive new dimension to the fire protection problem of warehouses. The pallet storage system provided as much as 36 times the surface area as boxes stacked solid. Initially, this problem was not universally recognized. Automatic sprinkler systems had such an excellent record that it was taken for granted that a sprinklered warehouse was safe.

Be aware that **idle pallet storage** (pallets without the product) is a dangerous practice inside of warehouses 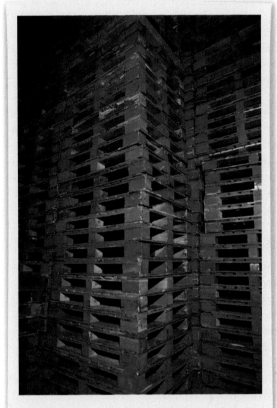. When they are piled up,

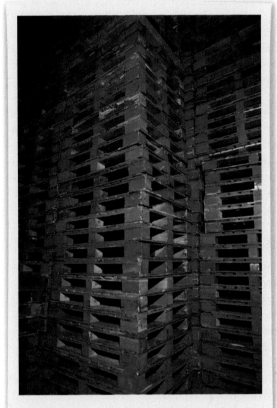

Figure 13-38 Dangerous idle pallet storage—plenty of fuel that is well ventilated.
© ThinkStock/age fotostock.

the pallets resemble the wood crib used to test fire extinguishers. They should be taken outside and away from the building.

Shelving

Although pallets are still used for storing some commodities, many other items require shelving. Lift trucks also make it possible to place stock on shelves. These layers—in effect, miniature floors—create a whole new fire problem.

An estimated $14 million loss occurred in a rack storage warehouse in Kernersville, North Carolina,

in March 1981. Sprinklers were not provided in the racks in a section almost 16 feet high. Although the fire had been burning for only 10 minutes when the fire department arrived, an interior attack was impossible. About 15 minutes later, the roof started to collapse. There was no FDC to the sprinkler system.

In the past, the FDC was optional. NFPA 13, *Standard for the Installation of Sprinkler Systems*, now requires the connection except for systems of 20 sprinklers or less or when permission has been obtained from the authority having jurisdiction (AHJ). Because the sprinkler system is a vital element of the fire department's suppression, the AHJ should be the fire department, and no such exception should be granted.

Modern Rack Storage Warehouses

Rack storage buildings—immense one-story warehouses in which the contents are stored on unprotected steel racks—are now found across the United States. Typically, such buildings are of noncombustible construction. The size can be unlimited. A Kmart warehouse that was destroyed by fire was as large as 33 football fields.

In these warehouses, merchandise is handled by mechanical equipment. In the most advanced warehouses, the operation is fully or partially automated. Computers keep stock records, and a signal from a computerized system operator can send the traveling elevator to the correct location. The elevator platform then ascends to the proper level, and the desired item—for example, a refrigerator, pallets of radios in polystyrene packaging, or automobile parts—is loaded onto the elevator. In some cases, the stock moves automatically all the way to the truck that will carry it to its destination.

A rack storage warehouse, depending on its height, is like a multistoried building without the fire resistance provided by even the poorest floor. Ordinary sprinklers located at the roof line are completely inadequate to control such a fire hazard. **In-rack sprinklers** must be provided at intermediate levels as well. Within certain limits, **early suppression/fast response (ESFR)** sprinklers are able to suppress a fire without in-rack sprinklers;

they allow warehouse owners flexibility to move the racks without having to repipe the in-rack sprinklers.

Some warehouses have interior rail tracks so cars can be unloaded within the warehouse. A railroad car represents a major unsprinklered area within a warehouse. A fire in a loaded railroad car within the building would be a major problem. Other warehouses have very tall storage arrangements (over 50 feet [15.2 m]) with automated product retrieval systems.

Dry Storage of Boats

A special type of rack storage warehouse stacks boats several levels high in open or partially enclosed rack structures. These warehouses are usually unsprinklered. The boats are of combustible fiberglass and many contain fuel. A hot, fast-growing fire with early structural collapse can be anticipated. Preincident plans should include immediate use of heavy-caliber streams in a defensive attack for a fire in such a facility.

Warehouse Concerns

There are several contributing elements in the potential for huge loss warehouse fires:

- Modern contents of warehouses are increasingly higher hazard materials such as plastics, flammable liquids and gases, and chemicals. Even noncombustible materials are packaged in combustible containers and packaging.
- Huge quantities of material in one fire area provide fire loads that can easily overcome the best defenses.
- Higher and deeper aisles limit access and encroach on the space over the tops of piles required for effective sprinkler or hose stream operation.
- The size and construction of the buildings often make even the heaviest caliber fire streams ineffective.
- Automatic sprinkler systems as installed are often inadequate for the potential fire problem.

- Automatic sprinkler systems that are adequate for the job as installed can be defeated by changes in the operation and storage patterns of the warehouse.

The Warehouse Building

Typically, warehouses are classified as noncombustible. This term is a code classification, not necessarily an accurate description. If the building is concrete, it is inherently noncombustible—it will not burn. The building is not inherently fire resistive. Unless it is fire resistive, the building may disintegrate and collapse in a fire.

In warehouses, tilt-slab concrete construction with precast wall panels is common. The roof is necessary to stabilize the walls. In addition, concrete T-beam roofs are subject to spalling or disintegration of the concrete below the tendons. Tendons exposed to 800°F (427°C) (less than the temperature produced by a self-cleaning oven) will lose their prestress capacity.

If the roof is a conventional metal deck built up from layers of asphalt and roofing paper, it can burn independently of the original fire, elongating the bar joists, which then push down the walls, possibly quite some distance from the original fire. At a later stage, the bar joists will collapse. In some warehouses, the racks themselves are the building's structural system.

Static Fire Defenses in Warehouses

Subdividing fire areas in a warehouse is accomplished by the use of fire walls. Fire walls in steel structures probably are not freestanding. Fire walls are considered to be **passive fire protection**, but there are few fire walls without openings. The integrity of the fire wall depends on the active closing of fire doors when a fire occurs. As many as 50% of fire doors in supposedly well-protected properties have been found to be inoperative. Overhead rolling doors are particularly susceptible to failure. In one case where 12 doors were trip tested, only 3 operated properly.

The solid masonry wall parapeted through the roof is the most dependable fire barrier in a warehouse. However, just because you can see a wall extending through the roof, don't assume that it is an effective fire wall. It may be decorative or it may be pierced.

In some cases, fire walls are made of combined elements. Consider a high bay steel-framed warehouse. A main girder is put in place with an upward camber. The parapet brick wall is built above the girder. Concrete block is built up to the girder. The girder is sprayed with fireproofing. There is no proof that these disparate elements will function together as a fire wall. If steel attached to the girder is heated, it may move the girder.

Dynamic Fire Defenses in Warehouses
Automatic Sprinklers

The principal active defense is automatic sprinkler protection. For more than 100 years, sprinklers have had an unparalleled record of suppression or control of incipient fires. Unfortunately, this superb record cannot be taken as an indication of what can happen in a high or dense storage warehouse. Many huge warehouse losses have occurred in sprinklered buildings. In too many cases, the sprinkler protection is inadequate and fails early, sometimes before the fire department arrives.

One of the causes of failure is the early distortion and collapse of the steel roof from which the sprinkler system is suspended. This can be caused by the exposure presented by a fire in stored pallets. Factory Mutual has developed recommendations to prevent the special problems caused by idle pallets from occurring, such as putting limits on pallet pile heights and keeping them outdoors.

As these fire hazards grow in size, the concept of sprinkler protection as we have known it may be superseded by or coordinated with fixed oscillating nozzles, similar to the water cannons used in the San Antonio Alamodome described in another chapter. Such systems are now used to protect huge outdoor hazards such as large lumber piles and refineries.

Foam System Protection

Some sprinkler systems are equipped to deliver low-expansion foam. These systems are used where there is a potential for flammable liquid fire. There is also some use of high-expansion foam. The *Chicago Tribune*'s rolled-paper warehouse is protected with a massive high-expansion foam system. Twenty-two massive foam generators are mounted on the ceiling. In 6 minutes, the entire warehouse could be covered or filled with foam, 50 feet (15.2 m) deep.

Attitudes Toward Warehouse Fire Protection

Management Attitudes

A distinction in attitude can be made between the management of a major national corporation operating a distribution warehouse, which probably has a loss control department, and entrepreneurial, currently fashionable bulk distribution businesses.

The fact that such tremendous risks for potential losses are developing seems to have caught American management unaware. In the minds of many managers, the possibility of a disastrous fire is the reason why insurance is purchased. The annual cost of insurance makes it easy to factor it into the cost of the operation. Thus, the problem is disposed of by financial problem solvers. Management attitudes can, in turn, range from "There is a serious fire potential here, and it's my job to see that it doesn't happen," to "These fire types are petty bureaucrats who are warts on the backside of progress."

It is unlikely that management is fully familiar with the details of serious fires. The fire department should be as familiar as possible with the circumstances of such disasters. The more closely the predisaster situation at a facility matches the fire experience, the more credibility it has for the warehouse manager. It takes a lot to overcome the attitude that "It can't happen here."

Many years ago, a major fire occurred in the Smithsonian Institution. At a meeting afterward, Frank Brannigan was seated next to an official who he knew had bitterly opposed the installation of a fire door recommended by Harold "Bud" Nelson. The fire door had saved from damage or destruction the flag that inspired the *Star Spangled Banner* national anthem and the national stamp and coin collection. Brannigan remarked on the fact that the unwanted fire door had performed admirably. The official said, "I'm the one who opposed that door. I've learned one thing here. If you design a major building, assume there will be a major fire." When we can get management to operate from the premise that there will be a major fire sooner or later, the rest is easy.

As a code official, I've had similar experiences dealing with warehouse owners, especially when the code was first enforced. Many had never seen the high-piled stock requirements of the fire code, so they were reluctant to upgrade their fire protection, including making costly smoke removal and upgrades to the sprinkler system when a new tenant was to move in. Most of the warehouses had been built to handle less combustible materials (Class I or II), yet were supposed to handle Class IV and plastic commodities.

Fire Department Actions

With regard to these warehouse structures, the fire department has several distinct areas of interest:

- Initial planning and plan review prior to construction
- Inspection of construction and witnessing of fire protection equipment tests
- Routine and special inspections during operations
- Regular liaison with the warehouse manager
- Adequate planning for fire suppression, "if and when"

These activities can be complicated if the warehouse property is located in two or more governmental jurisdictions.

Initial Planning and Plan Review

As soon as the municipality is aware that a warehouse structure is planned, a task force should be

assembled from all concerned agencies to address the many considerations. The first critical need is to establish what is being stored and how it is being stored. The San Antonio Fire Department instituted a "commodity letter" program. The letter asks the tenant questions about the materials themselves, where they are located in the building (diagrams), the method of storage (rack, solid pile, shelf), the aisle widths, rack dimensions, and the like. The tenant answers the questions on his or her own letterhead. This document then becomes a road map for inspectors to determine compliance and a permanent record of the facility. Fire departments are encouraged to develop such letters for code compliance and preincident planning efforts.

Water supply should be adequate for sprinklers and manual firefighting. Water supply for manual firefighting should not decrease sprinkler supply below that necessary for effective performance. Particularly in the case of tilt-slab buildings, hydrants should be located well out of the collapse zone to avoid loss of apparatus and to prevent subjecting fire fighters to a hazardous removal operation. Beware of yard hydrants in older industrial plants, which are intended to supply handlines only—they are not intended to supply pumpers.

Electrical service to fire pumps should be unaffected by a fire in the warehouse. If power must be cut to the warehouse, the pumps should continue to operate. Standby diesel power should be adequate to replace the entire electrical pump supply.

Often the only access to such a facility is through a single guarded gate. Additional road access and/or protected gates at distant points on the perimeter should be available for emergencies. Evacuation routes must be established in advance.

The potential for air contamination from fire smoke or other air toxics and the potential for groundwater contamination from fire suppression water should be planned for.

Insist on fire department Siamese connections being included in the sprinkler system. Although the on-site pumping system and mains may be adequate, the Siamese connection is the only practical method for the fire department to supplement the sprinklers. It is also necessary to supply the system and keep up the water flow and alarm feature if the pump or main is out of service.

Inspections During Construction
Whenever acceptance tests are conducted on the sprinkler systems; fire doors, particularly overhead rolling doors; or other fire protection features, fire department personnel should be present.

Routine and Special Inspections
Every effort should be made to assure management that the fire department's efforts are directed toward maintaining the warehouse as a vital contribution to the economy of the community. If this attitude is well received, a genuine cooperative effort may be achieved. On the other hand, management may be hostile, and perhaps require that an administrative search warrant be obtained before any inspection can be made. Maybe fire departments should adopt a policy that fire fighters will go no farther during a fire than is permitted during an inspection.

Fire prevention education commends neatness. However, neat, orderly storage on racks may create the impression that there is no hazard, when in fact the warehouse may be a potential disaster. During inspections make a particular effort to note when there is an increase in the hazard classification of the commodities stored.

Storage in the aisles can extend the fire and overwhelm the sprinkler system through fire jumping aisles. The usual excuse, "It's only temporary," can be countered with, "Right, the building may not be here in the morning."

Few people are aware of how fast a fire can develop. Make sure that warehouse personnel on all shifts, including contract cleaners and guards, are instructed by management that the fire department is to be called even if a fire is only suspected. First aid or brigade firefighting must not delay the alarm. Be familiar with the stories of major fires and use them in classes and casual conversation.

Preplanning

A specific member of the fire suppression forces should be assigned as liaison officer to be completely informed on all factors that affect firefighting operations in that building. It would be the liaison officer's concern to see to it that all vital information is disseminated to all those who should have it and that preincident plans are kept updated. The warehouse manager should designate a specific senior subordinate to maintain liaison with the fire department liaison officer.

On the Fireground

All personnel should be aware of the known weak points of the warehouse structure and construction. Particular emphasis should be placed on the potential for collapses or failures:

- A combustible metal deck roof fire
- Failure of pretensioned concrete T-beams
- Failure of truss roof
- Failure of connections or splices of heavy-timbered roof
- Potential fire collapse of tilt-slab walls outward and inward

Racks may be erected across the openings at the far end of aisles, making dead-end aisles. Furthermore, some warehouses have movable racks to create more storage room. When stock is to be handled, an aisle is made by moving racks, thus closing other aisles. It is conceivable that an electrical malfunction during a fire could cause racks to shift. The operation should be carefully studied to determine any hazards that might occur during a fire.

Note that fire codes call for fire fighter access doors every 100 feet (30.5 m) in a high-piled stock warehouse. Although there will not be a door at the end of every aisle, these doors (if present) should be opened/forced early in the fire to provide emergency egress for fire fighters **Figure 13-39**.

Solid Rack Shelves

After the Triangle Shirtwaist fire (New York, 1911) in which 142 people died in an unsprinklered high-rise

Figure 13-39 Not every aisle will provide egress from the building.
© Anton Gvozdikov/ShutterStock, Inc.

garment factory, the garment industry moved into sprinklered high-rise buildings in midtown New York City. Garment making generates huge amounts of combustible scraps that can accumulate under cutting tables. Fires can rage under these tables, unaffected by the sprinklers. The width of garment cutting tables is permitted to be 48 inches (61 cm). Some authorities having jurisdiction have interpreted the 48-inch concept to permit several levels of unsprinklered shelving no more than 48 inches (61 cm) wide. This interpretation is disastrously mistaken.

Many racks in retail big box stores have installed shelves (to prevent sales items from falling through). Solid shelves have also been found in warehouses. This blocks sprinkler spray. The only way around this problem is to place in-rack sprinklers under each shelf, a difficult and expensive but acceptable solution.

Ventilation

With warehouse fires, an argument rages as to whether it is better to close up the building and let the sprinklers do the job or to vent it and attempt a combined manual/automatic attack. The jury is still out on that one. Partisans of each approach present convincing arguments.

The fire code used in San Antonio and many western and southwestern cities during the late

1980s and early 1990s required smoke removal fans for some higher hazard commodities. Essentially, 30,000 cubic feet per minute (cfm) fans were positioned on the roof and covered 10,000 square feet (929 m²) areas of the building. Such a system was used successfully when a fire in a walk-in dumpster filled a Sam's Wholesale Club in San Antonio with smoke. The smoke removal system mechanically exhausted the smoke quickly.

Many warehouses use mechanical smoke and heat vents along with draft curtains to ventilate a building during a fire. The vents themselves may be fusible link-activated clamshell-type openings or rectangular plastic domes that shrink under the heat of a fire and drop into the warehouse below. Draft curtains—typically made of sheet metal—are used to contain the heat from a fire below, accelerating the operation of the vent Figure 13-40 . San Antonio, Texas, has permitted the use of manually operated smoke and heat vents in warehouses, with release cables positioned at the warehouse roof's edge, avoiding the need to send fire fighters up on to the roof itself.

There has been much debate over the interaction between vents and sprinklers, with some arguing that opening a vent will slow the opening of sprinkler. In any case, they are required in building and fire codes and are installed in warehouse type buildings, including big box retail stores.

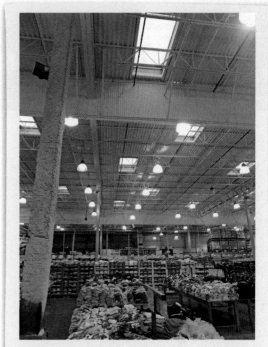

Figure 13-40 Warehouse stores with high-piled stock often incorporate smoke and heat vents with draft curtains. The curtains shown here are 4-foot-deep corrugated steel panels.

Courtesy of Glenn Corbett.

Handline Operations

In a number of warehouse fires, smoke has suddenly banked down to the floor, leaving fire fighters unsure of the way out. The hose line is the fire fighter's lifeline. "Follow the hose back to safety if lost" is one of the oldest instructions to new fire fighters. Remember, however, that a hose line fed from an interior hose outlet is not a lifeline. If interior outlets are used, lifelines should be strung to the exterior from the outlet. Use thermal imagers—they are essential.

Personal Safety

The art of warehouse protection is inexact to say the least. Impressive-looking systems have failed totally. Even if the management's planning is adequate, the execution is often faulty. All planning should place the safety of fire fighters first. This is the fire department's responsibility—no one else is going to take care of it.

TACTICAL CONSIDERATIONS

1. The evacuation of modern high-rise buildings has always been a challenging task. Fire operations primarily take place on the fire floor, the floor above, and two floors below the fire. Staging is usually two floors below the fire. Most critical is designating separate "attack" and "evacuation" stairwells; evacuation stairwells allow occupants to move safely to lower floors or out of the building, keeping them out of the "chimney" in the attack stairwell above the fire floor. Occupants should be moved to five floors below the fire, and then protected in place, if they are not evacuated completely out of the building. However, since the collapse of the World Trade Center Towers on September 11, 2001, occupants may no longer feel confident nor comfortable with anything less than total evacuation from the building. This can make evacuation an almost unmanageable problem. People will be anxious, perhaps even hysterical. A strong police presence may be required to provide the working space and order needed for fire department operations. Consider bringing in city transit buses to help shelter evacuees during inclement weather.

2. Water passes through tile arch floors easily, so a serious salvage problem should be anticipated with this type of construction. These floor units are actually arches and cutting out a tile may drop the arch. It's best to leave these floors alone. It is possible that the entire bay may collapse if the integrity of the floor is destroyed by saws. Even if personnel are protected against collapse, the collapse of a large floor section may cause an upward burst of fire.

3. Because of the falling glass hazard at high-rise fires, the preplan should include pertinent information on safe entry into the building, such as through an underground passage. Los Angeles, California, city fire fighters used an underground tunnel to enter the First Interstate building high-rise fire.

4. Interior personnel should be notified immediately if the windows blow out. This can dramatically change the situation on the fire floor for better or worse. In the Empire State Building fire, the windows failed on the windward side and pushed the fire down the hallway at the fire fighters. Winds in excess of 25 mph (40 kph) cannot be reversed with positive-pressure ventilation (PPV) fans. The situation created is called a *wind-driven fire*. (Two Fire Department City of New York [FDNY] fire fighters were killed in Brooklyn when the windows failed on the windward side of a residential high-rise, creating blow-torch-like conditions in the hallway. There was no reaction time for the fire fighters to escape.)

5. The need for search and rescue should never be overlooked. In a global economy, business employees can no longer be limited to the traditional nine-to-five work schedule. In addition to janitors, maintenance personnel, and computer technicians, a high-rise office building can have employees working around the clock. Often, they are working alone or in reduced staffing groups unknown to anyone else in the building. Don't assume a commercial high-rise building is unoccupied at 2:00 in the morning.

6. Fire fighter access to stairways may be impeded by the sheer number of people descending them. In residential occupancies, it is important to establish a firefighting stairway and an evacuation stairway. It depends on where the fire is; one objective is to get the nozzle between the fire and as many occupants as possible. Barring this strategy, if only one of the stairwells is pressurized, use that one for the evacuation stairwell. If possible, announce the designated stairway over the PA system to direct the occupants to safety.

7. There have always been two criteria for successful and efficient forcible entry: the door and locking system encountered and having the right tool for the job. It's not practical to go back to the apparatus for the right tool when you're on the 23rd floor. It may be much simpler to go through the wallboard and ignore the lock and the door altogether. This

may also be possible from apartment to apartment in upper levels of a high-rise or even in duplex and triplex apartments. The drywall is the weakest part of the wall. Wall studs that are toenailed into the bottom plate can be knocked out with any forcible entry tool.

8. Fire fighters forcing entry to a stairway door on to the fire floor may face a difficult, time-consuming task. It's easier to make the keys to the stairway door available to the attack team as they enter the lobby. This is the job of the unit assigned to lobby control. The attack crew must prepare for a burst of smoke, heat, and the possibility of fire as they open the door. Remember to keep control of the door. The stairway and landing above may suddenly become a hot, smoky chimney. Even with a designated "evacuation" stairwell, always check for occupants mistakenly descending the "attack" stairwell before opening the door to the fire floor. Make sure the standpipe connection for the attack line is made on the landing below the fire floor for the same reasons.

9. Some fire departments restrict the use of elevators by fire fighters going up to attack an upper floor fire. However, in high-rise fires, it is necessary to use them so you don't wear out your firefighting crews climbing 30+ flights of stairs while carrying heavy equipment. First-in companies will make the decision to use elevators long before a chief officer arrives on scene, so the practice should be spelled out in the department standard operating procedures (SOPs). Make sure the fire fighters assigned to run the elevators know how elevators operate in Phase I and Phase II, especially in Phase II. Elevators, especially in newer high-rise buildings, can have different Phase II sequences. Some elevators use different keys to place the car in Phase II and a totally different key to run the car from inside while in Phase II. The operator should initially stop every five floors to check the shaft for smoke. As fire fighters enter the elevator, they should all be wearing SCBA. The operator should stop well below the fire floor to allow the initial fire teams to get a quick visual orientation of a typical floor to help them maintain situational awareness once they enter a smoke-filled floor.

10. When an elevator shaft has a fire-resistive enclosure made of gypsum board, it is possible that hose streams striking the wall can disintegrate the gypsum board, making an open hole into the shaft. Holes can be unintentionally made into the shaft on the fire floor or exposure floors, without the elevator car in place. One Maryland fire fighter survived a multistory fall down an open shaft; however, two Chicago fire fighters did not and were killed when they fell through an open shaft.

11. When stack effect is a significant factor due to weather conditions, fire fighters should understand the situation created, for example, when the burn index is high. Company officers should be aware that the first smoke detector activated or the first call to the fire department may have come from a person many floors away from the scene of the actual fire. The IC should resist committing all his resources until the location of the fire is determined.

12. Stack effect can lead the IC to change tactical operations. For example, consider a fire in a high-rise building during winter. Let's say the fire is in a unit below the neutral plane and is also accessible from aerial ladders or platforms outside of the building. If conventional operations were used, the fire department would charge the standpipe and interiors crews would lay hose lines and enter the fire floor from the stairwell to attack the fire. Heat and smoke from the hallway would enter the stairwell and ride the available stack effect currents and contaminate the upper floors. An alternate attack might be to protect the interior exposures with charged hose lines but leave the door to the fire floor closed while the attack is made from the outside by aerial master streams. After the fire is knocked down, much of the smoke would vent out the windows by thermal pressure. The fire department can set up PPV fans and, (with a little luck) if the wind is blowing in the right direction, the fire floor could be quickly ventilated. The engine crew could enter with handlines and finish extinguishing the fire. The fire loss to the unit and floor would probably be the same with either attack, but the smoke damage and occupants endangered by the products of combustion throughout the building would be substantially reduced.

13. The fire department must train company officers so they are familiar with the many aspects of HVAC systems. The HVAC system can play a major role in the success of fighting a high-rise fire or turning it into a disaster. Company or chief officers who have the most experience with HVAC systems should be assigned to the command post to help the IC. All the necessary information and emergency contacts on

the structure's HVAC system should be included in the fire department's preincident plan. When a fire occurs, the knowledgeable HVAC technician should be called immediately to reduce response time and have that person report to the command post. A knowledgeable elevator technician should also be called at the same time.

14. If only certain stairwells are pressurized stairways, they should be reserved for evacuation purposes if possible.

15. Watch for signs like "Pardon our dust; we're remodeling." It may be a clue that structural components are temporarily being modified, which usually means they are not as strong as we would expect them to be. It may also mean that fire protection systems may be on hold and not readily available.

16. Rubbish is usually accumulated in the service lobby on each floor. Fire departments should be wary of using the service elevator to routinely approach all fires. Because of the prevalence of rubbish fires in high-rises, New York City fire fighters are not permitted to use the service elevator until it is declared safe.

17. Particular attention should be paid to the hazard created when the regular removal of rubbish is interrupted by a labor strike or other causes.

18. Beware of the obvious dangers of locked doors in prisons. Several fire fighters were temporarily trapped while extinguishing a fire in a county jail in New Jersey when a door closed and locked behind them.

19. Parking garages present a number of problems— especially underground garages. They are required to be sprinklered, but there are many areas where parked cars are outside the reach of the spray pattern. A car fire can go unchecked and even spread to adjacent vehicles outside the reach of the sprinkler heads. With low ceilings and no windows, smoke quickly banks down to the pavement. Finding the seat of the fire is difficult without a thermal imager. Extended hose lays are required and immediate ventilation is essential. Consider calling for extra PPV fans or a mobile ventilation unit (MVU).

20. The garage area may extend out beyond the exterior walls of the building. This was the case in Seattle where a motorcycle fire spread to the car next to it. The flames intensified and spread up the exterior B side of the building. The building had vinyl siding that allowed for rapid vertical spread. The fire broke out a window and started a full apartment fire on the fourth floor.

21. Overhead parking slabs may be designed only to carry the weight of cars. Heavier vehicles, like fire apparatus, may be restricted. One shopping mall in the Midwest consists of many buildings on a lightweight slab over the garage. No horizontal standpipe was installed. Fire apparatus cannot drive on the slab, so extremely long hose line stretches are required.

22. The high-volume fans used for PPV may be used to clear a stairway of smoke so that it can be used for evacuation. Consider the use of MVUs to pressurize stairwells beyond 20 floors. Don't worry about the production of carbon monoxide (CO) put out by the fan motors. The CO generated by fans is substantially less than the CO in the smoke being generated by the fire. Electric PPV fans are an excellent alternative, but still lack the total cfm output achievable with a motorized fan. Use an air monitor to check CO levels. After the smoke is cleared, change the ventilation plan to vent for CO. Consider using fan socks or RAM fans.

23. Expect that the rear doors of tenant spaces in a mall will be heavily secured to prevent burglary. Commit enough resources and equipment to force these doors, which are critical for horizontal ventilation. Crews need to remember to bring extra saw blades on the initial deployment. You don't want to waste time sending a fire fighter all the way back to the apparatus to get a new saw blade.

24. Opening the front display window in a retail store can provide substantial horizontal ventilation, but this must be coordinated with your interior attack hose team. Teach your fire fighters not to break windows for the sake of breaking windows. They must wait until the hose team is ready to enter. Remember, without pressurized ventilation, breaking a window will create the path of least resistance and the thermal pressure created by the fire will draw the fire right to the broken window. There are several videos showing fire fighters breaking display windows before the hose team is ready, causing a contents fire to turn into a structure fire. Don't be that fire fighter.

25. Fires can also spread horizontally below grade through a basement. Many strip mall basements do not have substantial separations between occupancies, allowing the fire to move unimpeded.

26. High-piled storage racks (rack storage) are being used in most big box warehouse stores throughout the country, even in areas that are protected by fire departments that have never faced such a fire problem. These warehouses present an extreme fire hazard and safety hazard to the life of the fire fighter. Fire officers and fire fighters alike must recognize that a completely different strategic situation exists. If the fire is attacked with standard tactics used for single-family dwellings and mercantile buildings, the fire fighters are going to be in for a big surprise, and a terrible loss of fire fighter lives may result.

27. Until the fire is knocked down, high-piled storage rack fires should be attacked with 2.5" (64 mm) hose lines or portable monitors. With the increase in fire load as well as the different types of fuels stored closely together, you need reach and gallons per minute (gpm) to safely extinguish this type of fire. Though 2.5" (64 mm) hose lines are needed for safety, the force of the streams hitting stored merchandise is going to cause storage containers or loaded pallets to fall off the shelves. Some of these containers are extremely heavy and can fall from a height of 30 feet (9.1 m). Falling merchandise can also fall against adjacent racks, causing them to collapse. The variables are great. High-piled storage racks that are involved in fire is no place for fire fighters to be working in close proximity. Use the reach of your hose line to operate from a safe distance.

28. The most important consideration in prefire planning is the safety of fire fighters. Operations at a fire in a rack storage warehouse require detailed advance study and preplanning. This must start when the structure is proposed. You cannot make it up as you go along. Fire departments whose experience is primarily in residential fires may have a "house fire" mindset, which can result in disaster in a warehouse fire.

29. Racks are simply lightweight steel structures. They have the same characteristics and weaknesses of steel when exposed to temperatures in excess of 800°F. They are subject to collapse due to the heating of steel. Fire fighters should never advance beyond the point where racks might collapse behind them. Attack the fire using 2.5" hose lines or portable monitors.

30. Due to the increase in fire load and mixed fuel classifications, venting big box warehouse roofs can be inadequate and dangerous. Outside of mill construction, most modern warehouse roofs now use lightweight steel or wood components. The amount of heat and Btu(s) produced require more than the standard 4- by 4-foot (1.2- by 1.2-m) hole, and these holes need to be made quickly while it can still withstand the weight of fire fighters. It should not be undertaken just as a routine to "get the roof." Experience limited to ventilating ordinarily single-family dwelling roofs may not be adequate. Consider using multiple PPV fans or an MVU. As more big box warehouse stores appear in communities, the need for MVUs is going to increase. It's simply a more efficient and safer way to ventilate a warehouse. An MVU can be operated by a single fire fighter.

31. No fire fighters should be permitted to move beyond the point where a defined path to safety exists. Hose lines can be repositioned and can also be confused with warehouse hose station lines that may have been deployed by employees attempting to fight the fire. When dealing with commercial structures and warehouses, it always increases the safety margin for fire fighters when lead lines or rescue ropes are deployed to help fire fighters find their way in and out of a burning warehouse. Battle lanterns and strobe lights placed at the entry/exit point also help fire fighters orient themselves toward the exits.

32. If a warehouse is heavily involved in fire upon arrival of the first fire unit, it is most likely lost. Use extreme caution in proceeding. It isn't worth getting a single fire fighter injured or killed for property that is probably insured and no longer has any salvageable value. Probably the best offensive tactic would be to supply the sprinkler systems to the maximum. Any systems or piping that has been destroyed in a collapse should be shut off at the main valve if possible to conserve water for defensive operations. Repeat, this is a defensive operation.

33. On March 27, 2007, Houston, Texas, Fire Captain Eric Abbt came within inches of his life when he became disoriented and lost. Houston fire fighters

Flames © Alma1968/Dreamstime.com; Texture © Yanas/Shutterstock, Inc; Steel © SharpShot/Dreamstime.com

responded to a six-story mid-rise fire in an open-concept atria building. The building was one giant open space from the roof down to the lobby. All the occupancies faced the atrium. The fire started in a dental office on the fifth floor and charged the entire building with smoke. Captain Abbt and his crew were assigned to search the fifth and sixth floors.

34. Upon entry to the atrium, there was zero visibility. The division chief assigned to the interior sector encountered zero visibility 10 feet (3 m) into the lobby. The chief had to crawl into the lobby to get some visual reference. Though there was a lot of smoke, there wasn't sufficient heat to fuse the sprinkler or to melt the fusible links to the automatic roof ventilation hatches. But even if there had been enough heat, the hatches wouldn't have opened. Six month prior, in preparation for an oncoming hurricane, the skylights and vent hatches were secured shut with sheet metal screws. After hurricane season, the screws were never removed.

35. Stairwells with roof access often have a narrow vertical ladder to a hatch that opens to the roof. The hatch usually has a padlock for security. Such was the case with the Houston mid-rise fire just discussed. But if you think a fully equipped fire fighter with SCBA and size 14 fire boots is going to be able to climb that ladder in zero visibility and use bolt cutters to clip a pad lock without falling off, think again. This might be possible during a prefire planning or a fire inspection, but it is impossible during a fire with thick black smoke. You have to find another way to access the roof. Through a series of hit-and-miss attempts, Captain Abbt was finally rescued when a window was broken out with the tip of the aerial ladder. He was semiconscious from the CO and had been out of air for over 20 minutes.

36. Local building and fire codes are sometimes more restrictive than the national codes because they are more specific to the hazards within the community.

37. Building codes usually don't require garden apartments of three stories or fewer to have a standpipe. Many apartment buildings are built to three stories to avoid this requirement. Therefore, if you come across a fire on the third floor of a three-story apartment, you will be laying hose lines from the apparatus to the fire room. If the fire is anywhere in a four-story apartment building or greater, you will be able to connect to a standpipe.

38. Three-story townhomes are becoming quite popular as single-family residences. Usually, there are two floors above the garage. The living area and kitchen are on the second floor, and the bedrooms are on the third floor. The main stair entry (one flight of stairs) is next to the garage. Many townhomes are built four to a plot. That means a fourplex of three-story townhomes. The space between them is narrow, which means the exposure problem is great, and fire apparatus have to remain in the street. It also means there is no aerial access to units in the back. Long lays are required to make the bedroom level in any of the townhomes. For the C side of the two townhomes in front and for the townhomes in the back, there is no aerial access. That means ground ladder evolutions—35-foot (10.6 m) ladders for the bedroom windows and 40- to 45-foot (12.2- to 13.7- m) ladders if you want to reach the roof.

39. Because of the height and close proximity of three-story town house structures on a single lot, many cities are adopting the NFPA 13R so the living area and bedrooms are sprinklered. That should help contain a room and contents fire. However, the garage may or may not be sprinklered, and usually the attic space is not sprinklered. This will require some preincident planning in your area where high-density neighborhoods are on the rise.

40. For a variety of reasons, some three-story townhomes have residential elevators. These elevators have a low load capacity and should *never* be used by fire fighters. Remember, during new construction, you have an open shaft: a fall hazard for fire fighters. After construction is complete, the elevator shaft is a vertical void space where fire can travel, so it needs to be checked after a fire.

41. Newer building codes require new multistory apartment buildings with exterior balconies to be sprinklered. Summer barbeques are an American tradition, but some have led to cooking fires on balconies where the overhang let flames quickly spread horizontally back into the apartment unit. If the overhang ceiling has holes, fire can enter into the space between the ceiling and the floor above. Fires on any type of unsprinklered balconies can easily spread to the floor above. Consider using a blitz attack with a 2.5" hose line or a deck gun to knock down a fully involved balcony fire before it spreads beyond your reach.

42. Vertical ventilation of a fire in an attic space with lightweight trusses is always a gamble and a race against time. Consider the indirect attack. There is no need for heroics when you're trying to extinguish an attic fire. You must realize the attic space is hot, oxygen deficient, and cannot support complete combustion. That means there is smoke and heat, but no fire—yet. Punching a hole through the roof large enough to insert a fog applicator or round nozzle is essentially creating a sprinkler system. With the expansion ratio from water to steam (1,700 times its volume), the fire will go out. Another consideration is using the gable ends of the attic space. Working off a ground ladder or aerial apparatus, this horizontal attack is putting the nozzle at the seat of the fire. The hose stream can reach to the opposite of the attic. This position is safer than roof operations. Of course, you can still try to attack the fire from the ceiling below, but remember, if you shouldn't be on top of it, you shouldn't be underneath it either.

43. Steel trusses need to be cooled. Consider an indirect attack from the safety of an aerial apparatus.

44. Bathrooms seem like the most benign rooms in a residence. In zero visibility, it's easy to figure out when you're in a bathroom and it's almost impossible to become lost or disoriented inside one, but they're a wild card. They must be thoroughly searched. For a variety of reason, including having a source for water, children and occupants hide in the bathroom as a place of refuge from a fire. On June 12, 2010, four children and one adult did just that; all five perished in that tragic apartment fire.

45. With regards to collapse, especially in older residential buildings, the bathroom floor is frequently more subject to collapse than the other floors for the following reasons:

 First, the bathroom fixtures create a heavy dead load or static fixed load. Cast-iron tubs, sinks, porcelain toilets, and tile floors can weigh up to a thousand pounds. This heavy weight is concentrated in the smallest room of the residence and, therefore, supported by fewer floor joists.

 Next, the bathroom has the most poke-through holes and concealed avenues for fire spread. With toilet drains; the holes for the hot and cold water lines for the sink, shower, and toilet; plus the lighting, heating, and ventilation systems, there are plenty of paths for vertical and horizontal fire spread.

Though a fire rarely originates in a bathroom, a fire seeking the path of least resistance will be drawn to the bathroom floor, weakening the nearby floor joists.

Finally, wood rot. Bathroom pipes leak or sweat on the outside. Water is spilled from sinks, showers and tubs, and toilets overflow. Over the years, moisture seeps down to the wood joists where they weaken from decay.

46. Bathroom collapses are usually localized collapses. That means the floor joists in the hallway should still be strong. During overhaul, sound bathroom floors with deliberate force.

47. Kitchens floors contain many of the collapse-potential problems associated with bathroom floors: poke-through holes for water lines, gas lines, and drains; and heavy concentrated loads from stoves, refrigerators, and tile floors. Before overhaul, sound kitchen floors with deliberate force or check floor joists from below when there has been extensive damage from fire.

48. Private single-family dwellings are not required by the building code to have fire-rated interior stairs. At best, traditional stairs are made of dimensional lumber but are still subject to fire and collapse. Very recently, a prefabricated stairwell manufacturer has created a wooden stair with risers and treads held together with sheet metal gusset plates. Wooden stairs without a fire-resistive soffit are vulnerable to fire damage from flames below the steps. Wooden basement or cellar stairs are often constructed without soffits and can result in early collapse. When arriving on scene, the initial company officer must determine if there is a basement. If the fire is in the basement, the crew has to be cautious with stairs leading to the front door. The porch also needs to be sounded. Don't take it for granted that you have a solid work platform.

49. When climbing sagging, fire-weakened wooden stairs, fire fighters should stay close to the inside wall. If the outer carriage beam has been destroyed and is not load bearing, the stair riser is supported on the wall side only. Weight, close to the wall, will put less stress on the stair than a size 14 fire boot walking near the outer edge—farthest from the wall.

50. Stairways and landings are the safest place to initiate an interior attack. However, any weaknesses or deterioration of its integrity may be impossible to

determine during the heat of the battle. Therefore, fire fighters should avoid bunching up on the stairway. The fire fighter on the nozzle and the company officer should be at point and the rest of the team should be spread out on the stairwell to help advance the hose line.

51. If stairs appear to be weakened or in danger of collapse, but the situation requires interior teams to remain (i.e., a rescue in progress), fire fighters could place a portable ladder or roof ladder over the weakened stair rise so fire fighters can use the ladder rungs instead of the steps. Ensure that the ladder spanning the stairs is supported by the floor below and the floor above so that if the stairs collapse, the ladder will remain in place. If this technique is not doable, exterior ground ladders must be placed at the windows of the involved floors so interior teams can have access and emergency egress.

52. Predicting a fireground outcome involves understanding gravitational forces and common sense. For example, three-story (or greater) burning wood-frame buildings collapse more frequently than burning wood-frame buildings that are one or two stories in height. If there is more weight supported by a fire-weakened support component, it will collapse. A serious fire that burns out the first floor of a two- or three-story wood-frame building presents an increased collapse potential than if the fire was on the upper floors.

53. Of the three types of wood-frame building collapses, the inward/outward collapse is the most dangerous because it gives no warning and can result in the simultaneous collapse of two or more sides of the structure.

54. Wooden structures located on a corner lot or standalone buildings are more susceptible to collapse when exposed to fire than wood-frame buildings in the center of a row of similar structures. Uninvolved buildings help shore up fire-weakened structures, supporting them if they begin to lean. This doesn't mean they are safe, so don't drop your guard.

55. Larger buildings that collapse on top of smaller buildings cause the smaller building to collapse. Try to predict this gravity-driven domino effect and prepare for it.

56. Older apartment complexes sometimes have the gas meters located in the basement. Relief valves sometimes release small amounts of natural gas, which spews the odor of mercaptan. Because natural gas is lighter than air, alarmed residents will call the fire department. To reduce the time it takes to determine the source, start the portable gas-detecting monitor while en route. It takes a few minutes for the unit to calibrate.

57. There are many places within urban areas where there are limited water supplies. These departments usually do not use water tenders. Keep in mind that modern big-city fire engines carry at least 500 gallons (1,893 L) of water. Depending on the fire situation, bringing in three engines gives you 1,500 gallons (5,678 L) of water. That's a lot of water to put out a fire. Most attack teams do not keep their nozzles fully opened for the duration of the fire; they spray water in bursts, shut down, advance, then open up again. For example, using the National Fire Academy fire flow formula, let's say you have a room fire that requires 100 gallons (379 L) of water to extinguish. With three engines, you have enough water to put this fire out 15 times without connecting to a hydrant. Again, depending on the situation, this tactic is quicker at extinguishing the fire than setting up a long lay or relay pumping operations to establish water supply.

58. City engine companies can also be rotated through at a working fire hydrant in the same way a tender fills up its tank. Depending on the situation, this may be a better option than setting up a long lay or relay pumping operations.

59. There are various styles of wood-frame houses with attached garages. Most contractors use gypsum wall board to meet the 1-hour (or greater) rating requirement for the walls. However, the nail heads and joints are not always properly taped over. This gives an avenue for a garage fire to spread into the house. When dispatched to an attached garage fire, some fire departments hesitate to attack the fire head on for fear they may push the fire throughout the house. Instead, they make entry into the house using the rule of thumb of attacking a fire from the uninvolved toward the involved. An attached garage fire is one exception to this rule. (Basement fires are another exception.) When fire fighters open the interior door into the garage, they immediately create a path of least resistance and the heat, smoke, and fire come toward them and enter the house. I have personally witnessed this tactical error more than once. What started out as a simple garage

fire escalated into a serious house fire. Despite the deficiencies of the gypsum board installation, the odds are on your side. When you arrive at a fully involved attached garage fire, attack the fire head on from the main garage door. Prop the door open with a baby ladder or a pike pole. Send the exposure line inside to protect the garage door leading into the house, but don't open it. Once the main body of fire is out, check the attic spaces and the rest of the house for extension. Using PPV fans to pressurize the house should keep smoke from the garage from entering the interior living space.

60. Fire walls are not always intact. Utilities sometimes pass through them, creating a path for extension. In strip malls or occupancies that use fire walls for separation, consider using PPV fans to pressurize the right and left exposure (exposure B and exposure D). The ceiling space must be opened in the B and D exposures to check for the extension of fire. If the attic spaces are clear, this is your first indication that the fire wall is intact. Continue to keep the exposures pressurized until the fire is extinguished. If there is light smoke, the positive pressure should push the products of combustion back toward the fire, provided the roof over the fire has been vented. This tactic is based on providing the path of least resistance for the fire. Once effective ventilation has been created over the fire (or as close as possible to the fire), the pressurization created by the fans should be more powerful than the horizontal thermal pressure created by the fire. The fans will help push the smoke back out the vent hole.

61. Row houses can extend for several hundred feet, encompassing numerous addresses along the same street. Some row housing can extend to the entire block. There are many challenges and hazards, but one of the paramount hazards is sharing a common cockloft. A trench cut is an effective ventilation tactic for cutting off the horizontal spread of the fire. The challenge for trench cutting is getting ahead of the fire. A trench cut is a labor-intensive, time-consuming tactic. The IC must send sufficient resources and equipment to complete the job in time (a minimum of six fire fighters). One of the errors truck crews commit is cutting an incomplete trench. A trench cut is not simply an elongated rectangle. A trench cut on the roof of a building can be compared to a fire break on a wildland fire. Leaving part of the

roof intact makes all the effort useless if fire can still extend beyond the trench. The trench must extend from one wall to another, for example, from side A to side C. Another error ICs make is not sending a hose line up on the roof to protect the trench or to the floor below to protect the trench from the ceiling. Though the flames and other products of combustion will vent through the trench, the exposures still need to be protected from radiant heat. That can be done only with water.

62. When fighting church fires, many fire departments are nervous about breaking large stained glass windows and rightly so. They should be protected when possible; however, to the fire, it's just a window. In an unscientific poll, 90% of the pastors surveyed would rather lose the stained glass windows than lose the building. If you need to use the large, round, rose-stained glass window to vent the fire or use the master stream, break it out.

63. Large shopping malls are difficult to pressurize and vent. Consider using an MVU. Some fire departments that do not have an MVU have been successful in requiring the corporation behind the construction project to purchase an MVU for the local fire department as part of the fire suppression/life safety system before signing off on the plans of the new monstrosity being built in their community. Remember, one function of ventilation is to direct the path in the direction you want the smoke to go. Many of the large indoor malls have glass skylights or roofs. Taking the precautions for falling glass, crews can create a trench by breaking out glass and the MVU can be positioned so it can force the smoke out of the trench.

64. Though hospital emergency plans call for sheltering patients in place or rolling hospital beds to another area behind fire door enclosures, there may be occasions when patients need to be placed as low as possible. This means getting them on the floor. The best way to do this is by sliding the mattress to the floor with the patient on it, then dragging the mattress across the floor to the safe area. The mattresses usually have fabric loops sewn on the sides. Fire fighters can hook a loop with a pike pole or a body loop strap (webbing) and drag the mattress and the patient like a wagon. This is an easy evolution. Many hospitals have an unoccupied wing where arrangements can be made to train on this procedure.

65. When fighting a fire in a nightclub, send a crew to go around the perimeter to force open doors and windows. Using chain saws or circular rescue saws, turn windows into doors. Remove doors from their hinges so they do not accidentally close again. Tying them off is another good alternative. Wedging doors may not be as effective—a rushing crowd can unknowingly kick a wedge out. A swinging door can injure people or it could close and relatch, narrowing the exit way. Putting the fire out solves myriad problems, but smoke still kills.

66. Most larger new "E" occupancies are required to be sprinklered per the new fire codes; however, there are still thousands of large two-story, brick exterior, traditional school buildings that do not have sprinklers. Remember, if the building is fewer than four floors, there will not be a standpipe. Longs hose lays will be required. If the fire is in a classroom or a lab on the second or third floor, lay a 2.5" (64 mm) hose line to the top of the stairs. Attach a 2.5" (64 mm) gated wye to the end. This can accommodate up to four 1.75" (44 mm) hand lines by attaching two-1.5" (38 mm) wyes.

67. Many house plans call for decorative dormers for the attic space. Though these look like bedroom windows, they are not; however, they can be used for horizontal ventilation; they can also be used for a horizontal attack point on an attic fire when there is no life hazard on the inside. Putting a nozzle through an attic dormer is, in essence, placing the nozzle at the seat of the fire. Use the stream reach to cover the entire attic space.

68. Warehouse fires are not residential fires. Don't pull a 1.75" (44 mm) handline because it's always the line of choice. You need gpm, stream penetration, and reach. There is the potential for long lays and for the line to get snagged around corners and around storage racks. Space fire fighters out along the stretch to keep the line moving forward. It is harder to pinch or kink a 2.5" (64 mm) hose line from fallen stock than a 1.75" (44 mm) handline.

69. Fire department ground ladders can be used to brace up exterior fire escapes when it appears it may be getting overloaded with a large number of panicked residents. Do not hesitate to make this decision. Many 26-foot (7.9 m) ground ladders and 12-foot (3.7 m) roof ladders mounted on top or on the sides of fire engines go untouched at fires.

70. Be prepared to initiate stairwell support operations. This consists of spreading fire fighters out throughout the stairwell. Place one fire fighter every two floors. All firefighting equipment is then shuttled up the stairwell two floors at a time. Once additional personnel arrive, you can place additional fire fighters in the stairwell so they only have to shuttle up equipment one floor, then return for another load. This is the "Old Bucket Brigade" relay system. If the stairwell is clear of smoke, fire fighters do not need to wear coats and SCBA while they are shuttling equipment. Make sure their PPE is on a landing within their work area in case conditions suddenly change.

71. If the strategy is to contain the fire on the fire floor by letting the fuel consume itself, firefighting efforts need to be concentrated on the floor below and the floor above the fire. There is a variety of fuels from carpeting to furniture that are being heated by the fire. They are emitting toxic fumes and gases. The need to be pulled away or wetted down to prevent fire spread. All poke-through holes and HVAC vents need to be continually monitored for fire spread.

72. High-rise buildings with Type II and (sometimes) Type III standpipes have a hose cabinet or hose station in the hallways. Make sure all your fire fighters are aware of the different terms used for this appliance. In Seattle, the term *house line* is used. Ladder Co. 6 was first-in at a high-rise fire with smoke and flames showing on the eighth floor. The first two fire fighters were sent to the fire floor with a 2.5 gallon (9.5 L) pressurized pump can, a TI, and the irons. The officer told them to "Grab the house line and take it!" Not wanting to appear ignorant, the two fire fighters nodded and quickly entered the lobby. On the eighth floor, they blew right past the hose station and tried to contain the fire with the extinguisher. They were not able to extinguish the fire and had to wait for the engine company. Because this was a room fire, had they grabbed the "house line," there is a good chance they would have completely extinguished the fire. The term was explained to them after the incident.

Chapter Summary

- Occupancy plays a role in how a building is constructed.
- Combustible multiple dwellings include garden apartments as well as modern row houses and townhouses and similar structures (three-deckers).
- From a fire protection standpoint, atria are large voids that pass through multiple floors, allowing smoke and heat to move vertically through the building.
- Houses of worship span the five basic types of construction and can present a myriad of challenges for fire fighters.
- A covered mall is a single building enclosing a number of tenants, which may include retail stores, drinking and dining establishments, entertainment facilities, offices, and other similar uses where the tenants have an opening onto one or more malls.
- The key to safe operations in a factory is first to understand what is to be made in the structure and how it will be made, and also how the structure has been designed/altered to meet these manufacturing needs.
- Most fire codes require the creation of an HMMP and an HMIS that fire fighters can consult when an incident occurs at a building where hazardous materials are produced or stored.
- It is also a serious error to consider all high-rise buildings as a single problem. There are fire-significant construction differences among high-rises.
- Hospitals and nursing homes have a common characteristic—they are home to numerous nonambulatory people. It is up to the building, the staff, and fire fighters to protect them.
- Although sprinklers are considered critical for larger hotels, small motels are not required to meet this requirement.
- Like hospitals, jails are considered institutional occupancies. The inmates of a jail are restrained and are incapable of getting out of the building to save their lives. They rely on prison staff and the building for their safety.
- Unfortunately, museums and libraries are not always protected with a fire suppression system. For years, many institutions mistakenly rejected sprinklers as causing too much water damage.
- Nightclubs are places where people go to have fun. The use of alcohol is a key element of that fun, resulting in patrons who are not fully aware of their surroundings. In addition, these clubs are often overcrowded beyond code-imposed limits.
- Office buildings are built in all five types of construction, large and small, high-rise or low-rise. These buildings have changed over the years, evolving from highly compartmentalized structures with offices for every worker to the open office plan.
- Many large open-area structures, such as churches and arenas, are built of wood or have an exposed wood plank roof that makes up the ceiling. Such a valuable building should be fully sprinklered.
- When preincident planning for a restaurant, pay particular attention to the location of the utilities, especially gas.
- A smoky fire in a self-storage building will make it very difficult to establish which units are involved.
- One notable issue with stadia is egress travel distance—most enclosed stadia cannot meet the traditional limits on travel distance from the most remote seat to an enclosed fire-rated stairwell.
- Fire fighters will be confronted with a variety of hazards as well as a variety of occupancy types in strip malls.

- Because underground buildings such as rail transit stations may house thousands of occupants at any given moment, moving these people quickly up stairwells to grade is nearly impossible.
- The advent of warehouse rack storage has brought potential major fire problems to Anyplace, USA.

Key Terms

Anchor stores Large stores (often department stores) attached to the mall that have all of their required exits independent of the mall.

Atria A large open space within a structure connecting two or more floors.

Brick nogging Brick and mortar filling between studs utilized as a makeshift fire barrier.

Calcination The deterioration of a product by heating to high temperatures.

Cold smoke Smoke that falls downward.

Control area A building or portion of a building within which hazardous materials are allowed to be stored, dispensed, used, or handled in quantities not exceeding the maximum allowable quantities.

Covered mall A single building enclosing a number of tenants, including retail stores, drinking and dining establishments, entertainment facilities, offices, and other similar uses where the tenants have an opening on to one or more malls.

Early suppression/fast response (ESFR) A type of fast-response sprinkler capable of providing fire suppression of specific high-challenge fire hazards.

Exit passageway Hallways, corridors, passages, or tunnels used as exit components and separated from other parts of the building in accordance with NFPA 101: Life Safety Code.

Friable Easily disintegrated.

Hazardous materials inventory statement (HMIS) Required in most codes, this statement lists the materials, hazards, and quantities of hazardous material products within a building.

Hazardous materials management plan (HMMP) Required in most codes, this plan explains how hazardous materials are to be stored and safely used within a building.

Idle pallet storage Pallets without product.

In-rack sprinklers Sprinklers that are placed within racks to control fires where overhead sprinklers are not adequate.

Inversion layer A layer of air that is warmer than the air below.

Landing zone In reference to an elevator, this zone is 18 inches above or below the landing floor.

Lapse The condition in which the atmospheric temperature is constantly decreasing as height increases.

Nonambulatory people Individuals who are not capable of self-preservation.

Occupancy Used in building codes to refer to the intended use of a building.

Open office plan A plan in which low-height partitions create cubicles for personal space; essentially one large open room.

Passive fire protection A material that is applied to a substrate and is designed to protect it from thermal effects.

Pause In reference to atmospheric conditions, the layer of air warmer than the air below it.

Platform A raised area in a building for presentations, wrestling matches, and the like where there are no hanging curtains, drops, or scenery other than lighting and sound effects.

Projected beam detectors Used in smoke control systems. These detectors can cover large areas with a single beam of light.

Proscenium arch and wall Found on a stage, it is the large ornamental opening and wall that separates the audience from the stage.

Pyrophoric gas Gas that ignites in air without the introduction of an ignition source.

RACE An acronym: R—remove all people in immediate danger to safety; A—activate the manual pull station and have someone call 911; C—close doors to confine the spread of smoke and fire; and E—extinguish the fire, if possible.

Racks Any combination of vertical, horizontal, and diagonal members that supports stored materials.

Shelves Storage on structures that are less than 2.5 feet (0.75 meters) deep, with shelves usually 2–3 feet (0.6–0.9 meters, respectively) apart vertically, and seldom exceeding 15 feet (4.5 meters) in total height.

Ship's ladder A completely vertical stair with a width that is not more than 24 inches wide.

Smoke barrier A continuous membrane, either vertical or horizontal, such as a wall, floor, or ceiling assembly, that is designed and constructed to restrict the movement of smoke. A smoke barrier might or might not have a fire resistance rating. Such barriers might have protected openings.

Solid pile Storage that is either box on box or pallet load on pallet load.

Stack effect The vertical airflow within buildings caused by the temperature-created density differences between the building interior and exterior or between two interior spaces.

Stage Performance area in a theater that has a proscenium arch and wall; hanging curtains, drops, and scenery; lighting; and support rooms (dressing room, etc.).

Tuned-mass dampers Heavy weights installed high up in a building that are adjusted by computers to counter wind-induced oscillations.

Case Study

Your crew has been assigned several preplans for the month of June. The first building on the list is an old multibuilding manufacturing facility. You and your crew know this facility well because many false alarms have been dispatched to this address. This building has changed hands many times over the years, but now manufactures and stores many different kinds of hazardous materials. Answer the following questions about the things to look for within this building.

1. True or False: This building will be required to have an HMIS and HMMP for the premises and available to the code official.

2. This structure has separating fire walls as dictated in the code. What type of protection best describes a fire wall?

 A. Active
 B. Dynamic
 C. Passive
 D. Static

3. Sprinkler systems are known as what type of defense?

 A. Active
 B. Dynamic
 C. Passive
 D. Static

4. You note that a yard hydrant is adjacent to a warehouse on the property. True or False: You should include this hydrant in your preplans to supply one of your engine companies.

Challenging Questions

1. Describe the difference between stages and platforms.

2. Describe the actions taken by nursing staff in a hospital with respect to each letter in the acronym RACE.

3. Describe the general characteristics of taxpayers and strip malls.

4. You are the lieutenant of an engine company and have been assigned to stretch the primary attack line to the seat of a fire in a tenant space in a covered mall. What things will you have to consider in completing this assignment?

5. Describe the phenomena of both winter and summer stack effect.

Collapse

OBJECTIVES

At the conclusion of this chapter, you will be able to:

- Explain the role of situational awareness in collapse.
- Identify the indicators used to anticipate collapse.
- Identify the various types of collapse.
- Apply risk analysis to fireground safety.
- Identify the specific basic competencies all fire fighters should have to perform at a structural collapse.
- Explain the essential elements of the collapse of the World Trade Center twin towers on September 11, 2001.

Case Study

Figure 14-1 The Provo Tabernacle fire.
© Jeffrey D. Allred/Deseret News/AP Images.

At 2:43 a.m. on December 17, 2010, Provo, Utah, fire fighters responded to the Provo Tabernacle for a report of a fire (**Figure 14-1**). Upon their arrival, smoke was observed coming from the massive, old heavy timber church. A quick interior investigation revealed that while little smoke was present at floor level (the interior of the church was a two-story space), fire was enveloping the piano and spreading to the choir seating area. A plaster and wood lath ceiling, suspended from iron rods attached to heavy timber roof trusses, blocked the view of the trusses and void space above the ceiling. On the exterior, however, smoke was seen coming from gable vents in the roof structure. Crews exited the building and reported the conditions to the incident commander. He immediately ordered a defensive exterior strategy, given the fire's spread, its inaccessibility, and real potential for roof failure. Exterior large-caliber streams were set up; the first portion of the roof collapsed at 4:30 a.m., followed by other sections of the roof over the next few hours.

1. How is the decision made to "go defensive" in a heavy timber building like this church?
2. What role does anticipating collapse play in an overall firefighting strategy?
3. As a fire fighter, what part do you play in the overall situational awareness of a fire scene?

Collapse and Situational Awareness

Individual fire fighters can be injured or killed in a variety of ways at a fire scene—heart attacks (the greatest killer today), smoke inhalation, falls, and more. It is building collapse, however, that has the potential not only to kill, but also to kill and injure in large numbers. In 2012, four fire fighters were killed by collapsing roofs and walls.

In the wake of the collapse of the World Trade Center on September 11, 2001, **situational awareness** has been widely discussed. According to one of the leading researchers in the field, Mica R. Endsley defines situational awareness as "the perception of the elements in the environment within a volume of time and space, the comprehension of their meaning, and the projection of their status in the near future." In other words, situational awareness is not only seeing and understanding what is happening, but also anticipating the future as the situation changes.

Everyone on the fireground must maintain a heightened sense of situational awareness, from the fire fighter all the way up to the incident commander (IC). From a practical standpoint, this means that fire fighters must be attuned to what is happening to the structure around them as they perform their duties, watching for signs of impending collapse. Company officers must maintain an even wider net of vigilance over the areas they are operating in. The IC must have a global understanding of what is happening in terms of the firefighting operations and the minute-by-minute destruction of the building.

Communication—the lifeblood of any fire response—is an essential element in maintaining situational awareness. Indications of the potential for collapse are critical for everyone and must be rapidly made known to all personnel. Many fire departments have specific protocols for collapse warnings; you must find out how your department handles such critical information on the fireground. Many use

a combination of radio warning signals (including urgent and emergency radio traffic) with specific numbers of air horn blasts to warn to evacuate a building.

Anticipating Collapse

No fire fighter or officer has a crystal ball through which to see the future outcome of a fire. Instead, fire fighters rely on training and experience to decide how to attack the fire and to rescue trapped occupants. Sometimes the wrong decision is made in the heat of battle when information is lacking and the incident is rapidly escalating.

It is imperative, then, that fire fighters do everything possible to anticipate collapse. In the previous chapters, we identified several indicators (signs that a collapse may occur) and precipitators (phenomena that cause a collapse) of collapse. Given their critical nature, a review of these indicators and precipitators is very important.

General Visual Indicators of Collapse

- Cracks in walls, especially cracks that develop or grow during a fire Figure 14-2
- Leaning walls
- Pitched or sagging floors
- "Racked" doorways (door stuck in shifted frame)
- The presence of building stabilization and bracing features, for example, cables and exterior "spreaders" (ornamental stars, plates, etc.), tie rods, and buttresses Figure 14-3

Audible and Physical Indicators of Collapse

- Moaning/groaning sounds
- Cracking noises
- Any type of movement
- Movement/shifting of water on floor
- Smoke pushing through cracks in the wall
- Vibrations
- Lack of water runoff from firefighting operations

Figure 14-2 A crack indicating serious building instability.

© monbibi/ShutterStock, Inc.

Figure 14-3 This old wood-frame commercial building has been retrofitted with cables and turnbuckles spanning a set of beams to hold the beams and their associated columns upright.

Courtesy of Glenn Corbett.

Environmental Precipitators of Collapse

- Heavy snow or rain load on roof
- Heavy wind conditions
- Earthquake
- Impact load (such as a vehicle striking a building)

Occupancy Precipitators of Collapse

- Overloading of floors and/or roof
- Concentrated loads
- Water-logged goods, soaked during firefighting operations

Existing Structural Instability Precipitators of Collapse

- Structural compromises of load-bearing members during renovations
- Previous fire damage
- Rotting or corrosion of structural members from water leaks/damage
- Compromised/broken structural elements
- Eccentrically loaded columns
- Buildings under renovation/construction and buildings being moved

Fire and Explosion Damage Precipitators of Collapse

- Sustained moderate to heavy fire conditions (measured from time of ignition, not arrival time)
- Explosion (such as from ignition of a natural gas leak)
- Loss of wood structural mass (beams) due to fire attack

Lightweight Construction Precipitators of Collapse

This is perhaps the greatest collapse danger to fire fighters today, given the complete proliferation of this type of construction throughout the country.

- Unprotected, bare steel members subjected to fire
- Wooden I-beams and lightweight wood trusses subjected to fire, particularly when directly exposed to basement fire or when burning in a truss void Figure 14-4

Figure 14-4 A new prefabricated "I-stair" using connections of galvanized steel in a manner similar to that of the gusset plates found in a wooden trusses. The steps (treads) and risers are made of oriented strand board. These stairs are particularly vulnerable on their underside, especially if unprotected within gypsum board such as in a basement.

Courtesy of Green Maltese, http://greenmaltese.com/.

Risk Analysis

When reports of fire fatalities from any type of building are studied, certain questions keep coming to the forefront: Why were those fire fighters where they were? What was the potential benefit from the risks undertaken? What were they doing? Too often the answers to these questions have been unsatisfactory. If the answers were completely truthful, it might be learned that the real reason was an antiquated macho attitude on the part of the fire fighters. This must be changed.

Fire fighters have been injured and have died unnecessarily in many fires. Taxpayers have been saddled with unnecessary expenses because fire fighters were operating in buildings that could not possibly be saved. It may provide a good feeling to say, "We made a fine stop," when fire fighters extinguish a fire in an already half-destroyed structure; however, fire departments are organized to provide a service, not to make people feel good.

If the fire service is to be regarded as truly professional, analysis of the relationship between risk incurred and benefits obtained must be undertaken. There are many unavoidable risks in firefighting, and they are all the more reason for the fire officer, who must consider the lives of the fire fighters, to be well informed about construction features of buildings that have been available for study for many years. In addition, all fire fighters must be continually mindful of the indicators and precipitators of collapse described in the previous section.

Types of Collapse

In previous chapters, we have studied collapse from a type of construction and building component perspective. For example, we have identified the reasons why brick bearing walls in ordinary construction fail as well as the dangers of lightweight wood trusses. There are, however, more general descriptions of the types of collapse, which can apply to all five types of construction and involve all types of building elements. The following is a list of descriptive collapse terms:

Figure 14-5 The result of the global collapse of the twin towers.
© Photos.com.

- **Global (total) collapse**: This type of collapse results in the complete failure of the building. The destruction of the World Trade Center twin towers and Building #7 are examples of global collapse **Figure 14-5**.
- **Partial collapse**: A type of collapse that results from a failure of a portion of a building. A situation in which a section of floor gives way without bringing down the rest of the building would be a partial collapse. A partial collapse (also referred to as a localized collapse) can be just as deadly as a global collapse.
- **Progressive collapse**: A type of collapse that occurs when a relatively small area of structural damage results in a loss of localized load-carrying capability, initiating a chain reaction of subsequent failures resulting in major structural failure, including global collapse.
- **Secondary collapse**: An additional collapse that occurs after the initial collapse. Secondary collapses often occur when loads shift after a primary collapse, causing additional portions

Figure 14-6 Searchers must be ever mindful of shifting loads, which can and will cause secondary collapses.

Courtesy of FEMA.

of the structure to fail **Figure 14-6**. Secondary collapses can and do happen during search and rescue operations.

- **Pancake collapse**: A collapse of a building's floors (and possibly roof) in a pancake stack fashion, with each floor (roof) laying flat on top of the one below **Figure 14-7**. This usually occurs when the connections between the floor and a load-bearing wall are lost, such as wood joists pulling out of a brick wall. Voids—debris-clear pockets where victims could possibly be found alive—are less prevalent in this type of collapse (compared with the other types that follow) and tend to be limited to areas surrounding furniture and other large objects.

- **Lean-to-floor collapse**: A collapse of a building's floor(s) in which one end of the floor is still supported, sometimes at or near the original point of connection to the wall **Figure 14-8**. A triangular void space is created.

- **V-shaped floor collapse**: A collapse in which the floor fails near its center, with the perimeter of the floor still partially supported by the walls **Figure 14-9**. Two voids are created in this type of collapse.

Figure 14-7 Pancake collapse.

Figure 14-8 Lean-to-floor collapse.

Figure 14-9 V-shaped floor collapse.

Figure 14-10 Cantilever floor collapse.

- **Cantilever floor collapse**: A collapse in which one end of the floor is still supported while the other end is unsupported Figure 14-10. Voids can be created in such situations.
- **A-frame floor collapse (tent collapse)**: A collapse in which one end of the collapsed floor is supported by an interior wall, creating two void spaces.
- **Inward outward collapse**: A collapse in which the exterior wall fails horizontally, with the interior wall "folding" horizontally. The top portion of the building fails inward while the bottom half fails outward. Braced-frame buildings are particularly susceptible to this type of collapse.
- **90-degree wall collapse**: A collapse in which an entire wall fails as one unit. Tilt-wall failure (which can fall in or out) is an example of a 90-degree collapse. Brick and other masonry walls can also fail in this manner Figure 14-11.

Figure 14-11 These insulating tilt-wall panels failed during a major fire in a food storage warehouse.

Courtesy of Glenn Corbett.

Figure 14-12 Establishing collapse zones saves fire fighter lives.

© Beth Keiser/AP Images.

- **Lean over collapse**: Usually limited to wood-frame structures, this type of collapse is characterized by a building that shifts at the upper levels, leaning into adjacent buildings or totally collapsing sideways in the absence of adjacent buildings.
- **Curtain wall collapse**: Often associated with brick veneer non-load-bearing walls, but also including other masonry walls. The wall falls like a curtain—straight down.

Establishing Collapse Zones

It is imperative that fire fighters and fire apparatus be kept out of collapse zones, especially when a fire attack has gone into a defensive mode (i.e., the building has been given up). This means staying far enough away from the structure so that any debris thrown (e.g., bricks, stones, cornices) during a collapse will not injure fire fighters surrounding the building Figure 14-12 .

It used to be said that fire fighters could stay back one-third the height of the wall and be safe. *This is total baloney!* Fire fighters must remain at least the full height of the wall away. Establish and enforce collapse zones with caution tape and cones. Safety officers must be wary of "creep up" by fire fighters as they attempt to get closer to the building in an exterior attack. Put away the handlines and go with the more effective master stream that has the reach and power to knock down the fire.

⊕ Branniganism

We must learn from tragedies so that those who died will not have died in vain. This is my self-appointed mission for the benefit of those who will listen. I get no pleasure from describing disasters—I do take pleasure in letters that say, "You saved my life."

Minimum Collapse Competencies for Fire Fighters

Although this text is not a training manual for technical collapse rescue, it is important to identify the minimum levels of competence that all fire fighters should have when dealing with a building collapse. National Fire Protection Association (NFPA) 1670, *Standard on Operations and Training for Technical Search and Rescue Incidents*, and 1006 identify the tasks the individuals should be capable of performing. The minimum competencies identified in this document are just that, the minimum. Higher competency levels—operations, technician, and specialist—require training in the use of specialized rescue equipment, the ability to perform shoring operations, and the capability to remove victims from voids.

Our Worst Collapse Tragedy

September 11, 2001, was a defining moment in American history. It was also a defining moment in the history of the fire service, when 343 fire fighters and 1 fire patrolman were lost in the 2 largest building collapses in human recorded history.

The twin towers of the World Trade Center were two of the tallest and largest buildings ever constructed. They were built by the Port Authority of New York and New Jersey, which was (and continues to be) exempt from all local building and fire codes. Each tower was 110 stories high (the North Tower/Tower 1 was 1,368 feet (417 m) high, while the South Tower/Tower 2 was 1,362 feet (415.1 m) high), and each floor was 40,000 square feet in area (an acre). Each side of the building measured 210 feet in length.

Four major structural components made up the buildings: the exterior walls (which were closely spaced columns on 3-foot, 4-inch (101.6 cm) centers, with 280 of them above the seventh floor), a series of columns in the center core (measuring 87 by 133 feet (26.5 by 40.5 m) and containing 48 columns), lightweight steel floor trusses forming a composite floor that tied the perimeter columns to the center core columns, and a "hat truss" (used at the very top of the building, between floors 106 and 110, to redistribute loads). With the structure supported at the perimeter and core columns, there were 60- and 36-foot (18.3- and 11- m) clear spans around the core *without any intervening columns*.

Although these facts about the construction of the building are undisputed by most individuals, the actual mechanism of collapse has been the source of debate. The National Institute of Standards and Technology (NIST), in a $16 million study, analyzed the collapse and arrived at conclusions as to how each building fell. The following is a brief synopsis of its analysis (the report consists of more than 10,000 pages of text), which is based on visual evidence of the exterior damage and computer modeling to simulate the fires and physical damage inflicted by the planes on the interior.

The North Tower (Tower 1)

At 8:43 a.m., the 767 aircraft impacted the north face of the tower between floors 93 and 99, cutting a large gash. Most damage on the interior of the building was done to floors 95 and 96. Thirty-five exterior columns were destroyed, and two were heavily damaged. Six center core columns were destroyed, and three were heavily damaged. Forty-three center core columns were stripped of fireproofing and 60,000 square feet (5,574.2 m²) of floor truss insulation was blown off. Multiple floor trusses on multiple floors were damaged or destroyed. Fifteen percent of the jet fuel carried on the plane was ignited inside the building, and subsequently ignited the combustible materials in the building.

NIST determined the following seven factors caused the collapse that occurred at 10:29 a.m.:

- The planes inflicted structural damage.
- Jet fuel sprayed on the interior, starting fires on multiple floors.

- Fireproofing was dislodged by the impact and subsequent fires, heating the unprotected steel.
- Open paths for fire spread were available because of the "open plan" of the impact floors and destruction of partitions by impact debris.
- Weakened core columns increased the load on the perimeter columns.
- Sagging of the "south floors" led to pull-in forces on the perimeter columns.
- Bowed south perimeter columns had a diminished capacity to carry loads.

The South Tower (Tower 2)

At 9:03 a.m., the 767 aircraft impacted the south face of the tower between floors 77 and 85, cutting a large hole. Most damage on the interior of the building was done to floors 80 and 81. Thirty-three exterior columns were destroyed, and one was heavily damaged. Ten center core columns were destroyed, and one was severely damaged. Thirty-nine center core columns were stripped of fireproofing and 80,000 square feet (7,432.2 m²) of floor truss insulation was blown off. Multiple floor trusses on multiple floors were damaged or destroyed. The jet fuel was ignited (also only 15% within the building as with the North Tower), and subsequently ignited the combustible materials in the building.

NIST determined the following seven factors caused the collapse that occurred at 9:59 a.m.:

- More direct damage was done by the plane to the structure than in the North Tower, including more severe damage to the center core columns.
- Jet fuel sprayed on the interior, starting fires on multiple floors.
- Fireproofing was dislodged by the impact and subsequent fires, heating the unprotected steel.
- Sustained fires on the east side of the building had an ample air supply.

- Weakened core columns increased the load on the perimeter columns.
- Sagging of the "east floors" led to pull-in forces on the east perimeter columns.
- Bowed east perimeter columns had a diminished capacity to carry loads.

There have been numerous debates as to the exact sequence of the collapse as well as the role of each of the structural elements. Some do not agree with the NIST conclusions about the collapse, especially about what was going on inside the structures, away from the cameras' views of the exterior. For example, exactly how many of the floor trusses were actually destroyed in the initial impact? How many failed during the subsequent fires? How much fireproofing was actually knocked off by the plane impact? (Photographic evidence prior to September 11, 2001, suggests some of it was already missing.) How many center core columns were damaged and destroyed? Which structural elements—the floors, the perimeter columns, and the center core's columns—played primary roles in the collapse and which played secondary roles?

Why is this important? Besides the obvious need to understand such a catastrophic event, there are other important reasons to analyze this disaster: Should lightweight trusses be used in high-rises of the future? Should spray-on fireproofing come under more scrutiny? Should any buildings be exempt from local building and fire codes? What role should the fire service play in the construction and regulation of a building?

This last point is a critical one for you to seriously consider. In this text, we have studied the failures of many types of buildings, including some that cost fire fighters their lives. Ultimately, it is better to take a proactive role in securing your safety and the safety of others by getting involved in the construction of a new building. Speak up about the hazards a building presents. Let the public know of the dangers that they will be faced with. Get involved in the code development and adoption process at both the local and national levels. Fire fighters are the

most knowledgeable people about the effects of fire on buildings—share your experiences with your brother and sister fire fighters.

The 344 fire personnel lost on September 11, 2001, left a legacy of heroism. So have the fire fighters lost in other building collapses. In the coming years, they also will have helped bring about positive changes in the fire safety of new buildings as new codes are created and obsolete practices are abandoned. What kind of legacy for fire fighter safety will you leave?

TACTICAL CONSIDERATIONS

1. Remember, the greatest threat on the fireground is collapse—not only because it can kill, but because it has the potential to kill in numbers. This should always be in the back of your mind when making tactical decisions. Many buildings today are "disposable," meaning they were built as quickly and as cheaply as possible, so that even moderate exposure to fire can cause significant damage. It is cheaper for owners to rebuild the structure than to repair it. Also, for liability reasons, it is better to knock these buildings down and build new ones than to risk repairs that were done improperly. Therefore, it is essential to perform a risk–benefit analysis. Case in point—on February 14, 2000, two Houston, Texas, fire fighters died in the line of duty while fighting a fire at a McDonald's restaurant. A few months prior, there was a similar fire at a McDonald's in Seattle, Washington. The Seattle fire fighters made a good stop. There was significant damage to a portion of the kitchen, but the eating area was relatively untouched. The average person would expect this restaurant to be repaired. The exterior of the building was intact except for the forcible entry damage to the door and one vent hole on the roof. Much to everyone's surprise, the McDonald's Corporation decided to tear the entire structure down and build a new restaurant, using a more modern design. Though no Seattle fire fighters were injured, their risks and firefighting efforts were all for naught. What we thought was valuable property worth protecting, the corporation saw as disposable. Perhaps it is time to "write off" such corporate properties that have a policy of disposing their fire-damaged buildings, no matter how slight the fire damage. If it's not important to them, should we be risking our lives for that type of property?

2. Building collapses are not always the result of a fire; they can occur for a variety of reasons: the age of the building, gas explosions, vehicle impacts, construction/demolition work adjacent to the building (especially at the foundation level), overloading, and earthquakes are all examples of precipitating events that can happen anywhere. Know what your department limitations are and who you can call for help, additional resources, and expertise.

3. If sufficient personnel are available, place "spotters" at the scene of a collapse for the sole purpose of watching for conditions leading to a secondary collapse.

4. When responding to a building fire, identifying hazards to fire fighters is a critical part of size up. Remember that burning building collapse is the fourth cause of fire fighter deaths. A building's life span is approximately 75–100 years. Beyond that (without diligent maintenance), the building becomes badly deteriorated. Many buildings throughout the country are reaching that 100-year mark, so structural collapse during firefighting should be expected to increase. There are three factors the company officer needs to consider: the age of the building, abandonment, and lightweight construction.

5. When a building is on fire, wall stability is based on the number of interconnected building components; the more interconnections, the more stable the wall is, the fewer number of interconnections, the less stable the wall is. A bearing wall is the most stable, a non-bearing wall is less stable, and a freestanding wall is the least stable and the most dangerous. A 90° vertical collapse of a freestanding masonry wall can run the entire length of the building. The minimum collapse zone should be equal to the height of the wall, but bricks or heavy masonry stone will fall and scatter with deadly force well beyond that. Even the sidewalk on the opposite side of the street can be within reach of flying shrapnel that can kill or injure fire fighters and spectators; therefore, it is recommended that the collapse zone be 1 ½ times the height of the wall. An IC cannot predict the type of collapse that may occur (i.e., a curtain fall or inward/outward) so the safest decision is to prepare for the worst case scenario—a 90° angle, vertical wall collapse.

6. If an interior roof or floor collapses as a result of the fire, the danger of an exterior wall collapse is greatly increased. Fire fighters operating on the outside of the building perimeter are exposed to the dangers of wall collapse. In addition to establishing the collapse, consider establishing a hot zone, a warm zone, and a cold zone. The command post and support positions should be located in the cold zone so it is safely outside the collapse zone and does not have to be relocated.

7. When the collapse zone is greater than the reach of the hose streams or the best angle for the hose stream is within the collapse zone, the company officer should reposition and operate the hose streams from a flanking position—that is, on either side of the wall in question—beyond the outer borders of the building's width. When you're performing this operation, it is usually a defensive fire. Don't waste your time (and water) on 1.75-inch (44 mm) handlines. Use 2.5-inch (64 mm) lines or monitors for reach and penetration from a flanking position.

8. Generally, high, ornate parapet walls collapse more frequently than lower, less decorative ones. The most dangerous parapet wall is the one constructed along the front wall of a one-story building above several large display windows. This type of parapet is supported and balanced on top of a steel I-beam that acts as a lintel, spanning above the show windows. The parapet wall is often secured with steel reinforcement tie rods embedded within the masonry and tied to a back wall or into the roof decking, visible only from the roof. In essence, this is a freestanding wall because it doesn't support any other part of the building; therefore, these types of walls are vulnerable to the impact lateral forces of high winds and master streams. A master stream striking a parapet wall from the rear of the building can collapse the wall onto the front sidewalk. Flanking streams or master streams deployed from the front side of the building can push the parapet wall back onto the roof, possibly causing a secondary roof collapse. By this stage of the game, it would be a defensive attack. There should be no interior crews entering or exiting the building from doors underneath a parapet wall, and there should be no crews working the interior underneath a rear collapse zone of a parapet wall.

9. Another type of hazard on the fireground is introduced by theater marquees. These are extremely heavy, cantilevered structures that are supported by a suspended beam. (The typical theater marquee is a suspended beam.) A suspended beam is a simple beam with one or both ends suspended on a tension member such as a chain, cable, or tie rod. These steel reinforced tie rods help secure the weight of the structure. The tie rods can be secured through the mortar in the walls, through a parapet wall, or to the roof decking. The author has seen these tie rods go through a parapet wall, and then down into the roof decking. Remember, these rods are unprotected steel and weaken or fail at temperatures between 800 and 1,000°F (538°C). When fire destroys this anchoring system, the marquee becomes an undersigned cantilever. One marquee collapse pulled the entire masonry wall down the entire length of the building, killing six fire fighters. One fire fighter was standing about 100 feet (30.5 m) from the initial collapse and was killed in the incident. The question a company officer needs to ask when dealing with a suspended beam (marquee) is, "What are the suspension rods tied to?"

10. A large commercial sign hanging or projecting out from the side of a building is an eccentric load and must be counterbalanced—usually by two steel-reinforced rods or lag bolts—tying it through the exterior wall and anchoring the sign into the roof rafters. If fire causes the steel cables or bolts to fail, the weight of the sign can cause it to release from the exterior wall and come crashing down like a guillotine. Worse, the wall can become outwardly overbalanced and will likely collapse. Again, the company officer needs to ask "What is this sign attached to? What's happening on the other end?" Also, remember: ice and snow can add tremendous weight to marquees, hanging signs, and canopies.

11. A parapet wall does not collapse during the early stages of a fire (a room and contents fire). If the interior attack is unsuccessful and the strategy is transitioning from offensive to defensive because the building is becoming involved, the truck officer should inspect the parapet for signs of failure. After the fire is out, don't drop your guard. Fire fighters often become complacent. The emergency is over but the hazards still exist. Stop and take a break. Let fire fighters rehab. During this time, the IC, the safety officer, and perhaps experienced company

officers should take a walk-around and identify all the safety hazards. They should be flagged, roped, or coned off. All fire fighters should be made aware of these safety hazards before overhaul starts. Parapet walls often collapse during overhaul because the structure has been weakened by the fire as well as master streams and other firefighting efforts. A parapet wall collapsed and killed six Fire Department City of New York (FDNY) fire fighters after the fire was declared under control.

12. Because the length of a parapet wall is usually greater than its height, the company officer must consider the horizontal distance of the wall and include this distance into the collapse zone. Consider using an uncharged fire hose to establish an unmistakable, clearly delineated, perimeter line.

13. A gable roof presents a collapse hazard and a fall hazard for fire fighters. Always take a roof ladder during vertical ventilation operations, no matter how low the angle of the pitch is. The roof ladder is not just for footing on a pitched roof. It's also the best tool to lower to a fire fighter who has fallen into the attic space below. The same goes for a flat roof. Take the roof ladder with you. It's for bridging and rescue.

14. Vertical ventilation typically results in extremely efficient removal of heat and smoke. The major risk is roof collapse. Fire fighters should understand that when they walk on the roof of a burning building, they risk the possibility of falling through the fire-weakened decking into a hot, fiery, attic space—or directly into the fire room from which they might not escape. If fire fighters or company officers aren't willing to risk rooftop operations, there is nothing wrong with venting top-floor windows with PPV. In rural areas, outside city limits or zoned areas, older roof construction may not conform to the local building code. Substandard roof construction may not support the weight of fire fighters. When in doubt, don't risk it.

15. Remember that pitched roofs are designed to shed water and snow and may be designed to support less of a live load than a flat roof. That means it may support fewer fire fighters than a flat roof will. Roof ladders also help distribute the live load weight of fire fighters on a pitched roof.

16. When exposure B and D are attached to the fire building and the roofs are separated only by fire walls, it is faster and safer to ladder the exposure buildings

first. This gives the rooftop ventilation team a safe platform to work from. They can evaluate the fire building and the condition of the roof before standing on it. The exposures also provide a secondary means of egress without placing extra ladders.

17. The dangers of bowstring trusses are now well recognized throughout the fire service. In unoccupied bowstring truss buildings where there is no life hazard, roof ventilation may be too dangerous to risk the lives of fire fighters. However, when roof stability is assured and roof ventilation is needed, fire fighters can sometimes open the skylights or scuttle covers quickly and leave the roof immediately. To quickly determine if a fire has spread to the roof space of a bowstring truss, a triangular cut can be made in the sloping front or rear of the roof from the safety of an aerial ladder or platform. If the web members aren't enclosed with plaster board, another cut can be made on the opposite sloping side, thereby providing horizontal cross-ventilation of the entire roof space. A hose line can also be aimed into this space on the windward side from the safety of the aerial and extinguish the fire in trusses.

18. This survival concept is essential for every fire fighter to learn. For example, consider a building that is 100 by 100 feet (30.5 by 30.5 m) with bowstring truss sections that are 20 feet (6.1 m) on center. In a building where the web areas of each truss are enclosed with firestopping and a serious fire burns in one bay between trusses, one truss is likely to fail first. A single truss collapse will open up a 4,000 square foot ($371.6 m^2$) section of roof (a 40- by 100-foot [12.2- by 30.5- m] area). When a bowstring truss roof appears weakened by fire, fire fighters ordered to evacuate the roof should retreat in a direction perpendicular to the roof trusses. That means the fire fighter will run 20 feet (6.1 m) in the direction perpendicular to the failing truss until he crosses into the space beyond the next truss, where he may temporarily be safe. If the fire fighter runs to the parapet wall (which is what we normally teach them to do), he must travel 50 feet (15.2 m) and still may go down with the collapsing roof.

The same principal applies to fire fighters operating inside the building, trying to get out of the path of a collapsing truss. If the exit is in-line with the burning truss, they may not escape. Fire fighters attempting an interior attack on the fire should be positioned behind

a truss section next to the one involved by fire. Make sure you have an escape route. This is also no place for 1.75" (44 mm) handlines. The reach of powerful 2.5" (64 mm) handlines or portable monitors will allow fire fighters to be out of the collapse zone of a single truss.

19. When a bowstring truss roof collapses, or any roof for that matter, it's not enough to avoid being on the roof or under it; there is a good chance this collapse could push a wall out, collapsing onto the sidewalk. Set up perimeter collapse zones—one and a half times the height of the wall.

20. When rapid intervention teams are sent in to rescue fire fighters trapped under a collapsed ceiling, it is important to sweep the collapsed area above the dropped ceiling with a hose line to knock down the flames and cool the area. This should happen at the same time rescue efforts are being initiated. If rescue attempts begin without extinguishing efforts, the trapped fire fighters may succumb to the products of combustion before they are reached. This is backed up with a case study where the trapped fire fighter needed to be cooled with a hose stream before rescue efforts were commenced because he was burning.

A fire fighter operating at a serious building fire with the possibility of a suspended ceiling collapse should attempt to stay near large pieces of furniture, stock, or equipment that can serve as shields or voids if the ceiling falls in on them. An open floor area or empty store is dangerous because if the ceiling collapses, a fire fighter operating below will receive the full blow of a heavy ceiling and will be pinned to the floor. The fire fighter will be unable to breathe and will quickly suffocate. Supermarket store shelves that extend down a shopping aisle have saved the lives of many fire fighters caught beneath a collapsing ceiling.

21. If fire fighters must work beneath a suspended ceiling that has partially collapsed or is sagging badly, consider using a portable ladder to shore up the weakened ceiling during operations. The vibration from tools used during overhaul could suddenly cause a partially collapsed ceiling section to pull down the entire ceiling.

22. When a suspended ceiling collapses, the grid system must first be cut apart with a power saw, and then the ceiling tiles must be cut or pulled apart. Fire fighters may then crawl beneath the ceiling into void spaces to search for victims. To access an area

below a collapsed ceiling quickly, a 12-foot (3.7 m) roof ladder can be shoved under a purchase point to lift a portion of the ceiling. The ladder can then be supported in the raised position with another solid object. This technique allows fire fighters to crawl beneath a fallen ceiling or break up a ceiling if the grid system is light enough.

23. If fire fighters are trapped in shelters or voids at the rear of a store beneath a collapsed suspended ceiling, rescue attempts can be made from adjoining stores. Partition walls separating stores in large one-story buildings can often be breached faster that rear entry through masonry walls or heavy steel doors. Hose lines should be placed into adjoining stores to prevent extension. Entry into a store beneath a ceiling collapse can sometimes be accomplished from a basement or cellar interior stairway. Below-grade access doors will be at the rear or the side of the building.

24. Exterior fire escapes are a thing of the past for modern construction—they are no longer permitted (exterior stairways, however, meeting all width, tread and risers requirements, and so on, are still permitted). Required maintenance makes them cost prohibitive as well as a liability. However, the many existing exterior escapes around the country will be here for decades to come. When climbing a fire escape, test the steps first before putting your full weight on it. Fire fighters should remember that a missing or broken step is a warning sign. A fire escape with one defective step is likely to have more. When a fire fighter is climbing a weakened fire escape stair, he should place his feet near the outer side of the step, where it meets the stringer. This method causes less deflection of the step and reduces the amount of stress caused by the fire fighter's weight.

25. A fire fighter on a fire escape landing preparing to advance a hose line into the fire building should always test steps, handrails, and side guardrails before leaning on it or applying his full weight. Test the vertical gooseneck ladder from the fire escape landing before ascending to or descending from the roof. Try to pull the ladder out from the wall to which it is attached to see if the connection is solid. If a fire fighter has any doubt about the integrity of a stair or a sliding drop ladder, use a fire department ground ladder. The extra time required to obtain and place the ladder is outweighed by the risk of using a fire escape that is ready to fail.

It is always safer to climb a fire department ladder than a fire escape.

26. When climbing or descending a fire escape ladder, weakened or otherwise, always face the ladder and always use the guide rails (beams) instead of holding on to the rungs. A rung will fail before the guide rail. Always keep at least one hand on the fire escape. Using the guide rails will minimize the risk of losing your grip should there be a sudden failure of the rung you're holding on to.

27. A gallon of water weighs 8.34 pounds (3.8 kg) and one ladder pipe can easily deliver 1,000 gallons per minute (3,785 Lpm). With multiple master streams operating on a "surround and drown" defensive operation, fire crews are dumping tons of water into a structure every minute. The building has been written off at that point and has no value. But forces applied to a building through massive water application can cause the same types of building collapses brought on by the destruction from a fire. Whether by fire or by water, the building can take only so much undesigned force before gravity takes over. The IC must prepare for a building collapse during defensive operations.

28. Here is a simple plan for rescue after a collapse. Whether the incident is as simple as a ceiling collapse or as complex as a high-rise building collapse, the tactical objectives can be broken down into the following sequences. Each sequence can take minutes, hours, or days.

 - Survey the collapse site. Perform a 360° walk-around. Get an aerial perspective from an aerial platform or from an adjacent rooftop. Determine potential dangers to rescue personnel.
 - Set up hose lines, monitors, and ladder pipes in case fires erupt.
 - Estimate the number of potential victims and set up a medical branch.
 - Shut off all utilities (gas, water, and electricity) that can injure trapped victims and rescuers.
 - Search for and remove surface victims first.
 - Search all voids and spaces created by the collapsed structure.
 - Start selected debris removal, digging to survivable areas where victims could be trapped.
 - Finally, start general debris removal to clear the entire collapse site.

WRAP-UP

Chapter Summary

- All fire fighters have the responsibility to know the conditions for collapse in different types of buildings as well as the signs of impending collapse in a given building. They must also communicate these signs to other fire fighters.
- It is imperative that fire fighters and fire apparatus be kept out of collapse zones, especially when a fire attack has gone into a defensive mode (the building has been given up). This means staying far enough away from the structure so that any debris thrown during a collapse will not injure fire fighters surrounding the building.

Key Terms

90-degree wall collapse A collapse in which an entire wall fails as one unit.

A-frame floor collapse A collapse in which one end of the collapsed floor is supported by an interior wall, creating two void spaces.

Cantilever floor collapse A collapse in which one end of the floor is still supported while the other end is unsupported. Voids can be created in such situations.

Curtain wall collapse Often associated with brick veneer non-load-bearing walls, but also including other masonry walls. The wall falls like a curtain—straight down.

Global (total) collapse Results in the complete failure of the building.

Inward outward collapse A collapse in which the exterior wall fails horizontally, with the interior wall "folding" horizontally. The top portion of the building fails inward while the bottom half fails outward.

Lean over collapse Limited to wood-frame structures, this type of collapse is characterized by a wooden building that shifts at the upper levels, leaning into adjacent buildings or totally collapsing sideways in the absence of adjacent buildings.

Lean-to-floor collapse A collapse of a building's floor(s) in which one end of the floor is still supported, sometimes at or near the original point of connection to the wall. A triangular void space is created.

Pancake collapse A collapse of a building's floors (and possibly roof) in a pancake stack fashion with each floor (roof) laying flat on top of the one below.

Partial collapse Results from a failure of a portion of a building. A situation in which a section of floor gives way without bringing down the rest of the building would be a partial collapse. A partial collapse can be just as deadly as a global collapse.

Progressive collapse Extensive structural failure initiated by local structural damage or a chain reaction of failures following damage to a small portion of a structure.

Secondary collapse An additional collapse that occurs after the initial collapse. Often occurs when loads shift after a primary collapse, causing additional portions of the structure to fail.

Situational awareness In firefighting terms, the perception of the surroundings in terms of time and space, the understanding and comprehension of these observations, and the projected outcome of a change in these conditions.

V-shaped floor collapse A collapse in which the floor fails near its center, with the perimeter of the floor still partially supported by the walls. Two voids are created in this type of collapse.

Case Study

While on light duty, you are reassigned to the training division. You have been asked to prepare a class on building collapse and the ever-present dangers associated with collapse. While studying the topic, you come across many different types and variations of collapse.

1. A type of collapse that is usually limited to wood-frame structures in which the upper floors shift would be considered a(n):

 A. 90-degree wall collapse
 B. Inward outward collapse
 C. Frame floor collapse
 D. Lean over collapse

2. Which type of collapse best describes a floor system that has failed and only one end of the floor is still supported?

 A. V-shaped floor collapse
 B. Cantilever floor collapse
 C. Frame floor collapse
 D. Lean-to-floor collapse

3. True or False: A pancake collapse usually occurs because the connection between the floor and the non-load-bearing walls is lost.

4. This particular type of collapse is a specific concern during search and rescue operations.

 A. Progressive collapse
 B. Global collapse
 C. Secondary collapse
 D. Partial collapse

5. Describe your department's procedures for setting up a collapse zone.

Challenging Question

Study the photo of the wood-frame church fire Figure 14-13. Identify the collapse indicators and the precautions you would take at this fire.

Figure 14-13 A wood-frame church fire.

Courtesy of Dave Traiforos.

Sprinkler System Design Occupancies

APPENDIX A

Light hazard occupancies include occupancies having uses and conditions similar to the following:

Animal shelters

Churches

Clubs

Eaves and overhangs, if of combustible construction with no combustibles beneath

Educational

Hospitals, including animal hospitals and veterinary facilities

Institutional

Kennels

Libraries, except large stack rooms

Museums

Nursing or convalescent homes

Offices, including data processing

Residential

Restaurant seating areas

Theaters and auditoriums, excluding stages and prosceniums

Unused attics

Ordinary hazard occupancies (Group 1) include occupancies having uses and conditions similar to the following:

Automobile parking and showrooms

Bakeries

Beverage manufacturing

Canneries

Dairy products manufacturing and processing

Electronic plants

Glass and glass products manufacturing

Laundries

Restaurant service areas

Ordinary hazard occupancies (Group 2) include occupancies having uses and conditions similar to the following:

Agricultural facilities

Barns and stables

Cereal mills

Chemical plants — ordinary

Confectionery products

Distilleries

Dry cleaners

Exterior Loading Docks
Exterior loading docks only utilized for loading and unloading should be classified as OH2. Where utilized for storage, exterior loading docks and all interior loading docks should be protected based upon the actual occupancy, the materials handled on the dock, as if the materials were actually stored in that configuration.

Feed mills

Horse stables

Leather goods manufacturing

Libraries — large stack room areas
It is not the NFPA 13 committee's intent to automatically equate library bookshelves with ordinary hazard occupancies or with library stacks. Typical library bookshelves of approximately 8 ft in height, containing books stored vertically on end, held in place in close association with each other, with aisles wider than 30 in.can be considered to be light hazard occupancies. Similarly, library stack areas, which are more akin to shelf storage or record storage, as defined in *NFPA 232, Standard for the Protection of Records*, should be considered to be ordinary hazard occupancies.

Machine shops

Metal working

Mercantile

Paper and pulp mills

Paper process plants

Piers and wharves

Post offices

Printing and publishing

Racetrack stable/kennel areas, including those stable/kennel areas, barns, and associated buildings at state, county, and local fairgrounds

Repair garages

Resin application area

Stages

Textile manufacturing

Tire manufacturing

Ordinary hazard occupancies (Group 2) include occupancies having uses and conditions similar to the following: (continued)

Tobacco products manufacturing

Wood machining

Wood product assembly

Extra hazard occupancies (Group 1) include occupancies having uses and conditions similar to the following:

Aircraft hangars (except as governed by *NFPA 409, Standard on Aircraft Hangars*)

Combustible hydraulic fluid use areas

Die casting

Metal extruding

Plywood and particle board manufacturing

Printing [using inks having flash points below 100°F (38°C)]

Rubber reclaiming, compounding, drying, milling, vulcanizing

Saw mills

Textile picking, opening, blending, garnetting, or carding, combining of cotton, synthetics, wool shoddy, or burlap

Upholstering with plastic foams

Extra hazard occupancies (Group 2) include occupancies having uses and conditions similar to the following:

Asphalt saturating

Flammable liquids spraying

Flow coating

Manufactured home or modular building assemblies (where finished enclosure is present and has combustible interiors)

Open oil quenching

Plastics processing

Solvent cleaning

Varnish and paint dipping

FESHE
Correlation Guide

Building Construction for Fire Protection FESHE Course Outcomes	Brannigan's Building Construction for the Fire Service, Fifth Edition Chapter Correlation
1. Describe building construction as it relates to firefighter safety, building codes, fire prevention, code inspection, firefighting strategy, and tactics.	1, 2, 3, 4, 5, 6, 7, 8, 9, 10, 11, 12, 13, 14
2. Classify major types of building construction in accordance with a local/model building code.	3, 4
3. Analyze the hazards and tactical considerations associated with the various types of building construction.	2, 3, 5, 7, 8, 9, 10, 11, 12, 13
4. Explain the different loads and stresses that are placed on a building and their interrelationships.	2, 3, 8, 9, 10
5. Identify the function of each principle structural component in typical building design.	2, 3, 7, 8, 9, 10, 11
6. Differentiate between fire resistance, flame spread, and describe the testing procedures used to establish ratings for each.	5, 6, 7, 11
7. Classify occupancy designations of the building code.	4, 13
8. Identify the indicators of potential structural failure as they relate to firefighter safety.	2, 7, 9, 13, 14
9. Identify the role of GIS as it relates to building construction.	1

90-degree wall collapse A collapse in which an entire wall fails as one unit.

A-frame floor collapse A collapse in which one end of the collapsed floor is supported by an interior wall, creating two void spaces.

Admixture A material other than water, aggregate, and cement used as an ingredient in concrete or mortar.

Adobe Large, roughly molded, sun-dried clay units of varying sizes.

Aggregate Any of a variety of materials, such as sand and gravel, added to a cement mixture to make concrete.

Air flow The use of high air velocity to stop smoke movement.

Aluminum A lightweight metal that is both malleable and nonmagnetic. This material has very good conductivity. This noncombustible material has a low melting point and little mass per unit of area, so it disintegrates rapidly in fire.

Americans with Disabilities Act (ADA) A set of regulations passed in 1990 that includes, among other things, regulations requiring areas of refuge for disabled people in multistory buildings.

Anchor stores Large stores (often department stores) attached to the mall that have all of their required exits independent of the mall.

Angles Steel members that have two legs at right angles to one another.

Arch Combines the function of a beam and a column.

Arched truss A truss with an arched upper chord and a straight bottom chord, with vertical hangers between the two chords.

Architect An individual who is engaged in the design of buildings and who often supervises construction.

Area of refuge An area of a building separated from other spaces by fire-rated smoke barriers in which a tenable environment is maintained for the period of time that such areas may need to be occupied at the time of a fire.

Asbestos cement shingle A "fireproof" roofing shingle that is composed of cement reinforced with asbestos fibers.

Ashlar masonry Stone cut in rectangular units.

Asphalt asbestos protected metal (AAPM) Asphalt coating that is combustible and used as a weather-protective coating on galvanized steel walls.

Asphalt felt siding A type of siding manufactured by saturating a dry felt with asphalt, then coating it with a fine glass fiber.

Atria A large open space within a structure connecting two or more floors.

Attitude Vertical or horizontal orientation.

Autoexposure The vertical spread of fire on the exterior of a multistory building, from one floor to the floor(s) above, via convection and radiation.

Autoignition A phenomenon in which a combustible material ignites spontaneously without the application of a flame or spark.

Automatic-wet standpipe system Standpipe system attached to a water supply capable of supplying the system demand at all times; it requires no action other than opening a hose valve to provide water at hose connections.

Axial load A load that passes through the centroid of a section under construction and is perpendicular to the plane of the section.

Backdraft The explosive burning of heated gaseous products of combustion when oxygen is introduced into an environment whose oxygen supply has been depleted due to fire.

Bagasse Low-density fiberboard made of wood fibers or sugar cane residue.

Balloon frame A wooden structure in which all vertical studs in the exterior bearing walls extend the full height of the frame from sill to roof. No firestops are present within the walls.

Bar joist A joist that generally runs in the same direction as a beam and forms a lightweight, long-span system, used as floor supports and built-up roofing supports.

Bars Plates fewer than 6 inches in width; may be square or round.

Batten A type of siding that uses vertical strips of wood to cover joints.

Batt insulation Fiberglass or rock wool insulation with various thicknesses; it may or may not have a paper covering.

Beam A structural member that transmits forces perpendicular to such forces to the reaction points.

Bent A line of columns in any direction.

Biopolymers In terms of construction materials, a naturally produced plastic from renewable, living sources (e.g., starches, wood), instead of nonrenewable sources such as oil.

Bottom plate A structural member laid on the sub-floor where the vertical studs are installed.

Box column A large hollow column built from steel plates.

Box girder A large girder, which is hollow, like a box column, and often used for highway bridges.

Braced frame A structural system that uses diagonal members to provide bracing against lateral wind and earthquake loads.

Bracket Diagonal member that supports what would otherwise be a cantilever.

Brick A solid or hollow masonry unit of clay mixed with sand, which is molded into a small rectangular shape while in a plastic state.

Brick and block composite wall Consists of an exterior wythe of brick directly mortared or parged to an inner wythe of concrete masonry unit (CMU).

Brick nogging Brick and mortar filling between studs utilized as a makeshift fire barrier.

Bridging Braces placed between parallel-frame members to prevent movement from their vertical axes. Also known as cross-bracing.

Btu British thermal unit; the quantity of heat required to raise the temperature of 1 pound of water 1°F at the pressure of 1 atmosphere and temperature of 60°F.

Building-integrated photovoltaics (BIPV) PV systems that are incorporated into the building itself in the form of either PV shingles or PV wall panel systems.

Building code Regulates the actual design and construction of new buildings, providing for minimum levels of health and safety. Regulates the level and amount of fire protection in a new structure.

Built-up girder Made of steel plates and angles riveted together, as distinguished from one rolled from one piece of steel.

Bulbtee A tee where the end of the cutoff is thickened.

Bulkhead An upright partition that divides a ship into compartments and is meant to prevent the spread of leakage or fire.

Buttress Mass of masonry built against a wall to strengthen it. Necessary when a vault or an arch places a heavy load or thrust on one part of a wall.

Cables Special high-strength, cold-drawn steel cables. Also referred to as strands or tendons.

Caisson A shaft of concrete placed under a building column or wall extending down to bedrock.

Calcination The deterioration of a product by heating to high temperatures.

Caloric value Measured in Btu; the amount of heat required to raise 1 pound of water 1°F.

Camber Upward rise.

Cantilever beam A beam supported at one end only, rigidly held in position at that end.

Cantilevered A type of construction in which an overhang is supported from only one end where one floor extends beyond and over a foundation wall.

Cantilever floor collapse A collapse in which one end of the floor is still supported while the other end is unsupported. Voids can be created in such situations.

Cantilever wall A freestanding wall unsecured at the top that acts like a cantilever beam with respect to lateral loads, such as wind or a hose stream.

Carbon dioxide system A fire protection system intended to protect materials that can be damaged by water; it uses carbon dioxide to suppress the fire.

Cast-in-place concrete Includes plain concrete, reinforced concrete, and post-tensioned concrete. This concrete is molded in the location in which it is expected to remain.

Cast-iron box An iron box built into a wall to receive the end of a girder.

Castellated beam A wide flange beam that has been cut in half in a zig-zag pattern and then welded back together in an offset manner, creating a new, deeper beam.

Casting A process of placing fluid concrete into molds, generally called forms, in which the concrete is permitted to harden to a certain shape.

Cavity or hollow wall A wall built of two wythes (a single vertical thickness of masonry) separated by a space for rain drainage or insulation.

Cavity wall Hollow wall in which wythes are tied together with steel ties or masonry trusses.

Ceiling layer The hot, buoyant gases that collect at the very top of a room.

Cells (concrete) The hollow portions of the core of a concrete block.

Cement-asbestos board Noncombustible material often used for friable construction.

Centroid The center point at which a body would be stable, or balance, under the influence of gravity.

Chairs Small devices designed to keep the rods up off the surface of the form so concrete will flow underneath.

Chamfered column A wooden heavy timber column that has been cut at an angle (beveled) on each of the corners to make it more difficult for fire to ignite the column at that location.

Chamfer To cut off the corners of a timber to retard ignition.

Channel Steel structural component that has a square U-shaped cross section.

Chipboard Wood chips that are glued together to make flat sheets. Often used in the floor construction of mobile homes.

Chord The outside members (top and bottom) of a truss, as opposed to the inner "webbed members."

Civil engineer An engineer who specializes in the design of parking lots, drainage areas, and roadways.

Clean agent system Gaseous fire extinguishing agent that does not leave a residue when it dissipates.

Cob A composite material composed of clay, straw, sand, and earth; similar to adobe; typically laid in courses.

Cockloft Void space between the top floor ceiling and the roof.

Cold-drawn steel A type of steel used in the construction of cables that are sometimes used to brace failing buildings or as tendons in tensioned concrete. It fails at 800°F (427°C).

Cold smoke Smoke that falls downward.

Column A structural member that transmits a compressive force along a straight path in the direction of the member.

Combustible acoustical tile Fiberboard in which holes have been punched.

Compartmentation Subdivision of a building into small areas so that fire or smoke is confined to the room or section in which it originates.

Composite Built up of different parts, pieces, or materials.

Composite and combination columns Columns that use steel and concrete combined into one unit.

Composite construction Buildings in which different load-bearing materials are used in different areas of the building.

Composite material A material built up of different parts, pieces, and materials intended to act as a single unit.

Composite wall Two different masonry materials, such as brick and concrete block, used in a wall and designed to react as one unit under load.

Compression Direct pushing force, in line with the axis member; the opposite of tension.

Concentrated load A load acting on a very small area of the structure's surface; the exact opposite of a distributed load.

Concrete masonry unit (CMU) Precast hollow or solid structural block. Sometimes referred to as cinder block.

Concrete topping Concrete placed over the first-floor wood floors for fire resistance or to provide sanitary floors.

Conduction The transfer of heat within an object or between objects through direct contact.

Conflagration Rapid involvement of a fire situation that goes beyond the normal attack situations that fire departments encounter.

Conflagration breeder A structure that presents severe exposure problems that are capable of initiating a conflagration—a large, multiple-building fire that is not easily contained.

Construction safety regulations Occupational health and safety regulations (enforced by state and/or federal authorities such as OSHA) that apply to construction sites. These include provisions to prevent falls (through the use of safety barriers) and protection from being hit or crushed.

Continuous beam A beam supported at three or more points. It is considered structurally advantageous because if the span between two supports is overloaded, the rest of the beam assists in carrying the load.

Continuous casting Process for casting or pouring concrete without interruption from start to finish.

Continuous slipforming Pouring concrete continuously as forms move upward so that continuous casting may be accomplished.

Control area A building or portion of a building within which hazardous materials are allowed to be stored, dispensed, used, or handled in quantities not exceeding the maximum allowable quantities.

Convection The transfer of heat through circulation within a medium such as a gas or a liquid.

Coping The masonry cap on top of a wall.

Corbelled A series of projections, each one stepped progressively outward from the vertical face of the wall as it rises up to support a cornice or overhanging member above.

Core construction No external braces involved; bracing is done within the core of the structure.

Cornice A projecting decorative (ledge) at the top of a masonry wall.

Corrugated metal siding Grooved panels utilized on industrial buildings of wood or steel framing; it poses an electrical hazard from stray electrical wires or lightning.

Corrugation Grooved ridged material, often metal.

Course A horizontal line of masonry.

Covered mall A single building enclosing a number of tenants, including retail stores, drinking and dining establishments, entertainment facilities, offices, and other similar uses where the tenants have an opening on to one or more malls.

Critical radiant flux (CRF) The amount of external radiant heat energy (measured in watts per square centimeter) below which a flame front will cease to propagate.

Crosslot bracing Internal bracing that transfers the lateral earth pressures between opposing walls through compressive struts.

Cross wall Any wall at right angles to any other wall; the walls should brace one another.

Cured (concrete) Concrete that is hardened to full strength.

Curtain wall A non-load-bearing wall that carries no weight other than its own. It is installed only to keep out the weather.

Curtain wall collapse Often associated with brick veneer non-load-bearing walls, but also including other masonry walls. The wall falls like a curtain—straight down.

Dead load The weight of a building; the dead load consists of the weight of all materials of construction incorporated into a building, including but not limited to walls, floor, roofs, ceilings, stairways, built-in partitions, finishes, cladding, and other similarly incorporated archi-

tectural and structural items, as well as fixed service equipment, including the weight of cranes.

Deflection The deformation or displacement of a structural member as a result of loads acting on it.

Demand area The "expected" area of sprinklers that will activate as anticipated by the sprinkler design standard; essentially the projected size of the fire.

Demising wall Wall bounding a tenant space.

Density The unit rate of water application to an area or surface; expressed in gpm/ft2 [(L/min)/m2].

Dewatering pump A pump that removes water from the ground or excavations that hinder construction from an area of the site.

Diaphragm floor A floor designed to stiffen a building against wind and other lateral loads such as earthquakes.

Dog iron Connects the girders and imparts some lateral stability under normal conditions; resembles a big staple.

Drop panel Thicker section of floor on top of columns to assist in resisting the natural tendency of the floor to shear off at the column.

Dry chemical system An automatic fire extinguishing system that discharges a dry chemical agent.

Early suppression/fast response (ESFR) A type of fast-response sprinkler capable of providing fire suppression of specific high-challenge fire hazards.

Eccentric load A force that is perpendicular to the plane of the section but does not pass through the center of the section.

Egress Adequate exits within a building.

Electrical contractor A subcontractor hired by the contractor to oversee all electrical installation.

Electrical engineer An engineer who specializes in the design of lighting, power telecommunications, and emergency power of a structure.

End matched A descriptor for lumber with tongues and grooves at the ends.

Engineered wood In the construction trade, laminated timbers. In this text, wood modified from its natural state.

English Tudor Balloon-frame structure finished to resemble post and frame construction.

Entrainment The process of oxygen (air) being drawn into a fire.

Excess flow valve Valve that senses a sudden increased flow, as from a broken line, and shuts off flammable gas.

Exit That portion of a means of egress that is separated from all other spaces of a building or structure by construction or equipment as required to provide a protected way of travel to the exit discharge.

Exit access That portion of a means of egress that leads to an exit.

Exit discharge That portion of a means of egress between the termination of an exit and a public way.

Exit passageway Hallways, corridors, passages, or tunnels used as exit components and separated from other parts of the building in accordance with NFPA 101: Life Safety Code.

Falsework Temporary shoring, formwork, beams, or lateral bracing to support the work in the process of construction; also known as formwork.

Finger joints Interlocking wood pieces that are glued together to create a longer piece of dimensional lumber.

Fire-rated The classification indicating in time (hours) the ability of a structure or component to withstand a standardized fire test. Does not necessarily reflect performance in an actual fire.

Fire alarm/security system contractor A subcontractor hired by the contractor to oversee all fire and security alarm installation.

Fire barriers Assemblies that are typically 1- to 2-hour rated and used to enclose shafts, exit stairwells, exit passageways, and horizontal exits, and to separate spaces from certain hazardous areas.

Fire code Regulates the activities that take place in existing buildings, including the maintenance of existing fire protection features such as automatic sprinklers and fire doors, hazardous processes such as spray finishing, the storage of hazardous materials, and general fire safety precautions.

Fire cut The end of a joist that is cut at an angle to permit the joist to fall out of a wall without damaging the load-bearing wall.

Fire damper A mechanical device installed in an HVAC system at the point at which it passes through a fire-resistive assembly so as to block the spread of fire through the fire-rated assembly.

Fire limit Older code provision that would not allow a structure to be built without the use of exterior masonry walls that would limit fire extension.

Fire load The potential fuel available for a fire in a building.

Fire partition A fire-rated assembly that subdivides a building to prevent the spread of fire such as between dwelling units in a multiple dwelling.

Fire plume The column of flames, smoke, and heated gases rising above the burning object.

Fireproof Material applied to structural elements or systems that provides increased fire resistance; usually serves no structural function.

Fireproofing contractor A subcontractor hired by the contractor to spray a layer of cementitious material onto steel for fire resistance purposes.

Fire protection engineer An engineer who specializes in the design of fire protection systems including sprinkler and standpipe systems, and fire alarm systems of a structure.

Fire resistance The ability of a material to avoid ignition, combustion, and the thermal effects of fire.

Fire resistance rating Quality ascribed to a wall, floor, or column assembly that has been tested in a standard manner to determine the length of time it remains structurally stable (or resists the passage of fire) when attacked by a test fire.

Fire resistant Construction designed to provide reasonable protection against fire.

Fire retardant Substance that helps delay or prevent combustion.

Fire service access elevators Elevators specifically designed for fire fighter use that are found in new buildings more than 120 feet (36 m) in height.

Fire wall Wall with a fire-resistive rating and structural stability that separates buildings or subdivides a building to prevent the spread of fire.

Fixed beam Beam supported at two points and rigidly held in position at both points. This rigidity may cause collapse of a wall if the beam collapses and the rigid connection does not yield properly.

Flameover A situation in which unburned fuel gases, having gathered in sufficient quantities at the ceiling layer, suddenly ignite; also called rollover.

Flameproof Resistant to fire.

Flammable Combustible that is capable of easily being ignited and rapidly consumed by fire. Flammables may be solid, liquid, or gas.

Flashover A transition stage in a fire in which exposed surfaces within the compartment ignite simultaneously and fire spreads throughout the compartment, resulting in full room involvement.

Flat plate structural system (continuous beam) Cast-in-place floor in which there are no beams supported by columns; the floor plate itself rests directly on the columns.

Flitch plate girder Composite of a steel plate or plywood sandwiched between two beams.

Flying buttress Masonry pier at a distance from the wall and connected to it that resists the outward thrust of the roof.

Foam system A firefighting system that provides for the delivery of a proportioned foam and water mixture for use in fire extinguishment.

Folded plate A structural element used to strengthen the roof of a structure over large areas.

Footing Thick concrete pads, usually heavily reinforced, which transfer the loads of piers or columns to the ground.

Formwork Temporary shoring, formwork, beams, or lateral bracing to support the work in the process of construction; also known as falsework.

Foundation The lower division of a building that serves to transmit and anchor the loads from the superstructure directly to its earth or rock, usually below ground level.

Framing A system of rough timber structural woodwork that is joined together to support or enclose, such as partitions, flooring, and roofing.

Friable Easily disintegrated.

Fuel A material that will sustain combustion.

Galvanized steel walls Walls made of weatherized steel. Can conduct heat easily.

Gang nail Another name for a gusset plate in a lightweight wood truss.

General contractor The prime contractor who oversees and is responsible for the overall work on the site.

Girder A beam that supports other beams.

Glass fiber–reinforced plastic A composite material made of plastic reinforced with glass fibers.

Global (total) collapse Results in the complete failure of the building.

Glued laminated timber Planks glued together to form a solid timber.

Gravity connection A connection that depends on the weight of the building to hold it in place.

Gravity resistance system All of the structural elements of a building and the connections that support and transfer the loads.

Gray water Wastewater from industrial processes, laundries, and other sources.

Grillage A series of closely spaced beams designed to carry a particularly heavy load.

Gusset plate (lightweight wood truss) Connecting plate made of a thin sheet of steel used to connect the components of the truss.

Gypsum A natural mineral used in the manufacture of drywall and plaster.

Halon system A fire protection system that provides for the transfer of halogenated agents between fire extinguishers, supply containers, and recharge and recovery containers so that none of the halogenated agents escape into the atmosphere.

Hazardous materials inventory statement (HMIS) Required in most codes, this statement lists the materials, hazards, and quantities of hazardous material products within a building.

Hazardous materials management plan (HMMP) Required in most codes, this plan explains how hazardous materials are to be stored and safely used within a building.

Header A masonry unit that overlaps two or more adjoining wythes of masonry to tie them together.

Header or bond course Bricks laid so that the end is visible.

Heat A form of energy that is the source of ignition.

Heat release rate (HRR) The rate at which the potential heat in a fuel is released.

Heavy timber Buildings constructed with noncombustible or limited-combustible exterior walls and floors made of large-dimension combustible materials. Also known as Type IV construction.

Hip The rafter at the angle where two sloping roofs or sides of a roof meet.

Hollow masonry wall Two connected wythes of masonry with an air space in between.

Hollow tile Tile unit composed of vertical hollow cells, utilized in early efforts to lighten concrete floors.

Homogeneous When describing wall construction, a wall that acts as one unit (good bonding exists between bricks blocks, and mortar).

Horizontal exit A 2-hour fire-rated wall typically running from exterior wall to exterior wall on an individual floor of a building, which is used when travel distances to a fire-rated stairwell or exit discharge are excessive.

Houseline Class II standpipe system that supplies 100 gpm at 65 psi. Intended for use by building occupants.

Hurricane bracing Components of a hurricane resistance system that prevent uplift of the components of a structure, including galvanized steel straps used to connect roof trusses to stud walls and anchor bolts used to connect stud walls with sill and sole plates.

I-beam Beam shaped like the letter "I".

Idle pallet storage Pallets without product.

Impact load The effect of a moving load upon a stationary structure.

In-rack sprinklers Sprinklers that are placed within racks to control fires where overhead sprinklers are not adequate.

Inherent fire resistance Flame resistance that is derived from an essential characteristic of the structural member.

Initiating device A system component that originates transmission of a change-of-state condition,

such as in a smoke detector, manual fire alarm box, or supervisory switch.

Insurance regulations The rules of an insurance company in regard to coverage. These regulations may require more fire protection for structures than local building and fire codes require.

Interior finish The exposed material on a wall, ceiling, or floor.

Interstitial space Void space made by utilizing deep parallel-chord trusses.

Inversion layer A layer of air that is warmer than the air below.

Inverted king post truss A truss incorporating a single compression member; it is inverted because the compression member extends downward.

Inverter A device used in a PV roof panel to convert DC into AC.

Inward outward collapse A collapse in which the exterior wall fails horizontally, with the interior wall "folding" horizontally. The top portion of the building fails inward while the bottom half fails outward.

Joist Wooden 2 × 8's, 2 × 10's, or 2 × 12's that run parallel to one another and support a floor or ceiling and are supported in turn by larger beams, girders, or bearing walls.

Joist hanger Metal angle or strap used to support an individual joist against a beam or a girder.

K-bracing An arrangement of braces between columns that resembles the letter "K."

Kilojoule (kJ) Metric unit approximately equivalent to one Btu.

Kilowatt (KW) A unit for measuring the energy release rate of a fire.

KIP 1,000 pounds force.

Knee wall A wall typically found in the top floor of a wood-frame home with a peaked roof. This short wall "squares off" the triangular area at the edge of the room where the sloping roof meets the floor.

Lally columns Steel pipes filled with concrete to increase their load-carrying capacity.

Landing zone In reference to an elevator, this zone is 18 inches above or below the landing floor.

Lapse The condition in which the atmospheric temperature is constantly decreasing as height increases.

Lateral impact load A force that acts on a structure from a horizontal direction, such as wind or seismic forces.

Lattice column Column made of vertical units connected with diagonal pieces.

Lean-to-floor collapse A collapse of a building's floor(s) in which one end of the floor is still supported, sometimes at or near the original point of connection to the wall. A triangular void space is created.

Lean over collapse Limited to wood-frame structures, this type of collapse is characterized by a wooden building that shifts at the upper levels, leaning into adjacent buildings or totally collapsing sideways in the absence of adjacent buildings.

Ledger board A wood board typically attached to a wall's studs that is used to support wood joists.

Left-in-place form Concrete floors that are cast onto corrugated steel.

Legacy code A source code that is no longer supported or printed.

Legacy construction In terms of wood-frame construction, the use of solid wood members as opposed to lightweight wood trusses and wooden I-joists.

Lift slab Type of building construction where concrete slabs are cast on the ground and lifted into place.

Light gauge steel-framed wall An exterior wall constructed with lightweight galvanized steel studs.

Lightweight truss A collection of lightweight structural components joined in a triangular unit that can be used to support either floors or roofs.

Lightweight wood truss A collection of lightweight wood structural components joined in a triangular configuration that can be used to support either floors or roofs.

Light well Small court commonly placed in large buildings to admit daylight into interior areas not exposed to an open view.

Lintel The horizontal beam that forms the upper structural member of an opening for a window or door and supports part of the structure above it.

Listing A compilation of fire-rated assemblies that have been tested by a testing laboratory and found to meet the minimum requirements of the test standard.

Live load The weight of the building contents.

Load Force or other action that results from the weight of all building materials, occupants and their possessions, environmental effects, differential movement, and restrained dimensional changes.

Load-bearing wall Any wall that carries a load in addition to its own weight.

Log cabin Structure constructed of entire tree trunks, as opposed to boards sawn from trees.

Low-density fiberboard Sometimes used as an interior finish, a product made of wood particles such as wood shavings and bound together with a suitable binder.

Lumber Wood that has been sawn and planed.

Malleable The property of a metal that allows it to be shaped by bending, hammering, or extruding without cracking or rupturing.

Manual-dry standpipe system Dry standpipe system that does not have a permanent water supply attached to the system.

Manual-wet standpipe system Wet standpipe system connected to a small water supply for the purpose of maintaining water within the system or sharing a water supply with an automatic sprinkler system, but not having a water supply capable of delivering the system demand attached to the system.

Masonry column Masonry bracing incorporated into unstable masonry walls; also called piers, buttresses, pilasters, or columns.

Masonry unit The type of masonry used on a given structure. Examples include bricks or concrete block.

Masonry walls The most common walls for unprotected steel-framed buildings; made of concrete block or a composite of concrete block and brick.

Masonry wire truss Wire truss embedded into the mortar in specified courses, making the header course no longer necessary.

Matchboarding Ceilings made of embossed steel and wooden boards.

Matched lumber Tongue and grooved lumber (usually lengthwise).

Mechanical engineer A person trained in a branch of engineering related to mechanical equipment. Usually designs the heating, ventilation, and air-conditioning systems and plumbing of the building.

Megastructure A very large structure.

Megawatt (MW) A unit for measuring the energy release rate of a fire.

Metal panel Prefabricated metal structure that is often made up in a sandwich construction to provide one unit combining thermal insulation and interior finish in a steel-framed structure.

Metal siding A type of siding that is often made to look like another type of siding—for example, stone or clapboards.

Mezzanine A low-ceilinged story located between two main stories; usually constructed directly above the ground floor, often projecting over it as a balcony.

Mill The earliest form of heavy timber construction; a type of factory that emerged during the Industrial Revolution.

Model code Complete regulatory documents that can be adopted by cities and states as they are written.

Modular construction A construction technique in which a set of prefabricated modules are transported to and assembled at a construction site, creating a building.

Modulus of elasticity A measurement of the ability of steel to distort and restore.

Moment The tendency of a force to rotate or twist a structural member.

Moment frame A structural system that utilizes special "moment" connections between columns and beams to resist rotation due to lateral loads such as earthquakes and wind.

Monolithic concrete A construction technique in which all successive poured concrete castings are

joined together so that the structure seems to be like one piece of stone.

Monolithic construction Method in which all the concrete in a building is properly bonded together and acts as one.

Mortise and tenon Joints that are fitted together to transfer loads properly.

Mudsills Planks on which formwork shores rest.

Mushroom cap Tapered extension at the tops of columns that assists in the transfer of loads from floor to column.

Needle beam When any change is to be made in the foundation of an existing wall, the wall must be supported. Often holes are cut through the wall, and so-called needle beams are inserted and supported on both sides. They pick up the load of the walls.

Neutral axis The line along which the length of the beam does not change.

Neutral plane The interface between the hot ceiling layer and the cooler air flowing into the compartment.

Non-load-bearing wall A wall supporting no load other than its own weight.

Nonambulatory people Individuals who are not capable of self-preservation.

Noncombustible A material that, in the form in which it is used and under the conditions anticipated, will not aid combustion or add appreciable heat to an ambient fire.

Novelty siding Combustible siding laid over the sheathing of a building. This type of siding comes in many different shapes and is laid horizontally.

Occupancy Used in building codes to refer to the intended use of a building.

Occupancy type The building code classification for the intended use or purpose of a building or building space.

Occupant evacuation elevators Special elevators intended for the self-evacuation of occupants in office buildings more than 420 feet (128 m) in height.

Off-grid systems PV systems that are not connected to a traditional public electrical grid; they are often located in rural areas.

One-way structural system System that incorporates floors that have beams running in one direction only.

Open office plan A plan in which low-height partitions create cubicles for personal space; essentially one large open room.

Ordinary construction Buildings in which the exterior walls are noncombustible or limited combustible, but the interior floors and walls are made of combustible materials.

Oriented flakeboard Another name for OSB.

Oriented strand board (OSB) A type of board made of layers of strands of wood cut from logs, with a fairly constant width-to-length ratio.

Overhanging beam A beam that projects beyond its support, but not far enough to be a cantilever.

Oxidizing agent Normally atmospheric oxygen, but may also be chemical compounds known as oxidizers that release oxygen as they react.

Pancake collapse A collapse of a building's floors (and possibly roof) in a pancake stack fashion with each floor (roof) laying flat on top of the one below.

Panel points The connection points joining ties, struts, and chords in a truss.

Panel wall (curtain wall) Non-load-bearing enclosing wall on framed buildings.

Parallel-chord truss A truss in which both the upper and lower chords are parallel to each other.

Parging (pargetting) Application of mortar to the back of the facing material or the face of the backing material.

Partial collapse Results from a failure of a portion of a building. A situation in which a section of floor gives way without bringing down the rest of the building would be a partial collapse. A partial collapse can be just as deadly as a global collapse.

Partition wall A non-load-bearing wall that subdivides spaces within any story of a building or room.

Party wall A load-bearing wall that is common to two structures.

Passive fire protection A material that is applied to a substrate and is designed to protect it from thermal effects.

Pause In reference to atmospheric conditions, the layer of air warmer than the air below it.

Peened Forcible attachment of a thermocouple to a steel column or beam.

Phase I operation Elevators are recalled to the first floor lobby of the building by the activation of a smoke detector located in an elevator lobby or on one of the upper floors. Upon arrival at the first floor, the doors open and allow the occupants to leave. The doors remain open for fire fighters to begin their use of the elevators.

Phase II operation A fire fighter selects one of the elevators in a bank to access the upper floors. Fire fighters can then control the elevators using a special key and set of procedures.

Pier A short column of masonry, usually rectangular in horizontal cross section, used to support other structural members.

Pilaster A masonry column projecting from one or both faces of the wall in which it is located.

Pile One of a series of large timbers or steel sections driven into soft ground down to bedrock to provide a solid foundation for the superstructure of a building.

Piloted ignition Ignition of a heated combustible material when a flame or spark is applied.

Pinned A description for structural elements that are connected by simple connectors such as bolts, rivets, or welded joints.

Pintle Square metal device used to transfer loads of columns on upper floors by passing the loads through intervening beams and girders to metal column caps on the column below.

Plain concrete Concrete that has no reinforcement, except possibly light reinforcement to resist temperature changes.

Plank and beam Used in modern construction, a wood-frame structure built with heavier beams that are spaced much farther apart than the traditional 16 inches.

Plastic Synthetic materials (typically hydrocarbon based) that are composed of organic polymers that can be shaped and molded.

Plastic design Design based on connections that redirect overloads to other sections of the building.

Plates Flat pieces of steel.

Platform A raised area in a building for presentations, wrestling matches, and the like where there are no hanging curtains, drops, or scenery other than lighting and sound effects.

Platform frame A structure in which subflooring is laid on the joists, and the frame for the first-floor walls is erected on the first floor.

Plumbing contractor A subcontractor hired by the contractor to oversee the installation of plumbing within a building.

Plumbing, mechanical, and electrical codes Plumbing codes apply to the installation of domestic water systems, sanitary systems, and the like. Mechanical codes apply to systems likes air-handling systems and smoke control systems. Electrical codes apply to the installation of electrical systems.

Plywood Wood panels manufactured with the grain of alternate plies laid at right angles to develop the approximate equal strength in either direction.

Plywood siding Four-foot-wide sheets that give the appearance of four-inch-wide strips about a half inch apart.

Polychlorinated biphenyl (PCB) Toxic chemicals used as nonflammable coolants in transformers that may be released during fires.

Polyvinyl chloride (PVC) A commercially produced organic compound used in transformers and capacitors due to its electrical insulator properties and low flammability rating.

Portal bracing Heavy riveting of girders to columns from the top to the bottom of the frame.

Portland cement mortar The most common type of cement in use today. The powder, when mixed with water, will become a hydrated solid over time.

Post and frame A wood-frame structure with an identifiable frame or skeleton of timber fitted together. Joints are constructed by mortise and tenon, fitted together to transfer loads properly.

Precast A concrete member that is cast and cured in a place other than its final position in the structure.

Precast concrete Concrete that has been cast at a location other than the place where it is to remain.

Precast concrete tilt slab wall A type of wall that acts as a vertical cantilever when it is being erected and is braced by tormentors or temporary bracing poles.

Precast prestressed concrete panels Concrete panels that are precast and brought to the construction site.

Prefire analysis A study of information that has been gathered as part of a prefire plan to identify specific construction issues/concerns and interactions.

Prefire planning A document developed by gathering general and detailed data used by responding personnel to determine the resources and actions necessary to mitigate anticipated emergencies at a specific facility.

Prestressing Engineered stresses placed in architectural and structural concrete to offset the stresses that occur in the concrete when it is placed under load.

Pretensioning and post-tensioning Processes by which steel tendons or rods are placed under tension, drawing the anchors together. Tensioned steel places the concrete in compression.

Progressive collapse Extensive structural failure initiated by local structural damage or a chain reaction of failures following damage to a small portion of a structure.

Projected beam detectors Used in smoke control systems. These detectors can cover large areas with a single beam of light.

Proscenium arch and wall Found on a stage, it is the large ornamental opening and wall that separates the audience from the stage.

Purge Flow of air or an inert medium at a rate that will effectively remove any gaseous or suspended combustibles and replace them with air.

Purlins Beams set at right angles to trusses or roof rafters to provide support for lightweight roofing.

Pyrolytic decomposition The chemical decomposition of a compound into one or more other substances by heat alone; pyrolysis often proceeds combustion.

Pyrophoric gas Gas that ignites in air without the introduction of an ignition source.

Quarried stone Natural stones such as granite, marble, limestone, and sandstone. Can be used in the construction of walls and foundations.

Queen post truss A truss with two compression members.

\dot{Q} A designation of the HRR; refers to the rate at which a fuel will burn.

RACE An acronym: R—remove all people in immediate danger to safety; A—activate the manual pull station and have someone call 911; C—close doors to confine the spread of smoke and fire; and E—extinguish the fire, if possible.

Racks Any combination of vertical, horizontal, and diagonal members that supports stored materials.

Radiation Heat transfer of heat through electromagnetic waves.

Rafters Wood members used to support the roof sheeting and loads.

Rakers Diagonal bracing columns.

Reaction The response in structures to the imposed loads, which are generally developed at the supports.

Reinforced concrete In concrete masonry construction, steel reinforcement that is embedded in such a manner that the two materials act together in resisting forces.

Reinforcement bar (rebar) Steel bar placed in concrete to increase its overall tensile strength.

Reinforcing bars or rods Steel rods or bars used to reinforce concrete.

Remote annunciator Device indicating an off-standard or abnormal condition by both visual and audible signals.

Repeated load A load that is applied intermittently.

Reshoring Shores that are put back into concrete to help carry the load of the still-curing concrete.

Ridge board A structural member placed on the ridge of the roof onto which the upper ends of rafters are fastened.

Rigid-foamed polyurethane Cellular foam plastic that is used as an interior finish.

Rigid frame Structural frame in which all columns and beams are rigidly connected. There are no hinged joints, and the angular relationship between beam and column members is maintained under load.

Rising roof A phenomenon in wood trusses in which differences in moisture levels between the upper and lower wood truss chords cause the truss to bend and create a rise in the roof.

Robertson protected metal Asphalt-coated steel.

Rolled or built-up members Steel structural members; rolled members are one solid piece of metal; built-up members are made up of different sections riveted, bolted, or welded together.

Rough carpentry All of the wood framing members and sheathing of a building. Also known as framing.

Rough lumber Lumber that is left as sawn on all four sides.

Rubble masonry Rough stones of irregular shapes and sizes, used in rough, uncoursed work in the construction of walls and foundations.

Rubble masonry wall A wall composed of an inner and outer wythe of coursed masonry. The space between is filled with random masonry sometimes mixed with mortar. Such walls are unstable to a lateral thrust.

Safety factor The ratio of the strength of the material just before failure to the safe working stress.

Sand-lime mortar A water-soluble mixture used in the past as mortar; when water is applied, the mortar can be washed away from the wall.

Sawn beam Wooden beam sawn out of a tree trunk. To some extent, these beams have been replaced by the wooden I-beam due to the weight and cost of the true wooden beam.

Scale (architectural) An instrument with uniformly spaced marks used to measure distance on a set of blueprints.

Scupper An outlet in a wall of a building for drainage of water from a floor or a flat roof.

Secondary collapse An additional collapse that occurs after the initial collapse. Often occurs when loads shift after a primary collapse, causing additional portions of the structure to fail.

Seismic protection Stabilization of outer walls, overhangs, floors, and walls to resist destruction by an earthquake.

Self-releasing floor A type of floor in which floor girders are set on anchor boxes in walls and caps attached to columns. A wood cleat or steel dog-iron similar to a big staple is used to provide minimal stability. Often used in heavy-timber construction.

Self-weight Another term for dead load.

Semiautomatic-dry standpipe system Standpipe system that is attached to a water supply capable of supplying the system demand at all times and that requires activation of a control device to provide water at hose connections.

Serpentine wall A curving wall.

Shear Forces occurring *within* a building member when opposing forces pull the member in opposite directions.

Shear wall A wall that counteracts the effects of lateral loads such as wind and earthquakes.

Shelves Storage on structures that are less than 2.5 feet (0.75 meters) deep, with shelves usually 2–3 feet (0.6–0.9 meters, respectively) apart vertically, and seldom exceeding 15 feet (4.5 meters) in total height.

Ship's ladder A completely vertical stair with a width that is not more than 24 inches wide.

Silence switch When flipped, the fire alarm is silenced but still active.

Sill Component found on the bottom of the frame of a wood structure. It rests on and is bolted to the foundation.

Simple beam A beam supported at two points near its ends. In simple beam construction, the load is delivered to the two reaction points and the rest of the structure renders no assistance in an overload.

Situational awareness In firefighting terms, the perception of the surroundings in terms of time and space, the understanding and comprehension of these observations, and the projected outcome of a change in these conditions.

Skewbacks Tiles shaped to fit around steel.

Slipforming Technique by which forms are moved upward as the concrete is poured.

Slow burning A characteristic of a building that should allow a fire in that building to be brought

under control before the building itself becomes involved.

Slurry wall A foundation wall in an excavation that is heavily reinforced with steel and temporarily filled with a liquid slurry mixture that is subsequently displaced by concrete.

Smoke The airborne solid and liquid particulates and gases produced by a fire.

Smoke barrier A 1-hour fire-rated assembly that has also been designed and tested to minimize the migration of smoke.

Smoke control System that utilizes fans to produce pressure differences so as to manage smoke movement.

Smoke developed In the Steiner tunnel test, a calculation of smoke spread made by measuring the obscuration as the smoke passes a photoelectric cell placed in the stack from the test tunnel.

Soffit False space above built-in cabinets, usually in a kitchen, or in the undersides of stairways and projecting eaves.

Solid masonry walls Masonry units (either solid or hollow) laid contiguously with the joints filled with mortar.

Solid pile Storage that is either box on box or pallet load on pallet load.

Spaceframe A three-dimensional pyramid-like truss.

Spall A small fragment or chip dislodged from the face of a stone or masonry unit.

Spalling Loss of surface material when concrete is subjected to heat.

Spandrel girder A girder that ties wall columns together in a framed building.

Spandrel space Distance between the top of one window and the bottom of the one above.

Specifications Manuals prepared by building designers that specify which particular type of device or equipment is to be used in the building.

Spliced timber Due to the shortage of solid timbers, these members are joined together by various metal connectors to transfer loads so that the spliced timber acts as a single member.

Spline Wooden strips that fit into grooves in two adjacent planks to make a tight floor.

Sprinkler contractor A subcontractor hired by a contractor to oversee the installation and design of sprinkler systems within a building.

Stack effect The vertical airflow within buildings caused by the temperature-created density differences between the building interior and exterior or between two interior spaces.

Stage Performance area in a theater that has a proscenium arch and wall; hanging curtains, drops, and scenery; lighting; and support rooms (dressing room, etc.).

Stairwell pressurization system System that introduces positive pressure into a stairwell to provide a tenable environment within the stair tower in the event of a building fire.

Standard time–temperature curve The profile of the furnace temperatures that are dictated by the test standards NFPA 251, UL 263, and ASTM E-119.

Standpipe system An arrangement of piping, valves, hose connections, and allied equipment installed in a building or structure, with the hose connections located in such a manner that water can be discharged in streams or spray patterns through attached hose and nozzles, for the purpose of extinguishing a fire, thereby protecting a building or structure and its contents in addition to protecting the occupants.

Steel expansion joints A metal connection that allows for movement of floors.

Steel joist An open web design used for the support of floors and roofs.

Steiner tunnel test A common term that refers to NFPA 255, *Standard Method of Test of Surface Burning Characteristics of Building Materials*.

Stiffness The capacity of a member or framework to resist imposed loads without excessive deflection.

Strain The actual percentage of elongation (deformation) when a material is stressed.

Strands Special high-strength, cold-drawn steel cables. Also referred to as cables or tendons.

Stress Force per unit area that produces a deformation.

Stressing the tendons After concrete is poured and reaches its specified strength, steel tendons are stretched and anchored at the ends of the unit. Also referred to as "jacking the cables."

Stretcher A masonry unit laid horizontally with its length in the direction of the face of the wall.

Stretcher course Bricks laid so that the long side is visible.

Stringers A component of a set of stairs used to support risers (vertical stair components between treads) and treads (horizontal stair components on which people place their feet).

Strip mall Modern one-story retail occupancy building that typically has a lightweight wood truss roof and concrete block walls (Type III construction) or steel bar joists and a metal deck roof with a masonry wall (Type II construction).

Structural elements Components of a structure that include beams, trusses, columns, arches, and walls.

Structural engineer An engineer who is consulted in large buildings to design the structural frame.

Structural frame All members of a structure that are tied together to carry the imposed loads to the substructure, and hence to the ground.

Structural insulated panels (SIP) "Stud-less" wall assemblies using a core of polyurethane.

Structural insulated sheathing (SIS) Installed as an exterior sheathing in wood-frame dwellings, providing structural stabilization (similar to plywood and particleboard sheathing over wood studs), insulation, and a barrier to moisture.

Structural steel Steel rolled in a variety of shapes and fabricated for use as load-bearing structural members or elements.

Strut A bracing column.

Stucco An exterior plaster finish made of Portland cement.

Stud A column in a frame building, usually nominal 2 by 4 inches (2×4) or 2 by 6 inches (2×6).

Subcontractor A specialist who undertakes part of a main building contract from a contractor.

Subflooring Laid on top of the joists, the structural member that serves as the base for the finished floor.

Substructure The foundation and footings.

Superstructure Any structure above the substructure.

Suspended beam A simple beam, with one or both ends suspended on a tension member such as a chain, cable, or rod.

Suspended load A hanging load supported from above.

Sustainable construction Construction that is deemed to be energy efficient and environmentally friendly; also known as green construction.

Tee A standard I-beam cut lengthwise through the web forms two such beams with T-shaped cross sections.

Temperature rods Thin rods installed near the surface of concrete, usually at right angles to the main reinforcing rods, to help the concrete resist cracking due to temperature changes.

Tendons Special high-strength, cold-drawn steel cables. Also referred to as strands or cables.

Tenement Multistory working-class apartment buildings constructed in the 1800s and early 1900s, often substandard in terms of fire safety and health.

Tension A pulling or stretching force in line with the axis of the body; the opposite of compression, which is pushing, crushing stress.

Terra-cotta tile Made of clay and fine sand and fired in a kiln.

Thermal pane windows Units composed of two or more glass panes in a hermetically sealed frame.

Thermocouple An electrical temperature-measuring device used in laboratory fire testing.

Thermoplastic Construction material that softens when heated and rehardens when cooled.

Thermoset Plastic that chars and burns, but does not flow.

Tie (truss member) The tensile connecting members of a truss web.

Tiebacks Cold-drawn steel cables inserted into holes driven into the rock and anchored with epoxy. Also refers to braced sheeting used in soil walls to protect against collapse.

Tied arch Arch in which a steel tension rod ties the ends of the arch together to eliminate the need for the masonry.

Tie rod A rod in tension; used to hold parts of a structure together.

Tin ceiling Embossed steel; will transfer heat in either direction.

Top plate Top horizontal member of a wood-frame wall that supports the ceiling joists.

Torque The measurable turning force applied to a structural member.

Torsion A force tending to twist a structural member.

Total flooding Act and manner of discharging an agent for the purpose of achieving a specified minimum agent concentration throughout a compartment.

Transfer beam A beam that typically carries a load around a large opening or over an area in order to avoid intervening columns.

Transmission The manner in which a load is spread from the point of application to the ground.

Travel distance The number of feet from any point in a building to a rated stairwell or outside.

Triage To evaluate and categorize.

Triangular truss A roof truss that is triangular in shape; it is used to create a peaked roof.

Trimmer A piece of lumber that supports a header and can be found in floor openings such as stairwells (runs perpendicular to floor joists).

Trombe wall A set of glass exterior panels and concrete block wall with a void space in between (used in smaller buildings).

Trunnel Wooden peg used to pin together mortise and tenon joints.

Truss A type of beam that is a framed structure consisting of a triangle or group of triangles arranged in a single plane in such a manner that loads applied at the points of intersections of the members will cause only direct stresses (tension or compression) in the members.

Truss frame Developed by the U.S. Forest Products Laboratory, a type of wood frame that substitutes sawn beams for truss members used in roofs and floors. It is held together with nails.

Truss void Found within a truss roof system, a void space located between the top floor ceiling and the roof. Also known as a cockloft.

Tube A steel structural member that is rolled in cylindrical, square, or rectangular shapes.

Tube construction Externally braced structure.

Tuned-mass dampers Heavy weights installed high up in a building that are adjusted by computers to counter wind-induced oscillations.

Two-way structural system System that incorporates floors that have beams running in two directions.

Type III construction Ordinary construction.

Type IV Heavy timber construction.

Ultimate strength The highest load that a member or structure can sustain before failure occurs.

Underwriters blocks Concrete blocks poured per Underwriters Laboratories, Inc.'s classifications.

Uniformly distributed load A load that is applied evenly over an area.

Uninhibited chemical chain reaction A sustained oxidation reaction that produces sufficient excess heat to continue heating unburned fuel, making it available for combustion.

Unreinforced masonry Ordinary masonry walls are not reinforced, so they have no resistance to lateral movement.

V-shaped floor collapse A collapse in which the floor fails near its center, with the perimeter of the floor still partially supported by the walls. Two voids are created in this type of collapse.

Valley The lower slope formed by the connection of two inclined planes of a roof.

Veneer wall A wall made up of a single vertical thickness of masonry that is designed to improve the exterior appearance of a building.

Vermiculite Mineral used as bulk insulation and as an aggregate in insulating.

Vierendeel truss A rectangular truss with very rigid corner bracing.

Vinyl siding A thermoplastic that will deform and drip when introduced to a fire situation. Often made to look like wood siding.

Voussoir A wedge-shaped block whose converging sides radiate from a center, forming an element of an arch or vaulted ceiling.

Waffle concrete Two-way structural system with closely spaced beams set at tight angles to one another in which unnecessary concrete is formed out; lower side resembles a waffle.

Waler A horizontal beam that ties rows of soldier beams together.

Wall A structural element that transmits to the ground the compressive forces applied along the top or received at any point on the wall.

Wall bearing A descriptor indicating a structure in which the entire structural load is carried on the walls; the opposite of a curtain wall.

Wallboard A large, rigid gypsum sheet that is fastened to the frame of a building and provides the surface finish.

Wallboard contractor A subcontractor hired by a contractor to oversee the surface finishes of walls that are fastened to the frame of the building.

Wall column A column of steel, reinforced concrete, or solid masonry (such as brick or solid block) in a block wall. Concentrated loads such as main girders are applied to the wall directly above the wall column.

Water flow switch A switch that detects the movement of water in a sprinkler system and transmits a signal to the alarm system.

Water mist system Distribution system connected to a water supply or water and atomizing media supplies that is equipped with one or more nozzles capable of delivering water mist intended to control, suppress, or extinguish fires and that has been demonstrated to meet the performance requirements of its listing.

Water spray system Special fixed pipe system connected to a reliable fire protection water supply and equipped with water spray nozzles for specific water discharge and distribution over the surface or area to be protected.

Watt (W) A unit for measuring the energy release rate of a fire.

Web The group of struts, ties, and panel points in a truss.

Weep hole A drainage hole in a masonry wall that allows water trapped inside the wall to escape.

Wet chemical system An automatic fire extinguishing system that discharges a wet chemical agent.

Wet joint Cast-in-place concrete that unites the rods projecting from precast sections.

Wide flange shapes I-beams that have flanges wider than standard I-beams.

Wind load The positive or negative force of the wind acting on a structure.

Wood A hard fibrous material forming the major part of trees. It is usually milled or otherwise processed for use in construction.

Wooden-walled building A wall-bearing building that carries the load of the structure and the contents.

Wooden I-joist Composite lightweight wood material used in an "I" cross section.

Wood framed Building with exterior walls, interior walls, floors, and roofs made of combustible wood material.

Wood lath Narrow, rough strips of wood nailed to studs. Plaster is spread on wood laths. Generally no longer used, wood lath is present in many existing buildings.

Wrought iron A commercially pure iron of fibrous nature; used for water pipes, rivets, and other ornamental forged work.

Wythe A single continuous vertical wall of bricks, one masonry unit in thickness.

Zees Members with a Z-shaped cross section.

Zone Defined area within the protected premises.

Zoned smoke control Smoke control system that includes smoke exhaust for the smoke zone and pressurization for all contiguous smoke control zones.

Zoning regulations Dictate land usage in specific cities (and are written by the cities themselves). Spell out which types of buildings can be built in specific locations and their permitted uses.

INDEX

Note: Page numbers followed by *f*, or *t* indicate material in figures, or tables, respectively.

U

V